Advances in Experimental Medicine and Biology

Volume 1101

More information about this series at http://www.springer.com/series/5584

Xiaoxiang Zheng
Editor

Neural Interface: Frontiers and Applications

Springer

Editor
Xiaoxiang Zheng
Qiushi Academy for Advanced Studies
Zhejiang University
Hangzhou, Zhejiang, China

ISSN 0065-2598 ISSN 2214-8019 (electronic)
Advances in Experimental Medicine and Biology
ISBN 978-981-13-2049-1 ISBN 978-981-13-2050-7 (eBook)
https://doi.org/10.1007/978-981-13-2050-7

This Springer imprint is published by the registered company Springer Nature Singapore Pte Ltd.
The registered company address is: 152 Beach Road, #21-01/04 Gateway East, Singapore 189721, Singapore

Preface

Neural interfaces interact directly with the central nervous system (CNS) or peripheral nervous system (PNS) to restore brain function impaired by neural diseases or trauma or to enhance motor or cognitive functions for the abled individuals. The technology interfacing the brain is also known as brain-computer interface (BCI) or brain-machine interface (BMI). While the cochlear prosthesis and deep brain stimulation for Parkinson's disease have become reality, other neural interface applications are yet fully matured for decades. However, in recent years we have witnessed the great advances in this field, and the advances in neuroscience and engineering are speeding up neural interface technology, opening the door to assist, augment, repair, or restore sensorimotor or other cognitive functions and thus improving the quality of life for the individuals with disabilities. Medical applications such as NeuroPace for seizure detection/suppression and spinal cord stimulation for pain management are now becoming increasingly common. Neural interfaces are now being explored in applications as diverse as rehabilitation, accessibility, gaming, education, recreation, robotics, and human enhancement.

Neural interfaces also represent a powerful tool to address fundamental questions in neuroscience. We also have witnessed tremendous advancements in the field of neural interface with high impact not only in the development of neuroprosthetics but also in our basic understanding of brain function. Neural interface technology can be seen as a bridge linking engineering and neuroscience.

This book focuses on frontiers of neural interface technology, including hardware, software, neural decoding and encoding, control system, and system integration. It also covers the respects of applications in neuroprosthetics, neural diseases treatment, neurorobotics, and also the research paradigms for basic neuroscience. This book aims at providing the researchers, graduate students, and upper undergraduate students in a wide range of disciplines with a cutting-edge and comprehensive summary of researches on neural interfaces.

Editing such a book is tough, yet delightful, challenging, and fundamentally a team effort. Thus, I am in gratitude to a bevy of individuals without whom we could not have hoped to succeed. First and foremost, I sincerely thank Dr. Peng Zhang, a senior editor from Springer Nature, who invited me to edit this book, for his understanding, support, and patience. I would like to thank all the authors who shared their works and wrote the manuscripts of their chapters. Their great contribution makes up this exciting work. I also would like to thank everyone who has helped me along the way but whom I did not mention explicitly.

Hangzhou, China Xiaoxiang Zheng

Contents

About the Editor

Xiaoxiang Zheng is a professor at the Qiushi Academy for Advanced Studies, Zhejiang University. She is currently a member of the academic board of Key Laboratory for Biomedical Engineering of Ministry of Education at Zhejiang University and was the director of the College of Biomedical Engineering and Instrument Science, Zhejiang University, and executive member of the council of the Chinese Society of Biomedical Engineering. Dr. Zheng's research interests include neural engineering, microcirculation, and physiology of the cell. She has published more than 200 articles in these fields, over 100 of which are indexed by SCI or EI.

Contributors

Mojisola Grace Asogbon CAS Key Laboratory of Human-Machine Intelligence-Synergy Systems, Shenzhen Institutes of Advanced Technology (SIAT), Chinese Academy of Science (CAS), Shenzhen, China
Institute of Advanced Integration Technology, SIAT, Shenzhen, China
Shenzhen College of Advanced Technology, University of Chinese Academy of Sciences, Shenzhen, China

Rajmohan Bhandari University of Utah, Salt Lake City, UT, USA
Blackrock Microsystems, Salt Lake City, UT, USA

Junjun Chen Qiushi Academy for Advanced Studies, Zhejiang University, Hangzhou, China
Department of Biomedical Engineering, Zhejiang University, Hangzhou, China

Weidong Chen Qiushi Academy for Advanced Studies (QAAS), Zhejiang University, Hangzhou, People's Republic of China
Department of Biomedical Engineering, Key Laboratory of Ministry of Education Ministry, Zhejiang University, Hangzhou, People's Republic of China
College of Computer Science and Technology, Zhejiang University, Hangzhou, People's Republic of China

Zhe Chen Department of Psychiatry, Department of Neuroscience and Physiology, New York University School of Medicine, New York, NY, USA

Peng Fang CAS Key Laboratory of Human-Machine Intelligence-Synergy Systems, Shenzhen Institutes of Advanced Technology (SIAT), Chinese Academy of Science (CAS), Shenzhen, China
Institute of Advanced Integration Technology, SIAT, Shenzhen, China

Robert Franklin Blackrock Microsystems, Salt Lake City, UT, USA

Jiping He Neural Interface and Rehabilitation Technology Research Center, School of Automation, Huazhong University of Science and Technology, Wuhan, China
Beijing Advanced Innovation Center for Intelligent Robot and System, School of Mechatronical Engineering, Beijing Institute of Technology, Beijing, China
Center for Neural Interface Design, School of Biological and Health Systems Engineering, Arizona State University, Tempe, AZ, USA

Dingyin Hu Neural Interface and Rehabilitation Technology Research Center, School of Automation, Huazhong University of Science and Technology, Wuhan, China

Jian Huang Neural Interface and Rehabilitation Technology Research Center, School of Automation, Huazhong University of Science and Technology, Wuhan, China

Sui Huang Nurotron Biotechnology Inc., Irvine, CA, USA

Paul Oluwagbengba Idowu CAS Key Laboratory of Human-Machine Intelligence-Synergy Systems, Shenzhen Institutes of Advanced Technology (SIAT), Chinese Academy of Science (CAS), Shenzhen, China
Institute of Advanced Integration Technology, SIAT, Shenzhen, China
Shenzhen College of Advanced Technology, University of Chinese Academy of Sciences, Shenzhen, China

Yile Jin Qiushi Academy for Advanced Studies, Zhejiang University, Hangzhou, China
Department of Biomedical Engineering, Zhejiang University, Hangzhou, China

Julia Körner University of Utah, Salt Lake City, UT, USA

Qiuxia Lai Neural Interface and Rehabilitation Technology Research Center, School of Automation, Huazhong University of Science and Technology, Wuhan, China

Farah Laiwalla School of Engineering, Brown University, Providence, RI, USA

Moritz Leber University of Utah, Salt Lake City, UT, USA
Blackrock Microsystems, Salt Lake City, UT, USA

Guanglin Li CAS Key Laboratory of Human-Machine Intelligence-Synergy Systems, Shenzhen Institutes of Advanced Technology (SIAT), Chinese Academy of Science (CAS), Shenzhen, China
Institute of Advanced Integration Technology, SIAT, Shenzhen, China

Wei Li Neural Interface and Rehabilitation Technology Research Center, School of Automation, Huazhong University of Science and Technology, Wuhan, China

Xiao Li Neural Interface and Rehabilitation Technology Research Center, School of Automation, Huazhong University of Science and Technology, Wuhan, China

Chuang Lin CAS Key Laboratory of Human-Machine Intelligence-Synergy Systems, Shenzhen Institutes of Advanced Technology (SIAT), Chinese Academy of Science (CAS), Shenzhen, China
Institute of Advanced Integration Technology, SIAT, Shenzhen, China

Xuan Ma Neural Interface and Rehabilitation Technology Research Center, School of Automation, Huazhong University of Science and Technology, Wuhan, China

Masaki Nakanishi Institute for Neural Computation, University of California San Diego, San Diego, CA, USA

Sandeep Negi University of Utah, Salt Lake City, UT, USA
Blackrock Microsystems, Salt Lake City, UT, USA

Arto Nurmikko School of Engineering, Brown University, Providence, RI, USA

Gang Pan College of Computer Science and Technology, Zhejiang University, Hangzhou, People's Republic of China

Christopher F. Reiche University of Utah, Salt Lake City, UT, USA

Oluwarotimi Williams Samuel CAS Key Laboratory of Human-Machine Intelligence-Synergy Systems, Shenzhen Institutes of Advanced Technology (SIAT), Chinese Academy of Science (CAS), Shenzhen, China
Institute of Advanced Integration Technology, SIAT, Shenzhen, China

Florian Solzbacher University of Utah, Salt Lake City, UT, USA
Blackrock Microsystems, Salt Lake City, UT, USA

Xiaoan Sun Nurotron Biotechnology Inc., Irvine, CA, USA

Jing Wang Department of Anesthesiology, Perioperative Care and Pain Medicine, Department of Neuroscience and Physiology, New York University School of Medicine, New York, NY, USA

Ningyuan Wang Nurotron Biotechnology Inc., Irvine, CA, USA

Yijun Wang Institute of Semiconductors, Chinese Academy of Sciences, Beijing, China

Yuanyuan Wang Neural Interface and Rehabilitation Technology Research Center, School of Automation, Huazhong University of Science and Technology, Wuhan, China

Yueming Wang Qiushi Academy for Advanced Studies (QAAS), Zhejiang University, Hangzhou, People's Republic of China
College of Computer Science and Technology, Zhejiang University, Hangzhou, People's Republic of China

Zhaohui Wu College of Computer Science and Technology, Zhejiang University, Hangzhou, People's Republic of China

Kedi Xu Qiushi Academy for Advanced Studies (QAAS), Zhejiang University, Hangzhou, People's Republic of China
Department of Biomedical Engineering, Key Laboratory of Ministry of Education Ministry, Zhejiang University, Hangzhou, People's Republic of China
Zhejiang Provincial Key Laboratory of Cardio-Cerebral Vascular Detection Technology and Medicinal Effectiveness Appraisal, Zhejiang University, Hangzhou, People's Republic of China

Qi Xu Neural Interface and Rehabilitation Technology Research Center, School of Automation, Huazhong University of Science and Technology, Wuhan, China

Ming Yin University of Utah, Salt Lake City, UT, USA
Blackrock Microsystems, Salt Lake City, UT, USA

Dan Zhang Department of Psychology, Tsinghua University, Beijing, China

Jiacheng Zhang Qiushi Academy for Advanced Studies (QAAS), Zhejiang University, Hangzhou, People's Republic of China
Department of Biomedical Engineering, Key Laboratory of Ministry of Education Ministry, Zhejiang University, Hangzhou, People's Republic of China
Zhejiang Provincial Key Laboratory of Cardio-Cerebral Vascular Detection Technology and Medicinal Effectiveness Appraisal, Zhejiang University, Hangzhou, People's Republic of China

Peng Zhang Neural Interface and Rehabilitation Technology Research Center, School of Automation, Huazhong University of Science and Technology, Wuhan, China

Shaomin Zhang Qiushi Academy for Advanced Studies (QAAS), Zhejiang University, Hangzhou, People's Republic of China
Department of Biomedical Engineering, Key Laboratory of Ministry of Education Ministry, Zhejiang University, Hangzhou, People's Republic of China
Zhejiang Provincial Key Laboratory of Cardio-Cerebral Vascular Detection Technology and Medicinal Effectiveness Appraisal, Zhejiang University, Hangzhou, People's Republic of China

Nenggan Zheng Qiushi Academy for Advanced Studies (QAAS), Zhejiang University, Hangzhou, People's Republic of China
College of Computer Science and Technology, Zhejiang University, Hangzhou, People's Republic of China

Xiaoxiang Zheng Qiushi Academy for Advanced Studies (QAAS), Zhejiang University, Hangzhou, People's Republic of China
Department of Biomedical Engineering, Key Laboratory of Ministry of Education Ministry, Zhejiang University, Hangzhou, People's Republic of China
Zhejiang Provincial Key Laboratory of Cardio-Cerebral Vascular Detection Technology and Medicinal Effectiveness Appraisal, Zhejiang University, Hangzhou, People's Republic of China

Chapter 1
Advances in Penetrating Multichannel Microelectrodes Based on the Utah Array Platform

Moritz Leber, Julia Körner, Christopher F. Reiche, Ming Yin, Rajmohan Bhandari, Robert Franklin, Sandeep Negi, and Florian Solzbacher

Abstract The Utah electrode array (UEA) and its many derivatives have become a gold standard for high-channel count bi-directional neural interfaces, in particular in human subject applications. The chapter provides a brief overview of leading electrode concepts and the context in which the UEA has to be understood. It goes on to discuss the key advances and developments of the UEA platform in the past 15 years, as well as novel wireless and system integration technologies that will merge into future generations of fully integrated devices. Aspects covered include novel device architectures that allow scaling of channel count and density of electrode contacts, material improvements to substrate, electrode contacts, and encapsulation. Further subjects are adaptations of the UEA platform to support IR and optogenetic simulation as well as an improved understanding of failure modes and methods to test and accelerate degradation in vitro such as to better predict device failure and lifetime in vivo.

Authors Moritz Leber, Julia Körner and Christopher F. Reiche have contributed equally to this chapter.

M. Leber (✉) · M. Yin · R. Bhandari · S. Negi · F. Solzbacher (✉)
University of Utah, Salt Lake City, UT, USA

Blackrock Microsystems, Salt Lake City, UT, USA
e-mail: mleber@blackrockmicro.com; florian.solzbacher@utah.edu

J. Körner · C. F. Reiche
University of Utah, Salt Lake City, UT, USA

R. Franklin
Blackrock Microsystems, Salt Lake City, UT, USA

X. Zheng (ed.), *Neural Interface: Frontiers and Applications*,
Advances in Experimental Medicine and Biology 1101,
https://doi.org/10.1007/978-981-13-2050-7_1

Keywords Utah electrode array (UEA) · Neural interface materials · Wireless technology · Accelerated aging · Advanced system integration

1.1 Application Areas and Challenges

Many neural interface electrode developments have been driven by government-sponsored research programs that focused on the restoration of motor or sensory function (neuroprosthetics). In recent years, increasing numbers of clinical research studies under FDA investigative device exemption (IDE) approvals have been carried out for a variety of interfaces. The Utah electrode array (UEA) is the only penetrating high-channel count device of its kind to have received FDA 510 (k) clearance and to have been used in such studies to date. Up until recently, the commercial sector however deemed this market to be too small and the technical hurdles and risks as too large to warrant investment. Subchronic monitoring and treatment of neurological disorders, such as epilepsy, and some peripheral and autonomic nervous system applications appear to have gained momentum in the commercial sector.

Key technological trends that can be observed are increasing channel count, enhanced biocompatibility through materials advances (porosity, coatings, etc.), the advent of devices for use in optogenetics and IR stimulation, advanced wireless (implantable and external) and novel system integration, and packaging concepts for these devices that, e.g., push the boundaries on electrode channel count into the 1000s of channels and beyond (the current Darpa NESD program targets 1 Mio channels for an implantable wireless device covering about $2 \times 2\ cm^2$).

The clinical trends are driven by key opinion leaders with clinical research studies in human patients at Brown University/Massachusetts General Hospital, Stanford, University of Pittsburgh Medical Center, University of Pennsylvania, Chicago Northwestern, Case Western Reserve University, Caltech, UCLA, Columbia, University of Washington, and a growing number of centers in Europe. To an increasing extent, epilepsy patients are being used as a recruiting group for a variety of clinical studies.

With increasing channel count on the hardware side, new challenges arise on the software and data analysis side if one takes into account that the current state of the art in analysis of the data is to search for specific features or carry out spike sorting of action potential events. Data amounts can easily exceed terabytes from a single subject within a 24-h period for devices in the 100-channel range. Assuming 50,000 channels, this would pose a serious challenge. Data collection, storage, management, and also visualization, search, and analysis become intractable with current methods. In addition, while the transition from nonhuman primates to human subjects has great advantages, it also comes with a more stringent regulatory framework and Health Insurance Portability and Accountability Act (HIPPA) regulations for protection of patient data. Future concepts for ultrahigh-channel count neural interfaces will thus likely have to mimic some of the architecture used in the human nervous system as well as in advanced sensor networks, where local "processors" can receive and process neural data at high resolution and depending on filters and parameters set by a central control unit can then either autonomously decide to pass on, compact, or discard data. Smart contextual visualization and search approaches could assume the role of a smart virtual assistant and become support tools for exploration of data.

1.2 Overview of Electrodes in the Field

There is a plethora of penetrating neural micro electrodes that has been published and demonstrated to varying degrees in vivo, mostly by academic institutions. Microwire electrodes were among the first implantable penetrating microelectrodes and were developed in the 1950s [25]. They are fabricated by cutting off the end of an insulated metal wire or by laser ablation of the insulation material around the sharpened tip of a wire [114]. With sufficient mechanical stiffness to withstand the insertion process into neural tissue, common materials for the 25–50 μm diameter wires are stainless steel, tungsten, iridium, and platinum-iridium alloys [25, 87, 114]. Several microwires can be assembled to an array, which is held together by a cast material such as biocompatible epoxy [114]. Further development led to the introduction of floating microelectrode arrays (FMAs), which are designed to move with the brain by "floating" on the surface of the brain and being mechanically decoupled from the skull. FMAs offer the advantage of being highly customizable regarding individual electrode length, position, and spacing [77]. The different electrodes are held together by lightweight substrates and are micro-welded to highly flexible insulated wires leading to the connector (Fig. 1.1a).

Besides microwire and FMA probes, silicon micromachined electrodes and arrays are prevalent in the field. The two main technologies include devices based on the Michigan style shank electrodes and the Utah electrode array (UEA). Both device types allow for high electrode density and electrode counts that are required for neural prosthetic applications [114], and they exhibit the advantage of micromachining batch process compatibility.

Michigan type microelectrodes consist of multiple electrode sites on a silicon shank and are fabricated using standard microfabrication techniques such as dopant diffusion, physical and chemical vapor deposition, photolithography, and etching processes [34, 125]. Furthermore, it is possible to assemble individual silicon shanks to three-dimensional array geometries [124].

In contrast, UEAs are immediately fabricated as a three-dimensional device. The UEA technology was developed at the University of Utah in the early 1990s [22] and is currently (as of 2019) the only FDA cleared microelectrode array for human

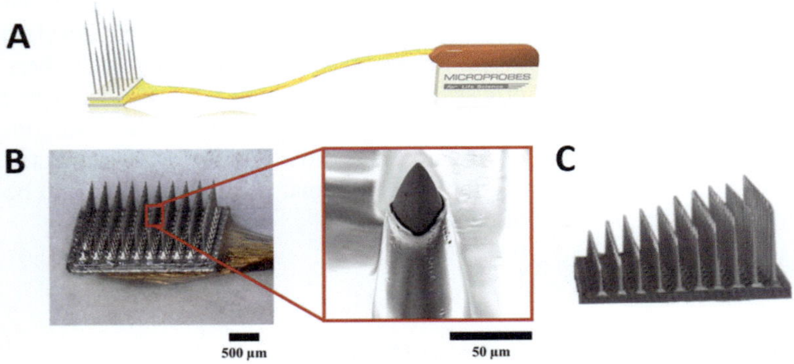

Fig. 1.1 (**a**) A floating microelectrode array architecture. The 18–36 microelectrodes of arbitrary lengths are held together by a lightweight ceramic platform and tethered to a connector by a thin and lightweight cable. (**b**) The Utah electrode array: a 10 × 10 grid of insulated boron doped silicon micro needles. Insulated gold wires are wire bonded before being encapsulated in medical grade silicone to form the assembled Utah electrode array. The magnification shows an active platinum coated electrode site at the tip of a parylene-C coated silicon shaft. (**c**) The Utah slant electrode array: a modification of the UEA with variable shaft lengths to access different depths in the tissue. (**a**. Image copyright and courtesy of Microprobes for Life Science. **c**. Image copyright and courtesy of Blackrock Microsystems LLC)

research use. The arrays consist of typically 10 × 10 electrodes, which are formed by cutting pillar geometries out of a highly p-doped silicon substrate with pre-structured insulation sites. The contact pads for each electrode are deposited and patterned on the back side of the wafer. A subsequent customized wet etching process sharpens the silicon pillars and forms a dense array of silicon needles with lengths varying from 500 to 1500 µm. The tips of the silicon needles are then metallized (typically with platinum or iridium oxide) in a sputtering process to improve the electrochemical charge transfer between the device and the surrounding biological media. To ensure biocompatibility and to protect the array from the foreign body response when implanted, the entire device is coated with parylene-C. Finally, the electrode sites are formed by removing the parylene-C encapsulation coating from the upper 30–50 µm of the metallized silicon tips. Insulated gold wires are wire bonded to the contact pads on the back of the UEA for each individual electrode and attach them to a connector. The bond sites are subsequently encapsulated with medical grade silicone.

Over two decades, the UEA was subject to multiple improvements and material changes [9]. The early UEA electrodes were insulated from each other by p-n junctions and the shafts were encapsulated with polyimide. The current UEA technology, however, uses glass lines between the shaft sites for insulation instead of p-n junctions and parylene-C instead of polyimide. As a modification of the standard UEA geometry with equal shaft lengths, the Utah slant electrode array (USEA) exhibits shaft lengths varying from 750 to 1500 µm. With the advantage of acting at different depths in the tissue, the USEA is commonly implanted into the peripheral nervous system [18]. Figure 1.1b displays an assembled standard UEA with constant shaft lengths of 1000 μm and Fig. 1.1c a USEA with varying shaft lengths.

In addition to the Michigan probes and the UEA technology, recent research efforts led to the introduction of high-density and high-channel silicon-based microelectrodes. Prominent technologies include the CMOS-based high-density microprobe arrays [100] developed at IMTEK/Germany and the neuropixels probes [56] developed at Janelia Research Campus/USA.

1.3 Key Developments of the UEA Platform

1.3.1 Array Shape Variations

Over time, adjustments have been made to the standard UEA geometry to allow for a variety of applications and to improve stimulation and recoding [98].

All shafts of the standard UEA have the same length (Fig. 1.2a). The arrays can be fabricated with a widely varying number of electrodes, ranging from small 4×4 (16 electrodes) arrays to larger ones with 10×10 (100 electrodes). High-density UEAs can comprise over 200 electrodes within the same dimensions as the standard UEAs [106].

A commonly used variation of the UEA is the Utah slant electrode array (USEA), which features electrodes with different lengths in one and constant length in the other direction (Fig. 1.2b). These arrays are favorably used in stimulation and recording in peripheral nerves as they allow to reach different axons in varying depths within the nerve [11, 18]. Furthermore, USEA with particularly long electrodes (1–9 mm compared to the usual 0.5–1.5 mm) have been reported (long USEA), allowing specifically deep penetration into neural tissue [106].

Another development, especially for applications in retinal implants and peripheral nerves, are convoluted shape UEAs. In that case, although all electrodes have the same lengths, they feature a varying penetration depth into the neural tissue due

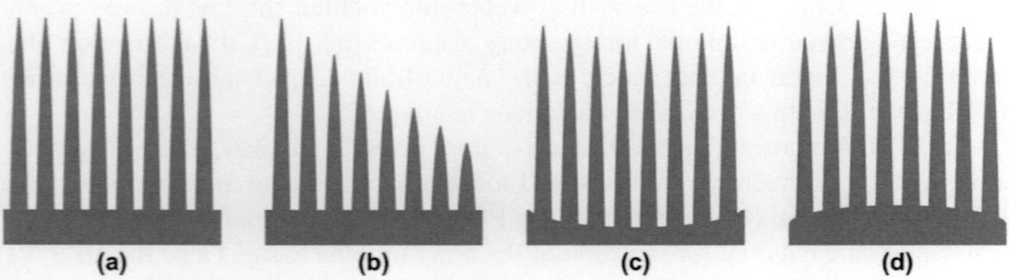

Fig. 1.2 Different geometries of the Utah electrode array: (**a**) standard UEA, (**b**) Utah slant electrode array, (**c**) and (**d**) convoluted shape UEA with concave and convex curvature, respectively

to the electrode base being curved, either in a convex or a concave fashion (Fig. 1.2c, d). According to Bhandari et al., this leads to a better integration of the array with the nerve geometry, consequently less extra-neural volume where fibrotic tissue can grow, and higher selectivity in stimulation as varying nerve sites can be reached [10].

1.3.2 Novel Architectures to Improve Tissue Response: The Natural Buoyancy Utah Electrode Array

Recording and stimulation performance of microelectrodes as well as foreign body response are critical factors limiting the longevity of neural interfaces and must be addressed in the design process [3, 19, 23, 62]. Shape, size, electrode material, tethering forces, and the implantation technique significantly influence the foreign body response to implanted neural electrodes [35, 77].

As an example, blood pulsation and head movements cause permanent relative motions of the brain with respect to the surrounding skull. Since the array must be implanted in the small space between skull and brain, it is subjected to these relative movements. Ideally, it would be advantageous for the electrodes to be mechanically decoupled from one another and to be able to float on the cerebral cortex. Minimizing the micromotions and resulting friction between the brain tissue and the implanted electrodes would reduce the risk of inflammation and the formation of an insulating sheath around the electrodes, leading to device insulation and failure. Numerical studies have shown that causes for severe foreign body response can be limited by flexible and/or soft implants [67, 109]. Hence, by eliminating the rigid backplane of the UEA, the mechanical mismatch between implant and tissue would be significantly reduced. Nevertheless, the backplane holding all 100 electrodes of the UEA in place is indispensable for the device handling and the high velocity insertion process during surgery. A dissolving and biodegradable backplane offering the required mechanical strength prior to and during the implantation process would therefore represent the ideal tradeoff solution. Once implanted, the dissolving backplane would decouple all electrodes from one another. Although the use of dissolvable implantation shuttles showed promising results by mitigating the foreign body response [60, 127], the fabrication of a high-channel count implant, such as the natural buoyancy Utah electrode array (NBUA) [65], with a dissolving backplane is unique.

The NBUA represents a modification of the conventional UEA, where the matrix of up to 10×10 microelectrodes is held together by the biocompatible dissolvable polyethylene glycol (PEG) material. The PEG provides a temporary rigid mechanical back-plane required for insertion of the array into the tissue. Once the 10×10 matrices of electrodes are inside the tissue, the PEG dissolves after it gets in contact with the biological fluid. At that point each electrode is floating independently from the others (Fig. 1.3).

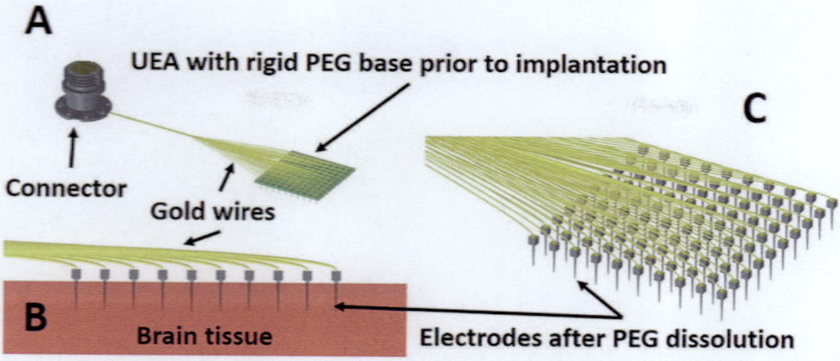

Fig. 1.3 Cartoon image of the dissolvable UEA prior to insertion (**a**). At this point all the 100 shafts are connected by PEG which represents the rigid backplane of the device required for the simultaneous insertion of all 100 electrodes. After insertion, PEG dissolves resulting in mechanically decoupled floating shafts (**b**, **c**). (Reprinted, with permission, from Leber et al. [65])

The fabrication of the NBUA comprises additional steps starting with a standard unwired UEA. Although the main modification is the substitution of the glass traces with PEG, the fabrication of single electrodes results in further challenges. The side walls of single electrodes, previously covered and protected by the glass traces, need to be encapsulated with parylene-C to avoid the exposure of silicon to the biological environment. Furthermore, the packaging process, which includes wire bonding and silicone encapsulation, needs to be adapted to decoupled single electrodes.

The first step in the fabrication process is the removal of the glass lines in between the different shafts of the standard Utah array while keeping the array in place. Glass lines are etched in hydrofluoric acid (HF). To keep the array in place and to protect the shaft and tip metal from HF, the UEA is entirely molded in photoresist with only the back side glass traces exposed. A polyethylene (PE) ring surrounds the array and serves as mold for the photoresist. After the HF etch, the photoresist is removed in acetone, releasing the singularized electrodes solely held together by a 3 μm thick parylene-C layer, which will subsequently be removed. Parylene-C is deposited again onto the array covering this time also the side walls of the electrodes, which were previously protected by the glass traces. Laser ablation disrupts the film connecting the single electrodes and forms the active electrode sites at the tip of each shaft.

Finally, all electrodes embedded in PEG and insulated gold wires are used to wire bond the bond sites on the back of each electrode to a connector. Silicone is applied using a thin platinum wire onto each bond site in order to provide additional mechanical strength to the bonds and to encapsulate the remaining exposed surface. The PEG is then removed from the tip side of the assembly by dipping the assembly into deionized water. The final device is shown in Fig. 1.4.

The PEG backplane has typically a mass of 300 mg, which dissolves in phosphate-buffered saline at room temperature in less than 45 minutes (Fig. 1.5). Electrical impedance spectroscopy experiments after dissolution yielded typical UEA impedance values for the single electrodes.

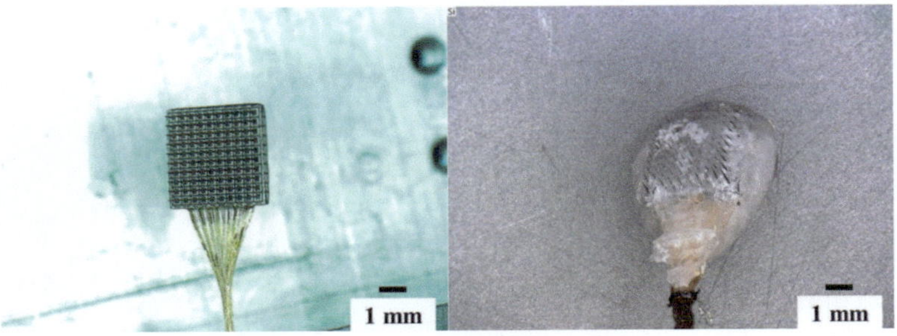

Fig. 1.4 Left: A standard UEA without dissolvable base for comparison. Right: The finalized NBUA device consists of 100 separated electrodes held in place by PEG and connected by gold wires to a connector. (© 2017 IEEE. Reprinted, with permission, from Leber et al. [65])

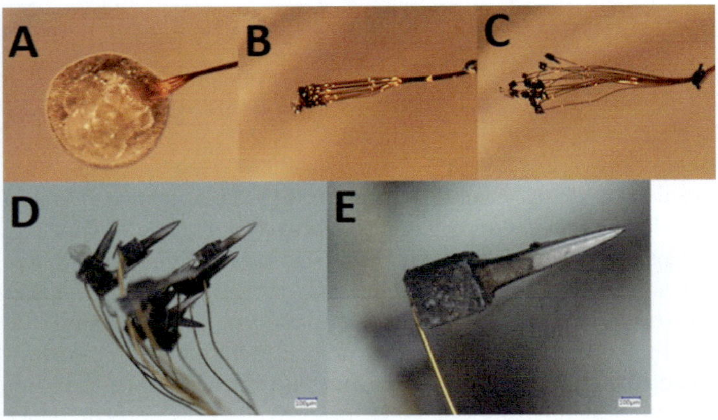

Fig. 1.5 A 4 × 4 device dissolving in PBS at room temperature. (**a**) Rear view of PEG-encapsulated device right after immersion in PBS (t = 0 min). (**b**) Rear view of device with dissolved PEG base (t = 45 min). (**c**) Same as (**b**) with deformed wire to demonstrate freely floating electrodes. (**d** and **e**) Close-up pictures of dissolved UEA and single electrode. (Reprinted, with permission, from Leber et al. [65])

It has not only been shown that the technology of a dissolving back plane is a promising evolution of the UEA, but it also offers the potential for many additional improvements. For once, PEG can be used as a carrier material for drug delivery to reduce inflammation or enhance neuronal growth. Furthermore, contacting the dissolving NBUA with highly flexible polymer-based ribbon cables will reduce the tethering forces induced by the currently used gold wires and provide even less mechanical mismatch between the implant and the tissue.

1.3.3 Ultrahigh-Channel Count Electrode Architectures

In order to reduce damage to and impact on the tissue in which stimulation and recording take place, it is beneficial for neural probes to have a small footprint. In case of the UEA, this requirement in combination with the 3D microelectrode geometry limits the number of active sites as only a certain number of electrodes fit within the device footprint due to the challenges of the dicing-based fabrication process. However, there is substantial interest to increase the number of active sites and their density to address larger fractions of the brain and nerve tissue as well as submillimeter neuroanatomical structures without increasing discomfort and risk of infection for the patient (Wark et al. 2013). Furthermore, the ability to simultaneously stimulate and record in a coordinated fashion at different locations is desired.

In order to enhance both coverage and resolution of the UEA architecture, different paths have been pursued:

1. Multiport devices that connect multiple UEAs to one percutaneous connector. This has allowed up to 1024 channels (sixteen 64-channel UEAs) to be attached to one connector (Fig. 1.6).
2. High-density (HD) UEAs that have a smaller pitch (200 μm, instead of 400 μm) between electrodes. These have been originally developed for the use in the peripheral nervous system as slant electrode arrays [118], but the technology is easily transferable to regular UEAs with constant shaft lengths.
3. Utah multisite electrode arrays (UMEAs) which are UEAs with multiple electrode sites per shank. This allows to increase the number of stimulation/recording sites while maintaining the same array footprint and form factor as the standard UEA [104]. Although initial demonstrators have shown up to 9 contacts per UEA

Fig. 1.6 Multiport UEA with 1024 channels (16 times 64 channel arrays attached to one connector). (Image copyright and courtesy of Blackrock Microsystems LLC)

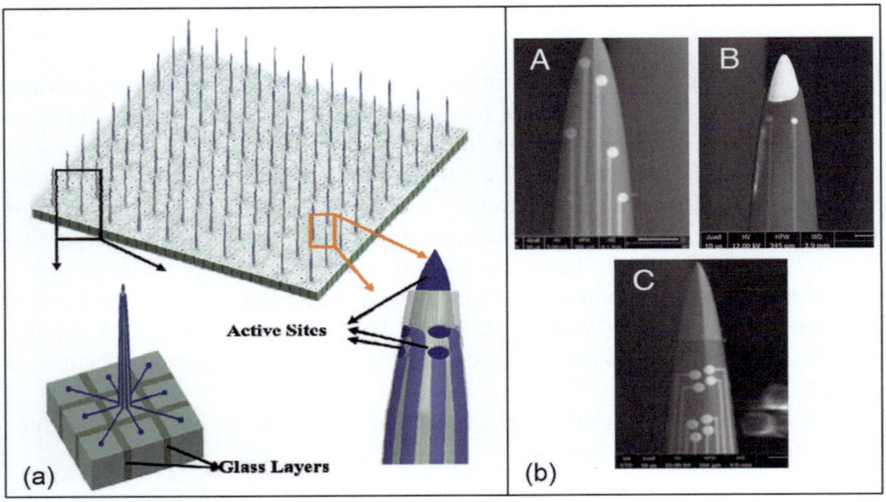

Fig. 1.7 (**a**) Concept of UMEA with up to nine active sites per electrode. (**b**) SEM micrographs of FIB fabricated UMEA with varying configurations: (A) 2 μm metal traces with multiple sites along shaft without tip electrode, (B) 5 μm diameter sites with 2 μm traces, (C) 2 pairs of 5 μm tetrodes. (A. Reprinted, with permission, from Shandhi et al. [103])

shank, practical implementations allow typically 3–4 contacts per shank (Fig. 1.7).

While the multiport and high-density devices are fabricated in the same way as the standard UEA with only small changes to the process, the creation of the Utah multisite electrode array requires significant modification and additional fabrication steps.

Each electrode of the UMEA has one active site at the electrode tip and up to eight active sites along the shaft. This requires the bottom of the electrode to be divided into nine isolated conductive squares which hold the contact pad for each site on the back (the shaft occupies one square). All active sites are connected to the contact pads via metal traces on the shaft's side, as seen in Fig. 1.7a [103]. In order to insulate all active sites from one another, a silicon nitride layer is deposited prior to metal sputtering [104].

Two approaches have been pursued to fabricate UMEAs: the serial direct writing by focused ion beam (FIB) patterning in combination with high-resolution scanning electron microscopy (SEM) [103] and a shadow mask process compatible with batch fabrication [104].

The FIB technique allows to deposit metal (platinum) patterns along the shafts of the UEA with precise control and reproducible resolution of <1 μm. It enables the creation of metal patterns of any shape (circles, square, ring, etc.), size, count, and location for neural recording and stimulation, on the tip of the electrode or along the shank of the Utah electrodes in various configurations [103].

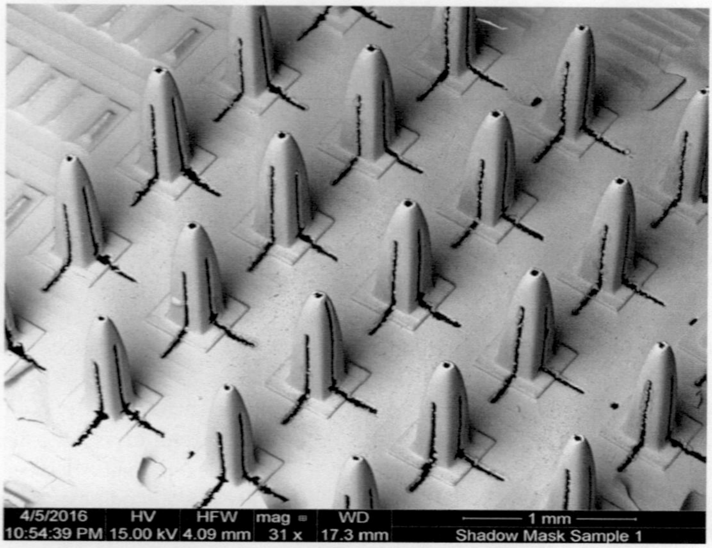

Fig. 1.8 SEM micrograph of the nickel shadow mask fabricated by electroplating and laser ablation on a sacrificial UEA. (Reprinted, with permission, from Shandhi et al. [104])

However, this approach is comparatively slow and expensive, as each active site needs to be written individually. Shandhi et al. have developed a process which addresses this challenge: first, a nickel shadow mask is created on a sacrificial UEA by electroplating and laser ablation (Fig. 1.8). After dissolution of the sacrificial UEA structure, this shadow mask can be used to deposit the metal traces and electrode sites by sputtering of platinum or another desired metal on another UEA during fabrication [104]. Although this approach requires additional fabrication steps compared to the standard UEA, it enables batch fabrication of arrays with an increased number of stimulation/recording sites while maintaining the footprint of a standard UEA.

This novel technology to pattern 3D UEA architectures opens new possibilities for commercial devices. For example, it enables the fabrication of different patterns (such as active sites in tetrode and laminar configuration) and/or in situ sensors for device and tissue health monitoring at the site of implantation. Figure 1.7b shows SEM micrographs of some of the fabricated UMEA configurations.

1.3.4 Improving the Stimulation and Recording Properties: Iridium Oxide and Roughened Electrodes

Neural stimulation and recording electrodes are generally characterized by their impedance, charge injection capacity (CIC), and surface area. It is common practice in the field to express the electrodes' impedance at the frequency of 1 kHz.

The CIC is defined as the highest amount of charge that can be delivered by an electrode within the leading phase of a stimulation pulse without inducing electrolysis of water, which is considered to have harmful effects on the surrounding tissue. Although the CIC is often assumed as a "safe" stimulation parameter, material degradation was observed at lower charge values [90, 91], and tissue damage can be induced by overexcitation of neural tissue regardless of the CIC [30].

The surface area can be differentiated into geometric surface area (GSA) and real surface area (RSA) that additionally considers surface roughness on a nanometer scale. Both impedance and CIC depend on the RSA: the impedance decreases while the CIC increases with increasing RSA [28]. Generally low impedance and high CIC values are desirable for neural stimulation and recording electrodes [119]. Microelectrodes like the UEA ensure high selectivity and spatial resolution to discriminate between single neurons. Consequently, they exhibit a rather small surface area (ca. 2000 μm^2 for UEA electrodes). Hence, a tradeoff must be made between small electrode sizes on the one hand and low impedance along with high CIC values on the other hand. As impedance and CIC depend not only on the GSA and RSA, but also on the material specific electrochemical charge transfer, two approaches to optimize impedance and CIC were pursued in research:

1. Exploration of various electrode coating materials regarding their charge transfer properties.
2. Creation of rough electrode surfaces to increase the RSA/GSA ratio.

Both approaches were investigated for the UEA technology by introducing sputtered iridium oxide as an alternative coating material to the well-established platinum and by roughening the silicon substrate material.

The Introduction of Iridium Oxide as UEA Coating Material

Over the years, iridium oxide has been shown to be one of the most promising electrode coating materials, especially due to its comparatively high CIC values [28]. Iridium oxide films can be created by using several techniques such as electrochemical activation by repeated oxidation and reduction in PBS to form activated iridium oxide films (AIROF) [90, 91], thermal decomposition of iridium salt on metal substrates [92], electrodeposition [72], or by sputtering processes [59]. Sputtered iridium oxide films (SIROF) have the advantage of the possibility to fine-tune process parameters to obtain the desired physical and electrical properties for neural stimulation and recording [80, 122]. Therefore, they were investigated as an alternative coating material to platinum for the UEA electrodes.

Negi et al. [80] studied the effect of pulsed DC reactive sputtering parameters on the resulting SIROF film quality [80]. The sputtering pressure was identified as the critical deposition parameter, which has a major impact on the film morphology, as well as the electrochemical properties of the coatings. At lower pressure values, the SIROF surface exhibits a granular morphology, transitioning to dendritic structures with increasing pressure. Furthermore, impedance and CIC values of the films were measured to be lowest and highest, respectively, at lower pressure values. Further studies confirmed the superior electrochemical properties of SIROF compared to AIROF and platinum. SIROF coated UEA electrodes yielded CIC values of 2 mC/cm^2 compared to 1 mC/cm^2 [83] and 0.3 mC/cm^2 [81] for AIROF and platinum coated electrodes, respectively. At the same time, the impedance values at 1 kHz for SIROF electrodes were measured to be lowest with 6–10 kΩ compared to 120 kΩ [83] and 125 kΩ [81] for AIROF and platinum, respectively. In addition to its superior CIC and impedance values, SIROF was shown to be more resilient against material degradation by analyzing the iridium content in PBS after continuous pulsing, using mass spectrometry [82].

Due to its superior material properties described above, iridium oxide has been validated as an alternative coating material for UEA electrodes and has demonstrated its usefulness in sophisticated human experiments [32].

Silicon Substrate Material Roughening for Improved Electrode Properties Although SIROF coatings have been shown to substantially increase the charge injection properties and to decrease the impedance values of UEA electrodes, platinum is the best studied material for neural electrodes to date, with numerous reported acute and chronic in vivo studies. As a result, much research was conducted on roughening platinum electrode surfaces for improved charge transfer properties. The efforts to create high RSA/GSA ratio platinum surfaces include laser processing [40, 41], electrochemical roughening [15, 115, 121], and electroplating of platinum gray [123, 142] as well as platinum black [38, 71, 85] films. Whereas laser processing is a time-consuming serial processing technique and presents difficulties regarding the three-dimensional UEA geometry, electrochemical roughening and deposition methods add to the process complexity and, in the case of platinum black, raise concerns regarding their mechanical integrity. This led to a substantially new approach to create high RSA/GSA ratio electrode surfaces by roughening the silicon substrate instead of the coating material, allowing the conformal deposition of platinum in a simple sputtering process.

Various physical and chemical roughening processes to alter the silicon surface of UEA tips were investigated. Despite well-described silicon roughening techniques in literature, the unusual high aspect ratio of the UEA architecture required a tailored roughening process. A customized reactive ion etching (RIE) and Bosch-type deep reactive ion etching (DRIE) recipe were developed to roughen the silicon tips of the UEA before coating them with a 500-nm-thick platinum layer [64]. The initial RIE step creates a shallow, well-defined microstructure on the silicon surface, whereas the second DRIE step deepens the pores from the first RIE etch until a minimum electrode impedance is reached, indicating maximum RSA increase. The process

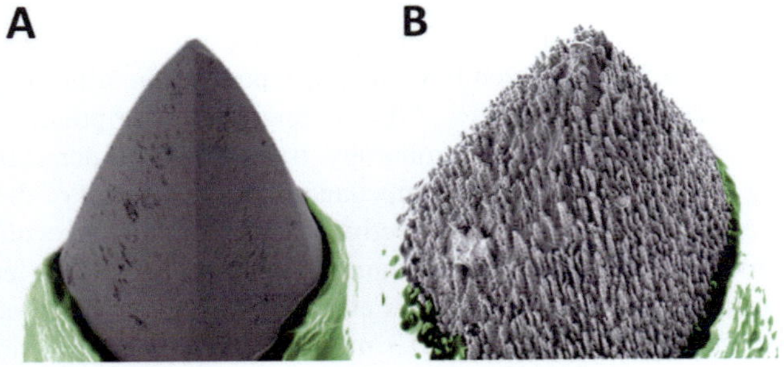

Fig. 1.9 Colored SEM images: (**a**) Standard platinum coated UEA electrode. (**b**) Roughened platinum coated UEA electrode

exhibits high controllability and has the advantage of batch fabrication compatibility, allowing to process the UEAs on wafer level. Fig. 1.9 displays a standard platinum coated UEA electrode and a roughened platinum coated UEA electrode.

Extensive soak testing in PBS over 500 days with regular impedance spectroscopy and CIC measurements demonstrated a decrease of impedance at 1 kHz by a factor of 0.32, and an increase of CIC by a factor of 1.8 of the roughened compared to standard UEA electrodes [63]. Furthermore, the roughened electrodes demonstrated an increased robustness toward stimulation-induced material failure. In a stress test, during which 1 million biphasic charge balanced pulses were applied to both roughened and standard electrodes, the roughened electrodes were found to be less susceptible to the degradation and delamination of the platinum coating.

Furthermore, first in vivo tests with the roughened UEAs were carried out in rodents. Although slightly higher platinum residuals in the neural tissue were detected for the roughened electrodes, a negative effect on the foreign body response and cytotoxicity could not be observed during a histological analysis of the neural tissue after 8 weeks of implantation into the cerebral cortices of rodents [63].

The substantially novel method of substrate material alteration toward increased RSA/GSA ratios of neural electrodes was only studied for platinum coatings to date. SIROF coated roughened UEA electrodes remain to be studied but are expected to even further increase the electrochemical properties such as the electrodes' impedance and CIC values.

1.3.5 Optrodes for Infrared and Optogenetic Stimulation

In the past, most devices that aim to interface with the nervous system and evoke action potentials in neurons relied on electrical stimulation. This is still the case, but there is an increasing interest in using electromagnetic radiation, both in infrared (IR) and visible wavelengths for neural stimulation as these optical methods carry the promise to address some of the challenges of electrical stimulation, such as neuronal specificity, spatial resolution, stimulation-induced tissue damage, and stimulation artifacts [86, 120].

IR stimulation-based approaches have the advantage that they can be employed without preceding alterations to the neurons or their surroundings as the tissue absorbs the radiative energy and the resulting rapid temperature change induces stimulation due to many different processes and effects in the neural tissue [112].

In contrast, for the stimulation with visible wavelengths, the neurons have to be modified via optogenetics. In this technique, chosen as Nature's method of the year 2010 [79], virus vectors are used to genetically target specific neuron subtypes to express light-activated proteins in their cell membrane. Depending on the protein chosen, the thus genetically modified neurons can either be stimulated or inhibited by light irradiation in a specific wavelength range of the visible and near-infrared spectrum with very high specificity [86].

In order to take full advantage of the potential of optical stimulation, the efficient delivery of radiation deep into the neuronal tissue with high spatiotemporal resolution is required in many use cases. Furthermore, a corresponding device should also be able to record the resulting electrochemical response of the neurons to monitor the effect of the stimulation. The simplest penetrating light delivery approach is given by implanting an optical fiber into the neural tissue, often together with a penetrating or surface electrode for electrophysiological measurements of the response of the neural cells [5, 6, 39]. In more advanced approaches of integrated devices, either Michigan-style penetrating electrodes are upgraded with optical waveguides [26] or even whole micro light emitting diodes (μLED) [95, 126]. Alternatively, the UEA platform is adapted to benefit the optical stimulation techniques as discussed in the following.

A way to integrate a UEA-type device with a simple optical fiber approach is given by removing one of the electrodes and creating a hole in the back plane of a UEA via laser ablation. The missing electrode can then be replaced by aligning and fixing a so-called optrode, a metal coated tapered optical fiber, in its place (Fig. 1.10). This combines the relatively simple single site optical fiber stimulation with the spatiotemporal multichannel recording capability of UEA-type devices. Successful optogenetic stimulation with such a device was demonstrated both in vitro and in vivo [117, 141].

Alternatively, the UEA geometry and batch compatible fabrication method can be modified to create novel devices for multichannel light delivery. To create a device for IR radiation transmission, undoped silicon is used as a base material in the fabrication process and the contact pads of the UEA are replaced with an aluminum mask with windows to facilitate the alignment of butt-coupled optical fibers to the silicon needles that are acting as waveguides. Additionally, other fabrication steps necessary for the electrostimulation such as the metal electrode material deposition and the electrical insulation of the individual silicon needles are omitted. The resulting IR optrode array (Fig. 1.10a) was able to efficiently and selectively deliver IR radiation into tissue for stimulation [2, 27]. In order to further enhance the efficiency of the IR transmission and to add the possibility of delivering visible light into tissue for optogenetic stimulation, the base material for the

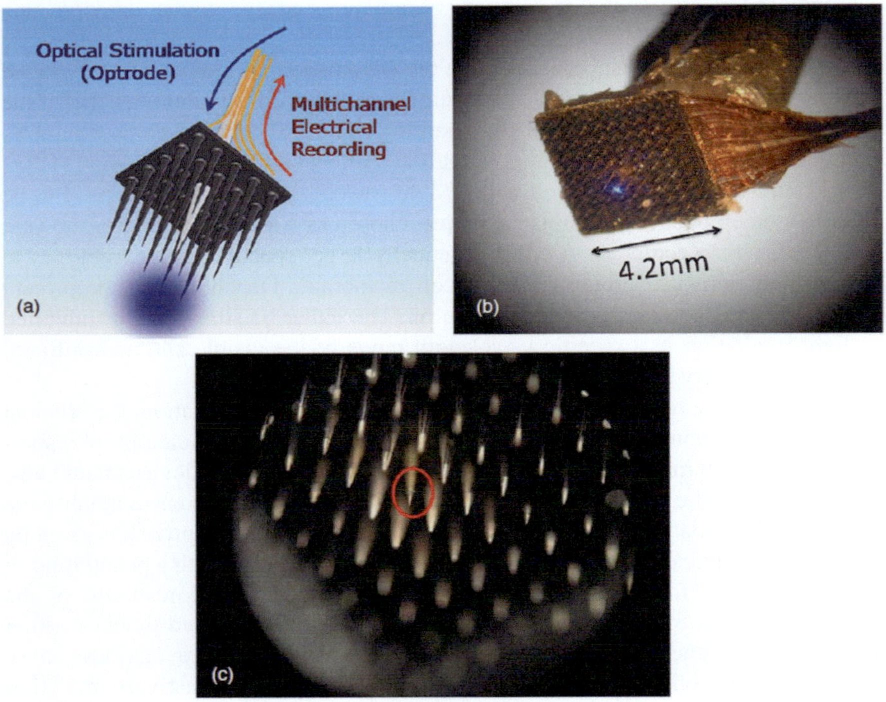

Fig. 1.10 Combination of a multi electrode array based on the Utah electrode array with a sharpened and metal coated optical fiber-based optrode. (**a**) Schematic of the device and (**b**) optical image with blue laser light being transmitted through the optrode. (**c**) Close-up optical image of the device. The position of the optrode between the electrodes is indicated. (Reprinted, with permission, from Zhang et al. [141])

device fabrication was changed from silicon to transparent silicate glass [1, 2]. However, glass behaves differently than silicon especially during the etching step of the UEA fabrication process and no sharp needles could be created this way. Therefore, several additional steps had to be added to the standard UEA fabrication process along with other smaller modifications to create a suitable device geometry. These additional steps include the use of a tapered dicing blade to pre-shape sharp optrode tips as well as a heat treatment step to smoothen the rough glass surface after etching [2, 17, 96].

The resulting passive glass Utah optrode array devices (Fig. 1.11b, c) can be used for light delivery using external stimulation sources. However, for neurological research and future clinical and neuroprosthetic applications, a fully integrated device, featuring both multichannel optical stimulation and electrical recording capacity, would be beneficial. Such a device is currently under development by a combination of the passive glass optrode array with a μLED array and an interposer layer to reduce cross-talk between optrodes (Fig. 1.12). By adding electrical

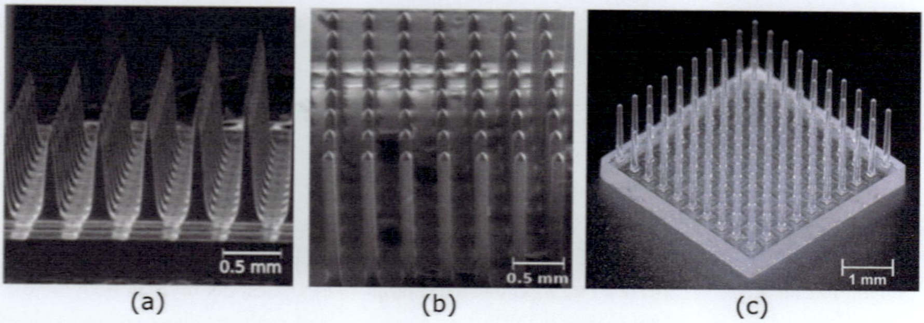

Fig. 1.11 SEM images of optrode arrays fabricated from (**a**) undoped silicon, (**b**) fused silica glass, and (c) optical image of an optrode array made from borosilicate glass. (**a**, **b** Reprinted, with permission, from Abaya et al. [2])

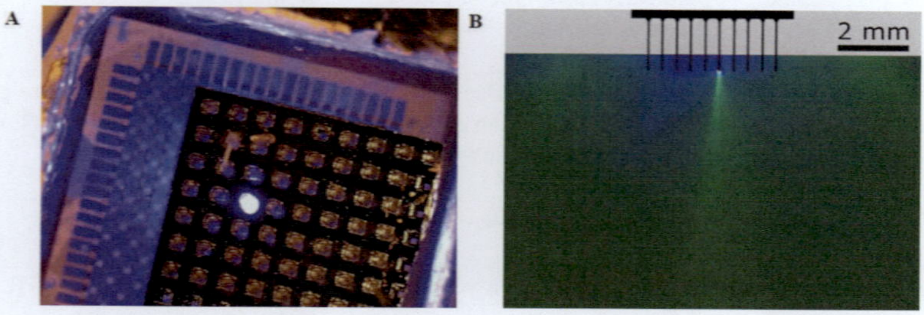

Fig. 1.12 Images of a glass optrode array with a special interposer to prevent cross talk, coupled to a μLED array. In (**a**), one of the μLEDs is turned on. (**b**) shows the emission profile of an optrode when immersed in fluorescein solution. (Reprinted, with permission, from Scharf et al. [96])

recording functionality to the array, a highly flexible device for optogenetic research applications will be created [96].

1.3.6 Advances in Encapsulation

An important requirement for implanted neural devices is the ability to perform their function in vivo for an extended amount of time to justify the risk and cost associated with implantation surgery. However, the physiological environment is hazardous to devices. Water and different ionic species dissolved in body fluids lead to corrosion and other electrochemical degradation processes of the various materials the device consists of, resulting in leakage currents or short circuits of electrical components. This will eventually lead to complete device failure. Therefore, the functional parts of the device, except the active electrode sites that are supposed to interact directly with the neural tissue, must be protected from the physiological environment by encapsulation.

Table 1.1 Methods and materials that can be used for the encapsulation of implantable neural devices along with their advantages and disadvantages

Method and material	Advantages	Disadvantages
Hermetic can [55]	Hermetic (airtight + waterproof), long lifetime	Large, assembly process complexity, need for feedthroughs
Silicon oxide ([43, 78])	Conformal, good insulation	Dissolves in body fluids (except for thermal SiO_2), needs to be defect free – > capacitively coupled electrodes
Silicon nitride ([97, 129])	Low WVTR, good dielectric properties, easy to deposit	Pin holes, dissolves in body fluids
Diamon-like carbon (DLC) [93]	Low friction, low wear, chemically inert, biocompatible, deposition at room temperature possible, doping can control electrical properties	Poor adhesion, tends to delaminate
Ultrananocrystalline diamond (UNCD) ([8, 128])	Chemically inert, biocompatible, high wear resistance, low friction surface	Requires high deposition temperatures
Silicon carbide ([29, 49, 66])	Very hard to dissolve/destroy, biocompatible, doping can control electrical properties, low water intake	Pin-holes, high temperature process, nonuniformity, cracks on polymer substrates
PTFE ([54, 69])	Low friction, biostable, biocompatible	Nonuniformity, byproduct incorporation
Silicone [116]	Biocompatible, cheap, widely used	Nonuniformity, high WVTR
Parylene ([51, 70])	Conformal, low deposition temperature, biocompatible, widely used	Relatively high WVTR, poor adhesion
Polyimide [88]	Biocompatible	Nonuniformity, high water absorption and WVTR

Any material that is used for encapsulation of implantable neural devices has to provide chemical resistance against biological media and their diffusion into the encapsulation material, biocompatibility, and long-term stability. It has to protect the inner parts of the device over extended periods of time in vivo by creating a near-hermetic isolation with a very low water vapor transmission rate (WVTR) [107]. Additionally, it has to ensure the proper electrical functionality of the devices, i.e. preferably should have high impedance and a low dielectric constant to reduce signal noise and ensure high selectivity for interaction with neural tissue [51].

Furthermore, the encapsulation steps need to be compatible with the fabrication or assembly processes, offer the potential for further miniaturization, and be compatible with wire-based or wireless data and power transmission. Identifying an encapsulation method and corresponding materials that fully satisfy all these requirements at the same time is very difficult [134].

Consequently, many different materials have been investigated for encapsulation purposes of implantable neural devices (Table 1.1). Methods can be divided into hermetic and near-hermetic encapsulation. The former can be realized as capsules or small implantable cans made of metal, glass, or ceramic and provide an airtight and waterproof environment [129]. They are rather large which is a drawback for miniaturized implantable devices and, furthermore, require feedthroughs for electrical data and power transfer. Therefore, near-hermetic encapsulation using thin films, which at least significantly slow down the ingress of body fluids, has significant potential [129, 134].

Thin-film encapsulation materials can be grouped into polymeric (e.g. polytetrafluoroethylene (PTFE), silicone elastomers, polyimide, and parylene) and non-polymeric materials (e.g. silicon dioxide, silicon nitride, silicon carbide, and diamond-like carbon (DLC)). Non-polymeric materials require chemical vapor deposition (CVD) processes with rather high temperatures, which is not suitable for neural devices with active electronics that would not sustain these high temperatures. Furthermore, some of them dissolve in saline solution over time, are prone to structural defects (pin-holes), form non-conformal layers, and tend to delaminate.

Polymeric materials are very flexible, inexpensive, and require low process temperatures. Disadvantages include a high WVTR, problems in creating conformal and pin-hole free films, and material degradation as well as poor adhesion [129].

For implantable neural interfaces, many different encapsulation materials have been studied and continuously developed for creating conformal and long-term stable encapsulation. In the following, the ones relevant and used for the UEA will be discussed, which include parylene-C, silicon carbide, and more advanced parylene-C/Al_2O_3 bilayer structures.

1.3.6.1 Parylene-C

Parylene, a semicrystalline polymer for the family of thermoplasts, is a widely used material for encapsulation and insulation of biomedical devices. It exists in several variants, among which parylene-C and parylene-N are certified as class VI polymers for use in biomedical devices by the Food and Drug Administration (FDA) [73, 108].

In the following, we will focus on parylene-C which is used for the UEA. Information on other variants may be found elsewhere [36, 61].

Parylene-C has many advantageous properties: it is highly biocompatible, non-toxic, chemically inert, an excellent ion barrier, has a low dielectric constant (~3.15 at 60 Hz), relatively low water absorption rate (~0.1% in 24 h), high resistivity, and dielectric stability in saline. Furthermore, due to the possibility of employing a CVD process, a very conformal deposition down to the nanoscale and also for complex geometries which include sharp edges and crevices is possible [70, 73, 102, 105, 107, 130, 133].

Due to parylene-C being a polymer, it exhibits some of the corresponding drawbacks as well, which include a relatively poor adhesion to inorganic and metallic substrates and a certain permeability to water [73].

Parylene-C is deposited according to the *Gorham* method, which is a three-step CVD process [140]. First, the precursor (di-para-xylene dimer) is sublimated at ~130 °C. Next, by elevating the temperature to ~680 °C, the dimer is cracked into monomers. It is important to achieve a complete dissociation of dimers into monomers to reduce the number of dimers in the deposited film. Reduction of the temperature (to at least <80 °C) allows the polymerization of the monomers on all available surfaces. The film grows due to a free radical chemical reaction process where first new carbon chains are formed (initiation) and then extended to higher molecular weight by attachment of monomers to the end of the chain [51, 101, 129]. Typical film thicknesses deposited on the UEA are 3–6 μm. Resulting film properties depend on the pressure, temperature, and deposition rate during the CVD process which provides a means to tailor the film's properties within a certain range [50].

The described CVD process has the additional advantage of not requiring any solvents or additives, which reduces the contamination of the film to a minimum and ensures almost defect free and conformal coverage, even of complex electrode geometries [51].

A problem which needs to be addressed during film deposition is the poor adhesion of parylene-C to metal and other inorganic surfaces. In case of the UEA, it has to stick well to silicon and platinum or iridium oxide, depending on the type of tip metallization used. Therefore, measures must be taken to increase the adhesion which include silanization, PECVD deposition of an interlayer, plasma surface treatment, heat treatment, and chemical modifications.

The use of silane (usually Silquest A-174) as adhesion promoter, which is vaporized onto the substrate before deposition, is a common method for the UEA [73, 105, 107]. The organo-silane creates a covalent bonding by functional groups between substrate and parylene-C [45, 135]. It is inexpensive and integrates very well in the process flow of UEA device fabrication.

Seymour et al. [102] developed a reactive parylene-C by adding a functional group and demonstrated that it improves adhesion and short-term insulation performance. Plasma surface treatment or chemical surface modifications of the substrate before parylene-C deposition increases the number of radical sites for covalent bonds, resulting in better adhesion [73]. This can even be enhanced by deposition of a plasma polymer as a thin interlayer between substrate and parylene-C [140]. Although that has been reported to be a very effective method, it requires another complex step and is therefore not used for the UEA [129]. Finally, the parylene-C coated UEA can be annealed to increase the physical interlock between substrate and thin film [73].

All of these methods address the adhesion problem but do not prevent moisture ingress over time.

The function of the UEA is based on the electrode tips being in contact with the neural tissue; hence, the encapsulation has to be removed from this area. Several methods have been proposed and studied, the most common one being the reactive ion etching (RIE) with oxygen plasma by using aluminum foil or photoresist as a mask [10, 107].

Yoo et al. employed laser ablation which is advantageous for mass production and allows to selectively treat individual electrodes. However, this process creates carbon debris and the amount thereof as well as the conformity of removal depend on the laser fluence. The debris can be removed by a subsequent short (1 min) oxygen plasma etch. However, the laser ablation requires a tip metal which can withstand the laser power [138, 139].

Many experiments have been carried out (both in vivo and in vitro) to investigate the long-term stability of parylene-C encapsulation to ensure the safe and reliable performance of chronically implanted neural devices. Test devices include flat interdigitated electrode structures and functional UEAs, which are submerged in phosphate-buffered saline (PBS) solution for prolonged amounts of time (several hundred days to more than a year). Typical temperatures range from 37 °C (human body temperature) up to 80 °C to achieve accelerated aging processes.

Although parylene-C coating can withstand very harsh conditions, such as the use of different sterilization methods and heat treatment [101], all experiments indicated an increasing amount of failed test devices after being subject to physiological saline solution for prolonged times. The failure mechanisms are not yet fully understood, but investigations of failed devices found blistering, microcrack formation, microscopic delamination, and thinning of the parylene-C film. One reason is believed to be the steady water uptake of the material. Furthermore, imperfections at the interface between the electrode and parylene-C coating may induce electrical stresses leading to microcracks and delamination of the film. Once moisture and possibly ions permeate though the encapsulation, chemical reactions degrade the metal film underneath and eventually induce failure of the electrode [73, 130, 133]. These effects are observed for all variants of parylene, although with slightly different characteristics. For example, parylene-N is prone to microcrack formation but does not show delamination [73].

Failing electrodes exhibit an increased leakage current, decreased impedance modulus for low frequencies, and an increased area under the cyclic voltammetry curve used for testing the devices which indicate exposure of underlying metal [73, 105, 130].

To overcome these problems, the single material encapsulation has been advanced to a bilayer of parylene-C and Al_3O_3 which will be discussed in the next section.

1.3.6.2 Parylene-C/Al_2O_3 Bilayer Encapsulation

The use of a bilayer encapsulation evolved due to the observed long-term degradation of parylene-C films in physiological environment. The bilayer encapsulation allows combining the favorable properties of both materials to form a better and more durable encapsulation for neural implants. With parylene-C still being the outer layer in direct contact with tissue, the ease of regulatory approval is ensured with its proven performance in patients for up to 6 years.

Al_2O_3 thin films deposited by atomic layer deposition (ALD) feature the formation of conformal and pin-hole free layers, high hardness and abrasion resistance, good biocompatibility, and perfect electrical performance and are a very good moisture barrier with very low WVTR. However, the major drawback of ALD Al_2O_3 is that it corrodes easily in water and more so in PBS solution [130].

Table 1.2 Relevant properties for the UEA for both individual materials (ALD Al_2O_3 and parylene-C) and the bilayer of these two materials (information taken from [[129], Xie et al. 2014, [20]])

Properties	ALD Al_2O_3	Parylene-C	Effects of bilayer
Water vapor transmission rate (WVTR)	~10^{-6} g/m^2·day [57]	~1 g/m^2·day (from SCS coating)	Isolates water vapor from the device
Deposition temperature	Room temperature to 200 °C	Room temperature	Low temperature coating process
Barrier to ions	Yes	Yes	Good ion barrier
Biocompatible	Yes	Yes	Biocompatible coating
Insulation	10^{14} ohm-cm	10^{15} ohm-cm [133]	High insulation impedance
Coating process	Plasma-assisted ALD	CVD	Pin-hole free, uniform coating
Dissolution in water	Yes [4]	No	Prevents alumina dissolution

By combining ALD Al_2O_3 and parylene-C, this disadvantage is mitigated as parylene-C provides a good ion barrier. It therefore prevents ions from reaching and dissolving the Al_2O_3 layer. Table 1.2 summarizes the properties relevant for use in the UEA for both individual materials and the bilayer. Typical thicknesses used for the UEA are ~50 nm Al_2O_3 and ~3–6 μm parylene-C [21, 131, 133].

Several studies have been carried out since the development of the bilayer encapsulation to investigate its long-term performance for test samples (flat glass IDEs) and actual UEAs soaked in PBS at body temperature as well as for elevated temperatures (for accelerated aging) for prolonged amounts of time (months to more than a year) [21, 73, 130, 131, 134].

These investigations indicated that the bilayer on flat IDE test samples shows a better long-term stability in physiological environment than a single parylene-C film. A comprehensive study from Minnikanti et al. found a mean time to failure (MTTF) of 8 months for single parylene-C encapsulation in saline solution at 37 °C, while the MTTF for the bilayer was 36 months – an increase by a factor of 4.6 and very close to that of other materials like liquid crystal polymers (LCP; up to 2 years) [73]. The MTTF for single parylene-C is consistent with values reported from other groups which are 6–12 months. However, most of these tests were only performed on flat IDE samples which does not reflect the complex geometry of the UEA.

Xie et al. and Caldwell et al. performed lifetime tests on actual UEAs encapsulated with the bilayer. While Xie et al. reported an improved long-term stability even under an increased bias voltage [134], the more comprehensive study (increased number of samples/improved statistics) from Caldwell et al. did not find conclusive evidence for an improved lifetime of the bilayer, neither on more complex flat test samples with additional features nor for the full UEAs [21]. Furthermore, they observed that, if the encapsulation fails, the bilayer is destroyed faster than the single parylene-C film [21]. The rationale in that case is that solution ingress does not have a strong effect for a single parylene-C film. For the bilayer, however, solution

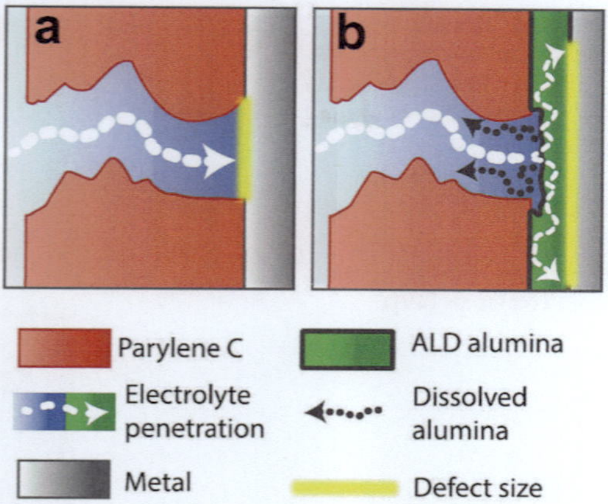

Fig. 1.13 Comparison of defect size for (**a**) single parylene-C encapsulation and (**b**) parylene-C/Al_2O_3 bilayer. (Reprinted, with permission, from Caldwell et al. [21])

permeating through the outer parylene-C layer reaches the underlying Al_2O_3 film which dissolves, leading to the creation of an undercut and effectively to a growth of the defect site (Fig. 1.13).

These results clearly show that imperfections of the parylene-C layer have a much more severe effect in case of the bilayer. Hence, although the bilayer could be an important improvement in the encapsulation of neural devices for long-term implantation, the current materials combination still gives rise to problems and needs to be developed further to allow for chronic implantation over several years.

Steps along that way have to include the development of failure models in order to understand the causes for formation of imperfections and defects of the films. Based on that, materials and deposition processes can be improved. A suggestion made by Caldwell et al. is the use of higher deposition temperatures for the ALD Al_2O_3 which would create less impurities, thereby reducing the probability of dissolving the film [21]. Furthermore, it may become necessary to investigate the feasibility of other materials such as silicon carbide or oxide-nitride-oxide layers in the future [21].

As in the case of a single parylene-C film, the bilayer has to be removed from the electrode tips to ensure contact between tip metallization and tissue. Xie et al. proposed a three-step process similar to that of single parylene removal: (1) laser ablation of the parylene-C layer, (2) reactive ion etching with oxygen plasma to remove carbon debris and parylene-C residue, and (3) buffered oxide etch (BOE) to remove the Al_2O_3 layer and expose the tip metal [132, 133].

1.3.6.3 Silicon Carbide as an Alternative Encapsulation Material

Another material which fulfills the criteria for an encapsulation material for implantable devices (i.e. high hardness, chemical inertness, biocompatibility, low dielectric constant. and low moisture intake) is silicon carbide [29]. Hsu et al. developed a low

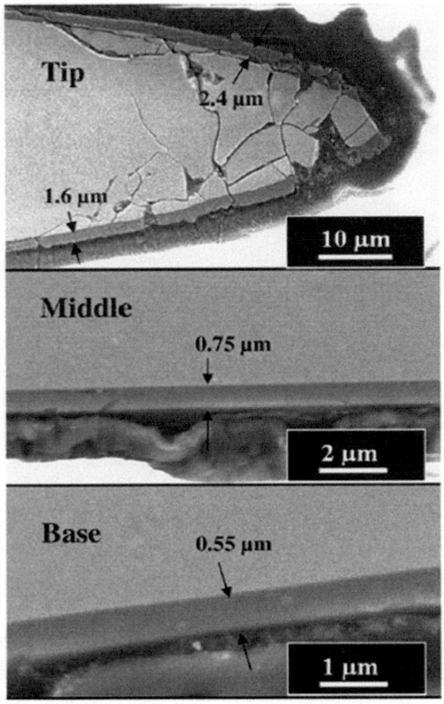

Fig. 1.14 Scanning electron micrographs from a-SiC_x:H films deposited on UEA. The thickness of the film was measured at different sites along the electrode shaft, and a wide variation is observable. (Reprinted, with permission, from Hsu et al. [49])

temperature process for the deposition of hydrogenated amorphous silicon carbide (a-SiC_x:H) on test samples (IDEs) and UEAs and investigated the properties of the resulting film with regard to stress, defect density, and dielectric constant. While for the actual UEAs only the conformity and coating uniformity were studied with electrode cross sections in scanning electron microscope (SEM), electrical tests with coated IDEs soaked in saline solution were performed [49].

The a-SiC_x:H films were deposited by plasma-enhanced CVD at temperatures below 200 °C and with hydrogen as dilution gas source and silane (SiH_4) and methane (CH_4) as Si and C precursors. Through variation of process parameters and subsequent characterization, it became evident that the quality of the deposited film (strength of Si-C bond, film stress) can be optimized and tailored within a certain range. However, while the results were promising for the flat IDE samples, the encapsulation of the UEAs showed a rather large thickness variation along the electrode shafts and a large pin-hole density (Fig. 1.14) [49]. Hence, the encapsulation of UEA with silicon carbide was not viable.

Another issue that has been observed is that silicon carbide is hard to remove which is crucial for the UEA as the metallized tips have to be exposed to get in contact with tissue. Laser ablation and mechanical grinding have been reported but led to alterations of the underlying metal film [29].

Despite these difficulties, silicon carbide stays in the focus as a possible material due to its good biocompatibility and promising results in IDE soak tests. Currently, it is being discussed if it is feasible to create an entire electrode from a-SiC allowing for greater flexibility of electrode structures.

1.3.6.4 Accelerated Aging

In order to evaluate the performance of any encapsulation, its short- and long-term performance in an in vivo environment and its possible failure modes have to be investigated. However, in the development of new processes and exploration of new materials, it is neither feasible nor reasonable to achieve this by chronic implantation studies due to the time and resources necessary for such studies. Therefore, one needs to test new encapsulation materials or processes in vitro in an experimental setup that mimics the physiological environment. This also avoids the ethical dilemma, high cost, and regulatory challenges associated with animal studies. However, these experiments are still time-consuming if the long-term performance has to be evaluated and therefore still pose a problem, especially since usually a large number of iterations are involved in device development. Therefore, accelerated aging methods have to be employed to estimate the lifetime performance of the encapsulation and the whole device in a reasonable amount of time.

A basic in vitro method to test the lifetime of biomedical devices is to immerse them in saline solution of physiological concentration at a temperature of 37 °C to mimic the human body temperature.

In order to induce accelerated aging of the device, a simple method is to increase the temperature of the solution above 37 °C as this will increase the rate at which chemical reactions proceed [46]. For polymeric encapsulation materials and assuming first-order reactions only, it can be shown by using chemical kinetics that increasing the temperature speeds up the reaction rate and therefore the aging process by a factor *f*:

$$f = \exp\left(\frac{R\left(T - 37^{\circ}C\right)}{E_{act}}\right)$$

where R being the universal gas constant (8.314 J mol^{-1}), T the elevated temperature, and E_{act} the activation energy of the aging reaction [52]. The application of this relation, however, has the challenge of determining the activation energy. This is avoided by employing the empirically determined "10-degree rule" which provides a conservative estimate about the accelerated aging process. It states that a temperature increase by 10 °C roughly doubles the rate of many polymeric reactions and therefore the aging process [46, 52]. For example, increasing the temperature by 20 °C allows to test a device for an effective lifetime of 12 months in only 3 months real time according to the "10-degree rule." However, this is also only valid for first-order chemical processes, and both approaches do not apply in cases where the

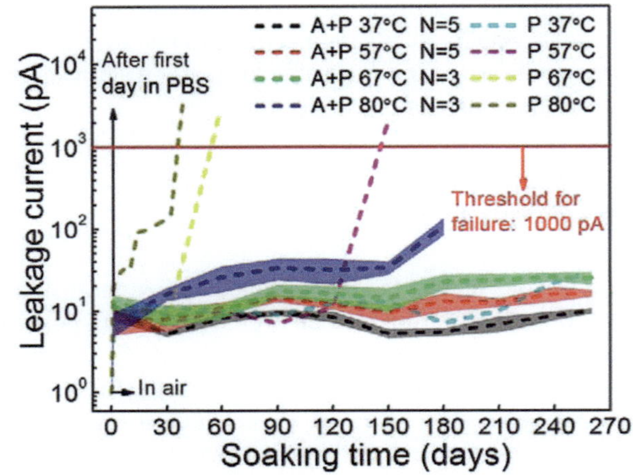

Fig. 1.15 Leaked current of planar interdigitated electrode test structures that were coated with either an Al2O3/parylene-C bilayer (A + P) or parylene-C only (P) and soaked in saline solution at various temperatures for an extended period of time. For the parylene-C only devices, the effect of the different temperatures resulting in different failure times and the improved performance of the bilayer are clearly visible. (Reprinted, with permission, from Xie et al. [131])

elevated temperature induces processes that are not present in normal aging, such as glass transitions or phase changes [52]. Consequently, any accelerated test protocol has to be designed with great care [46].

The application of either the chemical reaction formula or the “10-degree rule” enables the testing of encapsulation materials in a reduced time frame as this allows to state an effective time the material or device was aged.

However, it would be desirable to always test the whole device in accelerated aging studies. However, in case of complex devices with a high resource cost associated with every sample (like the UEA), this is not economically feasible if entirely new encapsulation materials or processes are studied. Therefore, one can employ more simple test structures as for example planar interdigitated electrodes (IDEs) that are coated with the materials to be studied and monitored with electrochemical impedance spectrometry or leakage current measurements in strongly accelerated aging experiments [131]. Figure 1.15 gives an example of how such test structures have been employed to evaluate the performance of an Al_2O_3/parylene-C bilayer compared to parylene-C only encapsulation in accelerated aging experiments.

While these simple planar test structures might give a first indication about encapsulation longevity and performance, respective results have to be taken with a grain of salt when generalizing them to the whole device. Even slightly increasing the complexity of the test structures might lead to different results and makes confirmatory accelerated aging testing of the whole device necessary [21].

Of course, the simple approach of evaluating device lifetime by soaking it in heated saline solution does not capture the multitude of possible chemical processes in an in vivo environment that involve the complex composition of body fluids and the immune system’s foreign body response. The insufficiency of simple immersion in (heated) saline solution is indicated by the inability to reproduce the failure modes observed on UEAs that were chronically implanted in vivo, as shown in Fig. 1.16.

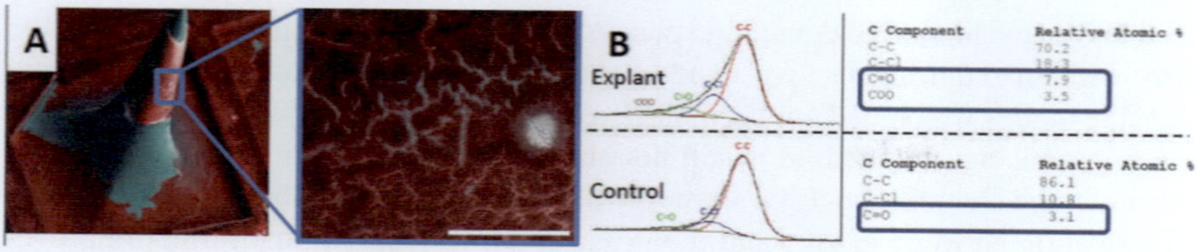

Fig. 1.16 Characterization of explanted UEA. (A) Electrode tine of UEA explanted after 3 years in feline nerve reveals silicon (green) exposed through the damaged parylene-C (red); inset shows detail of parylene-C degradation and cratering over IrO_x (white). Imaging XPS of inset area ("explant") shows evidence of C=O and COO bonds, while identical measurements made on an UEA aged in saline solution ("control") only show nominal levels of C=O binding energies. (Images and data courtesy of Ryan Caldwell)

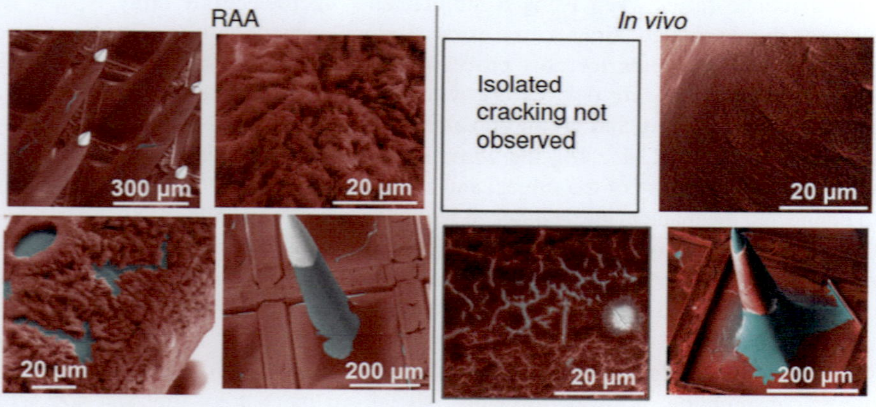

Fig. 1.17 Comparison of Utah electrode array that was subjected to RAA and a device which was explanted after more than 3 years from a feline peripheral nerve. The colors were added to highlight the different materials: red, parylene-C; white, IrO_x; green, Si. (Images and data courtesy of Ryan Caldwell)

In order to improve the predictive capabilities of accelerated aging experiments, it was suggested to use a reactive accelerated aging (RAA) method involving the regular addition of hydrogen peroxide to the saline solution [111]. This method tries to mimic the environment that is created by activated immune cells in response to foreign body intrusion, which includes the release of reactive oxygen species [7, 111].

This RAA method was applied to UEAs, and the damage to the arrays was assessed visually from scanning electron microscopy (SEM) images and chemically from energy dispersive X-ray spectroscopy (EDX) and X-ray photoelectron spectroscopy (XPS) measurements. It was found already from the visual examination that new failure modes had appeared as compared to heated saline solution only devices. Furthermore, comparison to respective measurements on UEAs that were removed after being implanted in a feline peripheral nerve for more than 3 years revealed that the visual damage modes looked very much alike for the explanted and the RAA devices (Fig. 1.17) [20]. This demonstrates the benefit of the RAA method which enables the evaluation of device and encapsulation performance in a fast and realistic way without requiring in vivo experiments. It allows the reproduction of in vivo damage modes which had not been observed before in aging experiments. However, while promising, the method is still not able to fully emulate in vivo aging. As an example, the chemical analysis showed a difference in both the bulk and the surface composition between in vivo and in vitro aged arrays, hinting that further improvements to the accelerated aging methods may be necessary.

1.4 Wireless Technologies for Use with the UEA

The challenge faced when using a multichannel microelectrode array is how to connect the implanted array to an external signal generator (for stimulation) or data evaluation systems for post processing (in case of recording). Most of the existing neural recording devices use wired connections for data communication. Despite the pioneering and ground-breaking results that tethered studies have produced in the past [31, 47, 48], the wired recording configuration faces limitations with regard to mobility of the subject and a significant inherent risk of infection due to connected mechanical stress induced by the constraining cabling [58]. A straightforward solution to overcome these challenges is the application of wireless technology. This, however, comes with its own obstacles: due to the size and power dissipation limitations of the application, the fabrication of neural recording devices with a high-speed wireless link for reliable wideband neural data transmission at very low power budget is extremely difficult. In device design, a tradeoff between power consumption, noise performance, signal quality, battery life, and size will have to be made. To achieve small device sizes, constraints consist of either limiting the number of recording channels [42] or conceding signal fidelity for increased channel count [99].

Over the years, these issues have been addressed one after another, leading to the development of headstage wireless connectors that are not fully implantable but mark an important milestone as they are already used in animals as well as investigative clinical applications.

Such wireless devices have been developed and tested for various animal species applications, such as insects [44, 94], rodents (Seung Bae [42, 68, 110]), sheep [89], and nonhuman primates (NHPs) [16, 24, 37, 53, 74, 99]. Some of these devices were even advanced beyond the simple headstage concept by employing an enclosure fit for subcutaneous implantation [16, 141]. Additionally, there are efforts to create and utilize application-specific integrated circuits (ASICs) to further advance miniaturization and power efficiency in these kind of devices [12, 33, 75, 136].

In order to advance the general idea of wireless implantable neurodevices even further, efforts have been made in parallel to create fully integrated neural interface (INI) devices. In corresponding developments that were directly tailored for the

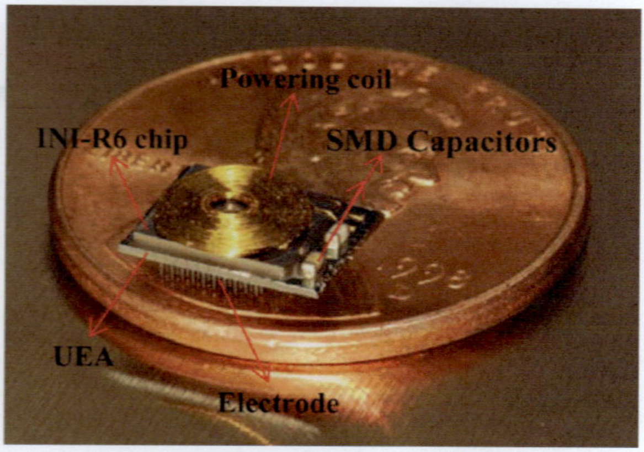

Fig. 1.18 Image of an integrated neural interface based on a modified Utah slant electrode array with flip chip bonded ASIC (in this case an INI-R6 chip) for amplification and spike thresholding with AuPd insulated wire coil and SMD capacitors. The device is inductively powered and transmits data wirelessly. (Reprinted, with permission, from Xie et al. [132])

UEA, an ASIC INI chip designed to fit the architecture and 400 μm interelectrode pitch of the UEA was directly flip-chip bonded to a modified array. The modification of the array allowed to use the backplane as a PCB substrate for direct chip mounting of active and passive components such as the ASIC INI chip, a power antenna, and SMD components (Fig. 1.18) [58, 105, 107, 132].

Due to the technical challenges of system miniaturization, power dissipation limits, and, most importantly, establishing a reliable wireless data link with high throughput in situations with a highly active and mobile patient, wireless recording systems have just recently been introduced on the market. Present examples for such systems are the W-Series recording system by Triangle BioSystems International [113]; the Cube2 system by NeuraLynx, Inc. [84]; the Wireless-Systems by Multichannel Systems [76]; and a series of wireless systems by Blackrock Microsystems LLC. The latter include the Brown Wireless Device (BWD), CerePlex W [14], and CerePlex Exilis [13], which are all headstage-based devices. Figure 1.19 depicts the roadmap of the wireless system development from Blackrock Microsystems LLC, with the first device in the series being the BWD.

Although the roadmap in Fig. 1.19 only covers a short time span of 2 years, the requirements and trends for the development are clearly visible. Preferable properties for wireless headstage-type devices are:

1. Small and lightweight (less than 10 g has been demonstrated for the CerePlex Exilis)
2. High-channel count
3. High data transmission rate
4. Range of few meters for reliable wireless transmission
5. Prolonged battery life-time
6. Offering the possibility for practical sterilization.

Reduced weight is ensured by using polymer-based encapsulation materials which not only allow protection of the electronics from mechanical and electrical impact but are also easily sterilizable. A common way to connect the wireless

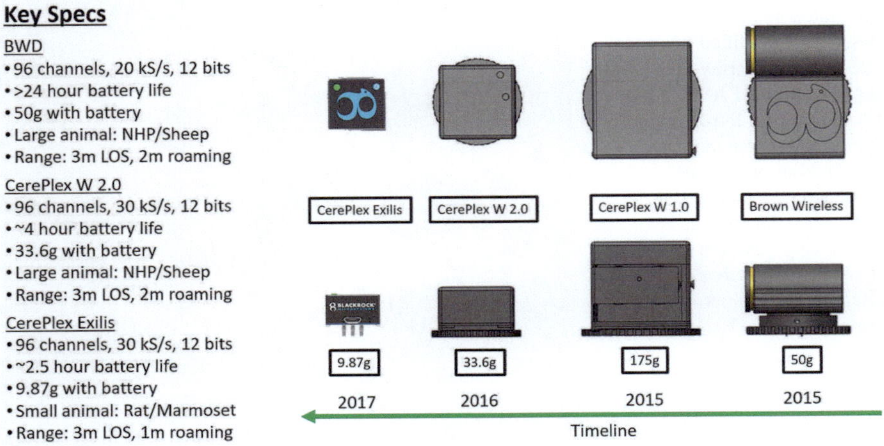

Fig. 1.19 The Roadmap of the 96-channel wireless headstage development at Blackrock Microsystems LLC: from BWD to CerePlex Exilis

transmitter to an implanted microelectrode array is via a "screw-on" attachment to a pedestal connector port on the patient's/subject's skull that is in turn connected to the neural implant. This is illustrated for the BWD from Blackrock Microsystems LLC in Fig. 1.20. The BWD has been extensively tested and is currently actively used in both animal and investigative clinical research.

Figure 1.21 gives an overview of the used wired and wireless components and the recorded neural signals in an experiment with a chair constraint monkey. The comparison clearly shows that the BWD wireless system produced a similar output signal as a typical commercial wired recording system. Other experiments indicated likewise performance. This validates the currently used setup for wireless recording which can be the basis for further developments toward a miniaturized and fully implantable chip.

Although the BWD's performance was promising, there were also drawbacks with regard to design and device performance which needed to be addressed. One of the concerns was that the size and weight of the device was still too big even for large animals and certainly not feasible for small animals such as rodents and marmosets. Motivated by that, the other two devices from the wireless series, CerePlex W and Exilis, were developed and commercialized by Blackrock Microsystems LLC to address these challenges. Compared to the BWD, the CerePlex W and Exilis are much smaller in size and lighter in weight, while providing 1.5 times higher sampling rates and lower noise levels. Furthermore, they both offer the possibility of electrode impedance measurement as an advancement.

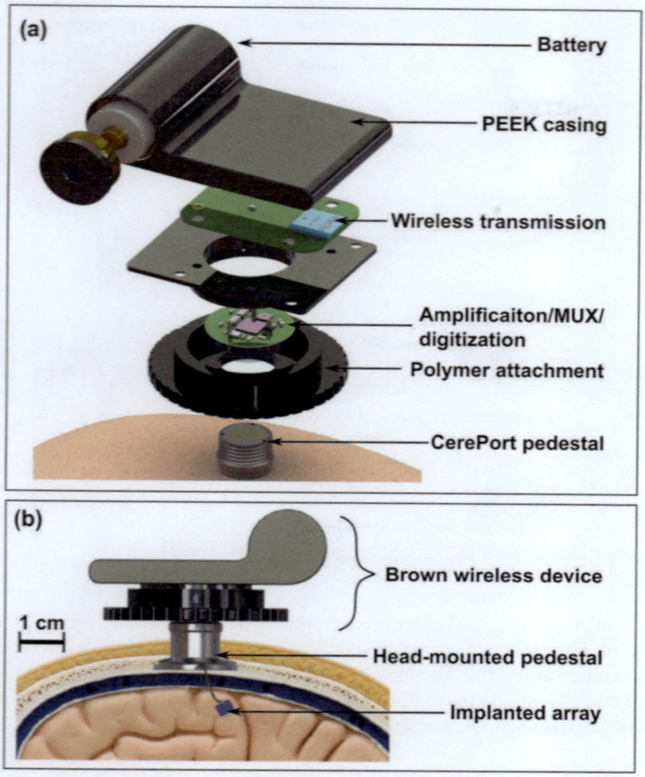

Fig. 1.20 Architecture of the Brown Wireless Device. (**a**) 3D computer-added design (CAD) model showing the complete assembly of the wireless neurosensor. (**b**) 3D CAD model showing the attachment of the wireless neurosensor to a head-mounted pedestal (side view). (Reprinted, with permission, from Yin et al. [137])

1.5 Vision and Concept for Integrated Systems and Implantable Electronics

The ultimate vision for advanced and fully implantable wireless neural interfaces are solutions that integrate the electrode and those electronics necessary to carry out rudimentary signal processing functions, e.g., for amplification and data compression and to facilitate wireless telemetry of data and power. While there is a multitude of options, a potential roadmap, applied to a penetrating electrode such as the UEA, is shown in Fig. 1.22. It has to be noted here that the developmental work on the engineering side is only one side of the medal and that on the other side, regulatory guidelines and constraints have to be met for implantable devices (especially for human use). The stepwise approach depicted in Fig. 1.22 is advantageous to address both parts by successive technological and regulatory iterations of the devices.

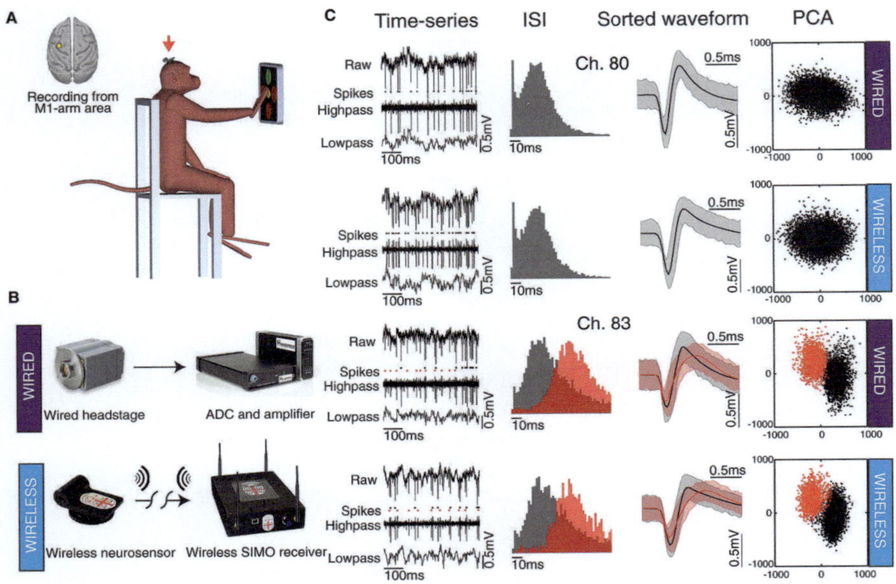

Fig. 1.21 Comparison of recorded neural data from a wired and a wireless setup in an experiment with a monkey. (**a**) Experimental setup and sketch of brain location where the recording took place (primary motor cortex, arm area). The red arrow marks the device location. (**b**) Images of the used devices: the commercial wired system (Cerebus) and the BWD system. (**c**) Recorded neural data for the wired and wireless device. (Reprinted, with permission, from Yin et al. [137])

The neural information system 1 (NIS-1) stage constitutes a conventional passive electrode with percutaneous connector, connected to an external headstage and cable. NIS-2 transfers the headstage function into an implantable device which then generates an amplified, digital signal and allows reduction of the number of wires, independent of the number of electrode channels. An example for this stage is the CerePlex-I system by Blackrock Microsystems LLC. NIS-3 constitutes a passive device with percutaneous connector, attached to a wireless headstage, allowing rigorous testing, market release, and regulatory clearance for a wireless module for high-resolution, high-channel neural data. This is the current development stage for the Utah electrode array by Blackrock Microsystems LLC. NIS-4 provides power through a percutaneous connection but allows data to be transmitted wirelessly from the implant. NIS-5 allows wireless power and data transfer (to a receiver hub in its direct proximity). The NIS-6 stage finally allows wireless power and longer distance transfer directly to, e.g., a beltpack receiver.

The abovementioned CerePlex-I system (Fig. 1.23) is an intermediate step toward a wireless fully integrated system, which transfers part of the headstage functionality to a digitization board which is implanted under the skin on the skull.

The device consists of three main parts: (1) a neural interface such as the UEA for stimulation/recording of neural tissue, (2) a microsystem composed of an

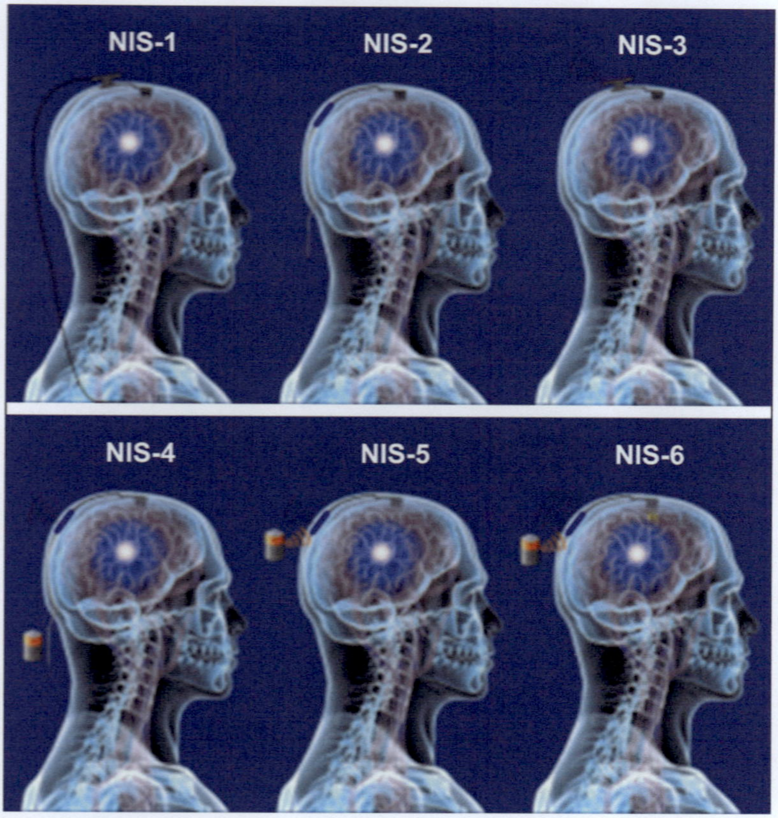

Fig. 1.22 Potential roadmap and concepts for implantable wired and wireless neural interfaces. (Image copyright and courtesy of Blackrock Microsystems LLC)

encapsulated printed circuit board (PCB) containing a 128 channel ASIC implanted under the skin, and (3) an electrode tail connector referred to as the pigtail, connecting the implanted part to the neural signal processing system. A combination of the CerePlex I and multiport approach allows further scaling of channel count.

The vision for the future is the complete miniaturization and integration of all components on the device. A first demonstration toward that goal was given by Sharma et al. [105]. Such a system would allow to fully implant the recording functionality by a microelectrode array in combination with an onboard first data analysis and wireless power and data transmission. Further iterations of the device designed for high-channel count would feature a modular design that amplifies and multiplexes neural signals through an ASIC mounted on the chip but connected with a wired connection to a pedestal connector or a chest-mounted pacemaker style implantable module with rechargeable batteries and wireless data link. With

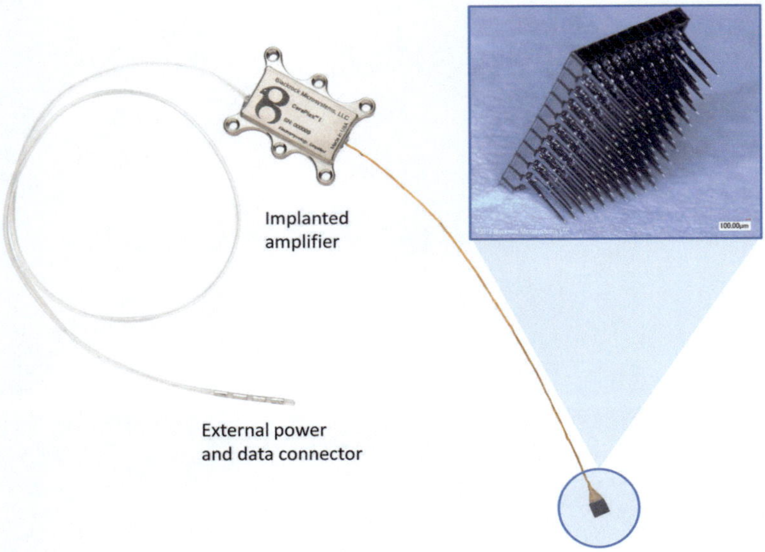

Fig. 1.23 Image of the Blackrock Microsystems LLC CerePlex-I system connected to a UEA. (Image copyright and courtesy of Blackrock Microsystems LLC)

constantly advancing microfabrication leading to decreasing footprint and ultralow power consumption of the necessary electronic components, big steps toward a realization of this vision are expected over the next years.

References

1. Abaya TVF et al (2012) 3D waveguide penetrating arrays for optical neural stimulation. In: 2012 international conference on optical MEMS and nanophotonics. IEEE, pp 216–217
2. Abaya TVF et al (2014) Deep-tissue light delivery via optrode arrays. J Biomed Opt 19 (1):15006
3. Abdo A et al (2011) Floating light-activated microelectrical stimulators tested in the rat spinal cord. J Neural Eng 8(5):56012
4. Abdulagatov AI et al (2011) Al_2O_3 and TiO_2 atomic layer deposition on copper for water corrosion resistance. ACS Appl Mater Interfaces 3(12):4593–4601
5. Adamantidis AR et al (2007) Neural substrates of awakening probed with optogenetic control of hypocretin neurons. Nature 450(7168):420–424
6. Aravanis AM et al (2007) An optical neural interface: *in vivo* control of rodent motor cortex with integrated fiberoptic and optogenetic technology. J Neural Eng 4(3):S143–S156
7. Badwey JA, Karnovsky ML (1980) Active oxygen species and the functions of phagocytic leucocytes. Annu Rev Biochem 49:695–726
8. Bajaj P et al (2007) Ultrananocrystalline diamond film as an optimal cell interface for biomedical applications. Biomed Microdevices 9(6):787–794
9. Barrese JC et al (2013) Failure mode analysis of silicon-based intracortical microelectrode arrays in non-human primates. J Neural Eng 10(6):66014

10. Bhandari R et al (2008) A novel method of fabricating convoluted shaped electrode arrays for neural and retinal prostheses. Sensors Actuators A Phys 145–146:123–130
11. Bhandari R et al (2009) Wafer-scale processed, low impedance, neural arrays with varying length microelectrodes. In: TRANSDUCERS 2009–2009 international solid-state sensors, actuators and microsystems conference. IEEE, pp 1210–1213
12. Biederman W et al (2013) A fully-integrated, miniaturized (0.125 mm^2) 10.5 μW wireless neural sensor. IEEE J Solid State Circuits 48(4):960–970
13. Blackrock Microsystems (2018a) CerePlex Exilis. Available at: http://blackrockmicro.com/cereplex-exilis/. Accessed 27 June 2018
14. Blackrock Microsystems (2018b) Cereplex W. Available at: http://blackrockmicro.com/cereplex-wireless-headstage/. Accessed 27 June 2018
15. Bolzan AE et al (1988) Smooth and rough platinum deposits resulting from the electroreduction of hydrous oxide platinum overlayers—a mechanistic approach. Electrochim Acta 33(12):1743–1751
16. Borton DA et al (2013) An implantable wireless neural interface for recording cortical circuit dynamics in moving primates. J Neural Eng 10(2):26010
17. Boutte RW, Blair S (2016) Maskless wafer-level microfabrication of optical penetrating neural arrays out of soda-lime glass: Utah optrode array. Biomed Microdevices 18(6):115
18. Branner A, Stein RB, Normann RA (2001) Selective stimulation of cat sciatic nerve using an Array of varying-length microelectrodes. J Neurophysiol 85(4):1585–1594
19. Buzsáki G (2004) Large-scale recording of neuronal ensembles. Nat Neurosci 7(5):446–451
20. Caldwell RB (2017) Strategies towards the mitigation of shunting in implanted neural arrays to improve device stability for chronic applications (Order No. 10688653). Available from ProQuest Dissertations & Theses Global. (2189849447)
21. Caldwell R et al (2017) Analysis of Al_2O_3 —parylene C bilayer coatings and impact of microelectrode topography on long term stability of implantable neural arrays. J Neural Eng 14 (4):46011
22. Campbell PK et al (1991) A silicon-based, three-dimensional neural interface: manufacturing processes for an intracortical electrode array. IEEE Trans Biomed Eng 38(8):758–768
23. Castagnola E et al (2014) Smaller, softer, lower-impedance electrodes for human neuroprosthesis: a pragmatic approach. Front Neuroeng 7:8
24. Chestek CA et al (2009) HermesC: low-power wireless neural recording system for freely moving Primates. IEEE Trans Neural Syst Rehabil Eng 17(4):330–338
25. Cheung K (2007) Implantable microscale neural interfaces. Biomed Microdevices 9 (6):923–938
26. Cho I-J, Baac HW, Yoon E (2010) A 16-site neural probe integrated with a waveguide for optical stimulation. In: 2010 IEEE 23rd international conference on micro electro mechanical systems (MEMS). IEEE, pp 995–998
27. Clark GA et al Selective, high-optrode-count, artifact-free stimulation with infrared light via intrafascicular utah slanted optrode arrays. In Photonic Therapeutics and Diagnostics VIII (Vol. 8207, p. 82075I). International Society for Optics and Photonics
28. Cogan SF (2008) Neural stimulation and recording electrodes. Annu Rev Biomed Eng 10:275–309
29. Cogan SF et al (2003) Plasma-enhanced chemical vapor deposited silicon carbide as an implantable dielectric coating. J Biomed Mater Res 67A(3):856–867
30. Cogan SF et al (2016) Tissue damage thresholds during therapeutic electrical stimulation. J Neural Eng 13(2):21001
31. Collinger JL et al (2013) High-performance neuroprosthetic control by an individual with tetraplegia. Lancet 381(9866):557–564
32. Davis TS et al (2016) Restoring motor control and sensory feedback in people with upper extremity amputations using arrays of 96 microelectrodes implanted in the median and ulnar nerves. J Neural Eng 13(3):36001

33. Denison T et al (2007) A 2 uW 100 nV/rtHz chopper-stabilized instrumentation amplifier for chronic measurement of neural field potentials. IEEE J Solid State Circuits 42(12):2934–2945
34. Drake KL et al (1988) Performance of planar multisite microprobes in recording extracellular single-unit intracortical activity. IEEE Trans Biomed Eng 35(9):719–732
35. Ersen A et al (2015) Chronic tissue response to untethered microelectrode implants in the rat brain and spinal cord. J Neural Eng 12(1):16019
36. Fortin JB, Lu T-M (2004) Chemical vapor deposition polymerization: the growth and properties of parylene thin films. Kluwer Academic Publishers, Boston
37. Foster DJ et al (2014) A freely-moving monkey treadmill model. J Neural Eng 11(4):46020
38. Franks W et al (2005) Impedance characterization and modeling of electrodes for biomedical applications. IEEE Trans Biomed Eng 52(7):1295–1302
39. Gradinaru V et al (2007) Targeting and readout strategies for fast optical neural control in vitro and in vivo. J Neurosci 27(52):14231–14238
40. Green RA et al (2012) Variation in performance of platinum electrodes with size and surface roughness. Sens Mater 24(4):165–180
41. Green RA et al (2014) Laser patterning of platinum electrodes for safe neurostimulation. J Neural Eng 11(5):56017
42. Greenwald E et al (2011) A VLSI neural monitoring system with ultra-wideband telemetry for awake behaving subjects. IEEE Trans Biomed Circuits Syst 5(2):112–119
43. Hämmerle H et al (2002) Biostability of micro-photodiode arrays for subretinal implantation. Biomaterials 23(3):797–804
44. Harrison RR et al (2011) Wireless neural/EMG telemetry Systems for Small Freely Moving Animals. IEEE Trans Biomed Circuits Syst 5(2):103–111
45. Hassler C et al (2010) Characterization of parylene C as an encapsulation material for implanted neural prostheses. J Biomed Mater Res B Appl Biomater 9999B(1):266–274
46. Hemmerich KJ (1998) General aging theory and simplified protocol for accelerated aging of medical devices. Med Plast Biomaterials 5:16–23
47. Hochberg LR et al (2006) Neuronal ensemble control of prosthetic devices by a human with tetraplegia. Nature 442(7099):164–171
48. Hochberg LR et al (2012) Reach and grasp by people with tetraplegia using a neurally controlled robotic arm. Nature 485(7398):372–375
49. Hsu J-M et al (2007) Characterization of a-SiCx:H thin films as an encapsulation material for integrated silicon based neural interface devices. Thin Solid Films 516(1):34–41
50. Hsu JM et al (2008) Effect of thermal and deposition processes on surface morphology, crystallinity, and adhesion of Parylene-C. Sens Mater 20(2):87–102
51. Hsu J-MJ et al (2009) Encapsulation of an integrated neural interface device with Parylene C. IEEE Trans Biomed Eng 56(1):23–29
52. Hukins DWL, Mahomed A, Kukureka SN (2008) Accelerated aging for testing polymeric biomaterials and medical devices. Med Eng Phys 30(10):1270–1274
53. Jackson A, Mavoori J, Fetz EE (2007) Correlations between the same motor cortex cells and arm muscles during a trained task, free behavior, and natural sleep in the macaque monkey. J Neurophysiol 97(1):360–374
54. Janting J, Branebjerg J, Rombach P (2001) Conformal coatings for 3D multichip microsystem encapsulation. Sensors Actuators A Phys 92(1–3):229–234
55. Jiang G, Zhou DD (2010) Technology advances and challenges in hermetic packaging for implantable medical devices. In: Zhou D, Greenbaum E (eds) Implantable neural prostheses 2: techniques and engineering approaches. Springer, New York, pp 27–61
56. Jun JJ et al (2017) Fully integrated silicon probes for high-density recording of neural activity. Nature 551(7679):232–236
57. Keuning W et al (2012) Cathode encapsulation of organic light emitting diodes by atomic layer deposited Al_2O_3 films and Al_2O_3/a-SiN_x:H stacks. J Vac Sci Technol A Vac Surf Films 30(1):01A131
58. Kim S et al (2009) Integrated wireless neural interface based on the Utah electrode array. Biomed Microdevices 11(2):453–466

59. Klein JD, Clauson SL, Cogan SF (1989) Morphology and charge capacity of sputtered iridium oxide films. J Vac Sci Technol A 7(5):3043–3047
60. Kozai TDY et al (2014) Chronic tissue response to carboxymethyl cellulose based dissolvable insertion needle for ultra-small neural probes. Biomaterials 35(34):9255–9268
61. Kuppusami S, Oskouei RH (2015) Parylene coatings in medical devices and implants: a review. Univ J Biomed Eng 3(2):9–14
62. Leach J, Achyuta AKH, Murthy SK (2010) Bridging the divide between neuroprosthetic design, tissue engineering and neurobiology. Front Neuroeng 2:18
63. Leber M (2018) Unpublished (in progress). *unpublished*
64. Leber M et al (2016) Different methods to alter surface morphology of high aspect ratio structures. Appl Surf Sci 365:180
65. Leber M et al (2017) Novel method of fabricating self-dissolvable and freely floating neural array. In 2017 19th international conference on solid-state sensors, actuators and microsystems (TRANSDUCERS). IEEE, pp 1726–1729
66. Ledermann N et al (2000) Sputtered silicon carbide thin films as protective coating for MEMS applications. Surf Coat Technol 125(1–3):246–250
67. Lee H et al (2005) Biomechanical analysis of silicon microelectrode-induced strain in the brain. J Neural Eng 2(4):81–89
68. Lee SB et al (2013) A wideband dual-antenna receiver for wireless recording from animals behaving in large arenas. IEEE Trans Biomed Eng 60(7):1993–2004
69. Limb SJ et al (1998) Growth of fluorocarbon polymer thin films with high CF2 fractions and low dangling bond concentrations by thermal chemical vapor deposition. Appl Phys Lett 68 (20):2810
70. Loeb GE et al (1977) Parylene as a chronically stable, reproducible microelectrode insulator. IEEE Trans Biomed Eng BME-24(2):121–128
71. Marrese CA (1987) Preparation of strongly adherent platinum black coatings. Anal Chem 59 (1):217–218
72. Meyer RD et al (2001) Electrodeposited iridium oxide for neural stimulation and recording electrodes. IEEE Trans Neural Syst Rehabil Eng 9(1):2–11
73. Minnikanti S et al (2014) Lifetime assessment of atomic-layer-deposited Al2O3–Parylene C bilayer coating for neural interfaces using accelerated age testing and electrochemical characterization. Acta Biomater 10(2):960–967
74. Miranda H et al (2010) HermesD: a high-rate long-range wireless transmission system for simultaneous multichannel neural recording applications. IEEE Trans Biomed Circuits Syst 4 (3):181–191
75. Muller R et al (2014) A miniaturized 64-channel 225μW wireless electrocorticographic neural sensor. In: 2014 IEEE international solid-state circuits conference digest of technical papers (ISSCC). IEEE, pp 412–413
76. Multi Channel Systems (2018) Wireless-systems. Available at: https://www.multichannelsystems.com/products/wireless-systems. Accessed 27 June 2018
77. Musallam S et al (2007) A floating metal microelectrode array for chronic implantation. J Neurosci Methods 160(1):122–127
78. Najafi K, Wise KD, Mochizuki T (1985) A high-yield IC-compatible multichannel recording array. IEEE Trans Electron Devices 32(7):1206–1211
79. Nature (2011) Method of the year 2010. Nat Methods 8(1):1–1
80. Negi S et al (2009) Effect of sputtering pressure on pulsed-DC sputtered iridium oxide films. Sensors Actuators B Chem 137(1):370–378
81. Negi S, Bhandari R, Rieth L, Solzbacher F (2010a) In vitro comparison of sputtered iridium oxide and platinum-coated neural implantable microelectrode arrays. Biomed Mater (Bristol, England) 5(1):15007
82. Negi S, Bhandari R, Rieth L, Van Wagenen R et al (2010b) Neural electrode degradation from continuous electrical stimulation: comparison of sputtered and activated iridium oxide. J Neurosci Methods 186(1):8–17

83. Negi S, Bhandari R, Solzbacher F (2012) Morphology and electrochemical properties of activated and sputtered iridium oxide films for functional electrostimulation. J Sensor Technol 2(3):138–147
84. NeuraLynx (2018) Cube2. Available at: https://neuralynx.com/hardware/cube2. Accessed 27 June 2018
85. Park S et al (2010) Nanoporous Pt microelectrode for neural stimulation and recording: in vitro characterization. J Phys Chem C 114(19):8721–8726
86. Pashaie R et al (2014) Optogenetic brain interfaces. IEEE Rev Biomed Eng 7:3–30
87. Prochazka A, Mushahwar VK, McCreery DB (2001) Neural prostheses. J Physiol 533 (1):99–109
88. Richardson RR, Miller JA, Reichert WM (1993) Polyimides as biomaterials: preliminary biocompatibility testing. Biomaterials 14(8):627–635
89. Rizk M et al (2009) A fully implantable 96-channel neural data acquisition system. J Neural Eng 6(2):26002
90. Robblee LS et al (1983a) Electrical stimulation with Pt electrodes. VII. Dissolution of Pt electrodes during electrical stimulation of the cat cerebral cortex. J Neurosci Methods 9 (4):301–308
91. Robblee LS, Lefko JL, Brummer SB (1983b) Activated Ir: an electrode suitable for reversible charge injection in saline solution. J Electrochem Soc 130(3):731
92. Robblee LS et al (1985) Charge injection properties of thermally-prepared iridium oxide films. MRS Proc 55:303
93. Roy RK, Lee K-R (2007) Biomedical applications of diamond-like carbon coatings: a review. J Biomed Mater Res B Appl Biomater 83B(1):72–84
94. Sato H et al (2009) Remote radio control of insect flight. Front Integr Neurosci 3:24
95. Scharf R et al (2016) Depth-specific optogenetic control in vivo with a scalable, high-density μLED neural probe. Sci Rep 6(1):28381
96. Scharf R, Reiche CF, McAlinden N, Cheng Y, Xie E, Sharma R, Tathireddy P, Rieth L, Mathieson K, Blair S (2018) A compact integrated device for spatially selective optogenetic neural stimulation based on the Utah Optrode Array. In Optogenetics and Optical Manipulation 2018 (Vol. 10482, p. 104820M). International Society for Optics and Photonics
97. Schmitt G et al (1999) Passivation and corrosion of microelectrode arrays. Electrochim Acta 44(21–22):3865–3883
98. Schwartz AB et al (2006) Brain-controlled interfaces: movement restoration with neural prosthetics. Neuron 52(1):205–220
99. Schwarz DA et al (2014) Chronic, wireless recordings of large-scale brain activity in freely moving rhesus monkeys. Nat Methods 11(6):670–676
100. Seidl K et al (2011) CMOS-based high-density silicon microprobe arrays for electronic depth control in Intracortical neural recording. J Microelectromech Syst 20(6):1439–1448
101. Selbmann F et al (2016) Deposition of Parylene C and characterization of its hermeticity for the encapsulation of MEMS and medical devices. In: 2016 IEEE 11th annual international conference on nano/micro engineered and molecular systems (NEMS). IEEE, pp 427–432
102. Seymour JP et al (2009) The insulation performance of reactive parylene films in implantable electronic devices. Biomaterials 30(31):6158–6167
103. Shandhi MMH et al (2015) A novel method of fabricating high channel density neural array for large neuronal mapping. In: 2015 Transducers – 2015 18th international conference on solid-state sensors, actuators and microsystems (TRANSDUCERS). IEEE, pp 1759–1762
104. Shandhi MMH et al (2017) Reusable high aspect ratio 3-D nickel shadow mask. J Microelectromech Syst 26(2):376–384
105. Sharma A et al (2010) Long term *in vitro* stability of fully integrated wireless neural interfaces based on Utah slant electrode array. Appl Phys Lett 96(7):73702
106. Sharma R et al (2011) Application-specific customizable architectures of Utah neural interfaces. Procedia Eng 25:1016–1019

107. Sharma A et al (2012) Evaluation of the packaging and encapsulation reliability in fully integrated, fully wireless 100 channel Utah slant electrode Array (USEA): implications for long term functionality. Sensors Actuators A Phys 188:167–172
108. Stadler S, Ajmera PK (2008) Sensors and materials: an international journal on sensor technology. Scientific Publ. Division of MYU, Tokyo
109. Subbaroyan J, Martin DC, Kipke DR (2005) A finite-element model of the mechanical effects of implantable microelectrodes in the cerebral cortex. J Neural Eng 2(4):103–113
110. Szuts TA et al (2011) A wireless multi-channel neural amplifier for freely moving animals. Nat Neurosci 14(2):263–269
111. Takmakov P et al (2015) Rapid evaluation of the durability of cortical neural implants using accelerated aging with reactive oxygen species. J Neural Eng 12(2):26003
112. Tan X et al (2018) Auditory neural activity in congenitally deaf mice induced by infrared neural stimulation. Sci Rep 8(1):388
113. Triangle Biosystems International (2018) W-Series. Available at: http://www.trianglebiosystems.com/w-series-systems.html. Accessed 27 June 2018
114. Troyk PR, Cogan SF (2005) Sensory neural prostheses. In: Neural engineering. Springer, Boston, pp 1–48
115. Tykocinski M et al (2001) Chronic electrical stimulation of the auditory nerve using high surface area (HiQ) platinum electrodes. Hear Res 159:53–68
116. Vanhoestenberghe A, Donaldson N (2013) Corrosion of silicon integrated circuits and lifetime predictions in implantable electronic devices. J Neural Eng 10(3):31002
117. Wang J et al (2012) Integrated device for combined optical neuromodulation and electrical recording for chronic *in vivo* applications. J Neural Eng 9(1):16001
118. Wark HAC et al (2013) A new high-density (25 electrodes/mm^2) penetrating microelectrode array for recording and stimulating sub-millimeter neuroanatomical structures. J Neural Eng 10(4):45003
119. Wellman SM et al (2017) A materials roadmap to functional neural Interface design. Adv Funct Mater 28(12):1701269
120. Wells J et al (2005) Optical stimulation of neural tissue in vivo. Opt Lett 30(5):504
121. Weremfo A et al (2015) Investigating the interfacial properties of electrochemically roughened platinum electrodes for neural stimulation. Langmuir 31(8):2593–2599
122. Wessling B, Mokwa W, Schnakenberg U (2006) RF-sputtering of iridium oxide to be used as stimulation material in functional medical implants. J Micromech Microeng 16(6):S142–S148
123. Whalen JJ et al (2006) Electrochemical characterization of charge injection at electrodeposited platinum electrodes in phosphate buffered saline. J Electrochem Soc 153(12):C834
124. Wise KD (2005) Silicon microsystems for neuroscience and neural prostheses. IEEE Eng Med Biol Mag 24(5):22–29
125. Wise KD, Angell JB, Starr A (1970) An integrated-circuit approach to extracellular microelectrodes. IEEE Trans Biomed Eng BME-17(3):238–247
126. Wu F et al (2015) Monolithically integrated μLEDs on silicon neural probes for high-resolution Optogenetic studies in behaving animals. Neuron 88(6):1136–1148
127. Xiang Z et al (2014) Ultra-thin flexible polyimide neural probe embedded in a dissolvable maltose-coated microneedle. J Micromech Microeng 24(6):65015
128. Xiao X et al (2006) In vitro andin vivo evaluation of ultrananocrystalline diamond for coating of implantable retinal microchips. J Biomed Mater Res B Appl Biomater 77B(2):273–281
129. Xie X (2013) Atomic layer deposited aluminum oxide and parylene C bi-layer encapsulation for biomedical implantable devices (Order No. 3607382). Available from ProQuest Dissertations & Theses Global. (1492997360)
130. Xie X et al (2012) Plasma-assisted atomic layer deposition of Al_2O_3 and parylene C bi-layer encapsulation for chronic implantable electronics. Appl Phys Lett 101(9):93702
131. Xie X et al (2013) Long-term bilayer encapsulation performance of atomic layer deposited Al2O3 and parylene c for biomedical implantable devices. IEEE Trans Biomed Eng 60 (10):2943–2951

132. Xie X, Rieth L, Williams L et al (2014a) Long-term reliability of Al2O3 and Parylene C bilayer encapsulated Utah electrode array based neural interfaces for chronic implantation. J Neural Eng 11(2):26016
133. Xie X, Rieth L, Negi S et al (2014b) Self aligned tip Deinsulation of atomic layer deposited Al2O3 and Parylene C coated Utah electrode Array based neural interfaces. J Micromech Microeng Struct Devices Syst 24(3):35003
134. Xie W, Kothari V, Terry BS (2015) A bio-inspired attachment mechanism for long-term adhesion to the small intestine. Biomed Microdevices 17(4):68
135. Yao Q et al (1999) Adhesion enhancement of underfill materials by silane additives. In: Proceedings – international symposium on advanced packaging materials: processes, properties and interfaces. IMAPS – International microelectronics assembly and packaging society, pp 27–30
136. Yin M et al (2013) A 100-channel hermetically sealed implantable device for chronic wireless neurosensing applications. IEEE Trans Biomed Circuits Syst 7(2):115–128
137. Yin M et al (2014) Wireless neurosensor for full-spectrum electrophysiology recordings during free behavior. Neuron 84(6):1170–1182
138. Yoo J-M et al (2012) Hybrid laser and reactive ion etching of Parylene-C for deinsulation of a Utah electrode array. J Micromech Microeng 22(10):105036
139. Yoo J-M et al (2013) Excimer laser deinsulation of Parylene-C on iridium for use in an activated iridium oxide film-coated Utah electrode array. J Neurosci Methods 215(1):78–87
140. Yu Q, Deffeyes J, Yasuda H (2001) Engineering the surface and interface of Parylene C coatings by low-temperature plasmas. Prog Org Coat 41(4):247–253
141. Zhang J et al (2009) Integrated device for optical stimulation and spatiotemporal electrical recording of neural activity in light-sensitized brain tissue. J Neural Eng 6(5):055007
142. Zhou D (2007) Platinum electrode surface coating and method for manufacturing the same

Chapter 2
EEG-Based Brain-Computer Interfaces

Yijun Wang, Masaki Nakanishi, and Dan Zhang

Abstract Brain-computer interfaces (BCIs) provide a direct communication channel between human brain and output devices. Due to advantages such as non-invasiveness, ease of use, and low cost, electroencephalography (EEG) is the most popular method for current BCIs. This chapter gives an overview of the current EEG-based BCIs for the main purpose of communication and control. This chapter first provides a taxonomy of the EEG-based BCI systems by categorizing them into three major groups: (1) BCIs based on event-related potentials (ERPs), (2) BCIs based on sensorimotor rhythms, and (3) hybrid BCIs. Next, this chapter describes challenges and potential solutions in developing practical BCI systems toward high communication speed, convenient system use, and low user variation. Then this chapter briefly reviews both medical and non-medical applications of current BCIs. Finally, this chapter concludes with a summary of current stage and future perspectives of the EEG-based BCI technology.

Keywords Brain-computer interfaces · EEG-based BCI · hybrid BCI · ERP · SSVEP

2.1 Introduction

Since Vidal first proposed the term "brain-computer interface (BCI)" in early 1970s [1], BCIs have received increasing attention in the fields of neuroscience, neural engineering, and clinical rehabilitation [2–4]. Especially during the past two

Y. Wang (✉)
Institute of Semiconductors, Chinese Academy of Sciences, Beijing, China
e-mail: wangyj@semi.ac.cn

M. Nakanishi
Institute for Neural Computation, University of California San Diego, San Diego, CA, USA

D. Zhang
Department of Psychology, Tsinghua University, Beijing, China

X. Zheng (ed.), *Neural Interface: Frontiers and Applications*,
Advances in Experimental Medicine and Biology 1101,
https://doi.org/10.1007/978-981-13-2050-7_2

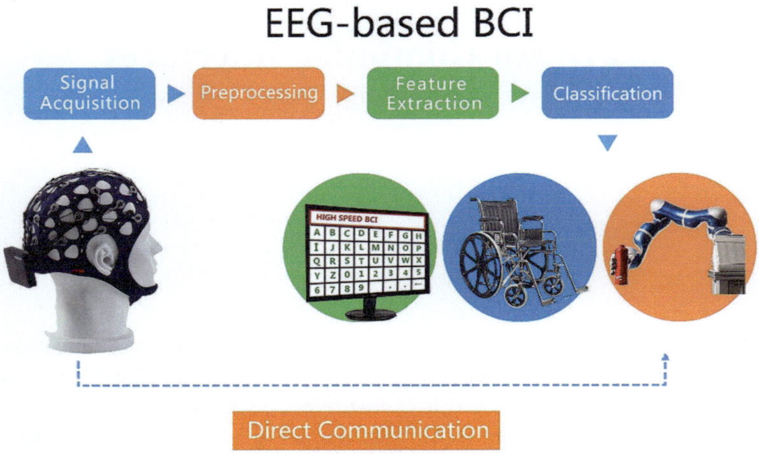

Fig. 2.1 System diagram of a typical EEG-based BCI system for communication and control

decades, BCI has rapidly become a hot topic in both research and practice. The primary goal of BCI research is to provide a non-muscular communication channel to help people with severe motor disabilities to communicate with their environments. Among different brain imaging techniques that have been applied to BCIs, electroencephalography (EEG) has become the most popular method due to its advantages such as non-invasiveness, ease of use, and low cost. Figure 2.1 shows the system diagram of a typical EEG-based BCI system for communication and control, which consists of three major components: signal acquisition, data processing, and device control. Although the scope of application of EEG-based BCIs has been significantly extended in recent years, this chapter mainly focuses on the traditional EEG-based BCIs for the purpose of communication and control.

In the twentieth century, early works by research pioneers laid important groundwork for the BCI field. In the 1970s, Vidal first proposed the term of BCI for the purpose of controlling external apparatus with brain signals and developed a BCI system based on visual event-related potentials (ERPs) [5]. The system converted ERPs corresponding to four retinotopic maps of a checkerboard stimulus into movement control commands to move a mobile through a maze. In 1988, Farwell and Donchin described a mental prosthesis system using the P300 component of ERPs [6]. The system consisting of a six-by-six matrix served as a keyboard and was operated by flash elicited P300 at a speed of ~2 characters per minute. In the early 1990s, BCI systems based on sensorimotor rhythms emerged. Woplaw et al. trained subjects to learn self-control of amplitude of mu rhythm [7]. The system realized one-dimensional cursor control by translating mu rhythm amplitudes into cursor movements. Pfurtscheller et al. performed a BCI system based on event-related desynchronization (ERD) of sensorimotor rhythms [8]. The system was operated by discriminating the spatio-temporal patterns of ERD during imagination of left and right hand movements. During this period, the ERP-based BCI paradigms were further extended. Sutter developed a brain response interface using pseudo-random sequence modulated visual evoked potentials (VEPs) in 1992 [9]. The eight-by-eight visual keyboard recognized the user's eye gaze direction using VEPs and obtained high communication rates. In the late 1990s, Birbaumer et al. developed a spelling device for the completely paralyzed patients based on slow cortical potentials (SCPs) [10]. The system allowed the subjects to select letters by driving a cursor with their SCPs using an imagery strategy.

Almost two decades into the twenty-first century, BCI research has achieved significant progress and shown promising prospects and applications in various fields. The progress mainly includes the following aspects: signal processing [11, 12], machine learning [13, 14], system paradigm [15, 16], and applications [17, 18]. Advanced EEG signal processing and machine learning algorithms have been applied to significantly improve the performance of the early developed BCI paradigms (e.g., BCIs based on P300 and sensorimotor rhythms). Meanwhile, new BCI paradigms have been proposed and achieved rapid development. Various types of visual and auditory BCIs based on ERPs have been developed and proven highly efficient in terms of information transfer rate (ITR) [19]. A hybrid BCI system that combines a traditional BCI with other physiological signals has shown advantages in BCI performance and system flexibility [20]. Over the past decade, a wide range of BCI applications have emerged, which considerably expanded the original goals. Assistive BCIs and rehabilitative BCIs, which are designed for communication and motor rehabilitation, have been widely investigated in experimental and clinical studies [21]. In addition, the potential applications of BCI go far beyond the medical applications. For example, passive BCIs have been developed for assessing and monitoring mental states in real time [22].

Although the performance of BCI systems has improved significantly in recent years, researchers still face a lot of challenges in promoting wide use of the BCI technology. On one hand, compared to traditional input technologies, the current levels of BCI performance are still very low and do not support fast and complex communication and control. The performance bottleneck is mainly caused by the lack of knowledge of the underlying neural mechanisms of information encoding in EEG signals. Besides, decoding EEG signals requires more efficient computational modeling approaches based on signal processing and machine learning techniques. On the other hand, transitioning BCI systems from a well-controlled laboratory setting to a real-life environment remains challenging due to the complications of EEG measurement [23]. A practical BCI system must meet the requirements of good user experience and robust system performance [24]. At present, there is a great need for a mobile EEG platform with user-friendly electrode technology.

This chapter gives a comprehensive overview of the current EEG-based BCIs for the purpose of communication and control. This chapter first provides a taxonomy of the EEG-based BCI systems by categorizing them into three major groups: (1) BCIs based on ERPs, (2) BCIs based on sensorimotor rhythms, and (3) hybrid BCIs. And then, this chapter describes challenges and potential solutions in developing practical BCI systems for real-life applications toward high communication speed, convenient system use, and low user variation. Next, this chapter briefly reviews both medical and non-medical applications of current BCIs. Finally, this chapter concludes with a summary of current stage and future perspectives of the EEG-based BCI technology.

2.2 Taxonomy of Current BCIs

Since many brain signals and system paradigms have been applied to BCIs, a taxonomy is useful for providing a summary of the current status of BCIs. A taxonomy not only provides a systematic description of brain signals and modulation methods used in system design but also aids building general frameworks in system implementation. In general, BCI systems can be classified by the brain signals employed in the system (e.g., P300, VEP, or sensorimotor rhythms) [2]. Alternatively, BCIs can be categorized by operation paradigms like dependent/independent BCIs or synchronous/asynchronous BCIs [15]. Recently, a taxonomy of visual and auditory BCIs was proposed according to the multiple target access methods used in information encoding [19]. To give a meaningful overview of the current BCI systems, this chapter provides a simple taxonomy of BCI based on the types of EEG control signals. By jointly considering brain signals summarized in recent reviews [19, 20, 25], this chapter categorizes the current BCI systems into three major groups: (1) BCIs based on ERPs, (2) BCIs based on sensorimotor rhythms, and (3) hybrid BCIs. Figure 2.2 illustrates the proportions of journal papers in ISI Web of Science for these three BCI groups in the last five years (2012–2016). The numbers of papers for the ERP-based BCIs and the sensorimotor BCIs are

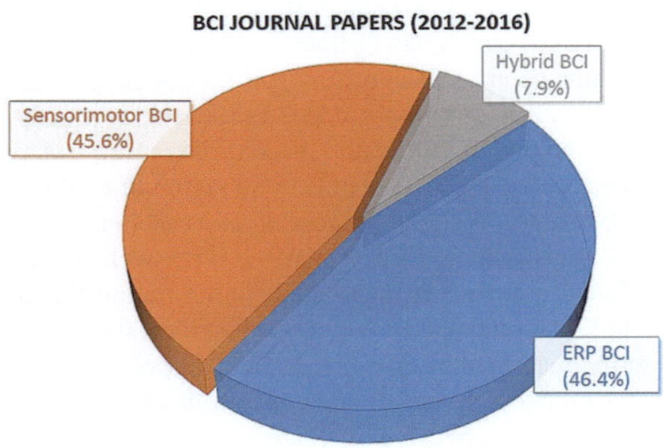

Fig. 2.2 Distribution of major BCI studies including ERP BCI, sensorimotor BCI, and hybrid BCI in recent years.

Notes: Data were obtained from ISI Web of Science using the corresponding topics of journal papers during the last five years (2012–2016). The analysis includes a total of 2658 papers (1234 papers for the ERP BCI, 1213 papers for the sensorimotor BCI, and 211 papers for the hybrid BCI).

comparable (1234 versus 1213, percentage: 46.4% versus 45.6%), which are significantly larger than that of the hybrid BCIs (211, percentage: 7.9%). Note that the hybrid BCIs also include other physiological signals such as electromyogram (EMG) and electrooculogram (EOG) [20]. Without loss of generality, this chapter focuses on the most popular brain signals and system paradigms used in recent BCI studies.

2.2.1 BCIs Based on ERPs

2.2.1.1 ERP Signals for BCI

ERPs, which were originally called evoked potentials (EPs), are measured electrical brain potentials that are evoked by stimuli, which display stable time relationships to a reference event (as opposed to the spontaneous EEG rhythms) [26]. Typically, an ERP waveform consists of multiple peaks and troughs, which are called ERP components. Major ERP components include visual and auditory responses, N2, P3, response-related components, and language-related components. Among these ERP components, two types of signals have been widely used in BCI systems since very early studies in the 1970s and 1980s [5, 6]. The first type is the VEP family, which includes subtypes of transient VEP (TVEP), steady-state VEP (SSVEP), motion VEP (mVEP), and code modulated VEP (cVEP). VEPs are ERPs elicited by visual stimuli, which can be measured with maximum amplitude over the occipital region [27]. Event-related P300 component is another type of commonly used ERP signal. P300, a positive deflection around 300–400 ms after a relevant stimulus, which can be measured over the central and parietal areas, reflects stimulus evaluation, selective attention, and conscious discrimination in an oddball task [28]. In addition to VEP and P300 signals, other types of ERP signals including auditory evoked potential, N2pc, error-related negativity, and N400 have also been investigated in a few BCI studies.

ERP signals are modulated by exogenous stimuli or endogenous mental activities. The exogenous stimuli in BCI systems can be visual or auditory stimuli (e.g., VEPs are elicited by exogenous visual stimuli), while endogenous activities include users' attention or mental tasks (e.g., P300 is an endogenous brain response to an oddball stimulus). In a BCI system, the ERP signals of sensation, perception, or cognition can be modulated by voluntary control (e.g., attention) of the subject. The information encoded in the modulated ERP signals can be decoded by signal processing and classification algorithms. In contrast to spontaneous EEG rhythms that do not require external stimuli, the ERP signals exhibit good characteristics including high signal-to-noise ratio (SNR), reliable spatio-temporal features, and low user variation. With rapid advances of signal modulation and demodulation techniques, ERP-based BCI systems have achieved very high communication speed and therefore become more popular in recent BCI research [19].

2.2.1.2 Modulation of ERP Signals

Signal modulation plays an important role in the ERP-based BCIs. First, to convey the user's intent, ERP signals must be modulated by the user's mental activities in an efficient way so that the modulated ERP signals can be reliably demodulated into the original messages. Second, toward high communication rate, a large bandwidth is required so that the modulated ERP signals corresponding to different intents can be easily discriminated. By introducing the multiple access (MA) techniques in telecommunications [29], the signal modulation approaches in ERP-based BCI systems can be classified by the major MA techniques: time division multiple access (TDMA), frequency division multiple access (FDMA), code division multiple access (CDMA), and space division multiple access (SDMA). This classification can facilitate the understanding of various ERP-based BCI paradigms, which can be considered under a unified signal modulation framework.

First, TDMA allows multiple users to share the same communication channel by dividing the signal into different time slots. In ERP-based BCIs, TDMA is the most popular method for multiple target coding. In these BCI systems, multiple targets appear at different time slots. For example, TDMA has been widely used in the visual P300-based BCI systems [30]. In the standard 'row/column' paradigm, in which each row and column are flashed in a random order, flashing of the row and column of the desired character will elicit P300 signals. Detection of P300 in different time slots can identify the target character. The other single character, checkerboard, and region-based P300 paradigms all follow the principle of TDMA. In addition to P300, the VEP-based BCIs using mVEP [31] and TVEP [32] also employ the TDMA method. Second, FDMA divides the entire bandwidth into multiple individual bands, each for a single user. The SSVEP-based BCI [33] is a typical FDMA system. In a typical frequency coding-based SSVEP BCI, all stimuli flash concurrently at different frequencies, and a target stimulus can be identified by analyzing the frequency of the elicited SSVEP. In ERP-based BCIs, FDMA has higher capacity than TDMA. The SSVEP-based BCIs can discriminate 40 different frequencies within a short stimulus duration (e.g., 500 ms) and thereby achieve very high ITRs [34, 35]. Third, CDMA assigns separate codes to modulate users' signals. The BCIs based on pseudorandom cVEP [9, 36] are typical CDMA systems. The CDMA BCI systems show high performance comparable to FDMA. Last, SDMA divides the geographical space into smaller spaces for multiple users. The SDMA BCI systems use retinotopic mappings of different stimulus locations in the visual field to elicit different spatial patterns of TVEP or SSVEP signals [5, 37]. The number of targets in SDMA is generally limited by the low spatial resolution of EEG-based retinotopic mapping.

2.2.1.3 Demodulation of ERP Signals

In an ERP-based BCI, information embedded in the modulated ERP signals needs to be demodulated and converted into control commands. Due to high trial-to-trial variability of ERP signals, advanced data analysis methods are required in the demodulation of ERP, especially in signal-trial analysis and classification of ERP [38]. ERP demodulation can be considered an EEG pattern recognition problem, which includes data pre-processing, feature extraction, and classification. Signal processing [11] and machine learning [13] are two major data analysis techniques. The signal processing algorithms have been widely used in the pre-processing and feature extraction procedures. The aim of data pre-processing is to improve the SNR of the ERP signals by removing task-irrelevant noises. Bandpass filtering and spatial filtering are two commonly used pre-processing procedures. Bandpass filtering only allows the frequency range of ERP signals to remain. Spatial filtering further enhances ERP signals by calculating a weighted linear combination of multiple EEG channels [39]. Besides, trial averaging is another efficient way to enhance the SNR of ERPs. After pre-processing, features that reflect the characteristic of ERP signals can be extracted. The features for ERPs generally exploit the spatio-temporal information of the signals (e.g., concatenation of ERP amplitude for all time points and all channels) [38]. In addition to time and space domains, features in the frequency domain (e.g., power spectral density) have also been used in the detection of SSVEP signals [40]. After feature extraction, the machine learning algorithms are applied to classification (e.g., discriminating target and non-target ERPs). Classifiers such as linear discriminant analysis (LDA) and support vector machine (SVM) have been widely used in the classification of ERPs with the extracted spatio-temporal features. In online BCI systems, classifiers are obtained from labelled training data collected in calibration sessions and then applied to predict the label of testing data during the system operation.

2.2.1.4 Dependent and Independent ERP-Based BCIs

EEG-based BCIs are categorized as dependent and independent systems [2]. An independent BCI does not depend on the brain's normal output pathways (i.e., peripheral nerves and muscles) to generate the brain activity that carries the user's intent. In contrast, a dependent BCI requires some activity from peripheral nerves and muscles to modulate the brain activity. The ERP-based BCIs include both dependent and independent systems. In BCIs, ERP signals are generally modulated by eye fixation or attention. BCIs based on visual ERPs majorly depend on the control of the muscles of the eye to gaze at desired targets and fall into the dependent BCI class. Most of VEP and visual P300-based BCIs use gaze-dependent paradigms. Different from direct eye gaze (i.e., overt attention), covert attention mechanisms such as spatial selective attention and feature selective attention have been investigated to develop gaze-independent BCI paradigms. There are three types of independent BCIs based on ERP signals: auditory BCIs, tactile BCIs, and independent visual BCIs [41]. For example, using standard oddball tasks, P300 signals can be elicited by auditory, tactile, or visual stimuli, which do not require gaze control to select a target. In addition, SSVEP BCIs that use spatial or feature selective attention to modulate the amplitude of attended/unattended SSVEPs are independent [42, 43]. Due to a larger number of classes and a higher level of signal modulation, the dependent BCIs can achieve much higher communication speed than the independent BCIs. However, for totally locked-in patients who even lose extraocular muscle control, only independent BCIs are useful.

2.2.2 *BCIs Based on Sensorimotor Rhythms*

2.2.2.1 Sensorimotor Rhythms for BCI

BCI systems based on sensorimotor mu/beta rhythms have been developed rapidly in recent years [25]. The neurophysiological studies of the sensorimotor cortex reveal that mu (8–12 Hz) and beta (18–26 Hz) rhythms are modulated by actual movement or preparation of movement. The event-related power change of brain rhythms in a specific frequency band is known as event-related desynchronization and synchronization (ERD/ERS) [44]. Specifically, in the sensorimotor cortex, ERD indicates power decrease of mu/beta rhythms during movement. In contrast, ERS indicates power increase that typically occurs after movement. ERD/ERS of sensorimotor rhythms occur not only with movement, but also with preparation for movement or motor imagery (i.e., imagined movement). During motor imagery, ERD/ERS of mu/beta rhythms display characteristic spatial patterns corresponding to different body parts (e.g., left hand versus right hand). Source imaging studies further show that the ERD sources are located in corresponding cortical areas somatotopically during motor imagery of different body parts. These findings suggest that motor imagery can provide a good way to implement an independent BCI based on self-regulation of the sensorimotor rhythms.

Motor imagery is the basic strategy to operate a sensorimotor BCI. Many studies have demonstrated that people can learn to control (increase or decrease) the amplitude of the sensorimotor rhythms using motor imagery. For example, with the Wadsworth BCI developed by Wolpaw and his colleagues, subjects were trained to learn self-control of mu/beta rhythm amplitude to realized 1-D, 2-D, or 3-D cursor control by translating mu/beta rhythm amplitudes into cursor movements [7, 45, 46]. In the initial training sessions, subjects mostly employed motor imagery (e.g., imagination of hand movements) to control the cursor movement. Instead of continuous movement control, another approach for sensorimotor BCIs is to classify the motor imagery states (e.g., left hand, right hand, feet) using pattern classification techniques. Spatial patterns of ERD/ERS of the mu/beta rhythms are used as features for classification. The Graz BCI developed by Pfurtscheller and his colleagues employed spatial filters and classifiers to train subjects to generate discriminable motor imagery states and therefore built a co-adaptive learning system, which allowed brain learning and machine learning at the same time [8, 47]. The system can discriminate more complex spatio-temporal patterns of ERD/ERS during imagination of movements of different body parts including left and right hands, feet, and tongue [48].

2.2.2.2 Feature Extraction and Classification

The most commonly used features for indexing ERD/ERS of the mu/beta rhythms include band power values, amplitude envelope by the Hilbert transform, and autoregressive model [47]. Time-frequency analysis can provide more detailed features by jointly considering EEG dynamics in frequency and time domains and thereby improve the classification performance [49, 50]. In addition to temporal and spectral features, since ERD/ERS during motor imagery exhibit characteristic spatial patterns, the extraction of spatial features is especially important for the discrimination of motor imagery states. In the space domain, various approaches have been proposed to enhance the extraction of spatial features. First, spatial filtering techniques can significantly enhance the SNR of the mu/beta rhythms by removing the task-irrelevant brain activities. The supervised common spatial pattern (CSP) algorithm and its extensions such as multi-class CSP and filter bank CSP algorithms have been proved highly efficient for extracting discriminative information cross-motor imagery states [51–53]. Unsupervised methods such as independent component analysis (ICA) have also been applied to extract independent sensorimotor components toward a calibration-free system [54]. Second, EEG source imaging methods have been proposed to extract spatial features in the source space, which show improved spatial resolution and specificity than scalp EEG, for classifying motor imagery states in BCIs [55]. Source imaging approaches with a high spatial resolution were able to enhance BCI performance of decoding complex right-hand motor imagery tasks (i.e., flexion, extension, supination, and pronation) [56]. Third, independent from spectral features, amplitude and phase coupling measures such as non-linear regressive coefficients and phase locking value have been applied to extract brain synchrony features that reflect connectivity between multiple cortical regions involved in motor imagery [57]. By enhancing the extraction of spatial features from the sensorimotor areas, these signal processing approaches significantly facilitate the classification of motor imagery states. After feature extraction, the widely used classifiers such as LDA and SVM convert the features to continuous or discrete control commands to operate output devices [14].

2.2.2.3 User Training

User training plays a key role in the sensorimotor BCI. In the BCI system that requires subject to learn self-control of mu/beta rhythm amplitude, subjects usually take multiple training sessions for weeks and months to realize multi-dimensional control of cursor movement [45]. For the system that discriminates motor imagery states, the selection of the kind of motor imagery affects the BCI performance. For example, kinesthetic motor imagery was found to be superior to visual-motor imagery with respect to distinct spatial patterns [58]. Therefore, user training should emphasize kinesthetic experiences instead of visual representations of movements to obtain reliable motor-imagery-based BCI control. In the BCI systems that employ machine learning approaches, the brain and the machine can adapt to each other simultaneously, leading to a co-adaptive learning paradigm, which can significantly reduce the user training time [59, 60].

In the sensorimotor BCI, "BCI Illiteracy" is a big challenge for user training and system applications [61]. An estimation of 10% of people was found not able to use the motor imagery BCI after a short training session around 10 min [47]. To avoid costly training of participants who might not obtain BCI control, a neurophysiological predictor of BCI performance based on resting state EEG was proposed and achieved a high correlation between the predictor and the BCI performance after online training [62]. Currently, a general concern of researchers is how a naïve user can learn to operate the sensorimotor BCI. There are few BCI studies focusing on the training of naïve users. By adopting advanced signal processing and machine learning algorithms, a motor imagery BCI system achieved accurate performance in naïve subjects even in the first session [63]. In addition, the co-adaptive learning approach was demonstrated helpful for subjects who suffer from the BCI illiteracy problem to successfully gain control of the system [59].

2.2.3 Hybrid BCIs

2.2.3.1 Hybrid Modalities

A hybrid BCI combines a BCI with other physiological or technical signals [20, 64]. The goal of hybrid BCIs is to integrate multiple input signals to improve the performance of normal BCIs, and enlarge the population of end-users. In a hybrid BCI, at least one brain signal should be employed to provide intentional input to the BCI for real-time communication or control. According to the type of the second signal modality, hybrid BCIs can be categorized into pure hybrid BCIs and mixed hybrid BCIs [65]. A pure hybrid BCI typically combines two EEG patterns together toward enhanced BCI performance. The widely used hybrid EEG patterns are the combination of ERPs and sensorimotor rhythms (e.g., SSVEP-motor imagery, P300-motor imagery), and the combination of two ERP signals (e.g., P300-SSVEP, N2pc-SSVEP). In addition to hybrid EEG signals, other types of brain signals can also be employed to build a pure hybrid BCI. For example, a hybrid near-infrared spectroscopy (NIRS)-EEG BCI was developed and significantly improved the classification accuracy of motor imagery [66]. A mixed hybrid BCI combines EEG with other non-neuronal control signals toward more robust and reliable control. The control signals can be physiological signals (e.g., EMG, EOG, or heart rate) or signals from other existing input devices like an eye tracking system. These control signals typically are generated by residual muscle functions. For example, EMG signals, which can be modulated by residual muscle activities of patients, have been used to implement hybrid BCIs with various EEG signals and achieved significantly improved accuracy and ITR [20, 65].

The hybrid BCI processes the multiple input signals simultaneously or sequentially. Accordingly, hybrid BCIs can be divided into these two types (i.e., simultaneous and sequential hybrid BCIs) [64]. Simultaneous hybrid BCIs fuse information from different input signals, which carry the same desired goal for intentional control, to improve the accuracy. For example, in visual ERP-based BCIs, SSVEP and P300 signals have been widely used to detect gaze direction, therefore, a hybrid SSVEP-P300 BCI can significantly enhance the performance of gaze-dependent target identification by combining the two independent EEG signals [67]. Data fusion plays an important role in a simultaneous hybrid BCI [68]. Information fusion of multiple input signals is required in the procedures of feature extraction, feature combination, and classification decision. Sequential hybrid BCIs operate two systems sequentially, typically with one system as a switch or selector and another system for regular BCI control. The input signal for a switch or selector generally requires high accuracy and reliability, but a small number of discriminable conditions. For example, a brain switch based on motor imagery has been widely used in sequential hybrid BCIs [64]. In a sequential hybrid BCI, multiple control signals are processed separately toward different outputs, and therefore data fusion is not required.

2.2.3.2 Hybrid EEG Signals

In hybrid BCIs, the simultaneous pure hybrid BCIs that only employ multiple EEG patterns are of great interest to BCI researchers due to the great challenges in system design and data fusion. There are two different ways to form hybrid EEG signals. The first approach is to design hybrid BCI paradigms with multiple mental tasks, which can be used to elicit different EEG signals for single or multiple dimensional controls. An example is hybrid BCIs based on SSVEP and P300 signals, which have been widely used in visual BCIs [19]. Frequency-coded SSVEP stimuli can be superimposed onto P300 stimuli coded with a standard row/column paradigm. The attended target stimuli can simultaneously evoke P300 and SSVEP, which can facilitate the discrimination between target and non-target stimuli [67]. Another diagram used the SSVEP blocking approach to evoke P300 and block SSVEP at the same time, which also significantly improved target discrimination [69]. The combination of motor imagery task and visual attention tasks was proposed to develop a hybrid BCI, which can be applicable for more users [70]. The concurrent motor imagery (i.e., imagination of left or right hand movements) and visual attention (gazing left or right flickering stimuli) tasks simultaneously produced independent ERD and SSVEP features for enhanced classification performance. In another study, motor imagery and P300 tasks were performed simultaneously and independently to control 2D cursor movement [71]. Subjects obtained independent mu/beta rhythms and P300 for horizontal and vertical movement controls, respectively. The second approach to form hybrid EEG signals only employs a single type of mental tasks, which can elicit multiple EEG signals. The combination of ERPs and oscillatory brain rhythms can improve the discriminability of EEG patterns. For example, the combination of ERD/ERS and the lateralized readiness potential facilities the classification of motor imagery tasks [72]. In the gaze-independent SSVEP BCI based on visual spatial attention, the combination of SSVEP and the parieto-occipital alpha rhythm resulted in improved classification accuracy [73]. In addition, with appropriate design, multiple ERP signals can be modulated by a common task. For example, hybrid N2pc and SSVEP features were combined to realize fast detection of covert visuospatial attention in a recent study [74].

2.3 Challenges in Developing BCIs for Practical Applications

Although the feasibility of various BCI paradigms has been well demonstrated in the past decade, translating BCI systems from well-controlled laboratories to complicated living environments still poses great challenges [19, 24]. First of all, the BCI performance in terms of ITR is a key factor in building practical applications for communication and control. Due to limits of the EEG technology, the bottleneck problem in communication speed remains unsolved. Especially in the noisy living conditions, the low ITR is still the key obstacle to BCI applications. Second, the usability of the system plays another key role in practical applications for daily routine use. The two major issues regarding user experience include system calibration and asynchronous system design. The third factor is the practicality of the hardware and software platform. Specifically, practical BCI applications require a convenient mobile EEG system with electrodes that are easy to apply. Besides, analysis of mobile EEG signals, which are more easily contaminated by various noises, requires further development on algorithms. These technical issues need to be addressed to develop BCIs that can satisfy requirements for widespread usage in daily life.

2.3.1 Improving Information Transfer Rate

2.3.1.1 ITR in EEG-Based BCIs

Various metrics (e.g., classification accuracy, ITR, sensitivity and specificity, efficiency and utility, and the like) have been proposed for evaluating the performance of BCIs [75]. Among them, the ITR has been the most widely used metric in BCIs for the purpose of communication and control. Currently, the most popular method for calculating ITR based on Shannon channel theory is defined as follows [2]:

$$\mathrm{ITR} = \left(\log_2 M + P\log_2 P + (1-P)\log_2\left[\frac{1-P}{M-1}\right]\right) * \left(\frac{60}{T}\right) \tag{2.1}$$

where N is the number of possible choices, P is the classification accuracy, and T (in seconds/selection) is the average time for each selection. Due to its capability to quantify communication speed, ITR in bits/minute (bpm) has been widely used to evaluate the performance of online BCI systems.

According to temporal control paradigm, BCI systems can be divided into synchronous and asynchronous BCIs [15]. Most current BCIs use synchronous control protocols where the system controls are periodically available and the timing of the operation is determined by the system. In general, ITR calculation in Eq. (2.1) can be directly used for synchronous BCIs. In contrast, the system controls are continuously available in the asynchronous control protocols. By discriminating non-control state and intentional control state, the asynchronous BCIs allow users to make self-paced decisions on when to start or stop the operation. Since the timing of system operation may vary dramatically depending on the user's control state, asynchronous BCI systems usually do not report ITR [76].

2.3.1.2 BCIs with High ITRs

In Wolpaw et al.'s review in 2002 [2], the BCIs showed maximum ITRs up to 10–25 bpm [2]. Over the past 15 years, advances in BCI research have dramatically increased the ITRs for BCIs in all categories. First, for the sensorimotor BCIs, the ITRs have been improved by applying efficient signal processing and machine learning algorithms. However, due to the small number of possible choices and the endogenous control property, ITRs reported for the motor imagery BCIs were generally lower than 15 bpm [77]. Second, for the ERP-based BCIs, the ITRs have achieved a number of breakthroughs. With recent advances in information coding and EEG decoding, the current ERP-based BCI systems have achieved multifold increases of ITR compared with the early systems. Notably in recent studies on visual BCI spellers, the ITRs above 100 bpm have been reported in BCIs using P300 [78], cVEP [36], and SSVEP [34, 35, 79–81]. Townsend et al. used temporal constraints to develop an asynchronous presentation paradigm for the P300 BCI and reported an ITR of 120 bpm with a subject using a 72-target matrix [78]. Bin et al. developed a 32-target BCI using cVEP and achieved an average ITR of 108 bpm [36]. In the past several years, the BCIs based on SSVEPs have demonstrated a series of systems with high ITRs. Chen et al. developed a 45-target BCI, which was calibration free, with frequency-based stimulus coding and EEG decoding methods and obtained an ITR of 105 bpm [79]. The ITR was further improved to 151 bpm by adopting a filter bank analysis algorithm [80]. More recently, the combination of advanced SSVEP detection and multi-target coding methods has been proved highly efficient for the SSVEP-based BCIs. Nakanishi et al. proposed to incorporate individual calibration data in SSVEP detection and reported an ITR of 166 bpm using a mixed frequency and phase coding method [81]. Chen et al. reached a spelling speed at 1 character per second by proposing a joint frequency and phase modulation method and pushed the ITR up to 270 bpm [34]. In a recent study, by adopting a new spatial filtering algorithm, Nakanishi et al. achieved an ITR of 325 bpm [35], which was the maximum ITR reported in current

BCIs. Note that, since the spelling speed of 1 character per second is close to the speed limit of human gaze control, there might be little room for improving the practical ITR of BCI spellers [34]. Third, for the simultaneous hybrid BCIs, the ITRs have been significantly increased compared with the single modalities. The ITRs of hybrid BCIs highly depend on the signals employed in system design. For example, the hybrid P300-SSVEP BCIs reported ITRs around 50–60 bpm [67, 82], while the hybrid SSVEP-EMG BCI reached an ITR above 90 bpm [65]. Compared with the ERP-based BCIs, there is still large room for improvement of ITRs in the hybrid BCIs.

2.3.1.3 Approaches for Improving ITR

The performance bottleneck of current BCIs is mostly attributed to the poor SNR of EEG signals. Therefore, approaches for improving ITR generally aim to improve the SNR of EEG by adopting advanced signal processing and machine learning algorithms [11, 13]. However, according to Eq. (2.1), the methods for improving ITR can be considered with respect to number of choices (N), classification accuracy (P), and target selection time (T) separately. Note that, these parameters cannot be considered independently because they always interact with each other in BCI systems. For example, there is always a tradeoff between classification accuracy and target selection time toward an optimal ITR.

Classification accuracy is crucial in ITR estimation. The methods to improve classification accuracy can be summarized into two directions [19]. The first direction is to enhance the SNR of task-related EEG signals. In this direction, advanced signal processing algorithms have been applied to artifact removal, spatial filtering, and feature extraction [11]. For example, in the sensorimotor BCIs, the CSP-based spatial filtering algorithms have been widely used to extract the task-related power change of mu/beta rhythms during motor imagery [52, 53]. In the SSVEP BCIs, the canonical correlation analysis-based spatial filtering algorithms significantly improve the SNR of SSVEPs [80, 81, 83]. Besides, the amplitude and SNR of task-related EEG signal can be enhanced by user training. In the sensorimotor BCIs, users learn to control the amplitude of sensorimotor rhythms through online training with feedbacks [45]. Alpha neurofeedback training has been shown efficient for improving the SNR of SSVEPs [84]. The second direction for improving classification accuracy is to maximize the separability of multiple classes. To this end, machine learning-based classification techniques have been widely used in feature selection, feature combination, and classification [13, 14]. For example, linear classifiers like LDA have shown good performance in the sensorimotor and ERP-based BCIs [38]. In addition, separability of multiple classes can be improved by encoding more informative features in EEG signals. For example, the hybrid frequency and phase coding methods have been applied in the SSVEP-based BCIs [34, 81]. The hybrid EEG signals used in the simultaneous hybrid BCIs can significantly improve the separability of different classes by combining multiple EEG signals.

The number of choices also plays an important role in ITR estimation. In the sensorimotor BCIs, the number of choices are generally very limited due to the low spatial resolution of EEG patterns. The discrete classification of motor imagery tasks typically includes imagination of hands, feet, and tongue movements [48]. The number of choices can be improved by enabling the continuous control of 2D and 3D movements [45, 46]. The ERP-based BCIs show good capability in implementing a large number of choices. For example, a P300 BCI using the method of flashing quasi-random groups of characters realized a 7×12 matrix speller [85]. The BCI system using cVEP could discriminate 64 classes [9]. In these systems, the MA methods from telecommunications facilitate the implementation of a large number of classes [19].

The target selection time requires to be optimized toward high ITRs, especially in the ERP-based BCIs. Due to the non-stationarity of EEG signals, the number of trial repetitions required for target identification should be able to vary over time. In the P300 BCI, the dynamic stopping method has been proposed to reduce the target selection time by adaptively adjusting the number of trial repetitions [86]. In addition, the target selection time can be largely reduced by employing more efficient target coding methods. For example, in the VEP-based BCIs, the FDMA and CDMA methods require a much shorter time duration to code multiple targets than the TDMA method.

2.3.2 Reducing System Calibration Time

The non-stationarity and individual variance of EEG signals pose great challenges to the implementation of BCIs using training-based machine learning algorithms [19]. A calibration session is typically required to collect training data for building a classification model before the operation of the BCI system. In practice, the repetitive calibration procedures could be boring and time-consuming. To facilitate system calibration, different methods have been developed to reduce the calibration time or even realize zero-training paradigms. To reduce the calibration time, the machine learning algorithms need to have good generalization ability and robustness even when the training dataset is small. In recent years, the zero-training methods have obtained increasing attention. These methods aim to transfer information across multiple sessions, subjects, or tasks to obtain generalizable models that can be directly used for operating BCIs. For example, for each individual user, the cross-session transfer of spatial filters and EEG templates was developed for detecting SSVEPs without new calibration [87]. Besides, the cross-subject transfer of models has been demonstrated feasible for SSVEP [88] and P300 [89] detection without individual calibration data. In addition, the cross-task transfer of spatial filters derived from resting EEG data was proposed to implement zero-training classification of EEG in motor imagery tasks [90]. Furthermore, to better address the non-stationarity issues, adaptive classification methods have been developed to automatically update the classifiers during online BCI operations [88, 91]. The combination of information transfer and adaptive classification methods provides a practical solution to facilitate the calibration procedure of current BCIs.

2.3.3 Asynchronous System

Asynchronous BCIs, in which users make self-paced operations, are more flexible and natural for routine use than synchronous BCIs [15]. In asynchronous BCIs, continuous detection of idle state is required for discriminating the intentional control state and the non-control state. The BCI system only sends control command when an intentional control state is detected. Accurate detection of the idle states is crucial for practical applications; however, it remains a challenging task due to the large variability of idle states, which might involve different mental activities. One way to address the control state detection issue is to design a switch, which can turn on/off the BCI control mode. This method has been applied in SSVEP-based BCIs. For example, the idle state and the control state were switched by selecting an on/off stimulus, which turned on/off the visual stimuli of the SSVEP BCI [92]. To implement a user-initiated asynchronous BCI system, the discrimination of EEG signals between the idle state and the intentional control state is important and of great interest. For example, in a sensorimotor BCI-based brain switch, EEGs corresponding to idle and imaginary movement states were classified sample-by-sample and the averaged results over time were used to control the switch [93]. In the ERP-based BCIs, the method for detecting idle states is generally based on the computational modeling of EEG signals under the idle and control states. In an SSVEP BCI, spontaneous EEG signals under different types of idle states (on-screen, off-screen, eye-closed) and SSVEPs were classified by SVM classifiers [94]. In an N200 BCI, a computational model of the mVEP response patterns, which integrated the spatial profile of the speller matrix, was used to detect the non-control state effectively [95]. In a P300 BCI, computational models for target P300, non-target P300, and non-control EEG signals were proposed for detecting control and non-control states based on likelihood [96].

2.3.4 Developing Mobile BCI System

Despite the significant performance improvement of EEG-based BCIs, real-life BCI applications have still been hindered by the lack of portability and cumbersome system settings such as skin preparation and gel application. Developing a mobile BCI platform enables the movement of BCI systems from well-controlled laboratory settings to real-world environments. To design a mobile BCI system, the following three major challenges need to be addressed.

First, a mobile BCI system requires advanced sensor technologies to facilitate the skin preparation and gel application, which may take around an hour for conventional high-density EEG systems. Recently, researchers have developed various types of EEG electrodes. For example, dry- and non-contact electrodes showed good performance in an online BCI experiment based on SSVEPs recorded from the occipital area without conductive gels [97]. In another study, a soft, foldable, and ultra-thin electrode constructing fractal mesh geometry was introduced for long-term recording of high-fidelity EEG signals for a persistent BCI [98]. Besides, placing electrodes over non-hair-bearing areas is another way to avoid the interference from hairs. For instance, several studies demonstrated that behind/around-the-ear electrodes can be an alternative solution to measure EEG for BCIs [99, 100]. In addition, the platform for ear-EEG, which can record EEG signals from electrodes placed in-the-ear, has been developed for brain monitoring [101]. Second, a portable platform using mobile hardware for stimulus presentation, signal acquisition, and data processing is required to expand the capability of BCIs. Recent rapid advances in mobile devices like smartphones made it possible to implement a truly portable BCI. For example, a cellphone-based phone-dialing BCI system was demonstrated using SSVEPs [102]. A head-mounted virtual reality (VR) platform also showed great potentials for implementing a mobile BCI system. Nakanishi et al. proposed a portable BCI, which integrated wireless and wearable EEG, dry EEG and EOG sensors, and a head-mounted VR display, for assessing visual field loss in glaucoma patients based on multi-focal SSVEPs [103]. Third, artifact removal techniques are necessary for a robust mobile BCI. In general, subjects are instructed to remain stationary and focus only on their tasks in laboratory experiments, resulting in clean EEG recordings. However, in real-world situations, motion artifacts and ambient noises can be a severe problem since users' behaviors are not restricted. To simulate realistic motion artifacts, dual-task experimental design, in which subjects needed to perform a primary motor task and a secondary cognitive task at the same time, has been proposed. For example, Gramann et al. recorded EEG from subjects standing or walking on a treadmill while performing a visual oddball response task [104]. Although the scalp recordings were contaminated with motion artifacts, source ERP activities decomposed by ICA did not differ across conditions [104]. Lin et al. performed an online SSVEP BCI using a wearable EEG system in human subjects under walking conditions [105]. These results suggest that subjects can use ERP-based BCIs even while moving by combining mobile EEG with efficient de-noising algorithms.

2.4 Applications of BCIs

Facilitated by the rapid development of modern information technology, BCIs are moving beyond laboratories into real-world applications. The majority of the available BCI applications has been focusing on decoding and outputting users' intention, in which the users actively perform certain mental tasks for the purpose of controlling external devices. The devices can be assistive or rehabilitative, used in both medical and non-medical scenarios. BCIs can also be applied to monitor users' mental states in real time without their active participation. Applications in this direction mainly focus on healthy people in non-medical scenarios. Below we briefly review the state-of-the-art BCI applications and organize them by their design purposes (i.e., medical or non-medical).

2.4.1 BCIs for Medical Use

ERP-based BCIs are probably the most popular BCI systems serving as an assistive tool for severely motor disabled people, for example, patients with amyotrophic lateral sclerosis (ALS) [106, 107]. Compared to sensorimotor BCIs, the major advantage for ERP-based BCIs is its relative ease of acquiring reliable signals with little training required [17]. Using the classical P300 speller, ALS patients have demonstrated the capability of operating necessary software such as word processing, painting, e-mailing, and controlling home appliances, in their home environment for a considerably long time (>2.5 years in [108] and 14 weeks in [109]). Nevertheless, challenges still remain for patients with complete locked-in syndrome, who are speculated to lack the contingency between a voluntary intention and its consequence [110].

Sensorimotor BCIs, by contrast, have been suggested as an effective approach for rehabilitative purposes (see [10, 45] for attempts to implement assistive BCIs). The most widely studied population is the stroke patients, in which rehabilitative BCIs are expected to help them regain the lost motor functions. Sensorimotor BCIs detect and translate patients' movement intentions into actual limb movements, with the help of a prosthesis or a direct functional electrical stimulation. Such rehabilitation training is believed to induce reorganization of neural circuits, thus facilitating motor function recovery [111].

2.4.2 BCIs for Non-medical Use

Both ERP-based and sensorimotor BCIs can be used in non-medical scenarios, as a novel human-computer interaction (HCI) approach, especially in gaming industry [112, 113]. BCIs are considered as an additional input modality (along with keyboard, speech, gestures, and so on) to control a game or a general HCI application (e.g., in VR environments), providing a unique and possibly useful way for interaction.

An emerging but more important BCI application direction for healthy users is real-time monitoring of critical mental states, such as attention, stress, cognitive load, emotion, and the like. Most of these BCIs, recently termed as 'passive BCIs'

[22], usually take the form of hybrid BCIs, in which other physiological or environmental signals are also included when necessary, with the aim of having a more comprehensive overview of the users' states. Passive BCIs provide an objective evaluation in the fields of driving safety [114, 115], education [116], user experience, or neuro-marketing [117]. Although still at an early stage, BCI development in this direction is believed to benefit a much larger population, beyond the scope of clinical population.

2.5 Summary

The past 20 years have witnessed unprecedented progress in the BCI technology. In addition to the original purpose to help the patients with motor disabilities to communicate with their environments, current BCI applications have been extended to various fields including brain state monitoring, neuro-rehabilitation, and human performance improvement in the last few years. More generally, BCI can be defined as a real-time platform that uses online brain signal analysis to influence human interactions with their environments. In the field of communication and control, beyond the initial definition of interface, the current BCI technology places more emphasis on the interaction between human brain and the output devices. The practical BCI applications will allow researchers to study the mutual interference between brain and devices in long-term interaction.

Despite significant progress in performance improvement, low communication rates remain key obstacles to BCI-based communication and control. The improvement of communication speed will continue to be a main challenge in BCI research. The following efforts can be made in this direction. First, by increasing our understanding of how sensation, perception, and cognition are encoded in brain signals, new BCI paradigms with wider communication bandwidths can be developed. Second, by considering BCI as a communication system, advanced technologies from telecommunications can be employed to improve information modulation and demodulation in BCIs. Third, more efficient signal processing and machine learning algorithms can facilitate the decoding of EEG. Fourth, brain stimulation methods such as electrical and magnetic stimulation can be employed to enhance the brain signals. Furthermore, the future performance breakthroughs of BCI require the collaboration between neuroscience, artificial intelligence (AI), and engineering. In the near future, the combination of BCI with AI can lead to new hybrid intelligence systems that can effectively fuse human and machine intelligence.

Acknowledgments This work is supported by the National Natural Science Foundation of China (61671424, 61335010, and 61634006), the Key project of Chinese Academy of Science (KJZD-EW-L11-01), Beijing S&T planning task (Z161100002616019), and the Recruitment Program of Young Professionals.

References

1. Vidal JJ (1973) Toward direct brain-computer communication. Annu Rev Riophys Bioeng 2:157–180
2. Wolpaw JR, Birbaumer N, McFarland DJ, Pfurtscheller G, Vaughan TM (2002) Brain computer interfaces for communication and control. Clin Neurophysiol 113(6):767–791
3. Birbaumer N (2006) Brain-computer-interface research: coming of age. Clin Neurophysiol 117(3):479–483
4. Lebedev MA, Nicolelis MAL (2017) Brain-machine interfaces: from basic science to neuroprostheses and neurorehabilitation. Physiol Rev 97(2):767–837
5. Vidal JJ (1977) Real-time detection of brain events in EEG. Proc IEEE 65(5):633–664
6. Farwell LA, Donchin E (1988) Talking off the top of your head – toward a mental prosthesis utilizing event-related brain potentials. Electroencephalogr Clin Neurophysiol 70(6):510–523
7. Wolpaw JR, McFarland DJ, Neat GW, Forneris CA (1991) An EEG-based brain-computer interface for cursor control. Electroencephalogr Clin Neurophysiol 78(3):252–259
8. Pfurtscheller G, Flotzinger D, Kalcher J (1993) Brain-computer interface – a new communication device for handicapped persons. J Microcomput Appl 16(3):293–299
9. Sutter EE (1992) The brain response interface: communication through visually-induced electrical brain responses. J Microcomput Appl 15(1):31–45
10. Birbaumer N, Ghanayim N, Hinterberger T, Iversen I, Kotchoubey B, Kubler A, Perelmouter J, Taub E, Flor H (1999) A spelling device for the paralyzed. Nature 398:297–298
11. Bashashati A, Fatourechi M, Ward RK, Birch GE (2007) A survey of signal processing algorithms in brain-computer interfaces based on electrical brain signals. J Neural Eng 4(2): R32–R57
12. Makeig S, Kothe C, Mullen T, Bigdely-Shamlo N, Zhang Z, Kreutz-Delgado K (2012) Evolving signal processing for brain–computer interfaces. Proc IEEE 100:1567–1584
13. Müller KR, Krauledat M, Dornhege G, Curio G, Blankertz B (2004) Machine learning techniques for brain-computer interfaces. Biomed Tech 49(1):11–22
14. Lotte F, Congedo M, Lecuyer A, Lamarche F, Arnaldi B (2007) A review of classification algorithms for EEG-based brain-computer interfaces. J Neural Eng 4(2):R1–R13
15. Mason SG, Bashashati A, Fatourechi M, Navarro KF, Birch GE (2007) A comprehensive survey of brain interface technology designs. Ann Biomed Eng 35(2):137–169
16. Lance BJ, Kerick SE, Ries AJ, Oie KS, McDowell K (2012) Brain computer interface technologies in the coming decades. Proc IEEE 100:1585–1599
17. Mak JN, Wolpaw JR (2009) Clinical applications of brain-computer interfaces: current state and future prospects. IEEE Rev Biomed Eng 2:187–199
18. Van Erp J, Lotte F, Tangermann M (2012) Brain-computer interfaces: beyond medical applications. Computer 45(4):26–34
19. Gao S, Wang Y, Gao X, Hong B (2014) Visual and auditory brain-computer interfaces. IEEE Trans Biomed Eng 61(5):1436–1447
20. Müller-Putz G, Leeb R, Tangermann M, Höhne J, Kübler A, Cincotti F, Mattia D, Rupp R, Müller KR, Millan JR (2015) Towards noninvasive hybrid brain–computer interfaces: framework, practice, clinical application, and beyond. Proc IEEE 103(6):926–943
21. Chaudhary U, Birbaumer N, Ramos-Murguialday A (2016) Brain–computer interfaces for communication and rehabilitation. Nat Rev Neurol 12(9):513–525
22. Zander TO, Kothe C (2011) Towards passive brain–computer interfaces: applying brain–computer interface technology to human–machine systems in general. J Neural Eng 8 (2):025005
23. Millán JR, Rupp R, Müller-Putz GR, Murray-Smith R, Giugliemma C, Tangermann M, Vidaurre C, Cincotti F, Kübler A, Leeb R, Neuper C, Müller KR, Mattia D (2010) Combining brain–computer interfaces and assistive technologies: state-of-the-art and challenges. Front Neurosci 4:161

24. Wang Y, Gao X, Hong B, Gao S (2010) Practical designs of brain–computer interfaces based on the modulation of EEG rhythms. In: Graimann B, Allison B, Pfurtscheller G (eds) Brain–computer interfaces. Springer, Heidelberg, pp 137–154
25. Yuan H, He B (2014) Brain–computer interfaces using sensorimotor rhythms: current state and future perspectives. IEEE Trans Biomed Eng 61(5):1425–1435
26. Luck SJ (2005) An introduction to the event-related potential technique. MIT Press, Cambridge
27. Regan D (1989) Human brain electrophysiology: evoked potentials and evoked magnetic fields in science and medicine. Elsevier, New York
28. Patel SH, Azzam PN (2005) Characterization of N200 and P300: selected studies of the event-related potential. Int J Med Sci 2(4):47–154
29. Rappaport TS (2001) Wireless communication, principle and practice, 2nd edn. Prentice-Hall, Englewood Cliffs
30. Fazel-Rezai R, Allison BZ, Guger C, Sellers EW, Kleih SC, Kübler A (2012) P300 brain computer interface: current challenges and emerging trends. Front Neuroeng 5:14
31. Guo F, Hong B, Gao X, Gao S (2008) A brain–computer interface using motion-onset visual evoked potential. J Neural Eng 5(4):477–485
32. Lee PL, Hsieh JC, Wu CH, Shyu KK, Chen SS, Yeh TC, Wu YT (2006) The brain computer interface using flash visual evoked potential and independent component analysis. Ann Biomed Eng 34(10):1641–1654
33. Vialatte FB, Maurice M, Dauwels J, Cichocki A (2010) Steady-state visually evoked potentials: focus on essential paradigms and future perspectives. Prog Neurobiol 90(4):418–438
34. Chen X, Wang Y, Nakanishi M, Gao X, Jung TP, Gao S (2015) High-speed spelling with a noninvasive brain-computer interface. Proc Natl Acad Sci U S A 112(44):E6058–E6067
35. Nakanishi M, Wang Y, Chen X, Wang YW, Gao X, Jung TP (2017) Enhancing detection of SSVEPs for a high-speed brain speller using task-related component analysis. IEEE Trans Biomed Eng. (in press)
36. Bin G, Gao X, Wang Y, Li Y, Hong B, Gao S (2011) A high-speed BCI based on code modulation VEP. J Neural Eng 8(2):025015
37. Maye A, Zhang D, Engel AK (2017) Utilizing retinotopic mapping for a multi-target SSVEP BCI with a single flicker frequency. IEEE Trans Neural Syst Rehabil Eng. (in press)
38. Blankertz B, Lemm S, Treder M, Haufe S, Müller KR (2011) Single trial analysis and classification of ERP components—a tutorial. NeuroImage 56(2):814–825
39. Parra LC, Spence CD, Gerson AD, Sajda P (2005) Recipes for the linear analysis of EEG. NeuroImage 28(2):326–341
40. Wang Y, Gao X, Hong B, Jia C, Gao S (2008) Brain–computer interfaces based on visual evoked potentials-feasibility of practical system designs. IEEE Eng Med Biol Mag 27 (5):64–71
41. Riccio A, Mattia D, Simione L, Olivetti M, Cincotti F (2012) Eye-gaze independent EEG-based brain–computer interfaces for communication. J Neural Eng 9(4):045001
42. Kelly SP, Lalor EC, Finucane C, McDarby G, Reilly RB (2005) Visual spatial attention control in an independent brain–computer interface. IEEE Trans Biomed Eng 52 (9):1588–1596
43. Zhang D, Maye A, Gao X, Hong B, Engel AK, Gao S (2010) An independent brain–computer interface using covert non-spatial visual selective attention. J Neural Eng 7(1):016010
44. Pfurtscheller G, Lopes da Silva FH (1999) Event-related EEG/MEG synchronization and desynchronization: basic principles. Clin Neurophysiol 110(11):1842–1857
45. Wolpaw JR, McFarland DJ (2004) Control of a two-dimensional movement signal by a noninvasive brain-computer interface in humans. Proc Natl Acad Sci U S A 101 (51):17849–17854
46. McFarland DJ, Sarnacki WA, Wolpaw JR (2010) Electroencephalographic (EEG) control of three-dimensional movement. J Neural Eng 7(3):036007

47. Pfurtscheller G, Neuper C (2001) Motor imagery and direct brain-computer communication. Proc IEEE 89(7):1123–1134
48. Pfurtscheller G, Brunner C, Schlogl A, Lopes da Silva FH (2006) Mu rhythm (de)synchronization and EEG single-trial classification of different motor imagery tasks. NeuroImage 31 (1):153–159
49. Wang T, Deng J, He B (2004) Classifying EEG-based motor imagery tasks by means of time-frequency synthesized spatial patterns. Clin Neurophysiol 115(12):2744–2753
50. Li J, Wang Y, Zhang L, Cichocki A, Jung TP (2016) Decoding EEG in cognitive tasks with time-frequency and connectivity masks. IEEE Trans Cogn Dev Syst 8(4):298–308
51. Ramoser H, Müller-Gerking J, Pfurtscheller G (2000) Optimal spatial filtering of single trial EEG during imagined hand movement. IEEE Trans Neural Syst Rehabil Eng 8(4):441–446
52. Blankertz B, Tomioka R, Lemm S, Kawanabe M, Müller KR (2008) Optimizing spatial filters for robust EEG single-trial analysis. IEEE Signal Process Mag 25(1):41–56
53. Ang KK, Chin ZY, Wang C, Guan C, Zhang H (2012) Filter bank common spatial pattern algorithm on BCI competition IV datasets 2a and 2b. Front Neurosci 6:39
54. Wang Y, Jung TP (2013) Improving brain-computer interfaces using independent component analysis. In: Allison B, Dunne S, Leeb R, Millan JR, Nijholt A (eds) Towards practical brain-computer interfaces: bridging the gap from research to real-world applications. Springer, Heidelberg, pp 67–83
55. Qin L, Ding L, He B (2005) Motor imagery classification by means of source analysis for brain-computer interface applications. J Neural Eng 2(4):65–72
56. Edelman BJ, Baxter B, He B (2016) EEG source imaging enhances the decoding of complex right-hand motor imagery tasks. IEEE Trans Biomed Eng 63(1):4–14
57. Wei Q, Wang Y, Gao X, Gao S (2007) Amplitude and phase coupling measures for feature extraction in an EEG-based brain-computer interface. J Neural Eng 4(2):120–129
58. Neuper C, Scherer R, Reiner M, Pfurtscheller G (2005) Imagery of motor actions: differential effects of kinesthetic and visual-motor mode of imagery in single-trial EEG. Brain Res Cogn Brain Res 25(3):668–677
59. Vidaurre C, Sannelli C, Müller KR, Blankertz B (2011) Machine-learning-based coadaptive calibration for brain-computer interfaces. Neural Comput 23(3):791–816
60. Acqualagna L, Botrel L, Vidaurre C, Kübler A, Blankertz B (2016) Large-scale assessment of a fully automatic co-adaptive motor imagery-based brain computer interface. PLoS ONE 11 (2):e0148886
61. Kübler A, Neumann N, Wilhelm B, Hinterberger T, Birbaumer N (2004) Predictability of brain-computer communication. J Psychophysiol 18:121–129
62. Blankertz B, Sannelli C, Halder S, Hammer EM, Kübler A, Müller KR, Curio G, Dickhaus T (2010) Neurophysiological predictor of SMR-based BCI performance. NeuroImage 51 (4):1303–1309
63. Blankertz B, Losch F, Krauledat M, Dornhege G, Curio G (2008) The Berlin brain-computer interface: accurate performance from first-session in BCI-naïve subjects. IEEE Trans Biomed Eng 55(10):2452–2462
64. Pfurtscheller G, Allison BZ, Bauernfeind G, Brunner C, Solis Escalante T, Scherer R, Zander TO, Müller-Putz G, Neuper C, Birbaumer N (2010) The hybrid BCI. Front Neurosci 4:30
65. Lin K, Cinetto A, Wang Y, Chen X, Gao S, Gao X (2016) An online hybrid BCI system based on SSVEP and EMG. J Neural Eng 13(2):026020
66. Fazli S, Mehnert J, Steinbrink J, Curio G, Villringer A, Müller KR, Blankertz B (2012) Enhanced performance by a hybrid NIRS–EEG brain computer interface. NeuroImage 59 (1):519–529
67. Yin E, Zhou Z, Jiang J, Chen F, Liu Y, Hu D (2013) A novel hybrid BCI speller based on the incorporation of SSVEP into the P300 paradigm. J Neural Eng 10(2):026012
68. Li Y, Pan J, Long J, Yu T, Wang F, Yu Z, Wu W (2016) Multimodal bcis: target detection, multidimensional control, and awareness evaluation in patients with disorder of consciousness. Proc IEEE 104(2):332–352

69. Xu M, Qi H, Wan B, Yin T, Liu Z, Ming D (2013) A hybrid BCI speller paradigm combining P300 potential and the SSVEP blocking feature. J Neural Eng 10(2):026001
70. Allison BZ, Brunner C, Kaiser V, Müller-Putz GR, Neuper C, Pfurtscheller G (2010) Toward a hybrid brain–computer interface based on imagined movement and visualattention. J Neural Eng 7(2):26007
71. Li Y, Long J, Yu T, Yu Z, Wang C, Zhang H, Guan C (2010) An EEG-based BCI system for 2-D cursor control by combining Mu/Beta rhythm and P300 potential. IEEE Trans Biomed Eng 57(10):2495–2505
72. Höhne J, Holz E, Staiger-Sälzer P, Müller KR, Kübler A, Tangermann M (2014) Motor imagery for severely motor-impaired patients: evidence for brain-computer interfacing as superior control solution. PLoS ONE 9(8):e104854
73. Kelly SP, Lalor EC, Reilly RB, Foxe JJ (2005) Visual spatial attention tracking using high-density SSVEP data for independent brain-computer communication. IEEE Trans Neural Syst Rehabil Eng 13(2):172–178
74. Xu M, Wang Y, Nakanishi M, Wang YT, Qi H, Jung TP, Ming D (2016) Fast detection of covert visuospatial attention using hybrid N2pc and SSVEP features. J Neural Eng 13 (6):066003
75. Billinger M, Daly I, Kaiser V, Jin J, Allison BZ, Müller-Putz GR, Brunner R (2013) Is it significant? Guidelines for reporting BCI performance. In: Allison B, Dunne S, Leeb R, Millan JR, Nijholt A (eds) Towards practical brain-computer interfaces: bridging the gap from research to real-world applications. Springer, Heidelberg, pp 333–354
76. Yuan P, Gao X, Allison B, Wang Y, Bin G, Gao S (2013) A study of the existing problems of estimating the information transfer rate in online brain–computer interfaces. J Neural Eng 10 (2):026014
77. Müller KR, Tangermann M, Dornhege G, Krauledat M, Curio G, Blankertz B (2008) Machine learning for real-time single-trial EEG-analysis: from brain – computer interfacing to mental state monitoring. J Neurosci Methods 167(1):82–90
78. Townsend G, Platsko V (2016) Pushing the P300-based brain-computer interface beyond 100 bpm: extending performance guided constraints into the temporal domain. J Neural Eng 13(2):026024
79. Chen X, Chen Z, Gao S, Gao X (2014) A high-ITR SSVEP-based BCI speller. Brain-Comp Interfaces 1(3–4):181–191
80. Chen X, Wang Y, Gao S, Jung TP, Gao X (2015) Filter bank canonical correlation analysis for implementing a high-speed SSVEP-based brain-computer interface. J Neural Eng 12 (4):046008
81. Nakanishi M, Wang Y, Wang YT, Mitsukura Y, Jung TP (2014) A high-speed brain speller using steady-state visual evoked potentials. Int J Neural Syst 24(6):1450019
82. Xu M, Chen L, Zhang L, Qi H, Ma L, Tang J, Wan B, Ming D (2014) A visual parallel-BCI speller based on the time–frequency coding strategy. J Neural Eng 11(2):026014
83. Lin Z, Zhang C, Wu W, Gao X (2007) Frequency recognition based on canonical correlation analysis for SSVEP-based BCIs. IEEE Trans Biomed Eng 54(6):1172–1176
84. Wan F, Da Cruz JN, Nan W, Wong CM, Vai MI, Rosa A (2016) Alpha neurofeedback training improves SSVEP-based BCI performance. J Neural Eng 13(3):036019
85. Jin J, Allison BZ, Sellers EW, Brunner C, Horki P, Wang X, Neuper C (2011) Optimized stimulus presentation patterns for an event-related potential EEG-based brain-computer interface. Med Biol Eng Comput 49(2):181–191
86. Schreuder M, Höhne J, Blankertz B, Haufe S, Dickhaus T, Tangermann M (2013) Optimizing event-related potential based brain–computer interfaces: a systematic evaluation of dynamic stopping methods. J Neural Eng 10(3):036025
87. Nakanishi M, Wang Y, Jung TP (2016) Session-to-session transfer in detecting SSVEPs with individual calibration data. In: Schmorrow DD, Fidopiastis CM (eds) Foundations of augmented cognition: neuroergonomics and operational neuroscience. Springer International Publishing, Cham, pp 253–260

88. Yuan P, Chen X, Wang Y, Gao X, Gao S (2015) Enhancing performances of SSVEP-based brain-computer interfaces via exploiting inter-subject information. J Neural Eng 12(4):046006
89. Xu M, Liu J, Chen L, Qi H, He F, Zhou P, Wan B, Ming D (2016) Incorporation of inter-subject information to improve the accuracy of subject-specific P300 classifiers. Int J Neural Syst 26(3):1650010
90. Wang Y, Wang YT, Jung TP (2012) Translation of EEG spatial filters from resting to motor imagery using independent component analysis. PLoS ONE 7(5):e37665
91. Vidaurre C, Sannelli C, Müller KR, Blankertz B (2010) Machine-learning-based coadaptive calibration for brain-computer interfaces. Neural Comput 23(3):791–816
92. Cheng M, Gao X, Gao S, Xu D (2002) Design and implementation of a brain-computer interface with high transfer rates. IEEE Trans Biomed Eng 49(10):1181–1186
93. Mason SG, Birch GE (2000) A brain-controlled switch for asynchronous control applications. IEEE Trans Biomed Eng 47(10):1297–1307
94. Zhang D, Huang B, Wu W, Li S (2015) An idle-state detection algorithm for SSVEP-based brain-computer interfaces using a maximum evoked response spatial filter. Int J Neural Syst 25 (7):1550030
95. Zhang D, Song H, Xu H, Wu W, Gao S, Hong B (2012) An N200 speller integrating the spatial profile for the detection of the non-control state. J Neural Eng 9(2):026016
96. Zhang H, Guan C, Wang C (2008) Asynchronous P300-based brain-computer interfaces: a computational approach with statistical models. IEEE Trans Biomed Eng 55(6):1754–1763
97. Chi YW, Wang YT, Wang Y, Maier C, Jung TP, Cauwenberghs G (2012) Dry and non-contact EEG sensors for mobile brain-computer interfaces. IEEE Trans Neural Syst Rehabil Eng 20 (2):228–235
98. Norton JJ, Lee DS, Lee JW, Lee W, Kwon O, Won P, Jung SY, Cheng H, Jeong JW, Akce A, Umunna S, Na I, Kwon YH, Wang XQ, Liu Z, Paik U, Huang Y, Bretl T, Yeo WH, Rogers JA (2015) Soft, curved electrode systems capable of integration on the auricle as a persistent brain-computer interface. Proc Natl Acad Sci U S A 112(13):3920–3925
99. Wang YT, Nakanishi M, Wang Y, Wei CS, Cheng CK, Jung TP (2017) An online brain-computer interface based on SSVEPs measured from non-hair-bearing areas. IEEE Trans Neural Syst Rehabil Eng 25(1):11–18
100. Pacharra M, Debener S, Wascher E (2017) Concealed around-the-ear EEG captures cognitive processing in a visual Simon task. Front Hum Neurosci 11:290
101. Looney D, Kidmose P, Park C, Ungstrup M, Rank ML, Rosenkranz K, Mandic DP (2012) The in-the-ear recording concept: use-centered and wearable brain monitoring. IEEE Pulse 3 (6):32–42
102. Wang YT, Wang Y, Jung TP (2011) A cell-phone based brain-computer interface for com-munication in daily life. J Neural Eng 8(2):025018
103. Nakanishi M, Wang YT, Jung TP, Zao JK, Chien YY, Diniz-Filho A, Daga FB, Lin YP, Wang Y, Medeiros FA (2017) Detecting glaucoma with a portable brain-computer interface for objective assessment of visual function loss. JAMA Ophthalmol 135(6):550–557
104. Gramann K, Gwin JT, Bigdely-Shamlo N, Ferris DP, Makeig S (2010) Visual evoked re-sponses during standing and walking. Front Hum Neurosci 4:202
105. Lin YP, Wang Y, Jung TP (2014) Assessing the feasibility of online SSVEP decoding for moving humans using a consumer EEG headset. J Neuroeng Rehabil 11:119
106. Sellers EW, Donchin E (2006) A P300-based brain-computer interface: initial tests by ALS patients. Clin Neurophysiol 117(3):538–548
107. Nijboer F, Sellers EW, Mellinger J, Jordan MA, Matuz T, Furdea A, Kübler A (2008) A P300-based brain–computer interface for people with amyotrophic lateral sclerosis. Clin Neurophysiol 119(8):1909–1916
108. Sellers EW, Vaughan TM, Wolpaw JR (2010) A brain-computer interface for long-term independent home use. Amyotroph Lateral Scler 11(5):449–455

109. Holz EM, Botrel L, Kaufmann T, Kübler A (2015) Long-term independent brain-computer interface home use improves quality of life of a patient in the locked-in state: a case study. Arch Phys Med Rehabil 96(3):S16–S26
110. Birbaumer N, Piccione F, Silvoni S, Wildgruber M (2012) Ideomotor silence: the case of complete paralysis and brain-computer interfaces (BCI). Psychol Res 76(2):183–191
111. Van Dokkum L, Ward T, Laffont I (2015) Brain computer interfaces for neurorehabilitation-its current status as a rehabilitation strategy post-stroke. Ann Phys Rehabil Med 58(1):3–8
112. Bos DPO, Reuderink B, van de Laar B, Gürkök H, Mühl C, Poel M, Heylen D (2010) Brain-computer interfacing and games. In: Tan DS, Nijholt A (eds) Brain-computer interfaces. Springer, London, pp 149–178
113. Ahn M, Lee M, Choi J, Jun SC (2014) A review of brain-computer interface games and an opinion survey from researchers, developers and users. Sensors 14(8):14601–14633
114. Lal SK, Craig A (2002) Driver fatigue: electroencephalography and psychological assessment. Psychophysiology 39(3):313–321
115. Healey J, Picard RW (2005) Detecting stress during real-world driving tasks using physiological sensors. IEEE Trans Intell Transp Syst 6(2):156–166
116. Poulsen AT, Kamronn S, Dmochowski J, Parra LC, Hansen LK (2017) EEG in the classroom: synchronised neural recordings during video presentation. Sci Rep 7:43916
117. Khushaba RN, Wise C, Kodagoda S, Louviere J, Kahn BE, Townsend C (2013) Consumer neuroscience: assessing the brain response to marketing stimuli using electroencephalogram (EEG) and eye tracking. Expert Syst Appl 40(9):3803–3812

Chapter 3
Invasive Brain Machine Interface System

Yile Jin, Junjun Chen, Shaomin Zhang, Weidong Chen, and Xiaoxiang Zheng

Abstract Because of high spatial-temporal resolution of neural signals obtained by invasive recording, the invasive brain-machine interfaces (BMI) have achieved great progress in the past two decades. With success in animal research, BMI technology is transferring to clinical trials for helping paralyzed people to restore their lost motor functions. This chapter gives a brief review of BMI development from animal experiments to human clinical studies in the following aspects: (1) BMIs based on rodent animals; (2) BMI based on non-human primates; and (3) pilot BMIs studies in clinical trials. In the end, the chapter concludes with a summary of potential opportunities and future challenges in BMI technology.

Keywords Brain-machine interfaces · Neural ensemble recording · Neural decoding · Neural stimulation · Electrocorticogram

3.1 Introduction

Spinal cord and limb nerve injury, amyotrophic lateral sclerosis, and other neuromuscular degeneration are common neurological disorders that cause hemiplegia and paraplegia. There are 130,000 new paralyzed patients due to spinal cord injury (SCI) per year in the world [1]. About half of these patients are injured above the sixth section of the cervical vertebra. Thus, the movement of the limbs will be affected. Except for spinal cord injury, some other diseases can cause a loss of motor

Y. Jin · J. Chen · S. Zhang (✉) · X. Zheng
Qiushi Academy for Advanced Studies, Zhejiang University, Hangzhou, China

Department of Biomedical Engineering, Zhejiang University, Hangzhou, China
e-mail: shaomin@zju.edu.cn

W. Chen
Qiushi Academy for Advanced Studies, Zhejiang University, Hangzhou, China

College of Computer Science, Zhejiang University, Hangzhou, China

X. Zheng (ed.), *Neural Interface: Frontiers and Applications*,
Advances in Experimental Medicine and Biology 1101,
https://doi.org/10.1007/978-981-13-2050-7_3

function, when patients deeply appreciate how important the motor function is. Most of these patients are unable to take care of themselves and have a poor quality of life. Meanwhile, the drug is not effective in their cases.

Brain-Machine interfaces (BMIs) establish a direct pathway for communication and control between the brain and external devices, which means utilizing neural signals to control external devices to accomplish the desired action [2–4]. This pathway enables direct interaction between the brain and external devices independent of spinal cord/peripheral neuromuscular system. Not only does it assist and enhance the interaction between the human and outside world, but also repair the sensory, motor, and other cognitive dysfunction, providing a new means of rehabilitation for the disabled.

BMI is mainly composed of signal acquisition, signal processing, signal decoding, and external device control module [3, 4]. Signal acquisition module utilizes various sensor brain signal acquisition to collect electroencephalogram (EEG), magnetic signals, image signals, and others which can reflect neural activities in the brain. Signal processing refers to amplifying and filtering the original signal, consequently enhancing the effective signals and inhibiting the invalid signals like noise. Signal decoding is a process mapping the neural activities to motion parameters, which can be further converted to commands to control external devices. External devices aim at assisting or enhancing the motor function of subjects. Some common ones are a computer mouse, electric wheelchair, robotic arm, and so on.

BMIs are divided into non-invasive BMIs and invasive BMIs according to whether the signal acquisition equipment invades the organism of subjects [2]. Invasive BMIs based on cortical neural signals can directly obtain spike and field potentials with the features of being an informative and high temporal-spatial resolution, so as to predict the subjects' behavior intention and achieve real-time and precise control of external devices in multiple degrees of freedom. Since this century, scientific journals such as Nature and Science have reported a number of significant research achievements about invasive BMIs. The related research also promotes people's awareness of the nervous system and establishes a large number of methods to process complex information, which greatly helps computer improve comprehension to complex perceptual information and increase the efficiency of processing mass heterogeneous information.

From the point of view of the development of BMIs studies, the current trend of development of BMI is gradually transformed from laboratory to clinical application. Experimental study of BMI is more and more advanced, from rodent animals gradually shifted to non-human primates, and has carried out preliminary clinical studies; The paradigms in studies are getting more complex, from the two-dimensional (2-D) cursor control to the development of multi-degree of freedom prosthesis for performing reach-to-grasp movements. The clinical study of these invasive BMI will provide a solid foundation for its clinical applications.

3.2 BMIs on Rodents

With the deepening understanding of the encoding rules of the cerebral motor cortex in the rodent and the continuous improvement of the accuracy of the off-line decoding algorithms, researchers try to bypass the natural neural pathway and establish the artificial information transmission pathways, which means direct extraction of neural signals of cerebral motor cortex and real-time control of external device. According to the difference of the direction of information transmission, the rodent BMI can be classified into one-way BMI and two-way BMI.

3.2.1 Neural Decoding Studies in Rodents

One-way BMI refers to the extraction of neural signal from the cerebral cortex of animals, the translation of neural signals into motion commands according to the mapping between them and controlling external device to complete the task. In one-way BMI, information is transmitted from the nervous system to the external device on the artificial pathway. The animal can only receive the feedback of external device through its natural sensory system, such as vision, audition, touch, etc. One-way BMI is the earliest brain-machine interface system.

According to whether the animal gets the feedback of external device in real time, one-way BMI is classified into open-loop BMI and closed-loop BMI. The world's first open-loop BMI system was built by Chapin in 1999, in which the rat pressed the lever by its neural signals [5]. As shown in Fig. 3.1, the rat was first trained to learn to press the lever with its forelimb (Fig. 3.1b). The displacement of the lever was proportional to that of a robot arm which delivers water to rats (Fig. 3.1c). Once the displacement exceeded the threshold, the rat could get the water reward. During the experiment, multi-channel neural signals of the primary motor cortex (MI) and the ventrolateral (VL) thalamus were recorded simultaneously. In behavioral control mode, the rat drank water by pressing the lever. In the BMI mode, the connection between the lever and the robot was cut off, and the displacement of the robot was controlled by control commands decoded from the 32-channel neural signals. Then, the control mode switches to BMI mode. In the early stage of BMI mode, the rat still tried to press the lever to get the reward. As time went on, the rat completed the task only by modulating its neural activity, and the association between the forelimb's displacement and neural activity was weaker and weaker. Finally, the rat drank water only by modulating neural signals without moving its forearm. The experiment first proved the feasibility of an online BMI system and opened the door to the BMI field.

In 2009, Lee et al. built a rat BMI system to control turntables [6]. In behavioral training, the rat was trained to control the direction and magnitude of a turntable to obtain water. The motion set C_q consisted of seven elements $\{-3, -2, -1, 0, 1, 2, 3\}$, where the sign of the element represented the rotation direction and the

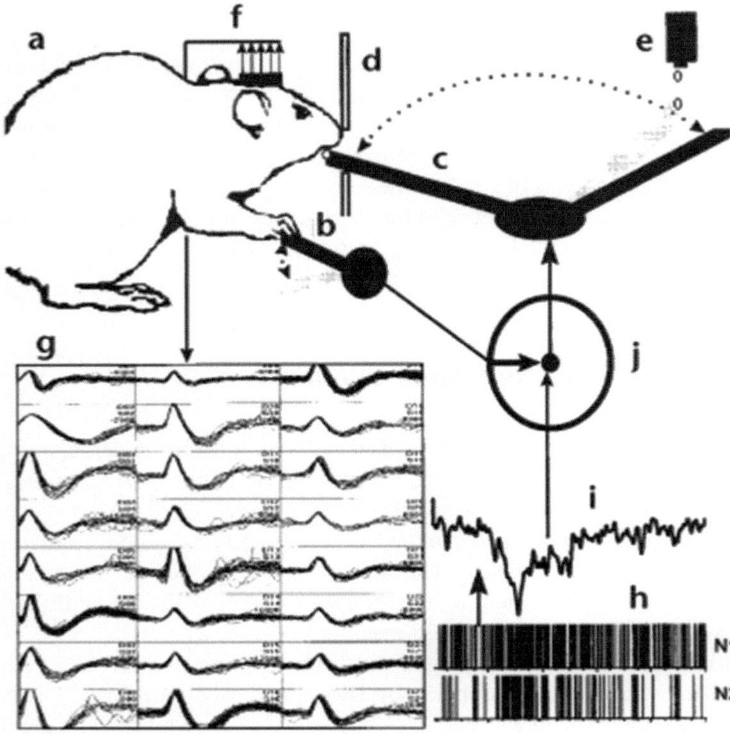

Fig. 3.1 The first BMI study in rodents [5]

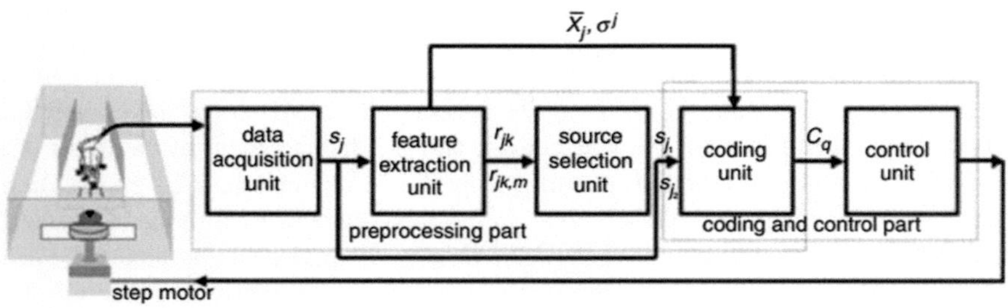

Fig. 3.2 A turning table paradigm in rodent BMI study [6]

magnitude represented the rotation amplitude. After behavioral training, multi-electrodes were implanted in the SI vibrissae area. As shown in Fig. 3.2, multi-channel spike train was sent into feature extraction and channel selection units which selected two channels s_{j1}, s_{j2} as decoding channels according to correlation. The coding unit obtained the mapping between the relative deviation of s_{j1}, s_{j2} and the motion data $C_q = f(s_{j1}, s_{j2})$. According to the mapping obtained above, the decoder generates the control command every 200 ms to make the turntable rotation.

In 2012, Manohar et al. built the BMI system of the rat hindlimb pressing lever under three different conditions [7]. The rat was trained to press the lever to get the reward in the behavioral control mode, which was followed by three neural control modes (NC). In the first mode, the mechanical connection between the lever and the reward delivery was cut off, but the rat could still press the lever. In the second mode, the lever was removed and the rat got the reward purely by modulating its neural signals. In the third mode, the spinal cord of the rat was completely transected and the rat relearned to modulate the BMI system. They compared the performance and neural firing pattern in the behavior control mode and three neural control modes. When the control mode was switched from the behavior control mode to the neural control mode, the response of the direct population was significantly enhanced, and the decoded information was significantly increased. When the lever was removed, the control performance was similar but the neural response changed significantly. When the spinal cord completely transected, the decoded information decreased by 40%. With a period of learning, the control performance gradually recovered, but still worse than before. The experiment demonstrated that the animals' state, such as the limitation of movement and the integrity of neural pathways, had a large influence on the BMI performance and the firing pattern of a neural population.

For some paralyzed patients, such as those with spinal cord injury, their cerebral motor cortex and effector (such as the arm) are intact and the lesion is located on the neural pathway, which breaks the transmission of information. For these patients, the best way to repair the motor function is to build an artificial neural pathway between the cerebral motor cortex and the effector. So researchers tried to combine traditional BMI with muscle electrical stimulation technique and put forward a new repairing technique of neural pathway. In 2014, Alam et al. reconstructed the neural pathway between the primary motor cortex and the hind limb of the rat with spinal cord transection [8]. They recorded the neural activity from the primary motor cortex of running rats before and after spinal cord transection in rats running tasks. The results showed that the association between neural activity and motion states decreased significantly after the transaction, but the stationary state and moving state could be still distinguished by decoding. After transection of the spinal cord, action potentials of the motor cortex were used to stimulate the muscles of the hindlimb. When the smoothed firing rate exceeds the threshold for the five consecutive time windows, the neural stimulator stimulated the hind limb muscle with a bidirectional pulse (biphasic pulse), which resulted in the synchronization movement between the hind legs and forelimb

The traditional BMI has to acquire a section of neural signal and movement data to train the decoder before translating the neural signals into control commands. However, for paralyzed patients, it is difficult to get the training data. A solution is based on motor imagination in which the paralyzed patients observe the automatic movement of external device while imagining themselves controlling the device. Then, the neural signals and the movement data are used to train decoder. With the in-depth study of brain plasticity, researchers try to set the mapping between the neural signal and motion states arbitrarily without a training set. Relying on the neural plasticity, the brain modulates its firing pattern to adapt to the artificial mapping and finally control the external device.

In 2005, without the advanced training and neural data for animals, Kipke et al. made use of the brain plasticity and adaptive decoder to build a rat BMI system to control an auditory cursor [9]. In each trial, the rat was given a target tone (10 Hz) and then required to control the auditory cursor to match the target tone. A Kalman decoder was used to translate the multi-channel neural signals into the pitch of an auditory cursor which was fed back to the rat in real time. The initial parameters of the decoder were randomly generated. A block estimation adaptive strategy was used to update the decoder parameters. Each block contained 10 trials. After each block, the pool of the neural signals and pitch of the auditory cursor during the previous 10 trials served as the training set to retrain the decoder. The new decoder was then used in the next block of trials. As the training went on, the success rate increased gradually from the random level.

The experiment carried by Kipke et al. harnessed the plasticity of the brain and adaptive strategy of the decoder to converge the encoding rules of both natural neural system and decoder. Thus, the final neural mapping was based on natural mapping, but not exactly the same. Considering the great role of neural plasticity in the BMI system, researchers have tried to build some BMI systems entirely relying on the plasticity of the neural system. In 2012, Carnema et al. proposed the operative conditioning experiment based on neurons, which artificially set new neuronal encoding rules which are completely different from natural ones. Then, the rat was trained to modulate the activities of neurons to control an auditory cursor [10]. The researchers successfully achieved the two-way control of the auditory cursor with this method and studied the functional coherence and plasticity of cortex and striatum. Specifically, two groups of primary motor cortex (MI) neurons were selected as conditioned neurons, and the rat was trained to control the auditory cursor to reach the high or low threshold by modulating neural firing patterns. The relationship between the pitch of the auditory cursor and the response of the two groups of neurons was opposite: the increased response of the first neural population increased the pitch of the auditory cursor while that of the second neural population reduced the pitch. Therefore, to reach the high pitch target, the rat had to activate the first neural population while suppressing the second one; to reach the low pitch target, the modulation was opposite. The results indicated that the firing patterns of striatal neurons changed as the rat learned to control the auditory cursor and gradually matched the preset encoding rules. Furthermore, the firing pattern of the motor cortex and striatum become more and more similar. The experiment further extended our understanding of neural plasticity and proved that through training the animal could master a new encoding rule which was completely different from natural ones and achieved control of the external device.

In 2014, Widge et al. made a similar operant conditioning experiment to control an auditory cursor [11]. Different from the Carnema's experiment, they only extracted one neuron from the prefrontal cortex (PFC) as the conditioned neuron, and its firing rate was positively related to the pitch of the auditory cursor. The experiment trained the rat to move the auditory cursor to the target within a period of time to and maintain for 500 ms. After training, all rats were able to complete the BMI tasks successfully. The results demonstrated the feasibility of a single neuron in the operant conditioning control of BMI. In 2013, Arduin et al. achieved to train rat to self-drink by modulating single neuron in the motor cortex to control the one-way movement of the water bottle [12]. In the same year, he further improved his experiment and achieved two-way continuous brain control [13].

3.2.2 Bidirectional Recording and Stimulating System for Free Moving Rats

Just as described before, the information pathway of the one-way BMI is from the brain to the external device whose feedback is transmitted back by natural sense such as vision, audition, or touch. The development of intracranial microstimulation (ICMS) provides a new approach to transmit information directly to the brain independent of the natural neural feedback pathways. Recently, the technique of guiding the animal to conduct a specific task by ICMS has become more and more mature. Therefore, researchers tried to build the two-way information interaction between the neural system and the external device, e.g., bidirectional BMI.

The first bidirectional BMI is a hybrid neuro-robotic system connecting the brainstem tissue of lamprey *in vitro* and a moving robot, which was built in Northwest University in 2003 [14]. After that, the bidirectional BMIs based on animal tissue *in vitro* have been widely studied.

In 2012, Vato et al. built a bidirectional BMI *in vivo* on an aesthetic rat. 16-channel microwire arrays were implanted into the primary motor cortex (M1) and primary sensory cortex (S1), respectively [15]. On the "read-out" pathway, the first two principal components of the M1 neural response were used to form a two-dimensional force vector (Fig. 3.3b, c), which drive a simulated point-mass moving in a viscous medium (Fig. 3.3d). On the "write-in" pathway, the position of the point-mass was transformed into one of four stimulus pattern by "Sensory Interface," and each stimulus pattern was produced by a pair of channels of S1 microwire array (Fig. 3.3a). The rat was trained to modulate its M1 neurons firing according to the stimulus feedback to move the point-mass from an arbitrary starting location to the selected equilibrium point.

In 2010, Kipke et al. built a bidirectional BMI based on the awake rat [16]. An abstract task was designed that required the rat to modulate its neural activity and gave the feedback by ICMS. The motor cortex and the visual cortex of the untrained

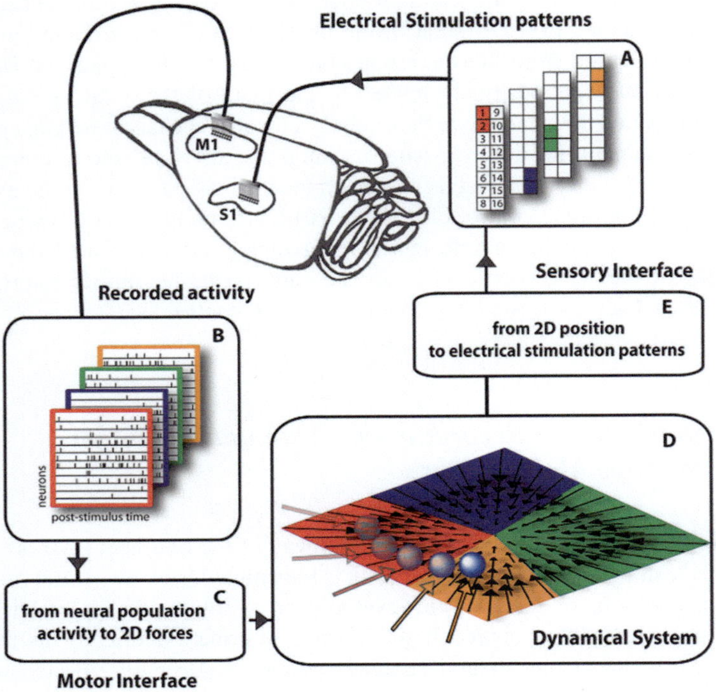

Fig. 3.3 A bidirectional BMI *in vivo* on an aesthetic rat [15]

rat was implanted with multi-channel electrodes. In each trial, the rat maintained its motor cortex ensemble baseline firing rates for 450 ms to begin a trial. Then, the "go" cue was given by ICMS and the rat had to modulate its motor cortex for 450 ms within the 4 s response period in order to receive the food reward. The rat was given continuous feedback via visual cortex ICMS during the response periods that was representative of the motor cortex ensemble dynamics. The results indicated that the rat could master an open-loop control task and build the association between the stimulus and the firing pattern of the motor cortex through a period of training.

Gage et al. designed a more systematic BMI to transform neurons in rats into sounds of different frequencies [9]. For a successful trial, rats need to regulate neuronal activity to reach the target frequency and maintain a certain time. DiGiovanna et al. proposed another BMI adaptation paradigm based on enhanced learning [17]. The goal of the intelligent system was to maximize the final reward after making the action in the intelligent system. The brain cortex uses the intrinsic reinforcement learning mechanism to regulate the relevant neurons, and the machine needs to simulate the corresponding reinforcement learning process. Mahmoudi et al. proposed the concept of symbiotic BMI, which obtained the reward information directly from the NAcc region of the rat brain as feedback to machine enhanced learning [18].

3.2.3 *New Exploration of BMIs Researches in Rodent Animals*

The BMI technique based on the neural signal recording, encoding, decoding, and ICMS could not only build the artificial open-loop information pathway between the brain and the external device and rebuild or repair some damaged motor function but also achieve the direct communication between the brain and external devices. In 2013, Nicolelis's team in Duke University built a brain-to-brain interface (BTBI) between two rats, which transmitted information from the brain of encoder rat to that of decoder rat [19]. Specifically, the two rats were placed in two separated boxes, and the encoder rat was trained to press one of two levels according to the visual stimulus to receive the water reward. Meanwhile, M1 neural activity was recorded from the encoder rat and transmitted to the decoder rat by ICMS. The decoder rat had to select the same lever pressed by the encoder to get the water reward. Once the decoder rat completed the trial successfully, the encoder rat got an extra reward. This suggested that BTBIs could enable networks of animal's brains to exchange, process, and store information and, hence, serve as the basis for studies of novel types of social interaction and for biological computing devices (Fig. 3.4).

Traditional BMI technique is based on electrophysiology, i.e., neural signal collecting and ICMS. Apart from this, researchers have tried to make use of other techniques to build BMIs. In 2014, Carnema's team in UC Berkley used two-photon imaging to perform a conditional operant experiment on awake rats and control an auditory cursor [20]. The fluorescence value was binned by 200 ms time windows and transmitted to the decoder to get the pitch of auditory cursor. The increase of the first neural ensemble (E1) increased the pitch while the increase of the second neural ensemble (E2) decreased the pitch. The rat was trained to modulate the activity of both E1 and E2 population to maintain the cursor within a baseline range to start a

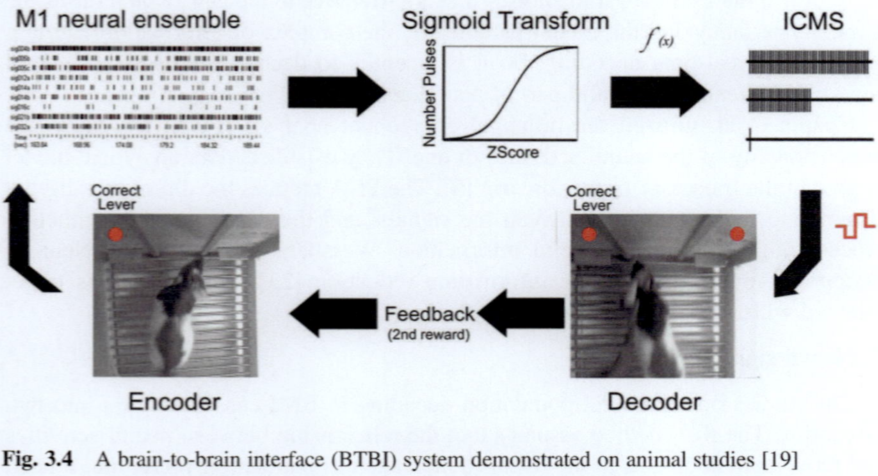

Fig. 3.4 A brain-to-brain interface (BTBI) system demonstrated on animal studies [19]

trial and then moved it to a high threshold within 30 s to complete a trial. This experiment expanded the technique of BMI further.

3.3 BMIs on Non-human Primates

3.3.1 Neural Signal Processing and Neural Decoding

- Neural signal processing

Carmena et al. started studying invasive BMIs in non-human primates (macaques) in 2003, which promoted the rapid development of this technology [21]. Researchers parsed a variety of movement-related parameters from recorded neural signals to control a robotic arm and accomplish reaching and grasp with the visual feedback.

In 2008, Schwartz and his colleagues implemented that the monkey could perform self-feeding in three-dimensional space with its arm fixed [22]. Researchers sent the food to a monkey by equipment during the experiment. While the neural signals of the monkey were decoded to control a robotic arm, the monkey could learn to grip the food and feed itself. The research was of great significance in that it reconstructed the motor function of arms in three-dimensional space.

The recording quality of neural activities is very important for neural signal analysis. The collecting and recording of neural signals are becoming more accurate with the development of the microelectrodes or probes. Nowadays, there exist commercial electrodes that were approved by the Food and Drug Administration (FDA) to be utilized even on human beings. The microelectrodes can be implanted into several brain cortex to collect multi-channel neural activities. One of the key questions in BMI is to decode the multi-channel neural activities efficiently and accurately.

Neural signals contain some noises that are irrelative to the task, which raises the difficulty of analyzing the model parameters, the complex of information learning models, and real-time decoding. So it is essential to decrease the scale of neural activities and extract the valid part of neural activities.

Chapin et al. utilized the principal component analysis (PCA) to reduce the dimensionality of the neural activities to one. They established an analytical model to predict the trajectory of the forearm [5]. The PCA reduces the dimensionality by eliminating the correlation between the signals, and the dimensionality reduction process might miss some useful information. Wessberg et al. used the Neuron-Dropping Analysis to achieve information reduction [23]. This method is target oriented with high computational complexity.

- Neural signal decoding

The studies on the neural population decoding in BMI can be divided into two categories. The first method assumes that the relationship between neural activities and movement conforms to the linear model, such as population vector and related algorithm [22, 24, 25], Wiener filter-based analysis method [21, 26, 27], Kalman filter-based analysis method [28–30], Fisher linear discriminant based method, and maximum likelihood estimation method. Although the linear analytic model has achieved good results in relatively simple decoding applications, recent studies have found that the nonlinear method has a more reasonable analogy to the real nervous system. In recent years, many attempts have been made to use non-linear methods for neural decodings, such as particle filter [31, 32] and unscented Kalman filters [33, 34].

The above researches have established a good foundation for neural population decoding. However, on account of the complexity of the brain system, it is necessary to achieve efficient and accurate neural population decoding. First, the brain cortex is a plastic system, so the time-varying neural activities raise great challenges to the traditional static information processing method; second, the neural activities show similar features in both time domain and frequency domain. Neural activities from limited brain cortex and frequency band can hardly represent the whole brain system. So the exploration of efficient, dynamic, and joint analytic models and algorithms is still an open question for BMI.

3.3.2 Closed-Loop Decoding for Online Control

The ideal BMI aims to achieve bi-directional information interaction between the brain and the external devices. In recent years, with the development of neural signal analysis technology and neural stimulation technology, researchers have focused their attention on bi-directional closed-loop BMI system research.

Novellino et al. combined *in vitro* cultured rat neurons and embedded signal processing platform to build a bi-directional BMI system prototype. The system used the signals of the neurons to control the movement of a car. There is stimulus feedback to the neurons when the car hits the obstacle. So the bi-directional interaction between neurons and the external environment was established [35]. Doherty et al. gave monkey direction instruction by electrical stimulation on the sensory cortex. Then, the direction selection task was achieved by real-time decoding of the motor cortex. The results showed that the monkey could obtain external information and output control sequences to the external devices [36]. In addition, Ye, Rolston, and Venkatraman et al. developed a number of stimulated neural signal acquisition systems that combine both neural stimulation and recording techniques to achieve a preliminary bi-directional communication system. Above studies showed that the bi-directional BMI has great application potential in developing disease treatment and understanding brain network mechanism [37–39].

Intracortical electrical stimulation, as a somatosensory information input, not only establishes a new information transfer path between brain cortex and external environment but may also improve the performance of the vision-based feedback BMI by information integration. Preliminary studies have shown that the electrical stimulation can not only instruct the movement of animals but also feedback the movement information to the brain cortex.

O'Doherty et al. trained monkeys to finish the movement direction selection task with tactile feedback and electrical stimulation [36, 40]. After two weeks, the result showed that the natural tactile feedback in the experiment and cortical electrical stimulation provide almost the same amount of information. At the same time, a considerable number of studies have confirmed that multi-modal feedback can provide more information and improve the performance of BMI. Nicoliles et al. found that visual feedback is very important for the motor control BMI and visual feedback coupled with tactile feedback can further improve the overall performance of the system; [4] Suminski et al. studied the natural somatosensory feedback and the mechanism of visual and found that somatosensory feedback is more likely to affect the neural activity of the motor cortex than visual feedback, and more importantly, the combination of two different modes of feedback can provide more information [41]. Further results showed that the dual-mode feedback could significantly improve the control performance of the BMI when the feedback of the two modes is consistent.

Several studies have shown that the plasticity of the brain cortex is closely related to BMI. In the process of neural decoding, neural activity is a dynamic learning process, and the mapping between the output signal (usually the neural signals) and the input signal (usually the kinematic data of the limb) is changing over time. In previous BMI studies, results have shown that the activity of a single neuron may change with the decoding algorithm. Ganguly et al. conducted a preliminary exploration of the adaptability and plasticity of the neural population in the context of BMI. Their findings suggest that the cerebral cortex has a very prominent learning ability and can quickly adapt to mutations in the decoding model [42].

Other researchers also found that cortical electrical stimulation can also affect the plasticity of the cortex. Moritz et al. have shown that functional connectivity can be established between the two neurons by cortical electrical stimulation, which provides a new idea for nerve repair [43]. Rebesco et al. have shown that cortical electrical stimulation can not only serve as an information input, but also can effectively enhance the functional connection of nerve, and eventually lead to behavioral changes, and this plasticity may be associated with natural stimulus caused by Hebby learning mechanism [44].

The study of the changes of the neuron population under the condition of invasive BMI would help us understand the plastic mechanism of the brain and construct the bi-directional closed-loop interaction. The use of the plasticity of the brain, combined with machine learning, can improve the performance of BMI. Taylor et al. used a mutual adaptation algorithm on the rhesus monkey's three-dimensional cursor control task and updated the mapping of neuronal activity to the movement of the cursor by iterating the changes in neuron during the learning process [24].

3.4 Pilot BMIs Studies in Clinic Researches

3.4.1 Invasive BMI System Based on Spike Signals from Human Brain

The ultimate goal of brain-machine interface development is to realize the practical application of clinical patients. Currently, there are mainly two types of invasive BMIs clinical studies. One type of BMIs employed fully implantable microelectrode arrays to collect the spike signals of neuron populations. This method has the advantage of acquiring high spatial and temporal resolution signals that contain rich information of movement and can be used to realize the high-precision motion control. However, due to the need of microelectrode array implanted directly into the cerebral cortex, the damage induced by implantation makes it difficult to achieve long-term application; the other type is a semi-invasive BMIs, which mainly used the electrodes placed on the surface of the cortex, collecting field potential signals from the cerebral cortex. Its damage to the brain is relatively small, but the spatial resolution of this type of signal is relatively lower than that of the invasive micro-electrode array. Each method has its advantages and disadvantages.

In the past two decades, many BMI system has also been developed for clinical transformations. In 1998, Philip Kennedy and Roy Bakay from Emory University started their researcher invasive BMIs [45]. Their study implanted a neurotrophic electrode into a patient who had severe amyotrophic lateral sclerosis (ALS) for 2.5 years prior to the implantation and was ventilator-dependent for 1.5 years. The electrode consists of a hollow glass conical tip about 1.5 mm in length which contains two gold recording wires. They recorded action potentials in the patient's brain over several months. By decoding the action potential, the patient ultimately was able to control the clicking of a computer mouse. This study was the first clinical study of invasive BMI which achieve the high-quality neural signal recording and movement output decoding, which laid the foundation for the development of invasive BMI.

With the development of the electrode technology, many researchers began to study the clinical BMI experiments based on the microelectrode arrays. In 2006, the group of Donoghue achieved a BMI for movement intent decoding for a tetraplegic patient [29]. The patient in this study was a 25-year-old male who had sustained knife wound, the spinal cord between cervical vertebrae C3–C4 of the patient was transected, which results in complete tetraplegia. During the research, a 96-ch microelectrode array was implanted in the arm area of the primary motor cortex of the patient for neural signal recording. The BMI system decoded the different direction movement imagination and was finally used to control a computer mouse in a two-dimensional plane. By making use of this BMI system, the patient was able to open the e-mail, draw figures using a paint program, play video games, and even adjusted the volume, channel, and power to his television. This study was the first attempt to make use of microelectrode arrays in clinical BMI use, which provides valuable research experience for the clinical transformation of invasive BMI. Then, the group further study the long-term stability performance of the BMI system in chronic application [46]. They found that the BMI system still kept a high system performance even 1000 days after the microelectrode arrays implantation. Though there were still many problems such as electrode aging and signal attenuation which may affect the long-term use of BMI system, this study proved that the BMI based on microelectrode array could keep a stable system performance for 2–3 years. In 2008, other two subjects participated in the BMI experiment. One participant was a 54-year-old woman who had thrombosis of the basilar artery and extensive pontine infarction 9 years prior to trial recruitment while another participant was a 37-year-old man who had been diagnosed with ALS for about 6 years. The two participants successfully achieved the online control of a computer mouse by using the BMI system after some training sessions. This result proved that the BMI system might be a benefit to the patients who suffer long-term motor dysfunction. Then, the further study of robot hand control by BMI was carried out in 2012 [29]. A stroke patient and the 54-year-old woman described above participated in this study. The system successfully decoded the three-dimensional trajectory of the hand movement. The brain activity of M1 was extracted to control the movement and grasp of a robot hand. The 54-year-old woman successfully used the robot hand to grasp a coffee cup and move to her own mouth to drink the coffee by absolutely brain control. This study shows us the prospects of invasive BMI system in daily life application.

The group of Andrew B Schwartz also had some research on clinical invasive BMI. They set up a clinical BMI system based on the microelectrode array in 2013 [46]. A 52-year-old woman who was tetraplegic participant in this study. Two 96-ch microelectrode arrays were implanted in her arm area of premotor cortex (PM), and the participant had to control a prosthesis using her brain signals by the BMI system after a 13-weeks' training. The participant had to perform a seven-dimensional sequence task including three-dimensional position, rotating the palm in one of six directions (possible orientations included pronation–supination $\pm 45°$, ulnar-radial deviation $\pm 20°$, and flexion-extension $\pm 45°$ of the wrist), or a grasp target (open or closed hand). The participant used the prosthesis under full brain control to do nine tasks (selected from 19 possible tasks) on the action research arm test (ARAT), which is an assessment of the unilateral upper limb function used commonly in patients who have had a stroke. This study showed a high-performance BMI system with multi-degree of freedom, which was difficult to achieve by non-invasive BMI. Then in 2015, the group of Schwartz further achieved the ten-dimensional anthropomorphic arm control in human BMI [47]. The ten dimensions contain translation (x,y,z), orientation (θx,θy,θz), pinch (p), scoop (s), finger ab/adduction (f), and thumb opposition/extension (t). This study found that individual motor cortical neurons encode many parameters of movement and that high-dimensional operation of prosthetic devices can be achieved with simple decoding algorithms, which shows the invasive BMI has the potential to control high-degree freedom prosthetic devices.

In recent years, more and more fully invasive BMI system has been studied. Richard A. Andersen et al. set up a BMI system based on the signals in posterior parietal cortex (PCC) [48]. PCC is regarded as being responsible for the process of high-level movement planning. This study implanted a microelectrode array to the PCC and successfully decoded the different object and trajectory of motor imagination. Bouton et al. study the hand motor function rehabilitation based on invasive BMI [49]. The study participant was a 24-year-old male with quadriplegia from cervical spinal cord injury sustained in a diving accident. The researcher decoded the neuronal activity and control activation of the participant's forearm muscles through a high-resolution neuromuscular electrical stimulation system. Through the system, the participant complete a Grasp-pour-and-stir functional movement task, which need the participant to opening his hand (a), grasping the glass bottle (b), pouring its contents (dice) into a jar (c), grasping a stir stick from another jar (d), transferring the stir stick without dropping it (e), and using it to stir the dice in the jar (f). These results have significant implications in advancing neuroprosthetic technology for people worldwide living with the effects of paralysis, which extended the application of BMI.

3.4.2 Invasive BMI System Based on Clinic ECoG Signals

We can obtain high temporal-spatial resolution brain signals by implanting microelectrode array to decode higher degrees of freedom and finer arm motion and achieve more natural and complex arm movement, which is the goal and direction of BMI. However, the fully implantable microelectrode arrays may cause damage to the blood-brain barrier, inflammatory response, and immune reaction, which may produce a series of biological and non-biological effects, leading to the attenuation and loss of the brain signals. It limits the application of implantable microelectrode array in the clinical BMI studies. On the contrary, the electrocorticogram (ECoG) electrode is located under the subdural and on the surface of cortical, which does not cause damage to the cerebral cortex. The neuronal damage can be completely avoided. Although the spatial resolution of the ECoG signal is slightly lower than that of the microelectrode array, the ECoG has a fewer surgical risk and better long-term stability of the signal. So the ECoG is relatively more meaningful for the clinical application of BMI. In the past few years, the study of clinical BMI based on ECoG signals had made great progress.

As early as in 2004, Gerwin Schalk, Eric C Leuthardt, Daniel W Moran ,and Jeffrey G Ojemann started the study of ECoG-based clinical BMI [49]. All the participants in the study were clinical epilepsy patients in whom subdural electrode arrays were implanted for three to eight days in preparation for surgery to remove an epileptic focus. The ECoG arrays located in sensorimotor cortex. Over brief training, the four participants then used these signals to master closed-loop control and to achieve success rates of 74–100% in a one-dimensional (1-D) binary task. Further analysis showed that ECoG also has a correlation with the direction of the two-dimensional joystick movement. Then in 2006, the same group set up a closed-loop BMI in paralyzed patients [50]. There are four participants, and the electrode covered their sensorimotor cortex and some speech cortex areas. In order to identify brain signals used for BCI control, the subjects were asked to perform paired blocks of repetitive hand/tongue, foot/shoulder movements, or repetitive speaking of the word "move" (screening procedure). From the spectral analysis, the researchers identified the locations and frequency bands in which amplitude was different between the task and rest, which will be selected in the subsequent closed-loop BMI. The participants received online feedback that consisted of the 1-D cursor movement that had shown correlation with tasks during the screening procedure. Their study showed that visual cursor feedback could effectively improve the success rate of the task. The work of Schalk was the first attempt to set up ECoG-based clinical BMI, which proved that ECoG was an available signal source in clinical BMI use.

The group of Schalk then used the ECoG signal to decode the trajectory of the cursor in two-dimensional plane [51]. The study included five patients with intractable epilepsy. The subjects were asked to use a joystick to move a white cursor in two dimensions to track a green target. By decoding the kinematic parameters from ECoG, the study showed that the correlation coefficients (r) between the actual and decoded kinematic parameters were as high as 0.7, which proved that ECoG-based BMI was feasible in two-dimensional trajectory decoding. Carsten Mehring et al. did similar work in 2008 [56]. Six patients suffering from intractable pharmaco-resistant epilepsy were included in their study, and the electrical stimulation through the electrode grid was performed to identify the functional mapping of the cortex. A Kalman filter was applied to decode the two-dimensional trajectory of the arm, and their result showed that the low-frequency component of ECoG has better performance for the two-dimensional trajectory decoding. In 2013, Yasuhiko Nakanishi and Takufumi Yanagisawa from Japan successfully decoded the three-dimensional trajectory of arm movement using ECoG signals [52]. There were three participants in this study; two of them had spastic paresis and weakness due to stroke, and the other was diagnosed with intractable epilepsy. The participants were asked to reposition three blocks one by one and clockwise at the corners of a square table, and an optical motion capture system was used to record the arm movement. Their work successfully predicted three-dimensional arm trajectories in time series from ECoG signals, which proved that we could extract the motion parameters of three-dimensional arm movement from ECoG signals.

In addition to continuous arm movement trajectory decoding, some teams have used ECoG signals to decode different finger movement and gestures. In 2009, Schalk et al. proved that it is possible to decode the flexion of individual fingers using ECoG signals [53]. A data glove (5DT) was used to record the flexion of each finger in this study. Their result showed that there was a different spatial pattern of ECoG signals for different finger movement and the LMP, and high gamma component of ECoG could decode the flexion of each finger with best decoding performance. Then in 2010, Nitish V. Thakor et al. found that ECoG signals correlated with finger positions [54]. Their study indicated that when the subjects engaged in slow and deliberate grasping motions, the amplitudes of the low-pass ECoG (LMP) over specific electrodes were correlated with the position of individual fingers. The work of Mehring and Shenoy both showed that ECoG signals could be used to decode different gestures [55–58]. Mehring's experiment requires subjects to grasp different objects with different gestures; and in Shenoy's experiment, the subject was asked to perform several specific gestures, which require different fingers to bend and stretch. Both experiments proved that ECoG signal was able to decode different gestures.

For that ECoG signals could be used to decode different motion parameters, some studies have implemented prosthesis control based on ECoG signals. In 2013, Matthew S. Fifer et al. trained two human subjects to achieve simultaneous neural control of reaching and grasping movements with a dexterous robotic prosthetic arm [59]. The group of Toshiki Yoshimine from Japan also used human ECoG signal to real-time control of a prosthetic hand [60]. Two decoders were trained to control the prosthetic hand. The first decoder was trained to classify the movement state R or M, and the second was trained to predict the types of performed movement with the state M. Based on the two decoders, the subject could asynchronously control of the prosthetic hand to perform three types of hand movement. In a further study, the subject could control the prosthetic hand to perform more tasks such as grasping, thumb flexion, elbow flexion, pinching, hand-opening, elbow flexion, and elbow extension [61]. This study tested the clinical BMI in a large number of subjects (12 subjects included) and achieved prosthetic hand control for multiple types of gestures. Wei Wang et al. used ECoG signals to control of 3-D cursor movement and 3-D prosthetic arm movement [62]. The participant in this study was a 30-year-old man with tetraplegia; he was asked to observe hand and arm movements of a virtual character on an LCD screen and simultaneously attempted the same movement. The participant successfully controlled the prosthetic arm hitting physical targets. This study indicated that we could extract the information of motion intent from ECoG signals. Another application of clinical BMI was a brain-control typing system. Ramsey et al. set up a BMI-based computer typing program for a locked-in patient with ALS [63]. The ALS patient was a 58 years old woman, and four subdural electrode strips were implanted in her brain motor cortex. The signal recording, conversion, and transmission devices of the typing system were fully implanted, which make the BMI system more simple and convenient for use. Portability and simplification have gradually become a key factor in the clinical transformation of the BMI system.

The key factor in invasive BMI is to improve the decoding performance of the system as much as possible. ECoG arrays usually cover a wide range, and ECoG signals have a wide frequency range, so it's meaningful to set up a BMI of ECoG based on multi-frequency band and multi-brain area. Ramsey et al. set up a BMI to decode gesture and compared the decoding performance of M1, S1, and whole sensorimotor cortex (M1 + S1) [63]. The result showed that decoding performance of M1 and S1 was similar, while the decoding performance of the whole sensorimotor cortex was significantly better, which means that the BMI of ECoG based on the multi-brain area may improve the system performance. Hotson et al. study the decoding performance of different frequency component of ECoG [64]. They divided the ECoG signal to seven frequency band, 0–4 Hz, 4–8 Hz, 7–13 Hz, 14–30 Hz, 30–50 Hz, 70–110 Hz, and 130–200 Hz, and used the signals of the seven single frequency band and multi-frequency band to decode the arm trajectory, respectively. Their result showed that multi-frequency band has the best decoding performance for all subjects. The above studies showed that take advantage of the ECoG signals of different brain area and different frequency band was of great benefit to improve the decoding performance.

3.5 Opportunities and Challenges

3.5.1 Opportunities

- Restoration of motor function in the paralyzed

One of the most important applications of BMIs is the restoration of motor function in the paralyzed. Two main reasons that lead to paralysis are descending corticospinal tract injury caused by spinal cord injury and neuromuscular dysfunction. The paralyzed suffer great physical and psychological trauma. One of the reasons for the rise of the technology of BMIs exactly is to help the paralyzed restore motor function. Researchers have conducted a series of experiments on non-human primates and human and have made some progress, laying a good foundation for the clinical practice of BMIs.

Capogross and his colleagues alleviated gait deficits after a spinal cord injury in non-human primates in 2016 [65]. First, unilateral spinal cord injury model was established by the side cutting the middle section of the monkey's spinal cord. Second, the monkeys were implanted with a microelectrode array into the leg area of the left motor cortex and with electrical stimulators into the lower spinal cord segment and the spinal cord segment associated with motion control of lower limbs. During the movement of monkeys, neural signals were used to decode the movement of lower limbs. Decoding results were then sent to control electrical stimulators to simulate spinal cord, consequently alleviating gait deficits of lower limbs.

- BMIs as a potential therapy for nerve recovery

Since the 1990s, BMIs have been mainly focused on two aspects: (1) The field of neuroscience, in which the physiological characteristics of the neural circuits are studied by designing new experimental paradigms. (2) The field of neural engineering, in which technologies and means of restoring motor function for the paralyzed are studied. Recent clinical studies have found that the BMIs used as one of the potential tools for neurorehabilitation. The results have shown that long-term BMIs training combined with physical therapy has a significant effect on the recovery of motor function. So does BMIs training combined with virtual reality system.

In 2016, Donati conducted 12 months training with a multi-stage BMI-based gait neurorehabilitation paradigm aimed at restoring locomotion in eight paraplegic patients [66]. This paradigm combined intense, immersive virtual reality training, enriched visual-tactile feedback, and walking with two EEG-controlled robotic actuators, including a custom-designed lower limb exoskeleton capable of delivering tactile feedback to subjects. Following 12 months of training with this paradigm, all eight patients experienced neurological improvements in somatic sensation. Patients also regained voluntary motor control in key muscles below the SCI level. As a result, 50% of these patients were upgraded to an incomplete paraplegia classification. Donati and his colleagues hypothesized that this unprecedented neurological recovery results from both cortical and spinal cord plasticity triggered by long-term BMI usage.

3.5.2 Challenges

- Biocompatibility and long-term stability of electrodes

Biocompatibility and long-term stability of electrodes is the key that determines whether BMIs technology can be widely applied to the clinic. Biocompatibility of electrodes is closely related to the long-term stability of recording. Currently, the biocompatibility of implanted microelectrodes is poor. And with the increase of implant time, it's easy to induce the rejection of biological tissue. The hyperplastic tissue covers the surface of the electrode, resulting in a reduction in the quality of the signal recorded by the electrodes. Simultaneously, the impedance variations caused by the corrosion and degradation of the electrode materials will also influence the long-term stability of recording.

Researchers have proposed some relevant strategies to enhance the stability of electrode recording, including electrode improvement in size, geometry, shape, coating, and materials. The results of the study confirm that smaller electrode size can achieve a relatively long signal recording time. At present, researchers have focused on the feasibility of extracellular or intracellular neural recording with carbon nanotube electrodes in that they have low impedance, and are capable of recording the activities of a single unit. Another focus is a bioactive coating, which has the effect of slowing the rejection and enhancing the stability of the signal recording. In addition, the bioactive coating can reduce the response of glial cells and increase the survival rate of peripheral nerve cells.

- The portability of BMIs system

In real life, a BMI operator may not carry a large number of signal cables and signal acquisition equipment. It is important to study how to simplify the BMI system or make it more portable. Ramsey et al. developed a highly integrated BMI system which could be fully implanted into a patient with late-stage amyotrophic lateral sclerosis (ALS) [63]. The patient could accurately and independently controlled a computer typing program with the assistance from this fully implanted BMI system. The future BMI system could be further integrated wireless communication and power. Another critical factor in the clinical application of invasive BMI is the decoding performance of the system. In practical applications, there might be no fault tolerant space so that we have to improve the decoding performance of a BMI system and minimize the decoding error rate as much as possible. The studies of Ramsey and Hotson et al. have proved that combining the neural signals from different brain regions and frequency components is of great significance to improve the decoding performance of a BMI system [63, 64]. In addition, looking for a better feature, exploring the neural mechanism underlying movement control, and developing decoding algorithm more in line with the neural characteristics will play a key role to improve the performance of the BMI system.

Clinical translation of invasive BMI is an essential process before a BMI system could be brought into clinical application and even real-life application. We still need to solve some critical issues when translating the BMI system from the lab to the bed of patients and form a stable and mature application system. When the BMI system could be applied to reallife application, the brain control as Avatar will no longer be fantasy.

References

1. Andersen RA et al (2014) Toward more versatile and intuitive cortical brain–machine interfaces. Curr Biol 24(18):R885–R897
2. Donoghue JP (2008) Bridging the brain to the world: a perspective on neural interface systems. Neuron 60(3):511–521
3. Hatsopoulos NG, Donoghue JP (2009) The science of neural interface systems. Annu Rev Neurosci 32:249–266
4. Lebedev MA, Nicolelis MAL (2006) Brain-machine interfaces: past, present and future. Trends Neurosci 29(9):536–546
5. Chapin JK et al (1999) Real-time control of a robot arm using simultaneously recorded neurons in the motor cortex. Nat Neurosci 2(7):664–670
6. Lee Y et al (2009) Classification of BMI control commands from rat's neural signals using extreme learning machine. Biomed Eng Online 8:29
7. Manohar A et al (2012) Decoding hindlimb movement for a brain machine interface after a complete spinal transection. PLoS One 7(12):e52173
8. Alam M et al (2014) A brain-machine-muscle interface for restoring hindlimb locomotion after complete spinal transection in rats. PLoS One 9(8):e103764
9. Gage GJ et al (2005) Naive coadaptive cortical control. J Neural Eng 2(2):52–63
10. Koralek AC et al (2012) Corticostriatal plasticity is necessary for learning intentional neuroprosthetic skills. Nature 483(7389):331–335
11. Widge AS, Moritz CT (2014) Pre-frontal control of closed-loop limbic neurostimulation by rodents using a brain-computer interface. J Neural Eng 11(2):024001
12. Arduin PJ et al (2013) "Master" neurons induced by operant conditioning in rat motor cortex during a brain-machine interface task. J Neurosci 33(19):8308–8320
13. Arduin PJ et al (2014) Bidirectional control of a one-dimensional robotic actuator by operant conditioning of a single unit in rat motor cortex. Front Neurosci 8:206
14. Kositsky M et al (2003) Dynamical dimension of a hybrid neurorobotic system. IEEE Trans Neural Syst Rehabil Eng 11(2):155–159
15. Vato A et al (2012) Shaping the dynamics of a bidirectional neural interface. PLoS Comput Biol 8(7):e1002578

16. Marzullo TC et al (2010) Development of closed-loop neural interface technology in a rat model: combining motor cortex operant conditioning with visual cortex microstimulation. IEEE Trans Neural Syst Rehabil Eng 18(2):117–126
17. DiGiovanna J et al (2009) Coadaptive brain-machine interface via reinforcement learning. IEEE Trans Biomed Eng 56(1):54–64
18. Mahmoudi B, Sanchez JC (2011) A symbiotic brain-machine interface through value-based decision making. PLoS One 6(3):e14760
19. Pais-Vieira M et al (2013) A brain-to-brain interface for real-time sharing of sensorimotor information. Sci Rep 3:1319
20. Clancy KB et al (2014) Volitional modulation of optically recorded calcium signals during neuroprosthetic learning. Nat Neurosci 17(6):807–809
21. Carmena JM et al (2003) Learning to control a brain-machine interface for reaching and grasping by primates. PLoS Biol 1(2):E42
22. Velliste M et al (2008) Cortical control of a prosthetic arm for self-feeding. Nature 453 (7198):1098–1101
23. Wessberg J et al (2000) Real-time prediction of hand trajectory by ensembles of cortical neurons in primates. Nature 408(6810):361–365
24. Taylor DM, Tillery SIH, Schwartz AB (2002) Direct cortical control of 3D neuroprosthetic devices. Science 296(5574):1829–1832
25. Wahnoun R, He J, Helms TS (2006) Selection and parameterization of cortical neurons for neuroprosthetic control. J Neural Eng 3(2):162–171
26. Kim S et al (2008) Neural control of computer cursor velocity by decoding motor cortical spiking activity in humans with tetraplegia. J Neural Eng 5(4):455–476
27. Hochberg LR et al (2006) Neuronal ensemble control of prosthetic devices by a human with tetraplegia. Nature 442(7099):164–171
28. Wu W et al (2003) Neural decoding of cursor motion using a Kalman filter
29. Hochberg LR et al (2012) Reach and grasp by people with tetraplegia using a neurally controlled robotic arm. Nature 485(7398):372–375
30. Gilja V et al (2012) A high-performance neural prosthesis enabled by control algorithm design. Nat Neurosci 15(12):1752–1757
31. Brockwell AE, Rojas AL, Kass RE (2004) Recursive bayesian decoding of motor cortical signals by particle filtering. J Neurophysiol 91(4):1899–1907
32. Shoham S et al (2005) Statistical encoding model for a primary motor cortical brain-machine interface. IEEE Trans Biomed Eng 52(7):1312–1322
33. Li S, Li J, Li Z (2016) An improved unscented Kalman filter based decoder for cortical brain-machine interfaces. Front Neurosci 10:587
34. Li Z et al (2009) Unscented Kalman filter for brain-machine interfaces. PLoS One 4(7):e6243
35. Novellino A et al (2007) Connecting neurons to a mobile robot: an *in vitro* bidirectional neural interface. Comput Intell Neurosci 2007:12725
36. O'Doherty JE et al (2011) Active tactile exploration using a brain-machine-brain interface. Nature 479(7372):228–U106
37. Ye X et al (2008) A portable telemetry system for brain stimulation and neuronal activity recording in freely behaving small animals. J Neurosci Methods 174(2):186–193
38. Rolston JD, Gross RE and Potter SM (2009) NeuroRighter: closed-loop multielectrode stimulation and recording for freely moving animals and cell cultures. Conf Proc IEEE Eng Med Biol Soc, 2009:6489–6492
39. Venkatraman S and Carmena JM (2011) Active sensing of target location encoded by cortical microstimulation. IEEE Trans Neural Syst Rehabil Eng, 19(3):317–324
40. O'Doherty JE et al (2011) Active tactile exploration using a brain-machine-brain interface. Nature, 479(7372):228–231
41. Suminski AJ et al (2010) Incorporating Feedback from Multiple Sensory Modalities Enhances Brain-Machine Interface Control. Journal of Neuroscience 30 (50):16777–16787

42. Karunesh Ganguly, Jose M. Carmena and James Ashe (2009) Emergence of a Stable Cortical Map for Neuroprosthetic Control. PLoS Biology 7(7):e1000153
43. Chet T Moritz and Eberhard E Fetz (2011) Volitional control of single cortical neurons in a brain–machine interface. Journal of Neural Engineering 8(2):025017
44. James M Rebesco and Lee E Miller (2011) Enhanced detection threshold for cortical stimulation produced by Hebbian conditioning . Journal of Neural Engineering 8(1):016011
45. P.R. Kennedy et al (2000) Direct control of a computer from the human central nervous system. IEEE Transactions on Rehabilitation Engineering 8(2):198–202
46. Collinger JL, Wodlinger B, Downey JE, Wang W, Tyler-Kabara EC, Weber DJ, McMorland AJC, Velliste M, Boninger ML, Schwartz AB (2013) High-performance neuroprosthetic control by an individual with tetraplegia. Lancet 381(9866):557–564
47. Wodlinger B, Downey JE, Tyler-Kabara EC, Schwartz AB, Boninger ML, Collinger JL (2015) Ten-dimensional anthropomorphic arm control in a human brain−machine interface: difficulties, solutions, and limitations. J Neural Eng 12(1):016011
48. T. Aflalo et al (2015) Decoding motor imagery from the posterior parietal cortex of a tetraplegic human. Science 348(6237):906–910
49. Eric C Leuthardt et al (2004) A brain–computer interface using electrocorticographic signals in humans. Journal of Neural Engineering 1(2):63–71
50. E.C. Leuthardt et al (2006) Electrocorticography-Based Brain Computer Interface—The Seattle Experience. IEEE Transactions on Neural Systems and Rehabilitation Engineering 14(2):194–198
51. G Schalk et al (2007) Decoding two-dimensional movement trajectories using electrocorticographic signals in humans. Journal of Neural Engineering 4(3):264–275
52. Yasuhiko Nakanishi et al (2013) Prediction of Three-Dimensional Arm Trajectories Based on ECoG Signals Recorded from Human Sensorimotor Cortex. PLoS ONE 8(8):e72085
53. J Kubánek et al (2009) Decoding flexion of individual fingers using electrocorticographic signals in humans. Journal of Neural Engineering 6(6):066001
54. Acharya S et al (2010) Electrocorticographic amplitude predicts finger positions during slow grasping motions of the hand. J Neural Eng 7(4):046002
55. Pistohl T et al (2008) Prediction of arm movement trajectories from ECoG-recordings in humans. J Neurosci Methods 167(1):105–114
56. Pistohl T et al (2012) Decoding natural grasp types from human ECoG. NeuroImage 59(1):248–260
57. Pistohl T et al (2013) Grasp detection from human ECoG during natural reach-to-grasp movements. PLoS One 8(1):e54658
58. Chestek CA et al (2013) Hand posture classification using electrocorticography signals in the gamma band over human sensorimotor brain areas. J Neural Eng 10(2):026002
59. Fifer MS et al (2014) Simultaneous neural control of simple reaching and grasping with the modular prosthetic limb using intracranial EEG. IEEE Trans Neural Syst Rehabil Eng 22(3):695–705
60. Yanagisawa T et al (2011) Real-time control of a prosthetic hand using human electrocorticography signals. J Neurosurg 114(6):1715–1722
61. Yanagisawa T et al (2012) Electrocorticographic control of a prosthetic arm in paralyzed patients. Ann Neurol 71(3):353–361
62. Wang W et al (2013) An Electrocorticographic brain Interface in an individual with tetraplegia. PLoS One 8(2):e55344
63. Vansteensel MJ et al (2016) Fully implanted brain–computer Interface in a locked-in patient with ALS. N Engl J Med 375(21):2060–2066
64. Hotson G, Fifer MS, Acharya S, Benz HL, Anderson WS, Thakor NV, Crone NE, Robin DA (2014) Coarse Electrocorticographic decoding of Ipsilateral reach in patients with brain lesions. PLoS One 9(12):e115236
65. Capogrosso M, Milekovic T, Borton D, Wagner F, Moraud EM, Mignardot J-B, Buse N, Gandar J, Barraud Q, Xing D, Rey E, Duis S, Jianzhong Y, Ko WKD, Li Q, Detemple P,

Denison T, Micera S, Bezard E, Bloch J, Courtine G (2016) A brain–spine interface alleviating gait deficits after spinal cord injury in primates. Nature 539(7628):284–288
66. Donati AR et al (2016) Long-term training with a brain-machine Interface-based gait protocol induces partial neurological recovery in paraplegic patients. Sci Rep 6:30383

Chapter 4
Peripheral Neural Interface

Peng Zhang, Xiao Li, Dingyin Hu, Qiuxia Lai, Yuanyuan Wang, Xuan Ma, Qi Xu, Wei Li, Jian Huang, and Jiping He

Abstract Peripheral nervous system, widely spread in the whole body, is the important bridge for the transmission of neural signals. Signals from the central nervous system (brain and spinal cord) are transmitted to different parts of the body by the peripheral nerves, while along the way they also feedback all kinds of sensory information. Certain level of information integration and processing also occurs in the system. It has been shown that neural signals could be extracted from the distal end of the stump, indicating that the bridge is still effective after limb damage or amputation, which is the neurophysiological basis for the research and development of peripheral nerve interface for the prosthetic system.

Keywords Peripheral nervous · Motor information · Sensory information · Nerve regeneration and repair

4.1 Introduction

The physical disability can be caused by many factors such as accidents, wars, and natural disasters besides the primary malignant tumors in the limb, such as thromboangiitis obliterans, diabetes, and other diseases caused by tissue infection

P. Zhang · X. Li · D. Hu · Q. Lai · Y. Wang · X. Ma · Q. Xu · W. Li · J. Huang
Neural Interface and Rehabilitation Technology Research Center, School of Automation, Huazhong University of Science and Technology, Wuhan, China

J. He (✉)
Neural Interface and Rehabilitation Technology Research Center, School of Automation, Huazhong University of Science and Technology, Wuhan, China

Beijing Advanced Innovation Center for Intelligent Robot and System, School of Mechatronical Engineering, Beijing Institute of Technology, Beijing, China

Center for Neural Interface Design, School of Biological and Health Systems Engineering, Arizona State University, Tempe, AZ, USA
e-mail: Jiping.He@asu.edu

X. Zheng (ed.), *Neural Interface: Frontiers and Applications*,
Advances in Experimental Medicine and Biology 1101,
https://doi.org/10.1007/978-981-13-2050-7_4

necrosis, which may also lead to amputation. According to the data from the National Bureau of Statistics in 2006, there are 24.12 million people with various physical disabilities in China, accounting for 29.07% of the total number of disabled people. For these persons with physical disability, the movement function restoration of hand and upper limb is particularly important. Developing novel intelligent prostheses and corresponding control methods can help them fetch food, grasp different objects, and restore important motor functions in their daily life, thus improving their life quality and even helping them return to work, which is of great significance to social development and stability.

Many amputees are not willing to wear and use traditional prostheses because of their limited functions and degrees of freedom (DoF). In recent years, some intelligent prostheses with excellent and reliable performance and high DoF have been developed, such as DLR/HIT prosthetic manipulator developed by the German Aerospace Center and Harbin Institute of Technology [1]; the i-limb prosthetic hand with multiple DoF made by British Touch Bionics; the SmartHand, dexterity prosthetic hand, developed in Italy; and so on. However, the outstanding performance of these prostheses will not be fully realized until the methods enabling the amputees to control the prosthesis with multiple DoF accurately and easily have been developed.

Extracting motor commands from neural signals to control the prosthesis by amputees' intention is a key breakthrough for improving the prosthesis performance and is also an interdisciplinary research focus. Electromyogram (EMG), electroencephalogram (EEG), cortical neural signal (CNS), or peripheral neural signal (PNS) all can be used as the source of prosthetic control instructions. However, they have different characteristics in signal acquisition technique and quality, and actual control performance while constructing interfaces with different degrees of invasiveness and feasibility, details are described in Table 4.1.

EMG can be recorded easily and noninvasively, and there are commercial EMG prosthetic products. However, limited control instructions can be extracted from EMG signals while a sufficient number of muscles without atrophy in the residual limb are required, which is not applicable for patients with high amputation and muscular atrophy. The target muscle rehabilitation technology (target muscle reconstruction, TMR), created by the American Chicago Rehabilitation Center, can effectively solve the problem of controlling in high limb amputation. To a certain extent, the DoF and flexibility of controlling can be improved by TMR, but its clinical application is limited for the complexity in implementation [2].

EEG can also be recorded noninvasively and has high temporal resolution. Brain computer interface (BCI) based on EEG has been widely applied in rehabilitation training and prostheses control. However, prosthesis control based on EEG requires users to concentrate on movement imagery without other actions (like talking or other movements and so on); otherwise, correct movement intentions cannot be extracted. In daily life, people always performed more than one action at the same time. So, this method is unnatural and is not consistent with user habits while having bottlenecks in rate of information transmission, control accuracy, and control commands decoding [3].

Table 4.1 Characteristic comparison among different prosthetic control interfaces

Interface type			Advantages	Disadvantages
EMG	Surface EMG (sEMG)	Normal sEMG	Noninvasive, simple procedure and convenient adjustment	Low DOF, unnatural
			High real-time in controlling	Sufficient normal muscles in residual limb
			Existing commercial products, mature technology	Not suitable for high level amputation
				Electrode movements, channel crosstalk, Low long term stability
		TMR	Suitable for high level amputation	Not suitable for distal amputation
			Higher flexibility of control	
			Some sensory feedback can be restored by functional electrical stimulation	Complexity and a large number of electrodes are required
	Invasive EMG (iEMG)		High signal quality	A lot of electrodes are required
			Small channel crosstalk	
			High selectivity	
EEG			High temporal resolution	Electrode placement is complex
			Noninvasive	Unnatural
			Have been widely used in disease detection and rehabilitation treatment	Low information transfer rate
				Low control command type
				Susceptible to interference
CNS (Local field potential and spike)			Directly decode the high-level control command	Large invasive
PNS			Natural	Special electrode is required
			More commands can be decoded from one electrode channel	Lack of decoding method
			High reliability	Immature technology

A large amount of movement and sensory information can be accurately extracted from cortical neural signals in particular cortex areas for controlling assistive devices precisely. However, electrode implantation in the cortex has to be performed via craniotomy, and this surgery has relatively larger risk. Moreover, long-term reliability and biocompatibility of the electrodes are also a problem because the patient can only undergo such electrode implantation surgery once [4].

Comprehensive consideration of invasiveness, expected effects, and ease of implementation, the best way for constructing interface for prosthesis control, is based on PNS. G. S. Dhillon et al. in the University of Utah have demonstrated that residual peripheral nerve keeps the original function and can transmit control commands for the amputated limb even after amputation for a long time [5]. Making full use of the signals from the residual peripheral nerve, amputees may use the prosthesis freely as with their own arms. PNS can be acquired from special electrodes implanted into the peripheral nerve with relatively simple surgery, while CNS recording requires craniotomy and is more complex with greater risk. In addition, both movement commands and sensory information are transmitted in the peripheral nerve and make closed-loop neural interface with sensory feedback possible.

There are three important research topics about PNS which will be presented in this chapter. Sections 4.2 and 4.3 present the research about PNS decoding and sensory feedback, respectively. Movement signals decoding and sensory signal feedback construct a closed-loop system, while nerve regeneration and repair in Sect. 4.4 provide the foundation for this closed-loop system. How to acquire peripheral neural signals, which were transmitted from the central nervous system, from residual peripheral nerve will be introduced in Sect. 4.2; how to provide sensory feedback by peripheral neural interface will be introduced in Sect. 4.3; some findings in neural cell culture and artificial stretch will be presented in Sect. 4.4.

4.2 Peripheral Neural Signal Acquisition

PNS has captured wide attention from international research community. As early as 2007, Chen et al. of Zhongshan Hospital affiliated to Fudan University conducted an experimental study on the control of electronic prostheses by the amputee for the first time in China [6]. They inserted the intrafascicular electrodes directly into the three main nerve bundles of the upper arm, and the electrodes were connected to a prosthetic device at the distal end. It was found that the signal derived from the radial nerve (simple motor nerve bundle) could trigger the prosthetic simulation device reaching out the fingers when the patient imagined the movement of the hand, but the fingers could not be closed. On the other hand, signals from other nerves (mixed nerve bundles) couldn’t effectively drive the prosthetic device. This study validates the possibility of using PNS to control prostheses in a further step; however, the neural signals are only used as a control switch for prosthetics other than being fully analyzed. In 2010, PM Rossini and S. Micera et al. implanted four intrafascicular electrodes into the ulnar nerve and the median nerve of an amputee to perform a long track recording [7]. Slight hold, tight hold, and flexion and extension of little thumb are extracted from the recording signals to develop commands for the robotic arm’s performing online. However, there is still a lack of research on the use of PNS to control prosthesis to complete more complex reaching and grasping tasks. For peripheral neural electrode, S. Micera et al. developed a variety of recording or stimulation electrodes, such as cuff electrode, the flat interface nerve electrode (FINE), longitudinal intrafascicular electrode (LIFE), etc. As to the PNS processing, they applied wavelet denoising method for preprocessing and the support vector machine (SVM) or artificial neural network (ANN) for classification corresponding to different tasks. Clinical trials have been conducted to verify the feasibility of the PNS on controlling the external machines. S. Micera pointed out that their future PNS research would focus on decoding more control instructions from the electrode recordings to achieve a higher degree of freedom for prosthesis control; higher user autonomy; developing bio-PNS interface by providing the users with the sense of feedback; and artificial limb control algorithm with higher robustness and adaptability.

Overall, the earlier carried-out research on EMG and EEG identification and analysis comes out with many practical applications. The modeling research on neural signals in the cerebral cortex and the corresponding limb motor function gained much progress, which provides lots of references for the study of PNS. In recent years, PNS shows the unique advantages and great promotion potential in the field of brain machine interface, thus attracting great interest from international research groups. However, great breakthroughs are expected in the PNS identification and other issues, as with the specialized analysis software and practical algorithms. Although the use of PNS to control robots is already realized by some teams, there is still a long walk to achieve flexible movement patterns with more degrees of freedom by accurate control. Before applying the neural interface constructed by PNS to clinical practice, a large number of animal studies are needed. Primates (such as macaques) become ideal animal test subjects since the anatomical and physiological functions of the nervous system as well as the motor system are the closest to humans. Moreover, PNS animal experiments are related to some key techniques such as the implantation of multichannel nerve electrodes, neural signal acquisition, and monkeys' training, which are challenges for the research institutions.

4.2.1 *Surgery and Data Collection*

There are mainly three peripheral nerves involved in the transmission of control commands and sensory signals for upper limb, including the ulnar nerve, the radial nerve, and the median nerve. In order to acquire meaningful neural information and to perform electrical stimulation for sensory feedback, we implanted special designed electrodes into these nerves of the monkey. All surgery and data recording procedures were strictly in compliance with the guidelines for the care and use of laboratory animals and were approved by the Institutional Animal Care and Use Committee in Huazhong University of Science and Technology.

4.2.1.1 Electrode Implantation

The electrodes we used in this experiment included floating microelectrode arrays (FMAs), which can be inserted into the nerve fasciculus and cuff electrode, which wrapped the nerves. One of the key concerns for electrode implantation is to meet the requirements for stable recording while allowing monkeys to perform various behavior tasks. To prevent the monkey scratching the connectors or the wires, the connectors must be mounted on its skull firmly. Since the electrodes were implanted into the nerves in the arm, long wires had to be used for linking the electrodes and the connector. Here

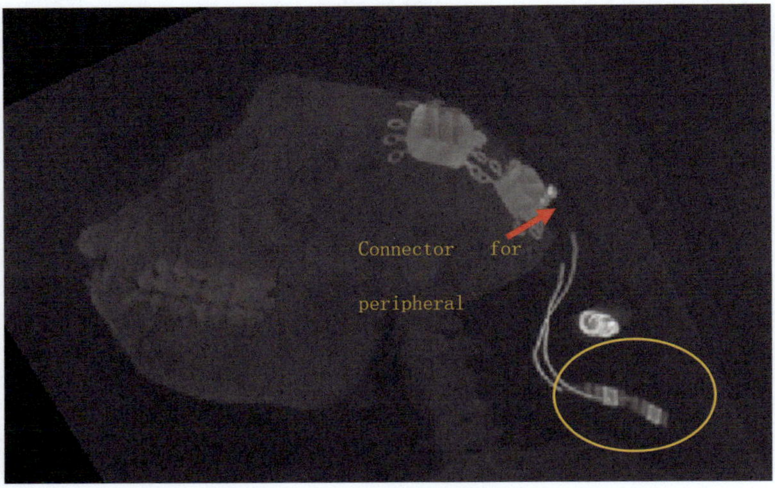

Fig. 4.1 The CT image of the monkey with peripheral neural electrodes implantation. The electrodes were implanted into the median nerve and ulnar nerve of the monkey and the electrodes were connected to the connector with wire through a subcutaneous tunnel. The connector was fixed on the skull with a custom-made pedestal

the wires we adopted in this experiment had a length of 30 cm, as shown in Fig. 4.1, which were connected with the connector and electrode through a subcutaneous tunnel.

The surgery procedures are shown in Fig. 4.2. The first step was to incise the muscles and expose the nerves. According to the relative positions of the nerves, different nerves can be identified clearly, mainly were median nerve, ulnar nerve and radial nerve. For FMA implantation, the array was held by an inserter, which can generate negative pressure, and then inserted fast into the target nerve. For firm fixing, a silicone film was used to wrap the array and sewed up with medical sutures. The surgery was finished after sewing the cuttings, and the total surgery time expended was about 3–5 h.

One FMA electrode and one cuff electrode were implanted in the median nerve and ulnar nerve of a monkey. The two electrodes implantation was achieved through a single surgical site at the elbow. The FMA electrode was inserted into the median nerve with its pins and the cuff electrode wrapped on the ulnar nerve. Moreover, the FMA electrode and that part of the nerve were wrapped by a silicone film.

4.2.1.2 Acute Electrode Implantation test

Movements like hand grasping can be generated by current pulse stimulation on given electrodes after peripheral neural electrode implantation, as shown in

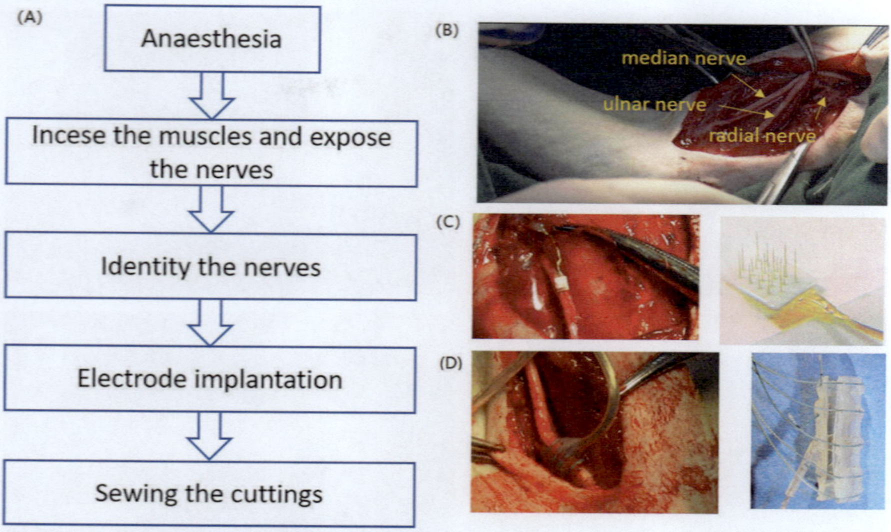

Fig. 4.2 Procedure of the surgery for peripheral electrode implantation. (A) is the flow chart of the surgery. (B) shows the surgical stie and these three exposed nerves, median nerve, ulnar nerve and radial nerve. (C) shows the implantation of the FMA electrode. (D) shows the implantation of the cuff electrode

Fig. 4.3 [8]. The hand shape at static state, when no current pulses were delivered to the electrodes, was shown in Fig. 4.3a. At the beginning phase of electrical stimulation, clear movement tendency could be observed, as indicated in Fig. 4.3b, c. The termination of a movement induced by current pulse was shown in Fig. 4.3d. The experiments described here demonstrate the feasibility of recruiting motor units to generate certain movements by electrical stimulation. However, the details about the patterns of stimulation parameters require further researches.

4.2.2 *Experiment*

4.2.2.1 Experimental Protocols

When performing the tasks, the animal was comfortably seated in a primate chair with its left arm restricted. The monkey was required to reach and grasp one of the three target objects indicated by the target LED in each trial, as shown in Fig. 4.4. For investigating the neuronal activity corresponding to different grasping gestures, three types of target objects were used, including cylinder, plate, and ring. It is obvious in Fig. 4.4b that the monkey adopted three kinds of hand shapes with large discrepancy with regard to the three types of target objects.

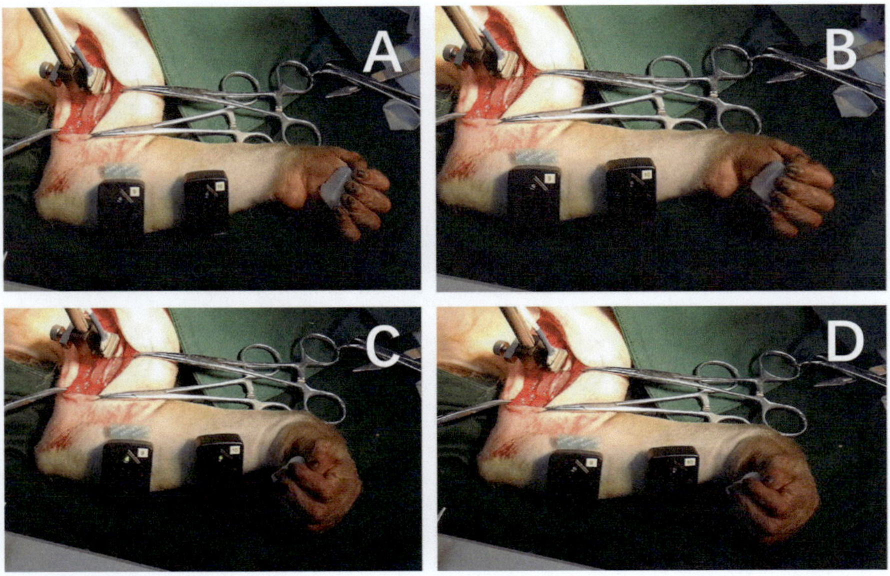

Fig. 4.3 Stimulation of median nerve recruits specific digit and wrist movements. A microelectrode array was implanted into the median nerve of the monkey and pulse-trian was delivered by a electrical stimulator (Plexstim, Plexon, Inc.), meanwhile sEMGs of the forearm were recorded by the Delsys surface EMG system (Delsys, Inc.) to evaluate the electrical stimulation performance. Hand grasping can be generated with specific electrical stimulation parameters and different parameters can evoke different movements

The procedures of the task in Fig. 4.4c show a typical successful trial. One successful experimental trial began with the center LED lighting (center on); then the monkey rapidly put its hand on the center pad and kept for 500 ms at least. Then, one of the three target LEDs was on (target on), cueing the monkey to release from the center pad (center release) and reach to the target indicated by that LED. The monkey should maintain a proper hand gesture according to the shape of the object and made a powerful grasp so as to trigger the FSRs (target hit). After keeping a firm grasping for about 200 ms, it was allowed to release the hand (target release) and then received several drops of water as reward. After an inter-trial interval of about 5–10 s, the monkey returned its hand to the center pad and waited for the beginning of the next trial.

The definition about several important periods during the tasks, are the baseline, the reaction period, and the moving period. During the reaction period, the monkey was in preparation for the future movements, but no movement occurred. During the early stage of the moving period (0–200 ms), hand grasping gestures have not been fully formed.

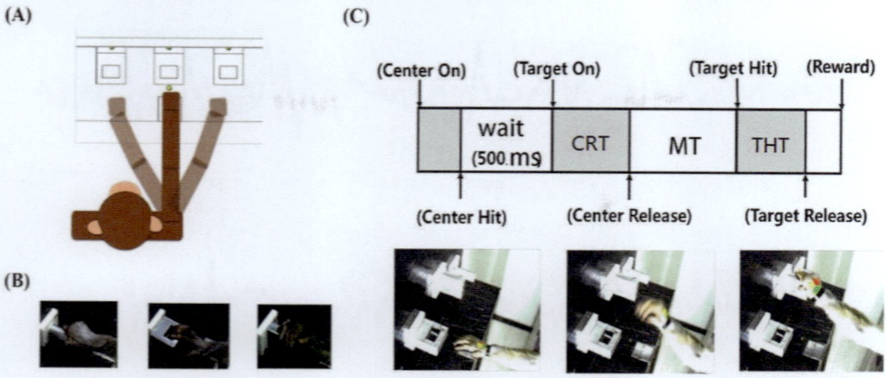

Fig. 4.4 Paradigm design of grasping multiple targets. (A) is the schematic diagram of monkey behavioral experiment. The monkey was guided to reach and grasp three different target objects. (B) is the pictures that the monkey was grasping these three target objects, cylinder, plate, and ring. (C) shows the sequence of the behavioral task. Each trial began with a cueing of center light on, instructing the monkey to fixate on the center pad. After a center holding time of 500ms, the center light went out meanwhile one arbitrary target light went on, cueing the monkey to reach for the corresponding target and make a whole-hand grasp movement. After a target holding time (THT), the target light went out. The monkey would return the hand and a reward of a few drops of water would be delivered

4.2.2.2 Peripheral Neural Data

The acquisition and recordings of neural data were finished by the OmniPlex system (Plexon, USA). The field potentials acquired by the cuff and the FMA on the median nerve were shown in Figs. 4.5 and 4.6 correspondingly.

4.2.3 Neural Decoding Algorithms

Over the decades, various practical decoding algorithms have been developed to analyze neural signals, with continuous improvement in their applicability to neural experimental data (see [9, 10] for recent reviews). A typical Bayesian framework of recursive estimation is introduced here [11], known as a Bayes filter, where the EMG signals from each muscle and the limb kinematics from the lower limb movement were regarded as hidden states of a dynamic network system, while neuronal spiking signals were recorded from motor cortical cortex using the implanted microelectrode arrays as the measurements of the dynamic system, in which the value updates constantly (see [12] for more details about the experiment).

In our experimental data analysis, the x_n denoted the hidden state x value at the n-th time instant. In EMG signal decoding, x_n is given by $x_n = [u_1(n), u_2(n), \ldots, u_M(n)]^T$, where $u_m(n)$, $m = 1, \ldots, M$ is the value of EMG signal from m-th muscle

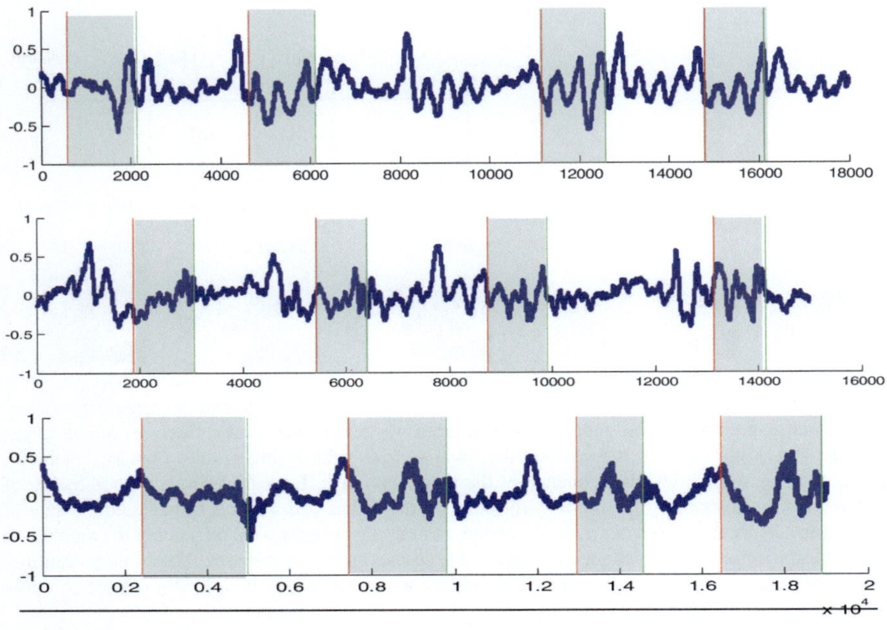

Fig. 4.5 The field potentials acquired by the cuff electrode

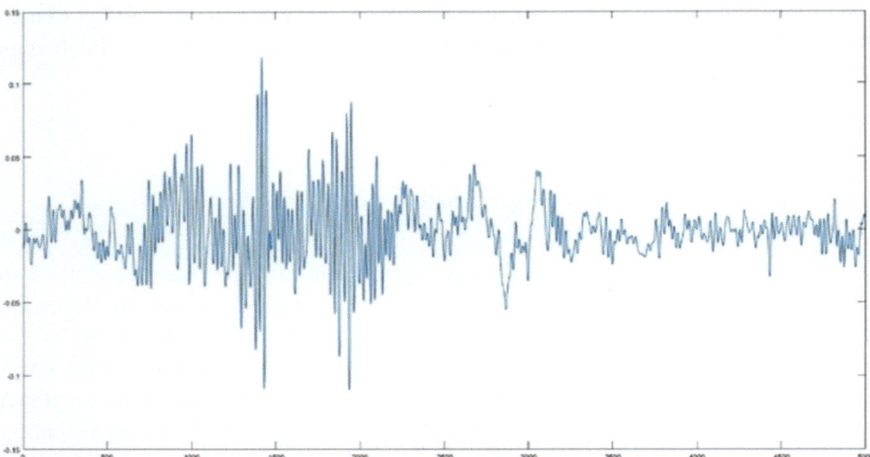

Fig. 4.6 The field potentials acquired by the FMA

activity at the time n and M denotes the total number of muscles where we implanted EMG recording electrodes. When decoding lower limb kinematics, the x_n given by $x_n = [Z(n), V(n), A(n)]^T$, where the $Z(n)$, $V(n)$, $A(n)$ separately respect the position, the velocity, and the acceleration of the marker at the time n. At the same time, the y_n denotes the constantly updated measurement values of the dynamic system at the time n, which is a vector data containing the spike counts of neurons in the n-th time bin.

The estimation process consists of two steps: (i) prediction and (ii) update. In the prediction step, the prior distribution of x_n is estimated from the previous states, then use a Markov model to generate x_n , so we have:

$$p(x_n|y_{1:n-1}) = \int p(x_n|x_{n-1})p(x_{n-1}|y_{n-1})dx_{n-1} \tag{4.1}$$

where $p(x_n| x_{n-1})$ represents the state of dynamic system changed from current time instant to the next moment and $p(x_{n-1}| y_{n-1})$ corresponds to the posterior probability value for the previous states.

The update step, the posterior prediction $p(x_n| y_{1:n})$ is calculated by updating the prior prediction with the new measurement values y_n:

$$p(x_n|y_{1:n}) = \frac{p(y_n|x_n)p(x_n|y_{1:n-1})}{p(y_n|y_{1:n-1})} = \alpha p(y_n|x_n)p(x_n|y_{1:n-1}) \tag{4.2}$$

where $p(y_n| x_n)$ represents the mapping relation between the dynamic system and the spike measurements and the normalization constant α in the formula can usually be ignored. Utilizing Eqs. (4.1) and (4.2), we can calculate the $p(x_n| y_{1:n})$ recursively. For practical data process, there are two widely used algorithms, the Kalman filter method [13] and the unscented Kalman filter method [14], that could be used to estimate the EMG signals and kinematics recursively from the neuronal signals recorded from motor cortex; for more detailed information, refer to [15].

4.3 Sensory Signal Feedback

For people who lost a limb after amputation, the concomitant loss of motor and sensory functions results in enormous inconvenience in activities of daily living and deterioration in quality of life [16–19]. To regain ability for grasping and manipulating objects requires appropriate tactile information in addition to motor-driven grippers [20–22]. To better restore amputees' limb function and enable them to better manipulate delicate objects in daily life, adequate sensory feedback is necessary, especially tactile information.

Open-loop myoelectric control via surface EMG once was the most sophisticated bio-control strategy clinically available since powered limb prostheses were developed to functionally restore arm/hand function in 1960s. However, this kind of control approach did not improve significantly grasping/manipulation due to lack of intuitive user interaction and/or sensory feedback involved in task control [23]. In order to achieve closed-loop control in grasping tasks, several feedback approaches have been proposed in the development of neural interface. For example, force feedback is often used in the hierarchical manner of shared-control for the manipulation tasks involving grasp [24]. Visual and/or auditory feedback has frequently been introduced to assist users in the control of prosthetic limb for reaching and grasping [25]. For sensory substitution, vibrotactile or electrotactile stimulations play a crucial role in providing sensory perception as feedback [26–29]. More recently direct stimulation of the peripheral nerves becomes more focused research corresponding to signals generated by the artificial sensor systems on the prosthetic limb [30–32].

As the complexity of the prosthetic hardware advanced significantly over the last decade, effective control strategies utilizing such high degrees of freedom in more dexterous prosthetic limbs become notably inadequate. For example, the sophisticated prosthesis, SmartHand [33–35], contains so many degrees of freedom, yet it is difficult for a patient to control it naturally. Furthermore, versatile sensors on new prostheses provide much valuable feedback information that substitute naturally the missing sensory information, and if transmitted to patient in a more natural or intuitive way [36], such feedback can be perceived as originating from the missing limb while without overwhelming the concentration of the user [37].

Recently, a study [38] has presented an evident improvement in the quality of sensation and showed that experience of stable touch perceptions can lead to more precise and reliable control of the prosthetic limb. While too many afferents are synchronously stimulated through these electrodes, it will be feasible to achieve more selective stimulation in order to generate more naturalistic tactile sensations. This is important to make the conversion of the prosthetic sensor outputs into patterns of electrical stimulation on peripheral sensory nerves [39].

In this section, we describe a neural prosthesis based on electrical stimulation to provide sensory feedback for closed-loop control. A Schunk Dexterous Hand (SDH) with array sensors on every segment of its fingers provides sensory feedback through a fully implantable electrical stimulation device, which can generate time-variant stimulation parameters according to the sensor measurements and is promising for the generation of natural, graded intensity of grasp force perception [38]. Particularly, the implanted part of this stimulation device is wirelessly powered via magnetically resonant coupling so that there's no need for an internal battery [40].

4.3.1 Materials and Method

The prosthetic part of the experimental neural prosthetic system consists of a 6-degree-of-freedom (DoF) Schunk Powerball Lightweight Arm 4P (LWA 4P), with a 7-DoF three-fingered SDH attached as its end effector. Each finger is

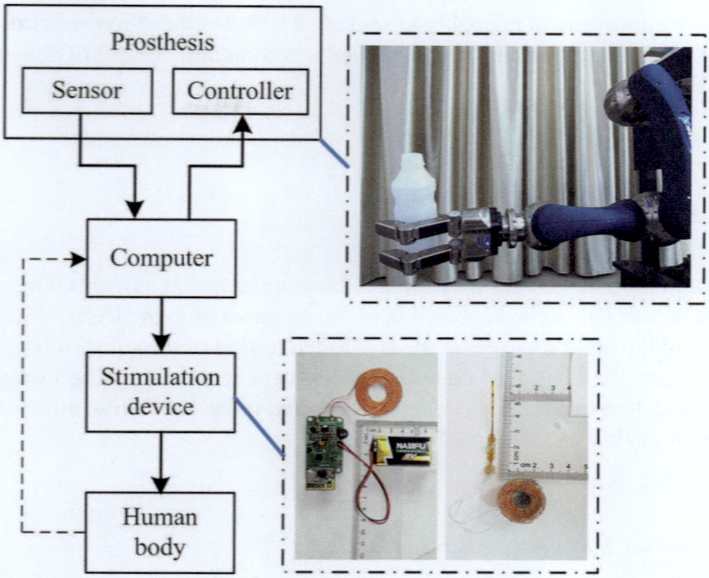

Fig. 4.7 The prosthesis is a combination of a LWA 4P and a SDH, which can provide sensory feedback information as well as moving according to the commands delivered by the computer. The stimulation device consists of the external controller, the implantable pulse generator (IPG), the lead extension and electrode. (Adapted from Qiuxia Lai; Dingyin Hu; Ang Ke; Jiping He, Providing Sensory Feedback Using Electrical Stimulation for Neural Prosthesis, Neuroscience and Biomedical Engineering, 2014. Spring, 2: 99~104)

equipped with two tactile array sensors to provide primitive contact pressure distribution as well as the synthetic force information for the calculation of the parameters needed for electrical stimulation, namely, the pulse width, the pulse amplitude, and the stimulation frequency. The robotic arm and hand communicate with a computer through CAN (controller area network) bus, and the tactile sensors are connected through serial port.

The stimulation system is based on wireless power transfer technology and contains the external controller, the implantable pulse generator (IPG), the lead extension, and electrode, as shown in Fig. 4.7. The controller receives stimulation parameters from the computer via a USB port after the real-time processing of the sensor data and then delivers the parameters to the IPG via wireless communication. The IPG generates time-variant stimulation pulses and delivers them to the bipolar electrode by the lead extension to elicit continuous "natural" sensations on the human body according to the stimulation patterns. Due to its tiny size and wireless power transfer this stimulation system is easily suited for implant operation.

We intend to provide sensory feedback signals that mimic the body's original biological sensors according to the tactile sensor data via the electrical stimulation device. The improvement in quality of sensation and the closed-loop control strategy is promising to lead to more precise, reliable, and natural control of the prosthetic limbs in daily life tasks.

4.3.2 Experiment

At current preliminary stage, we conducted two separate experiments. One aimed at collecting sensor data to be analyzed later for the generation of electrical stimulation patterns, and the other focused on the functional testing of the stimulation device. In the near future, the integrated complete experiment combining these two parts will be conducted, first in animal models for safety and reliability verification before into clinical trial having patients' participation.

4.3.2.1 Sensor Data Acquisition

In daily life, drinking water from a bottle is a simple but frequent behavior, yet most current prostheses cannot provide satisfactory assistance due to their innate shortage in utilizing natural sensory information.

In order to study the sensory information and regulation in this daily life activity as well as to find out a general research approach for similar issues, we conducted the experiment to collect the outputs of these array sensors during the task execution process when the prosthesis took up a bottle of water, moved to another place, poured water out, held the bottle upright again, and then put down the emptied bottle.

The recorded contact sensor data were visualized using pseudo-coloring scheme to present intuitive information of the pressure distribution and its dynamic variation during the experiment, as shown in Fig. 4.8. Detailed explanation is presented next.

In Fig. 4.8 four phases of the water bottle grasp and take-up, tilting for water-pouring procedure in the experiment task, are shown on the top panel, and the corresponding visualized contact pressure presentations are placed below. The water level is indicated by dash line on the bottle. The meaning of pseudo-color picture used to visualize the acquired contact pressure data is illustrated by the color bar and the unit of sensor readouts is millivolt. Each of the three fingers contains two tactile array sensors consisting of 78 and 86 tactile cells, respectively, and each cell can perceive the pressure applied on it.

All sensor data were delivered to the computer and recorded for offline analyses. In the pseudo-color picture, each figure under the representing bottle orientation shows three dynamic pressure distributions for the three fingers: the left color column represents the right-lower finger seen in the real object picture, the middle column represents the left (opposing) finger, and the right represents the right-upper finger. The primitive readouts are the voltages measured on each cell.

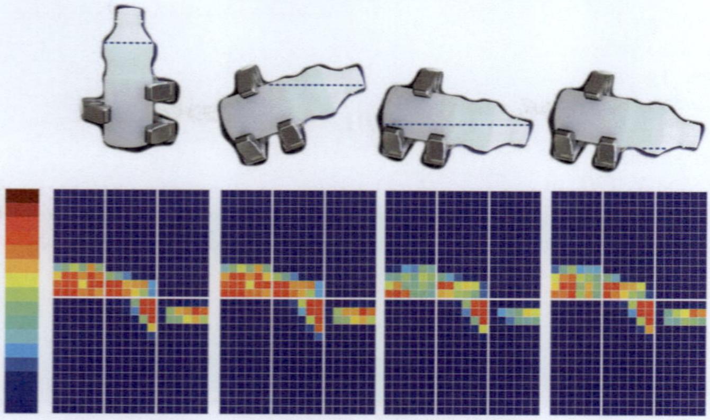

Fig. 4.8 Several typical time phases during the water-pouring procedure in the experiment task. The top panel shows the real scenes and the bottom panel shows the corresponding pseudo-color picture drawn according to the sensor readouts. The dash line indicates the water level inside the bottle, while the water outside isn't shown. The color bar demonstrates the meaning of the pseudo-color picture. Each of the color square in the pseudo-color picture represents a cell in the array sensor and its color demonstrates the voltage value which reflects the force strength exerted on it at that time point. Qiuxia Lai; Dingyin Hu; Ang Ke; Jiping He, Providing Sensory Feedback Using Electrical Stimulation for Neural Prosthesis, Neuroscience and Biomedical Engineering, 2014. Spring, 2: 99~104

4.3.2.2 Stimulation Pattern Generation

The stimulation device was designed to be capable of altering stimulation parameters in real time during its working procedure. The external controller receives commands from a computer via a USB port connection [41]. Then the external controller transmits the stimulation parameters to the IPG by radio frequency telemetry. The IPG generates pulses through the electrode to elicit certain neural responses [42].

In order to prevent corrosion and irreversible electrolyte reactions, the waveforms used in clinical stimulation should be charge-balanced, while compared with biphasic charge-balanced waveforms, asymmetric charge-balanced waveforms can provide better control of electrochemical reactions and may suppress undesired reactions [43].

To ensure charge balance at the electrode site, we monitored the raw output signals applied to a pair of electrode contacts under different specified stimulation parameters with the active electrodes immersed in 0.9% saline. Two waveforms from each electrode contact form together and generate the final stimulation pulse, as shown in Fig. 4.9.

We also altered the stimulation parameters during voltage observing procedure in order to test the capability of our stimulation device to generate different stimulation waveforms real time according to the computer commands. The voltage waveforms were recorded via the oscilloscope, as well. Some examples are demonstrated in Fig. 4.9. The stimulation parameters tested in the experiment were 50 Hz, 2.64 V, and 1200 μs and 10 Hz, 1.32 V, and 400 μs, respectively.

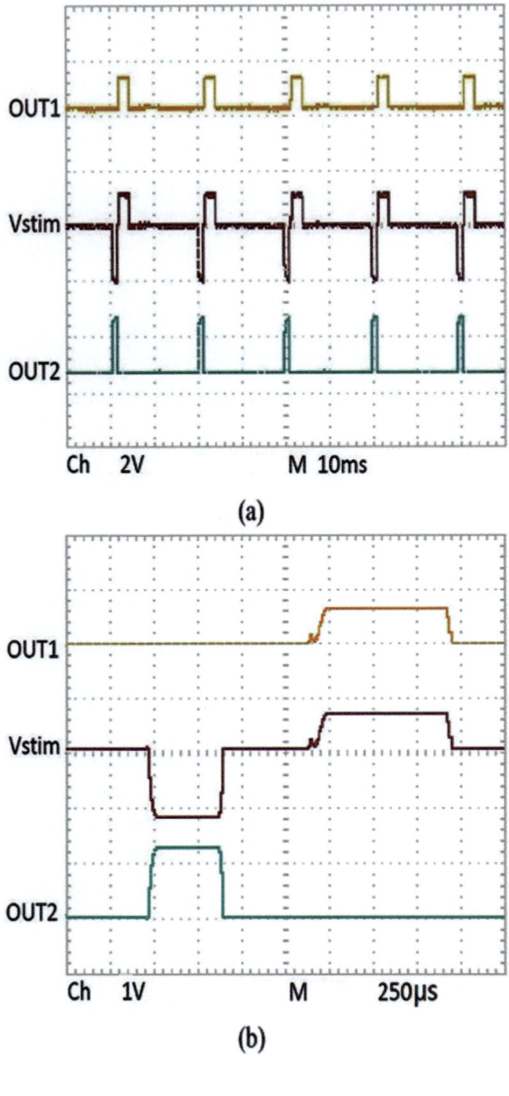

Fig. 4.9 Voltage waveforms observed on the oscilloscope. The active electrodes were immersed in 0.9% saline. The specified stimulation parameters are (**a**) 50 Hz, 2.64 V, 1200 μs and (**b**) 10 Hz, 1.32 V, 400 μs, respectively

4.3.3 Results and Discussions

4.3.3.1 Sensor Data Analysis

In the sensor data acquisition experiment, we observed an evident variation of the measured voltage. The pressure remained zero before the SDH touched the water bottle, and as can be seen in Fig. 4.8, the pressure later reached its desired maximum value when the fingers firmly held the bottle without squeezing hard to break it, and

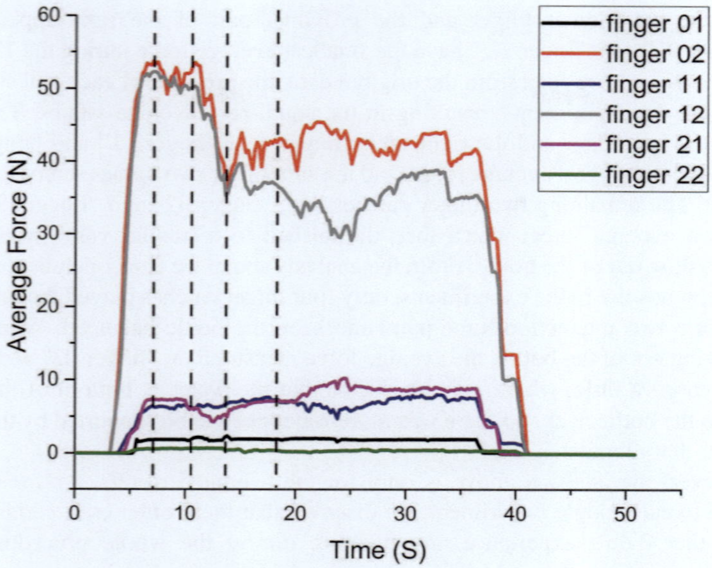

Fig. 4.10 Average forces of the 6 array sensors averaged from 12 experiments. In the "finger xy," x with the value 0–2 represents the three fingers from left to right in the bottom panel of Fig. 4.8, and y with value 1 represents the proximal part, and 2 represents the distal part of the finger

then decreased when water flowed out of the bottle. Finally, it returned to zero after putting down the bottle. We can also find out that each array sensor experienced different amounts of pressure, and this pressure distribution may result from the relative position and contact area between the bottle and the SDH, i.e., the holding posture affected the sensor output value. However, the variation rules of the pressure values on all six array sensors are similar during the experiment.

The array sensor data are initially in the form of voltage, and we calibrated the voltage into pressure using the formula:

$$\mathrm{p} = \mathrm{v}\, P^{*}/V^{*}. \tag{4.3}$$

where the $P*$ = 0.000473 N/(mm × mm) and the V^{*} = 592.1 mV according to the documentation of SDH.

After acquiring the pressure value on each cell, the average force on every array sensor can be calculated. The calculated force of the 6 array sensors averaged from 12 experiments is displayed in Fig. 4.10, and the time points corresponding to the given phases in Fig. 4.8 are marked by the dash lines. In Fig. 4.10, the "finger xy" indicates the six tactile array sensors embedded in the SDH, in which x ranging from 0 to 2 indicates the finger and y with values 1 and 2 indicates whether the sensor is proximal or distal.

We can see that there exists high-frequency jiggering in the force curves, indicating that the bottle slid a little during the whole experiment procedure. The distal part of the right-lower finger and the proximal part of the right-upper finger, i.e. “finger 01” and “finger 22,” have the smallest average force during the 12 experiments, and as can be seen from the original data, the pressure of each cell stays 0 in most of the 12 experiments, resulting in the equal-zero average values. The distal parts of the left finger and the right-upper finger, i.e., “finger 11” and “finger 21,” have the second largest average force, and the sum of the two values seems to remain the same. The remaining two finger patches, i.e., “finger 02” and “finger 12,” have the largest average force, which then diminished to a smaller value when water started to flow out of the bottle. From the analysis above we can conclude that as for the holding posture in the experiments, only four finger patches played the main role, which form two interaction force pairs and keep the bottle balanced. When water was flowing out of the bottle, the average force measured on “finger 02” and “finger 21” increased a little, which may be due to that the water at bottom of the bottle flowed to the bottle neck. So there was more water need to be supported by these two fingers at that moment.

However, the analysis above is only for the averaged results. As for the data acquired in each single experiment, we discover that there often exist certain finger patches that didn’t experience any pressure during the whole procedure. This phenomenon may not be ideal for shared load of fingers, but it is not unusually seen in real life because of the uneven surface or shape when grasping an object. This observation suggests that, for larger, heavier or irregular shaped/structured objects, grasping postures may need pre-planning to distribute pressure more evenly among all fingers of a robotic hand to prevent slippery or damage of robot hand or weak parts of the objects.

In further research, we plan to extract more information from these sensor data to generate proper electrical stimulation parameters and elicit sensations in human body real time. The type of stimulation represents the objects’ contact properties detected by the array sensors on the robotic hand and is promising to contribute a lot in the control of neural prosthesis because the sensory feedback information provided can help to reach a closed-loop control effect and will greatly improve the performance of the neural prosthesis when carrying out daily life tasks. What we need to do next is to propose more effective algorithms and make better use of these force data to acquire motor-related information and apply it to the controller designing.

4.3.3.2 Stimulation Pattern Generation

In the stimulation pattern generation experiment, we observed a charge-balanced waveform between the bipolar electrodes from oscilloscope. The active electrodes were immersed in 0.9% saline and the raw output signals were monitored under different specified stimulation parameters. As illustrated in Fig. 4.9, stimulation parameters are 50 Hz, 2.64 V, and 1200 μs (a) and 10 Hz, 1.32 V, and 400 μs (b), respectively. One channel of the electrode contact outputs a square pulse OUT1 and the other electrode contact outputs a delayed square pulse OUT2 whose amplitude is one-half and the duration is double of the OUT1 waveform. The charge-balanced biphasic stimulation waveforms Vsim are formed by these two square pulses and have a negative phase followed by a positive phase with an inter-pulse delay. We can observe a significant difference when stimulation parameters changed.

We conclude that both changes in the pulse width and the stimulation frequency can result in different stimulation waveforms, demonstrating the capability of the stimulation device to provide different stimulation patterns for eliciting variable sensations.

In future research, we will utilize the calculated force information to generate corresponding stimulation parameters (pulse width, amplitude, and frequency) that are promising to elicit natural sensory responses. The sensation elicited on sensory nerves can provide sensory feedback different from other kinds of feedback such as visual and vibratory feedback, and can induce users' intuitive response in the control of neural prosthesis, which is destined to realize a more natural control performance. Currently the exact relationship between the stimulation parameters and the elicited sensations is still unclear; we need to do further carefully designed research to investigate and acquire this kind of information and find out the proper way to provide users with natural sensations.

In this section, we report a novel experiment to measure the tactile sensor information during the process of taking up a bottle or cup of water, moving to a designated location while holding the bottle/cup upright, pouring the water out, returning the bottle back to upright position again and then putting down the empty bottle, and monitored the output stimulation waveforms of the stimulation device under specific stimulation parameters. These are preliminary experiments to explore approaches to restore the motor function for paralyzed people or amputees by introducing sensory feedback via electrical stimulation on peripheral nerves and controlling the motion of prostheses according to the motor intention decoded from user consciousness.

We now have successful access to acquire the data used for sensory feedback as well as the approach to generate various stimulation patterns; what we need to do in successive stages include conducting more experiments combining these two subsystems and developing an effective algorithm to convert the output of tactile array sensors on the prosthesis into patterns of electrical stimulation of sensory nerve. Moreover, we will also do investigative work to use decoded motor intention into real-time control commands to directly manipulate the prosthetic limbs as the natural sensory feedback are elicited via electrical stimulation synchronically, thus reaching a natural closed-loop control performance.

There are many scientific and technical challenges to be conquered before we can achieve the ultimate research goal. Along the way, the development effort should bring significant medical benefits to the patients and society. If the natural sensory feedback is achieved, the amputees will be able to perform naturally functional control on the prostheses as if they were just extension parts of their body.

4.4 Neural Regeneration and Repair

Autologous nerve grafts remain the "gold standard" for repair of peripheral nerve lesions. However, this approach is plagued by permanent loss of harvested nerve function and the potential formation of painful neuromas. Furthermore, there are obvious limitations in the supply of donor nerves, making autografts can only be used to repair short nerve lesions. Neural repair technology based on tissue engineering has shown promising prospects in the treatment of nerve injury. However, this treatment needs autologous nerve to stimulate axonal regeneration and extension into target tissues. Therefore, how to produce plenty of living nerve constructs in a short time is full of significance.

Fine-motor control of an artificial arm will ultimately require a different kind of link between living tissue and the prosthesis. Except for muscles, nerve fibers can also be the intermediary between the severed axons in a stump and the electrical wiring of a prosthetic device. D. Kacy Cullen and Douglas H. Smith believed this kind of neural interface has some advantages, such as low invasion, high sensitivity, convenient for signal separation, and short rehabilitation period [44]. To create such a neural bridge, we firstly have to figure out how to grow nerve fibers that are long enough to span the gap between the host axons and the electronics.

Although it has been well known that axon growth can be guided by chemical cues and electrical stimulation, both of these two methods couldn't produce regular neural tissues. In nature, the actual growth of the axon also appears to be dependent on mechanical stimuli. By the proper stretch speed, we can achieve regular nerve constructs. The nervous tissue constructs consisting of living axons, which provides an enticing target for host axon ingrowth and synaptic integration. Compared with original neural interface, our method shows some advantages. Firstly, living axonal constructs can provide regenerative support for the proximal stump where target is cut off. Secondly, the array and neurons can make stable connection in vitro. Thirdly, most host nerves will never connect directly with the electrical devices which usually cause damage to the respective tissue. Finally, the process of mechanical stretch can produce more regular nerve fascicles (details in Table 4.2).

Table 4.2 Comparison of different methods for the directed nerve growth

Methods	Advantages	Disadvantages
Chemical	The inherent characteristics of the organisms	Long period
	The growth speed and direction are controllable	Unquantifiable
	Both direction: Attraction and rejection	Irregular
	Kinds of neurotrophic factors for varied nerves	
Electrical	The growth speed and direction are controllable	Unquantifiable
	Noninvasive	Irregular
	Neural regulation and functional reconstruction	
Mechanical	The inherent characteristics of the organisms	Easily cause nerve damage
	The growth speed and direction are controllable	
	Regular nerve fibers	Few studies
	Quantifiable	

4.4.1 Axon Stretch Growth

In the past, a few reported studies have attempted gradually lengthening the axons by towing growth cone. In 1984, Bray successfully stretched the growth cones of individual chick sensory axons more than 100 μm within a few hours by a microelectrode [45]. In 1993, Heidemann et al. performed a series of studies. The growth cones of single axons were lengthened at precise increments via stretching glass pipettes using an external device. The rate of towed growth showed a linear dependence of growth rate with applied tensions between 25 and 560 micro dynes. He found the axons could be towed at a maximal limit of 1 mm every day without thinning down [46]. In 1977, S. Chada found the elongation rate and magnitude of applied tension have linear relationship in both central and peripheral neurons [47].

When the stretch time is too long or the speed too fast, the axon will be disconnected. Therefore, we can't produce long neural tissues by stretch growth cone. Actually, integrated axon tracts without growth cones could also undergo a form of more rapid and sustained growth under stretch. The majority of cell types in nature increase their volumes by division during development. However, neurons are non-mitotic, so the expansion of nervous tissue depends on another process. For example, during development of blue whale, the spinal axons need to undergo continuous mechanical tension as their spinal cord grows in length. Although without growth cones, these integrated axons could reach 30 m in the end.

However, the elongation of integrated axon tracts was not experimentally demonstrated until 2001 [47]. Smith et al. found integrated axon tracts without growth cones could be stretched at a faster rate. Two large populations of neurons were seeded on adjacent membranes within a bioreactor, and then the top membrane was slided across the bottom membrane at set rates by a programmable microstepper motor system. When the microstepper motor moved the top membrane, the axonal bundles crossing between the two membranes were stretched to a new length. In the end, the axons had grown up to 1 cm after 10 days of stretch. In the following years, they optimized the stretch growth paradigm, and dorsal root ganglion (DRG) axon tracts initially only 100 μm in length could be extended up to 10 cm by stretch for 2 weeks [48]. Despite this rapid stretching, sodium channel activation, inactivation, and recovery or potassium channel activation were demonstrated no change according to the whole cell patch clamp technique.

4.4.1.1 Principles of Axon Stretch Growth

In order to stretch axon bundles, as depicted in Fig. 4.11, one stationary substrate membrane as the bottom of the culture chamber. Another towing membrane made by the same material was positioned over the bottom membrane. Both of the two membranes would need to be thin and transparent enough so that adherent cells on their surfaces could still be within the working distance of a high magnification microscope objective. The towing membrane was fixed to a machined plastic block

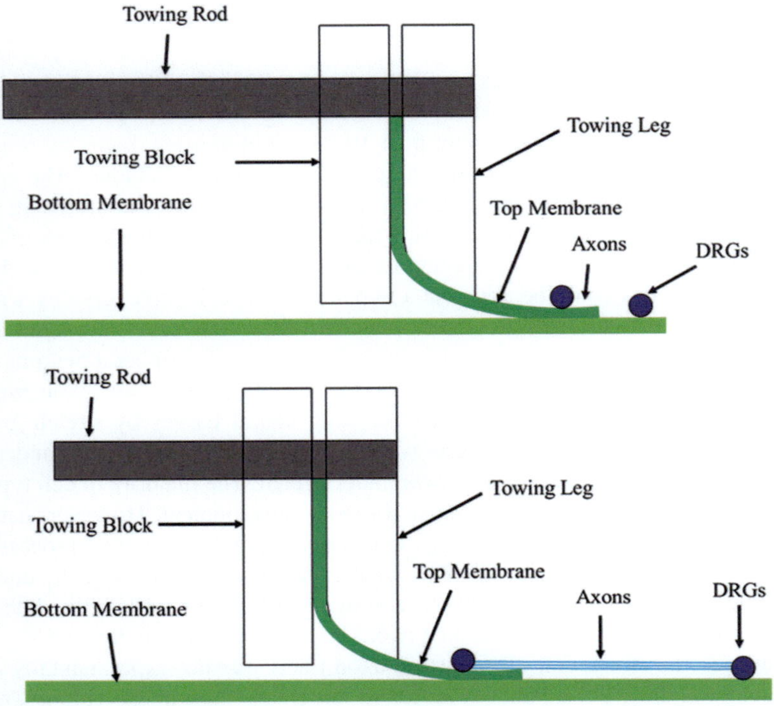

Fig. 4.11 Schematic diagram of axon stretch-growth

so that it could be moved over the bottom membrane by a microstepper motor system. First, the neurons were placed on the two adjoining substrates and formed new synaptic connections between each other over 5 days. Then, the axon bundles across the border between the top and bottom membranes were stretched in a stepwise fashion. After several days' stretch, the axon tracts could reach several centimeters in length.

4.4.1.2 Fabrication of Bioreactor

In order to stretch neural cells, axon expansion chamber will not only need to culture cells like a common culture vessel, but also need to stretch axons. The design needs to meet the following requirements:

1. Conditions for culture of neural cells in vitro. A. It can store and replace culture medium. B. It can be sterilized and maintained in a sterile environment. C. To exchange gas and maintain appropriate pH value medium. D. The material used in the incubator has no toxic gas under special temperature and humidity conditions. E. The bottom and top are transparent to facilitate observation during culture.

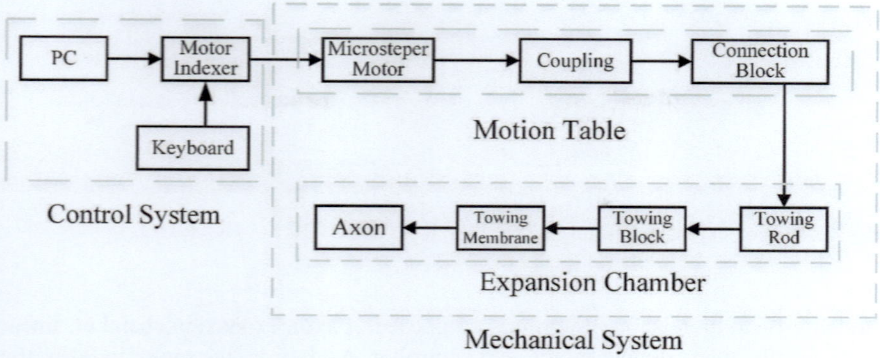

Fig. 4.12 Control structure diagram of axon stretch system

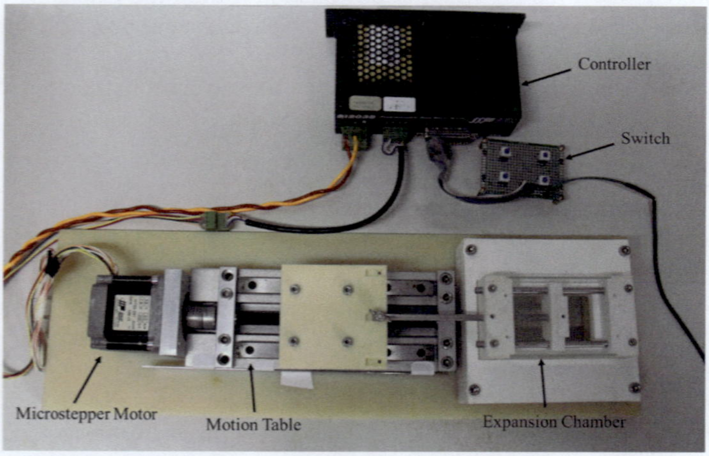

Fig. 4.13 Axon stretch growth bioreactor system. (Adapted from Li, Xiao, et al. "Development of a new miniaturized bioreactor for axon stretch growth." Journal of integrative neuroscience 15.03 (2016): 365–380)

2. Conditions for axon stretch growth. A. The materials used under special temperature and humidity conditions of the incubator will not deform and ensure the stability and accuracy during stretch growth. B. Stepper motor needs to work well in the incubator; each displacement distance is small enough to not affect the cell growth. C. The wire passes through the incubator, but does not destroy the air pressure and the sterile environment in the incubator.

A miniaturized bioreactor (Fig. 4.12) was engineered to gradually apply tension to axon bundles spanning two separate membranes. As showed in Fig. 4.13, the axon

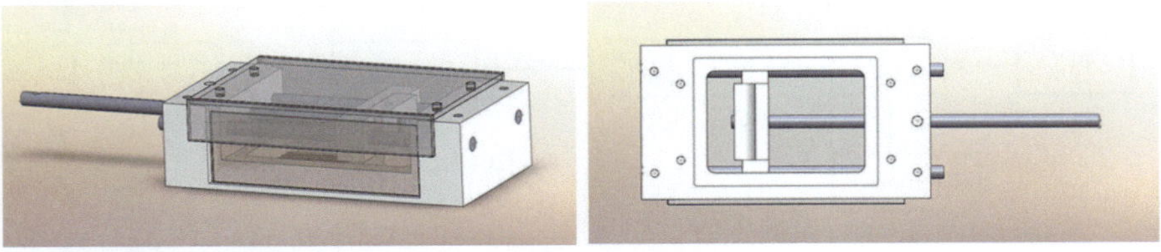

Fig. 4.14 Axon expansion chamber

stretch growth bioreactor system was composed of axon expansion chamber, linear motion table, microstepper motor, and controller. Aside from the motor indexer, the bioreactor was located in a CO2 incubator during neural culture and axon stretch growth experiments. The axon expansion chamber was seated within the chassis underlying the linear motion table, meanwhile the towing rod extending from the axon expansion chamber was fastened to the table using an acrylonitrile-butadiene-styrene (ABS) adaptor.

As is shown in Fig. 4.14, the axon expansion chamber is composed of the stretching frame, towing block, towing leg, and so on.

4.4.2 Axon Stretch Growth

When the axons sprouted from their cell body and formed synaptic connections with target cells, the axon stretch growth will begin. As the distance between the neuronal somata and the target cells increased, the axons will undergo the continuous tension. These forces increased axons' growth rate by stimulating axons to add cytoskeleton, axolemma, and other cellular building materials somewhere along their central length to minimize strain.

4.4.2.1 Key Factors in Axon Stretch Growth

1. Traction Mode.

Small and frequent stretches are important to the survival of axons undergoing stretch growth, particularly at the start of stretching. As shown in Fig. 4.15, if high elongation rates are applied early, growth cannot be sustained, and axons will rupture within the first 24 h. But if the rate increases slowly, the cell can accommodate the rate and grow healthily. Moreover, if the stretch rate increased

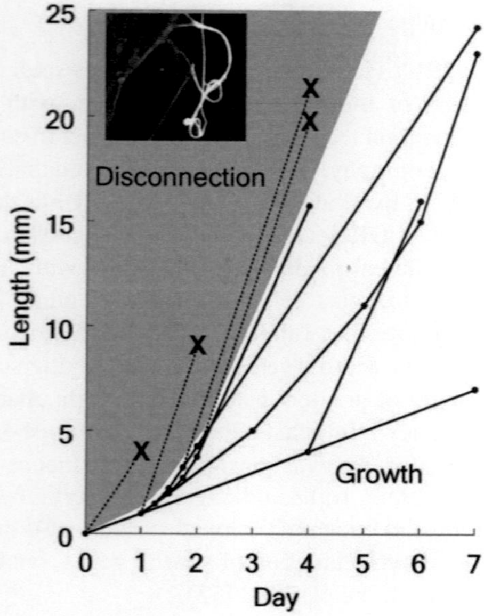

Fig. 4.15 Graphic representation of stretching conditions that define the boundaries of axon growth or disconnection

faster than the neuron, the tension will accumulate and cause the axons rupture. Therefore, two sequential factors that defined the boundaries of long-term stretch growth of integrated axons: strain and acclimation.

2. Substrate Attachment.

Another key factor is the neuronal adhesion with the substrate. The elongator substrates coated by both PDL and rattail collagen had been proved it was the best coating method. PLL had been widely used in the cell culture experiment. However, it could be taken in through adsorptive endocytosis by neurons which may provoke inflammatory responses either directly or indirectly through its necrosis-inducing abilities, most of axons had not grown out from the DRGs. Although collagen could provide an optimal condition for cell adhesion, it will rehydrate in the culture medium, which would often cause axon stretch growth experiments to fail. The axons grew very well after 3 days' culture, while they were seemed to be ill at the sixth day. High-molecular-weight PDL could provide enough adhesive force during axon stretch growth, but some residual PDL could not be digested by lysosome which would cause inflammatory responses when the nervous tissue constructs comprised of stretch-grown axons were used to repair nerve damage. By combining soluble PDL and collagen, we can get the most ideal result.

3. Slope Substrate.

In 1911, Harrison cultured cells on a spider web and found the cells grew along the fibers of the web [49]. Since then, with the development of micro- and nano-fabrication techniques, a large number of studies have shown that cells react strongly to topography. Rajnicek et al. reported that central nervous system neurites could be guided by shallow grooves with 14 nm deep and 1 mm wide [50]. Stepien grew chicken DRG neurons on single scratches. He found that direction often increases with increasing depth, but decreases with increasing groove width. If the grooves are >20 mm, most cell types will lose guide [51]. Goldner et al. describe an unusual capability of a subpopulation of DRG neurons could span across the grooves with 200 mm width, even with no underlying solid support [52]. Fricke et al. designed a variety of gradient patterns with slight changes in slope to control axon directionality. They found that reduction in the slope of structure from 0.04 (0.3 mm/7.5 mm) to 0.01 (0.1 mm/7.5 mm) strongly influence neurite growth. Actin and microtubules align along walls and edges, the microtubules being the first element to be aligned, followed by actin. Grooved surfaces also induce changes in transcription and the up and down regulation of several genes, but the explicit mechanism for cell guidance has yet to be clarified [53].

In the axon stretch growth system, there is a topography between the top membrane and the bottom membrane, which will affect the growth direction of the axons. The top membrane need to be polished on one edge to create gradual slope to the border of the exposed underlying membrane.

4.4.2.2 Experimental Design of Axon Stretch Growth

1. Modification of the Membranes.

DRG were isolated from 1 day-old infant Sprague–Dawley rats. The culture surfaces were respectively treated with high-molecular-weight PDL and collagen.

2. The Process of Axon Stretch Growth.

The neurons are placed on the two adjoining substrates and formed new synaptic connections between each other over 5 days. When the axon bundles are across the border between the top and bottom membranes, we can begin to stretch. The following axon stretch growth began at a rate of 0.864 mm/day by taking 1-μm steps every 100 s over 864 iterations. Alternatively, in order to achieve several centimeters' axon tracts rapidly, the stretch rate could begin slowly and gradually accelerate to the desired rate. As is shown in the Fig. 4.16, the important steps are as follows:

Fig. 4.16 Flowchart of axon stretch growth

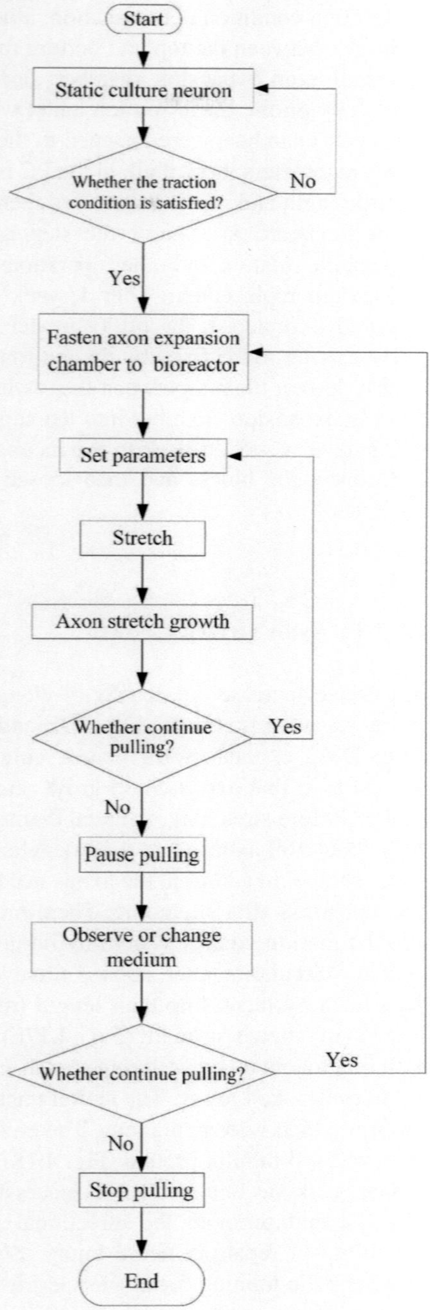

1. Traction condition determination: a large number of axon bundles are across the border between the top and bottom membranes.
2. Install axon expansion chamber: fasten the culture chamber within the chassis underlying the linear motion table, whereas the towing rods extending from the culture chambers were fastened to the table using an ABS adaptor.
3. Set parameters: first of all, in the PC machine, set the traction parameters such as: step length and waiting time, and then downloaded to the controller, you can use the keyboard to achieve the stepper motor to start, pause, positive rotation, negative rotation and other operations.
4. Medium replacement: after 1 week's culture, replace the medium. When the liquid is replaced, the utility model is firstly suspended by the keyboard, and the towing rod is fixed by the nut to prevent the towing rod from moving. After that, loosen the nut between the towing rod and the connecting block, and move axon expansion chamber into the super clean table. Finally, the axon expansion chamber is moved back to the incubator, and the towing rod is fixed with nuts. Connect the block, and then loosen the nut on the pull link, you can restart the pull.

4.4.2.3 Axon Stretch Growth

In order to increase the density of elongating axons in each culture, the elongator substrates were first treated by PDL and then followed by rattail collagen coating. After DRG explants were plated, cultures were maintained for 10 days in the incubator so that axons could grow across the interface of the two adjoining substrates. Before stretching, the total diameter of the axon bundles was about 100 μm, with about 100 axons (Fig. 4.17b). According to the above test, 1 μm step was the better choice. In addition, the axons need sufficient acclimation time to decrease the residual stress after stretching. Therefore, the stretch growth rate was programmed into the motion control system by implementing a displacement step and a resting time in a circular manner, about 1 μm every 100 s. These bridging axons adapted to the stretch by increasing their length from 500 μm to become 5.94 mm long over 7 days of stretch growth (Fig. 4.17h). Because the neighboring axon bundles preferred to join together during stretch-induced growth, the number of axon bundles in the end looked fewer. The neural tracts were stained with β-tubulin antibodies to confirm which were real axons. The entire length of stretch-growing axons labeled positive for β-tubulin protein (Fig. 4.18).

Our work has obtained some successful experience in verifying that mechanical stretch could promote the directional growth of axons, and demonstrating the possibility of repairing nerve injury. Some recent reports highlight the necessary of synergistic training for axons and myelin. This will be our next research goal.

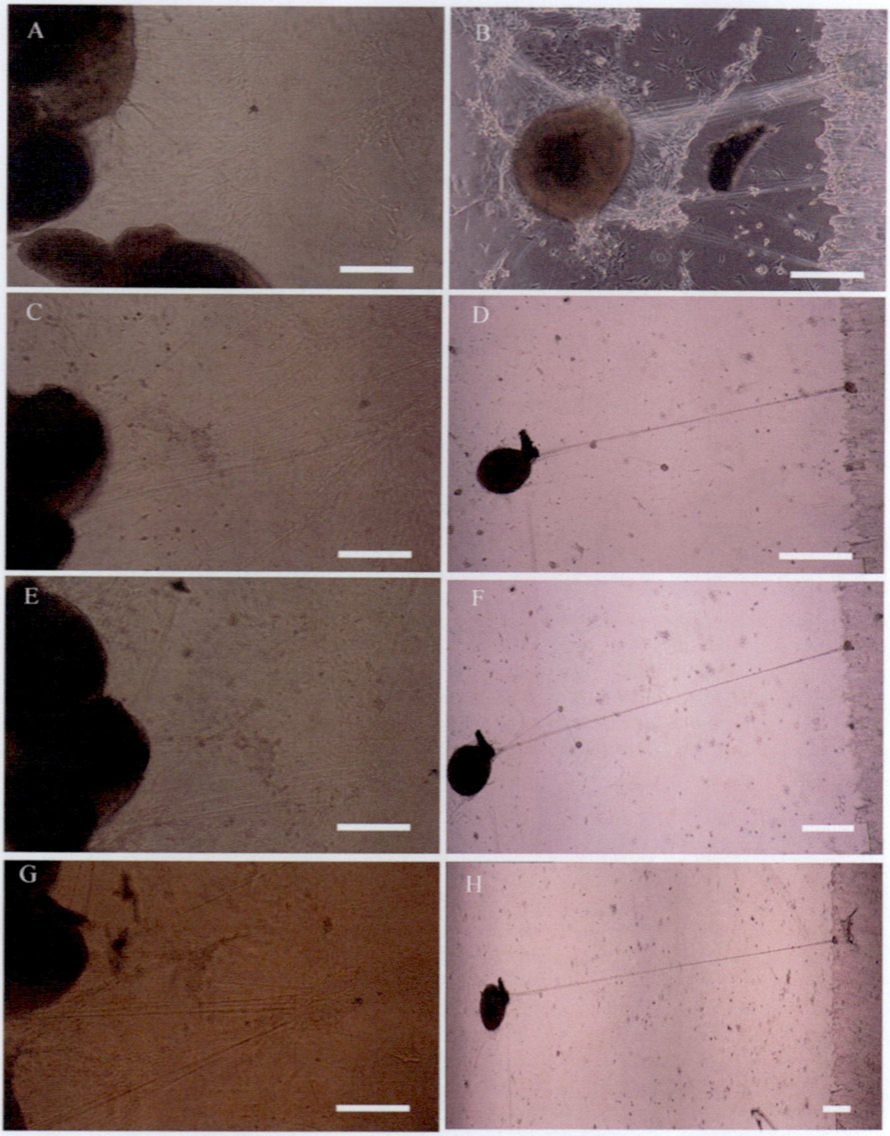

Fig. 4.17 Comparison of DRGs static culture group (left) and stretch-grown group (right) in the axon stretch growth bioreactor. (**a**) DRGs were cultured 3 days. (**b**) DRGs were cultured 10 days. (**c**) DRGs were cultured 10 days. (**d**) DRGs were stretched growth 3 days. (**e**) DRGs were cultured 13 days. (**f**) DRGs were stretched growth 5 days. (**g**) DRGs were cultured 17 days. (**h**) DRGs were stretched growth 7 days. Scale bars: (**a**)–(**c**), (**e**), and (**g**) 200 μm; (**d**), (**f**), and (**h**) 500 μm. (Adapted from Li, Xiao, et al. "Development of a new miniaturized bioreactor for axon stretch growth." Journal of integrative neuroscience 15.03 (2016): 365–380)

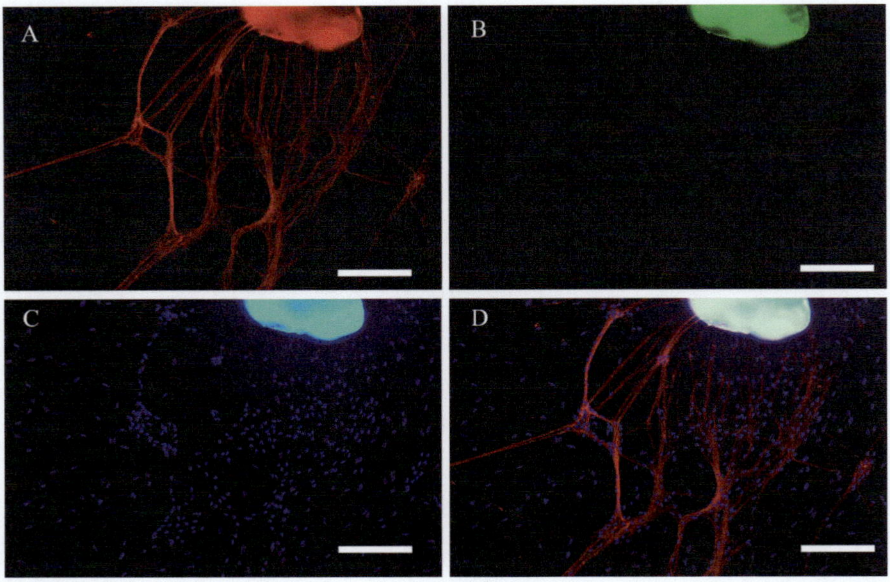

Fig. 4.18 Immunohistochemistry analysis of axon. (**a**) Antibodies against β-tubulin (red) were utilized to determine that elongating processes were axons. (**b**) Antibodies against myelin basic protein (green) were utilized to determine whether there was myelin around the axons. (**c**) Cell nuclei were labelled with DAPI (blue). (**e**) overlay. Scale bars: 100 μm. (Adapted from Li, Xiao, et al. "Development of a new miniaturized bioreactor for axon stretch growth." Journal of integrative neuroscience 15.03 (2016): 365–380)

References

1. Liu H et al (2008) Multisensory five-finger dexterous hand: the DLR/HIT hand II. IEEE/RSJ international conference on intelligent robots and systems, 22–26 September 2008, Nice, France
2. Micera S, Carpaneto J, Raspopovic S (2010) Control of hand prostheses using peripheral information. IEEE Rev Biomed Eng 3:48–68
3. Reuderink B, Poel M, Nijholt A (2011) The impact of loss of control on movement BCIs. IEEE Trans Neural Syst Rehabil Eng 19(6):628–637
4. McFarland DJ, Wolpaw JR (2008) Brain-computer Interface operation of robotic and prosthetic devices. Computer 41(10):52–56
5. Dhillon GS et al (2005) Effects of short-term training on sensory and motor function in severed nerves of long-term human amputees. J Neurophysiol 93(5):2625–2633
6. Xiaofeng Jia, Matthew A. Koenig, Xiaowen Zhang, Jian Zhang, Tongyi Chen, Zhongwei Chen (2007) Residual motor signal in long-term human severed peripheral nerves and feasibility of neural signal-controlled artificial limb. J Hand Surg Am 32(5):657–666
7. Rossini PM et al (2010) Double nerve intraneural interface implant on a human amputee for robotic hand control. Clin Neurophysiol 121(5):777–783
8. Ledbetter NM et al (2013) Intrafascicular stimulation of monkey arm nerves evokes coordinated grasp and sensory responses. J Neurophysiol 109(2):580–590
9. Quian Quiroga R, Panzeri S (2009) Extracting information from neuronal populations: information theory and decoding approaches. Nat Rev Neurosci 10(3):173–185

10. Rolls ET, Treves A (2011) The neuronal encoding of information in the brain. Prog Neurobiol 95(3):448–490
11. Bergman N (1999) Recursive Bayesian estimation, Doctoral dissertation, Department of Electrical Engineering, Linköping University
12. Chaolin Ma XM, Zhang H, Jiang X, He J (2015) Neuronal representation of stand and squat in the primary motor cortex of monkeys. Behav Brain Funct 11:15
13. Grewal MS (2011) Kalman filtering. Springer, Berlin
14. Julier SJ, Uhlmann JK, Durrant-Whyte HF (1995) A new approach for filtering nonlinear systems. In: Proceedings of the American control conference, 21–23 June 1995, Seattle, WA, USA
15. Ma X et al (2017) Decoding lower limb muscle activity and kinematics from cortical neural spike trains during monkey performing stand and squat movements. Front Neurosci 11:44
16. Antfolk C et al (2013) Sensory feedback in upper limb prosthetics. Expert Rev Med Devices 10 (1):45–54
17. Deans SA, McFadyen AK, Rowe PJ (2008) Physical activity and quality of life: a study of a lower-limb amputee population. Prosthetics Orthot Int 32(2):186–200
18. Whitehouse F, Jurgensen C, Block MA (1968) The later life of the diabetic amputee: another look at fate of the second leg. Diabetes 17(8):520–521
19. Østlie K, Magnus P, Skjeldal OH, Garfelt B, Tambs K (2011) Mental health and satisfaction with life among upper limb amputees: a Norwegian population-based survey comparing adult acquired major upper limb amputees with a control group. Disabil Rehabil 33:17–18, 1594–1607. https://doi.org/10.3109/09638288.2010.540293
20. Johansson RS, Flanagan JR (2009) Coding and use of tactile signals from the fingertips in object manipulation tasks. Nat Rev Neurosci 10(5):345–359
21. Campbell CS et al (1999) What you feel must be what you see: adding tactile feedback to the trackpoint. In: Proceedings of IFIP INTERACT'99, p 383–390
22. Mugge W et al (2009) Sensory weighting of force and position feedback in human motor control tasks. J Neurosci 29(17):5476–5482
23. Ortiz-Catalan M, Hakansson B, Branemark R (2014) An osseointegrated human-machine gateway for long-term sensory feedback and motor control of artificial limbs. Sci Transl Med 6(257):257re6
24. Cipriani C et al (2008) On the shared control of an EMG-controlled prosthetic hand: analysis of user-prosthesis interaction. IEEE Trans Robot 24(1):170–184
25. Childress DS (1980) Closed-loop control in prosthetic systems: historical perspective. Ann Biomed Eng 8(4–6):293–303
26. Pylatiuk C et al (2004) Progress in the development of a multifunctional hand prosthesis. Conference on engineering in medicine and biology society, 1–5 September 2004, San Francisco, CA, USA
27. Sasaki Y, Nakayama Y, Yoshida M (2002) Sensory feedback system using interferential current for EMG prosthetic hand. Conference proceedings of second joint Embs-Bmes conference, 23–26 October 2002, Houston, TX, USA
28. Kaczmarek KA et al (1991) Electrotactile and vibrotactile displays for sensory substitution systems. IEEE Trans Biomed Eng 38(1):1–16
29. Riso RR, Ignagni AR, Keith MW (1991) Cognitive feedback for use with FES upper extremity neuroprostheses. IEEE Trans Biomed Eng 38(1):29–38
30. Navarro X et al (2005) A critical review of interfaces with the peripheral nervous system for the control of neuroprostheses and hybrid bionic systems. J Peripher Nerv Syst 10(3):229–258
31. Micera S et al (2006) Hybrid bionic systems for the replacement of hand function. Proc IEEE 94 (9):1752–1762
32. Dhillon GS, Horch KW (2005) Direct neural sensory feedback and control of a prosthetic arm. IEEE Trans Neural Syst Rehabil Eng 13(4):468–472
33. Cipriani C, Controzzi M, Carrozza MC (2011) The SmartHand transradial prosthesis. J Neuroeng Rehabil 8(1):29

34. Cipriani C, Controzzi M, Carrozza MC (2010) Objectives, criteria and methods for the design of the SmartHand transradial prosthesis. Robotica 28(06):919–927
35. Antfolk C et al (2010) SmartHand tactile display: a new concept for providing sensory feedback in hand prostheses. J Plast Surg Hand Surg 44(1):50–53
36. Micera S, Carpaneto J, Raspopovic S (2010) Control of hand prostheses using peripheral information. IEEE Rev Biomed Eng 3:48–68
37. Almström C, Herberts P, Körner L (1981) Experience with Swedish multifunctional prosthetic hands controlled by pattern recognition of multiple myoelectric signals. Int Orthop 5(1):15–21
38. Tan DW et al (2014) A neural interface provides long-term stable natural touch perception. Sci Transl Med 6(257):257ra138
39. Talan J (2014) Neural Interface activates a sense of touch in prosthetic limbs. Neurol Today 14 (21):10–11
40. Xu Q et al (2013) A novel mat-based system for position-varying wireless power transfer to biomedical implants. IEEE Trans Magn 49(8):4774–4779
41. Zhou H et al (2012) A fully implanted programmable stimulator based on wireless communication for epidural spinal cord stimulation in rats. J Neurosci Methods 204(2):341–348
42. Xu Q et al (2011) A fully implantable stimulator with wireless power and data transmission for experimental use in epidural spinal cord stimulation. Annual international conference of the Ieee engineering in medicine and biology society, 30 August to 3 September 2011, Boston, MA, USA
43. Qi X et al (2005) A versatile microprocessor-based multichannel stimulator for experimental use in epidural spinal cord stimulation. In: IEEE proceedings of first international conference on neural Interface and control, 28–26 May 2005, Wuhan, China
44. Cullen DK, Smith DH (2013) Bionic connections. Sci Am 308(1):52–57
45. Bray D (1984) Axonal growth in response to experimentally applied mechanical tension. Dev Biol 102(2):379–389
46. Heidemann SR, Buxbaum RE (1994) Mechanical tension as a regulator of axonal development. Neurotoxicology 13(1):95–107
47. Smith DH, Smith DH, Wolf JA, Meaney DF (2004) A new strategy to produce sustained growth of central nervous system axons: continuous mechanical tension. Tissue Eng 7(2):131–139
48. Pfister BJ et al (2004) Extreme stretch growth of integrated axons. J Neurosci 24 (36):7978–7983
49. Harrison RG (1911) On the stereotropism of embryonic cells. Science 34(870):279–281
50. Rajnicek A, Britland S, McCaig C (1997) Contact guidance of CNS neurites on grooved quartz: influence of groove dimensions, neuronal age and cell type. J Cell Sci 110(23):2905–2913
51. StEpieŃ E, Stanisz J, Korohoda W (1999) Contact guidance of chick embryo neurons on single scratches in glass and on underlying aligned human skin fibroblasts. Cell Biol Int 23 (2):105–116
52. Goldner JS et al (2006) Neurite bridging across micropatterned grooves. Biomaterials 27 (3):460–472
53. Fricke R et al (2011) Axon guidance of rat cortical neurons by microcontact printed gradients. Biomaterials 32(8):2070–2076

Chapter 5
Brain-Machine Interface-Based Rat-Robot Behavior Control

Jiacheng Zhang, Kedi Xu, Shaomin Zhang, Yueming Wang, Nenggan Zheng, Gang Pan, Weidong Chen, Zhaohui Wu, and Xiaoxiang Zheng

Abstract Brain-machine interface (BMI) provides a bidirectional pathway between the brain and external facilities. The machine-to-brain pathway makes it possible to send artificial information back into the biological brain, interfering neural activities and generating sensations. The idea of the BMI-assisted bio-robotic animal system is accomplished by stimulations on specific sites of the nervous system. With the technology of BMI, animals' locomotion behavior can be precisely controlled as robots, which made the animal turning into bio-robot. In this chapter, we reviewed our lab works focused on rat-robot navigation. The principles of rat-robot system

---

J. Zhang · K. Xu (✉) · S. Zhang · X. Zheng
Qiushi Academy for Advanced Studies (QAAS), Zhejiang University, Hangzhou, People's Republic of China

Department of Biomedical Engineering, Key Laboratory of Ministry of Education Ministry, Zhejiang University, Hangzhou, People's Republic of China

Zhejiang Provincial Key Laboratory of Cardio-Cerebral Vascular Detection Technology and Medicinal Effectiveness Appraisal, Zhejiang University, Hangzhou, People's Republic of China
e-mail: xukd@zju.edu.cn

Y. Wang · N. Zheng
Qiushi Academy for Advanced Studies (QAAS), Zhejiang University, Hangzhou, People's Republic of China

College of Computer Science and Technology, Zhejiang University, Hangzhou, People's Republic of China

G. Pan · Z. Wu
College of Computer Science and Technology, Zhejiang University, Hangzhou, People's Republic of China

W. Chen
Qiushi Academy for Advanced Studies (QAAS), Zhejiang University, Hangzhou, People's Republic of China

Department of Biomedical Engineering, Key Laboratory of Ministry of Education Ministry, Zhejiang University, Hangzhou, People's Republic of China

College of Computer Science and Technology, Zhejiang University, Hangzhou, People's Republic of China

X. Zheng (ed.), *Neural Interface: Frontiers and Applications*, Advances in Experimental Medicine and Biology 1101,
https://doi.org/10.1007/978-981-13-2050-7_5

have been briefly described first, including the target brain sites chosen for locomotion control and the design of remote control system. Some methodological advances made by optogenetic technologies for better modulation control have then been introduced. Besides, we also introduced our implementation of "mind-controlled" rat navigation system. Moreover, we have presented our efforts made on combining biological intelligence with artificial intelligence, with developments of automatic control and training system assisted with images or voices inputs. We concluded this chapter by discussing further developments to acquire environmental information as well as promising applications with write-in BMIs.

Keywords Brain-machine interface · Rat-robot · Behavior control · Brain stimulation

5.1 Introduction

Brain-machine interface (BMI) is a bidirectional pathway between the brain and external facilities. Besides its ability to read out information from the brain, providing an available communication method for the disabled, the BMI also has further applications in "writing" information into the brain. With encoded stimulations, researchers can send artificial information into targeted brain area, interfering neural activities and even generating sensations.

The using of physical stimulation to affect the brain function, mainly by applying electrical current into the brain, has been applied for clinical treatments for more than a century. Since electrical stimulation targeted at motor cortex can elicit limb movements [1], it was well adopted for improving neurosurgical procedures from early times [2]. More recently, deep brain stimulation (DBS) has become a common therapy to alleviate symptoms of some brain disorders such as chronic pain [3, 4] and the Parkinson's disease [5, 6]. Besides, electrical stimulation was also introduced as a tool for regenerating sensory feedback. Cochlear implants [7, 8] and retinal prostheses [9, 10] are two of such applications, replacing the impaired peripheral nerves with artificial devices. Moreover, recent works have shown that intracortical micro-stimulation in primary somatosensory system can generate discriminable tactile percepts from lab animals [11, 12] to human beings [13]. These approaches draw bright prospects of the "write-in" BMI applications.

However, it is not until the mid-1990s that this therapeutic stimulation application had found a new potential in the field of BMIs, when the concept of "bio-robot" first appeared. The idea of the BMI-assisted robotic animal system is accomplished by stimulations on specific sites of the nervous system. Studies in functional mechanisms of the nervous system gave rise to the potential of this animal locomotion control system with focal brain stimulations. Taking advantages of the artificial-evoked locomotion changes, researchers attempted to control animals with virtual perception cues elicited by electrical stimulations. Compared to traditional robot or biomimetic robot, robotic animals have shown superior performances due to their flexibility, self-sufficient energy supply, and adaptability to complex environments.

One of the earliest attempts on direct animal controlling was made by a Japanese group. In their experiment, an American cockroach was successfully controlled to follow the forward and/or turning commands with electrical stimulations on the antenna [14]. Since the advent of the remarkable work, the researches on robotic animals, or bio-robots, have been developing rapidly in the following decades. Winged insects were one of the most interesting subjects. Artificial stimulation-controlled insects were thought to be an alternative to the insect-mimetic micro-air vehicles (MAVs) for their fascinating flight performances. An expanded variety of insects have been engaged in this cyborg insects study, including months [15, 16], beetles [17], and honeybees [18]. By implanting microelectrodes into antenna and optic lobes, the insects can be controlled to fly according to preset routes or even switch the wing oscillations.

Another hot spot in the field of bio-robot is behavior control of vertebrates [19–22]. Compared to insects, vertebrates have much more complicated brain structural and functional divisions and thus can be controlled to finish more complex tasks. Among various animals, rodents are always elusive with high agility, indicating the possibility of utilizing them for investigations and rescues. Furthermore, the rodents have an acute sense of smell, which exceeds most of the man-made sensors in both sensitivity and accuracy, making it a promising subject for explosive or drug detections. However, these small mammals are hard to be tamed with traditional training protocols. With the help of electrical stimulations, locomotion commands can be directly "written" into the animal's brain, saving training time and simplifying training paradigms at the same time. The initial idea of rat-robot control was accomplished by evoking virtual reward stimulation and whisker touching feelings to indicate moving forward or provide turning cues, respectively. Encouraged by this work, some other groups have made advances in control improvements.

In this chapter, we firstly introduced how rat-robot be controlled with write-in BMI. We then described some methodological advances developed for better locomotion modulation, such as optogenetic modulation methods. We also introduced our implementation of combining "write-in" BMI with "readout" BMI, as mind-controlled rat navigation system. Besides, we presented our efforts made on optimizing control strategies for automatic control and combining machine intelligence with biological intelligence for cognition enhancements. We concluded this chapter by discussing further developments to acquire environment information as well as promising applications with write-in BMIs.

5.2 Basis of Rat Navigation System

In 2002, Dr. Chapin and his colleagues managed to navigate rats through a complex field with electrical stimuli instead of external turning cues and rewards used in operant learning paradigms [23]. To accomplish the locomotion control, the rats

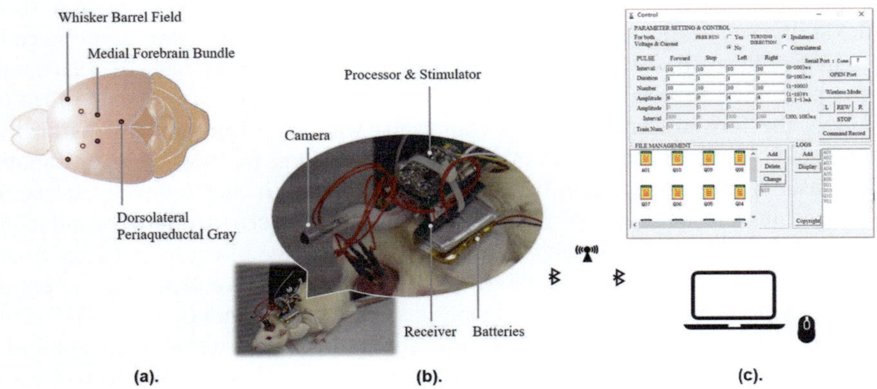

Fig. 5.1 Overview of the rat navigation system. The whole system is consisted of three parts: animal preparation, stimulator backpack, and remote control system. (**a**) Target brain sites for rat locomotion control, including turning cues (whisker barrel field/ventral posterior medial nucleus of thalamus, indicated in black dots and red circles, respectively), reward stimulation (medial forebrain bundle), and force holding (dorsolateral periaqueductal gray). (**b**) The stimulation backpack equipped with a video camera, which turning the received commands into electrical stimulations into corresponding brain sites and relaying captured videos back to operators at the same time. (**c**) GUI for operators to adjust stimulation parameters and deliver locomotion commands. The rat-robot can be controlled with a wireless mouse, and the information is exchanged wirelessly with Bluetooth

were firstly implanted with bipolar microelectrodes. After a period of training for the rats to correlate the electrical stimulation commands and corresponding behaviors, the rat's locomotor behavior can be controlled by the operators. In this model, electrical stimulations in different sites of the brain evoked different behavioral responses in rats, enabling accurate movement control in real time. Besides stimulation electrodes implantation, a backpack containing microprocessor and micro-stimulator needs to be mounted onto the rat for electrical stimuli delivering. The commands are then transferred wirelessly between computer and backpack. Figure 5.1 shows an overview of the whole rat-robot navigation system.

5.2.1 *Principles of Rat-Robot Control*

The BMI-assisted rat-robot locomotion control is mainly managed by evoking sensations and perceptions that lead to behavior changes caused by habits or reflections rather than directly activating targeted muscles. For movement control strategy design, the most classical method is to use electrical stimulation-evoked reward mechanism as an alternative for operant conditioning training. Other methods based on fear or punishment were also brought up as training-free solutions for rat-robot system improvement.

5.2.1.1 Virtual Reward-Based Training Method

To navigate a robot, at least two commands are needed for basic movement control: moving forward and turning left or right. Similarly, for typical rat-robot navigation control system, electrical stimulations are delivered into medial forebrain bundle (MFB) and somatosensory cortex (barrel field, S1BF) as virtual reward and turning cues, respectively, for movement control.

Medial forebrain bundle (MFB) has been proved to be involved in study and reward mechanisms. Electrical stimulation of MFB not only reinforces forward locomotion but also motivates further locomotion. Thus, MFB stimulation can be served as a virtual reward cue for rats to move forwards [24]. The effects of MFB stimulation can be tested and evaluated by behavioral tasks. Two typical behavioral designs are free-roaming task and lever-pressing task. The free-roaming task takes place in circular field or eight-arm maze. Rats receiving electrode implantation in MFB are allowed free roaming in the field. MFB stimulation with appropriate intensity will increase long-term locomotor activity of rats, inducing the rats to approach toward centered regions more often. In lever-pressing task, the rats are trained for a self-stimulation reward task. Once the lever is pressed by rat, a train of MFB stimulation will be delivered immediately to the rat as virtual reward. In the beginning of the first training session, the rats may touch the lever accidentally. After several tries, the rats learn the rule of self-reward stimulation and start pressing the lever frequently for reward seeking. After several days of reward training, the rats will show active movements with MFB reward stimulation.

The turning cues are induced by stimulating bilateral somatosensory cortex. Rat's somatosensory cortex barrel field (S1BF) is related to whisker sensing process. Stimulations in S1BF produce virtual whisker "touch" perceptions, inducing head movements and turning behaviors [25]. With this strategy, the rats need to receive a reinforcement training section of about 5–7 days. During this section, the animals are trained to link electrical cues with appropriate turning directions. At the beginning of each training session, the rats are motivated to keep running forwards in an eight-arm maze with MFB stimulations. Either left or right turning cues will be given when the rats entering the central area of the maze. Once the animal turns correctly after turning cue delivers, the rat will receive a train of MFB stimulation as reward. By associating MFB rewards to each correct turn, the rats can learn to perform highly accurate turning behavior following electrical stimulations. Generally, the initial turning directions are not uniform across untrained rats. While some animals would turn ipsilateral to the stimulated site, others show contralateral turning behavior. As the training procedure associated turning cues with rewards, the rats could finally turn to consistent direction according to turning instructions. Particularly, the turning direction is largely depended on the association of MFB rewards, especially in those initial trainings. When MFB rewards are given upon ipsilateral turns, the rat could learn to turn ipsilaterally to the S1BF stimulation, and vice versa [26].

5.2.1.2 Optimization of the Turning Behavior

With the combination of MFB reward and S1BF turning cues, the rats can be navigated over three-dimensional structures by operators. Motivated by MFB rewards, the animals are able to run forwards and even climb or descend steps. However, the control of turning behavior of the reported rat-robot exhibits a large variability in the results produced. Although the training procedure ensures a high success rate of turning behavior in rats after corresponding stimulation cue, the trained rats are observed to have no response to the electrical stimuli occasionally. Recent studies suggested that deep brain stimulation in thalamus might elicit more stable sensations in animals [27]. In rat's whisker sensing pathway, deflections of whiskers first activate neural responses in the brainstem. The information is then transferred into ventral posteromedial (VPM) nuclei of thalamus before it finally projected to the somatosensory cortex (S1BF).

We have tested the effects of VPM stimulation for turning behavior control [28]. The behavioral study showed robust turning reactions following VPM stimulation without training (see Fig. 5.2a). Parameter modulation tests display a near-linear relationship between the turning angle of the rat and stimulation intensity, suggesting the possibility of quantitative control of turning behavior [29]. A later micro-PET imaging study has shown that stimulation in VPM induces significant glucose uptake changes in S1 [30]. Further optical imaging studies display a larger activated area in S1BF with stimulation in VPM thalamus rather than directly in S1BF area (see Fig. 5.3), suggesting an amplifying effect through thalamocortical pathway. These behavioral and imaging results indicated that VPM stimulation may provide a more efficient alternative method for turning control in rat-robot system.

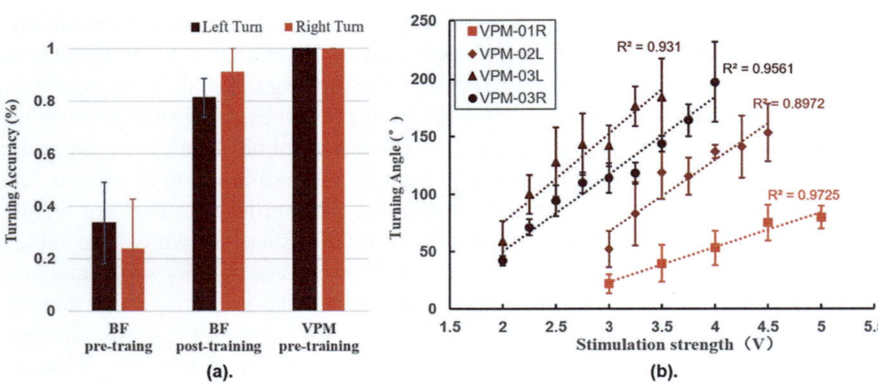

Fig. 5.2 Performances of turning behavior with VPM stimulations. (**a**) Turning accuracies calculated from both S1BF stimulation group and VPM stimulation group. VPM stimulation resulted in highly successful turning behavior without training. (**b**) The effect of stimulation strength on rat's turning angle. With the increase of stimulation voltage, the turning angle of rat showed near-linear trends. (**a** and **b** are modified from [28], Table 1 and Fig. 2A, respectively.)

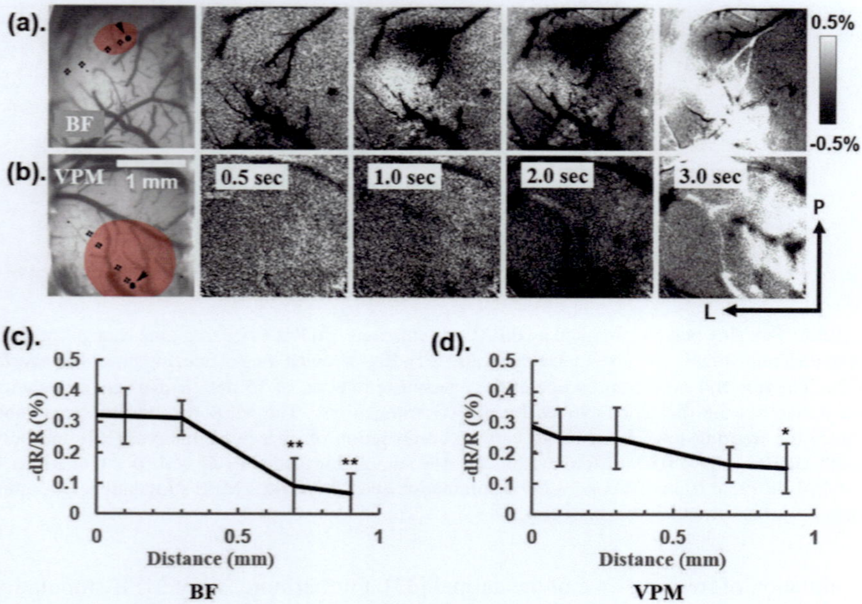

Fig. 5.3 Optical imaging responses to electrical stimulations. (**a** and **b**) Intrinsic optical imaging responses to S1BF stimulation and VPM stimulation, respectively. The darkened area indicated cortical activation spread. After a 500 ms stimulation (between 0.5 s and 1.0 s illustrated), the cortical area imaged over time responded differently to the two stimulation methods. The red-colored areas in the vessel maps illustrated significance ($p < 0.01$, t-test) between stimulation trials and control trials. *P* posterior, *L* lateral. (**c** and **d**) The corresponding response magnitudes as function of distance from the arrowed ROIs (regions of interest). S1BF stimulation resulted in a quicker drop of response magnitudes at distant spots. (Student's t-test, ∗, $p < 0.05$; ∗∗, $p < 0.01$). (**b** and **d** are modified from [29], Fig. 2.)

5.2.1.3 Punishment-Driven Behavior Control

With a combination of forward and turning commands, the rat can be controlled to move in a complex environment. Unlikely to traditional mechanical robots, one of the limitations of this intelligent robot is that the rats can hardly stay still without commands. Therefore, it is important to find a way to keep the rat immobile. One of the solutions is to utilize the distinct feel of fear and defense. Researches have shown that electrical stimulation in dorsolateral periaqueductal gray (dlPAG) induces defensive behaviors [31]. With the increase of stimulation strength, the animals show series of behavioral changes from alertness to freezing and finally escape [32]. In order to maintain immobility, stimuli should be carefully controlled in a certain range. We have tested the impacts of stimulation parameters and found that variation of both stimulation frequency and pulse train duration is efficient in

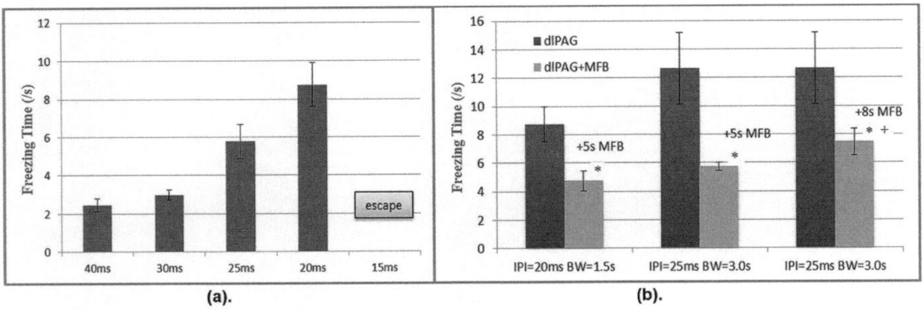

Fig. 5.4 Freezing behavior evoked by dlPAG stimulations. (**a**) Rat's freezing time changes resulted from different stimulation inter-pulse intervals (IPI). Rats showed longer freezing time with shorter pulse intervals and even start escape at the inter-pulse interval of 15 ms. (**b**) Reward stimulation effects on freezing behavior evoked by dlPAG stimulations. The MFB reward can remarkably reduce the freezing time durations at varying combination of inter-pulse interval (IPI) and burst width (BW). ∗, $P < 0.01$ different from dlPAG stimulation and dlPAG +MFB stimulation; +, $P < 0.01$ different from dlPAG+5 s MFB stimulation and dlPAG +8 s MFB stimulation. (Modified from [29], Figs. 1 and 3, respectively)

modulation of freezing time of the animal [33]. Furthermore, since MFB stimulation has reward effects, an MFB stimulation following dlPAG stimulation can remarkably reduce the freezing time, bringing the animal back from the holding state more quickly (as illustrated in Fig. 5.4b). Hence, by stimulating dlPAG or MFB, the rat's behavior can be switched between hold still and move on, like a real robot.

This "fear-caused" freezing reaction is more instinctive and can be evoked directly without any training. Similarly, a new method was proposed as a "virtual punishment"-driven solution to rat-robot navigation control [34]. This study investigated the responses of rats with stimulations in amygdale nucleus (AMY) and thalamic ventral posterolateral (VPL) as the alternatives to forward and turning cues, respectively. These two sites are involved in nociception or fear/anxiety information transmitting, providing a potential for learning-free rat-robot developments. More lately, another team also reported a direct turning behavior control method accomplished by stimulating nigrostriatal pathway [35]. This newly adopted brain site exhibits an immediate contralateral turning behavior in response to a single stimulation.

5.2.2 *The Remote Control System*

During locomotion control of rat-robot, cable connections for stimulation delivery may be a major limitation for animal's movements while roaming in large area. Thus, telemetry systems have been designed to overcome this problem. Feng et al. have designed a remote control stimulation system for controlling free-roaming animals [26]. The control system includes a remote-controlled micro-stimulator, a control program, and the Bluetooth modules for wireless communication between stimulator and control program. The control program firstly initiates the hardware and opens a protocol file for a specific rat. When user orders an instruction such as move forward or turn left/right, the program sends commands to the transmitter with preset stimulation parameters. The commands are finally translated by the receiver, and corresponding stimulations are delivered into the corresponding brain sites through electrodes implanted.

The stimulator is designed to be mounted on the rats for commands receiving and stimulation delivering. To fit small animals, the remote-controlled micro-stimulator should be kept small and light. At this point, one of the main concerns is the power consumption of the micro-stimulator. Thanks to the development of microelectronics, microprogrammed control unit (MCU) with high speed and low power consumption can be adopted as the main processor for control of receiving commands and delivering stimulations. The whole backpack mounted on controlled rat contains a receiver and a stimulator and weights only 20 g, with a total size of 36 mm $\times$ 22 mm $\times$ 15 mm. The stimulation commands received are translated by the MCU and then electrical pulses are delivered into corresponding sites of the rat's brain. Generally, the electrical stimuli are trains of biphasic, charge-balanced pulse trains. The stimuli delivered are effective with either constant voltage or constant current.

The control program is designed for navigation orders delivering as well as stimulation parameter setting. A basic instruction set for rat-robot navigation includes "Forward", "Left", and "Right", corresponding to the virtual reward and turning cues. For some cases, the "Stop" command can be added for the holding behavior control. With preset communication protocols, the instructions made by operators can then be transmitted to the stimulator and be finally translated into electrical stimulation pulses. For convenience, a wireless mouse or other portable devices can be employed as remote controller, providing a real mobile solution to working in complicated circumstances. As to optimize the animal locomotion control, the stimulation parameters are adjusted individually and can be saved and retrieved for reuse or edition in file management panel.

A Bluetooth module is applied for wireless communication in this system. The module can work properly within a range of 100 m with communication speed up to 70 kb/s, which is even sufficient for multi-channel bidirectional communication. Therefore, an improved version has been developed realizing a bidirectional telemetry system with both electrical stimulation and signal recording for free-roaming rats [36]. In this version, besides producing four-channel stimulus, the system can also acquire two-channel neural signals in real time. The recording module is capable of recording both neural action potentials (APs) and local field potentials (LFPs). This bidirectional communication system takes step towards the final goal of developing a closed-loop BMI. By analyzing neural signals in real time, stimulations can be triggered when necessary, forming a closed-loop brain-machine interface. Such closed-loop system is flexible in stimulation time, providing better treatment effects in clinical therapy with deep brain stimulation (DBS).

5.2.3 Rat Navigation Guided by Remote Control

After a few days of adaptation and training, the rats can learn to link the S1BF turning cues with MFB stimulation rewards, showing the feasibility of electrical stimulation-evoked "virtual" learning. With proper combinations of these two stimulation commands, the rats can be well steered towards targeted locations by operators. Notably, during navigation control, the MFB stimulation not simply acts as a reward feedback whenever the rat responding to turning cue correctly but also serves as a driving force to keep the rats moving continuously. In fact, frequent MFB stimulations during rats' locomotion reinforce their motivation of moving forward. Free-roaming rats with MFB stimulations show an increased locomotion activity and take more approaches toward the central area of the field. Therefore, with the help of this "virtual reward" stimulation, the rats can be induced to climb or descend from ladders/stairs and steep ramps or even jump off platforms.

Together with turning cues, well-trained rat-robots can then be adapted to complex three-dimensional obstacle courses quickly. As shown in Fig. 5.5, with the

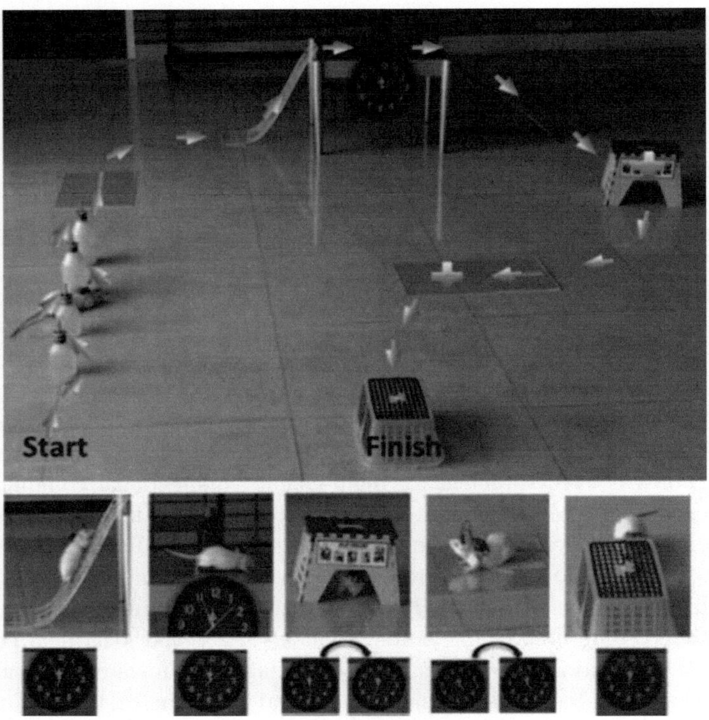

Fig. 5.5 An example of guided rat navigation using brain micro-stimulation. Well-trained rat-robot can be guided to pass through slalom courses, climb steep ramps, and even hold still for 3–5 s at target spots. (From [33], Fig. 4.)

"holding" command of dPAG stimulation, the rats can even be controlled to stay still for several seconds at any locations that be instructed to. Moreover, we have managed to guide the rats in some real conditions, such as passing through offices and corridors, crossing narrow pipelines and clumps of grass, and circumventing obstacles. While configuring the rat's backpack with a wireless micro-camera, the environmental images can be captured in real time, enabling the operators to navigate the rat-robots at beyond-visual ranges.

5.3 Novel Rat-Robot Control Method Based on Optogenetics

Optogenetics is one of the newly developed neural modulation techniques in recent years. By genetically expressing light-sensitive channel protein on specific type of neural population, targeted opsin-transduced neurons can be activated via optical stimulations with certain wavelengths. Compared with traditional electrical stimulation, optogenetic technique has the advantage of activating specific types of neurons, thus narrowing the activated range to desired areas. Besides, it also provides an opportunity to record neural electrophysiological activities during stimulation. For precise locomotion control, we developed a rat-robot system based on optical modulation.

5.3.1 Model of Optogenetics-Based Rat-Robot

The basic idea of optogenetic technique is to utilize the light-sensitive membrane ion channels to selectively activate neural populations. Different opsin channels can be activated with stimulation lights at different wavelengths, enabling either excitation or inhibition of the selected neural populations. Neural modulation by optogenetic technique requires transduction of light-sensitive opsin, optical stimulus with specific wavelengths, and proper stimulation intensities. In our current optical modulated rat-robot model, opsin gene of channelrhodopsin-2 (ChR2) is chosen for exciting neural populations. This opsin carried by adeno-associated virus (AAV) is transduced into excitatory neurons under the guidance of calcium-calmodulin-dependent kinase II type-FC; (CaMKIIFC;) promoters. After 3–4 weeks of recovery and ChR2 expression, the targeted areas in the rat's brain can then be activated by delivering blue light at a wavelength of 473 nm.

Similar to electrical stimulation control, optogenetic modulation of locomotion is also based on evoking virtual reward and defense behavior. For the "holding" behavioral modulation, optical stimulation is targeted at dPAG, the same brain area as used in electrical stimulation model. However, as to reward delivery, MFB

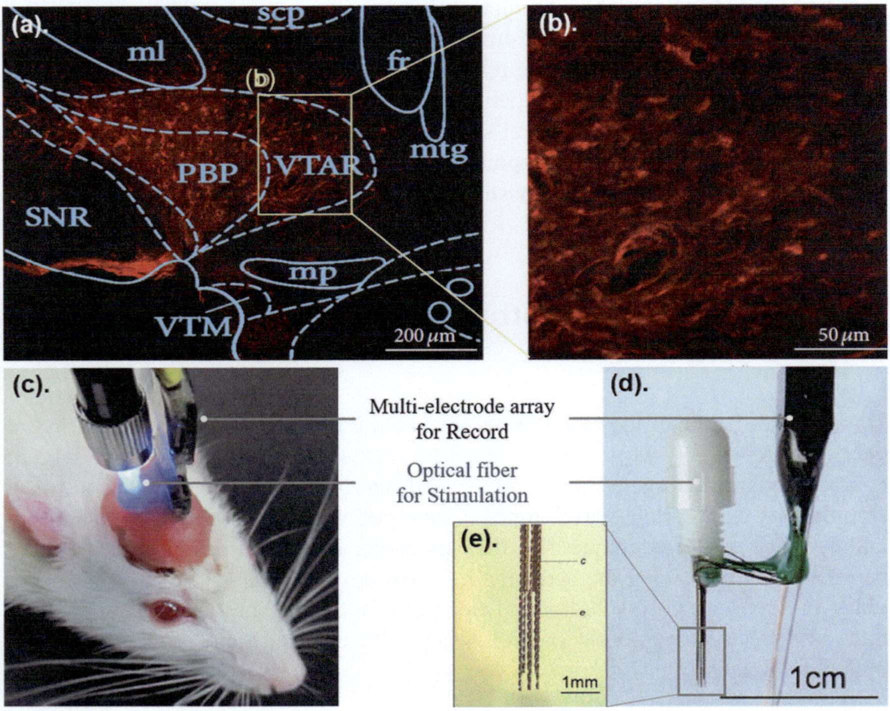

Fig. 5.6 Setup of the optogenetic model. (**a**) A typical view of the ChR2-mCherry expression on the neurons expressing CaMKIIα in and around the region of VTA (ventral tegmental area). An enlarged region of interest (ROI, squared in **a**) is illustrated in (**b**). (**c**) ChR2-transduced rat with implanted devices for optogenetic modulation. (**d**) The implanted device ("optrodes") with both optogenetic stimulations and electrophysiological recordings. (**e**) Enlargement of the electrode tips. The centered guide cannula is prepared for virus injection and fiber implantation, and the surrounded electrodes are for neural signal recording, with two different depths. (**a** and **b** are modified from [37], and Fig. 1. (**c–e**) is modified from [38], Figs. 1 and 3.)

is not an ideal target for optogenetic model since this bypassing axon bundle is lacking of cell bodies to be transduced with opsin genes and activated by light. Actually, MFB is one of the projections from the ventral tegmental area (VTA) with ventral striatum. VTA is believed to be involved in natural reward circuitry and motor activity, as well as drug addictions, providing an ideal site for opsin gene expression.

Successful expression of opsin genes can be verified by histological studies. The ChR2-mCherry expression traces can be observed on frozen brain slices with fluorescent images taken under a microscope. Figure 5.6a and b shows a typical view of the ChR2-mCherry expression image captured with fluorescent microscope, revealing a high density of transduced neural populations in the targeted brain regions with normal cell morphologies [37].

5.3.2 Optical Stimulation and Electrophysiology

Similar to electrical stimulations, during optical stimulations, the light intensity is of crucial importance. Previous researches showed that light intensity lower than the threshold of 1 mW·mm^{-2} can hardly activate ChR2-transduced neurons [39, 40], while overlarge stimulation intensity may damage the brain tissue around the stimulation site. Therefore, to activate enough range of brain tissue safely and efficiently, the appropriate stimulation intensity needs to be carefully calculated. Besides, due to diffusion and scattering, the light transmission distance in brain tissue is limited. According to the Kubelka-Munk model for diffuse scattering media [39], the light transmission fraction of an ideal model expresses relationships with both light penetration in the brain and scatter coefficient of penetration distance. For rats we have used in rat-robot control model, the appropriate distance for light delivery over the targeted brain area is 0.5 mm [38].

Besides the main advantage of activating neural populations selectively and precisely, the optical stimulation also enjoys the benefit of being free from stimulation artifacts. We have designed a multi-electrode array coupled with fiberoptic for simultaneous optical stimulation and electrophysiological recording [38]. As shown in Fig. 5.6c–e, this integrated optrode array device is consisted of a centered fiber guide cannula and 16 surrounding recording electrodes with 2 layers for different recording depths. In vivo electrophysiological recording demonstrates that optical stimulation can successfully evoke an increased neuronal activity with time specificity. Figure 5.7a and b illustrates the electrophysiological responses to optical stimulation [37]. During optical stimulation, neural spike firing activities show obvious increases in recorded area. At the end of optical stimulation, neural activities drop back to normal background level. Similar changes are also observed in local field potential signals, where the magnitude recorded increases during optical stimulation period. With the help of simultaneous electrophysiological recording, we have further tested the effects of stimulation parameters. As shown in Fig. 5.7c, optical stimulation with low frequency (5 Hz) can hardly induce significant increase in neuronal activities. While stimulated with higher frequencies, the spike firing rates recorded display significant increases from background level. The spike firing rate changes can be modulated by both frequency and stimulation intensity, showing increasing firing activities with higher frequency or intensity.

5.3.3 Behavioral Modulation of Optogenetic Rat-Robots

Optical stimulation with proper parameters also exhibits stable behavioral modulation effects on optogenetic rats. Optical stimulation in dPAG area can induce typical freezing and escape responses in rats, similar to that observed on electrically induced behavior change of rats. Track navigation tests showed precise freezing

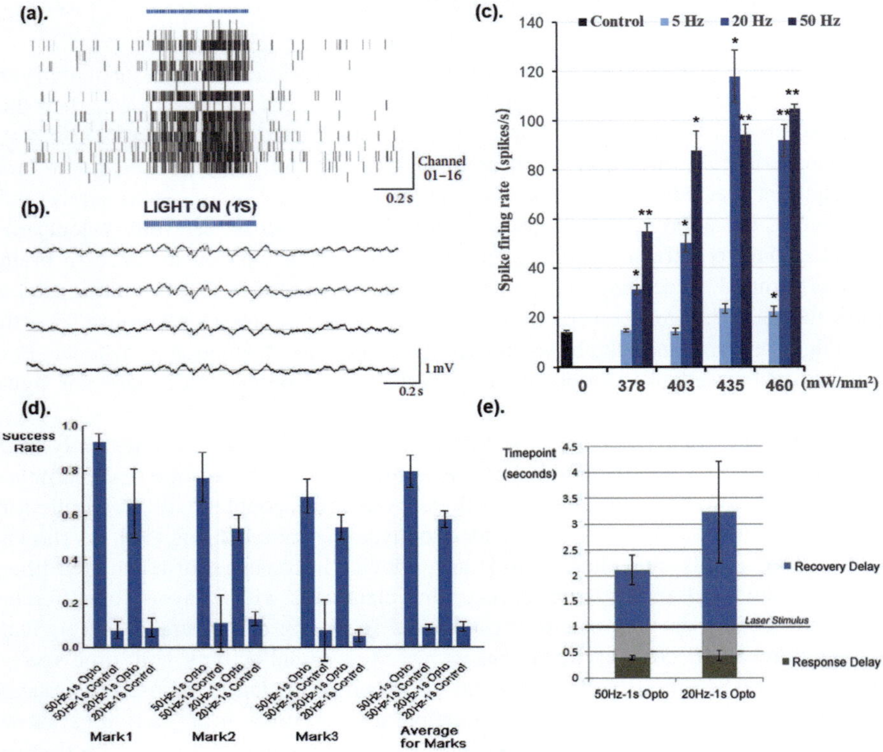

Fig. 5.7 Optogenetic modulation effects on rat-robots. (**a** and **b**) Typical in vivo spike and LFP signals recorded from the implanted optrode device, respectively. The blue bars indicate the onset of stimulation light, a train of 1.0 s pulses, with 15 ms pulse width and 50 Hz frequency. (**c**) The optical modulation effects on electrophysiological recordings. Both stimulation intensity and frequency have modulation effects on spiking firing rates (∗, $p < 0.05$; ∗∗, $p < 0.01$; one-way ANOVA, between stimulation and control group). (**d** and **e**) The modulation effects on rat holding behaviors. The optical simulations have obvious and stable modulation impacts on rat's behaviors at different mark points. (**a** and **b** are modified from [37], Fig. 1c is modified from [41], Fig. 4, and (**d**) and (**f**) are modified from [42], Fig. 1.)

behavior modulation both temporally and spatially. The optogenetic rats can be modulated to hold on any location along the trace, with short response delays (to start freezing) as well as recovery delays (to end freezing after stimulation) [42]. In accordance with electrophysiological recordings, stimulations with higher frequency display better modulation performances on both success rate and response delays (shown in Fig. 5.7d and e). Compared with electrical stimulation, this optical stimulation in dPAG induces more stable and moderate defensive responses, exhibiting shorter recovery periods after stimulation [41]. These results show the superiority of optogenetics over the traditional electrical stimulations.

As to stimulation in VTA, the optogenetic rats exhibit conditionings on reward learning. With optical stimulation, the optogenetic rats exhibit an increased, long-term locomotion activity in a free-roaming task. Further self-reward lever-pressing task shows an ascending tendency of lever presses over training sessions, demonstrating the "virtual reward" effect induced by optical stimulation in VTA neurons [37]. These results indicate that optical stimulation in VTA is an alternative way to induce reward-seeking behavior in rats rather than electrical stimulation in MFB. Taken together, the technique of optogenetics provides a novel method to regulate rat's locomotion switching between move and stop more precisely than traditional electrical stimulations both spatially and temporally.

5.4 Rat-Robot Navigation System Controlled by Human Brain

As mentioned above, the BMIs are bidirectional communication pathways which are able to modulate brain activities as well as to interpret thoughts in living brains through computers. Therefore, by combining the two directional BMI pathways, a "Brain-to-Brain Interface" (BBI) can be established where neural activities of two or more individual brains can be linked by external devices [43]. Experiments have proved that BBIs have the ability to translate locomotor intentions in real-time among brains [44–48] and even across species [49]. The application of BBI may augment the mutual coupling of brains, coupling neural processes of one brain to another, and thus may have a positive impact on human social behavior. Another potential application of BBI system is to implement locomotor control of cyborgs directly with human thoughts. Zhang's group has demonstrated a cyborg cockroach navigation system controlled by steady-state visual evoked potential (SSVEP) with a noninvasive BMI device. Since the SSVEP-based BMI decoding relies on visual stimulations, the operator's attention may be distracted from cyborg's location feedback by real-time images. To overcome this shortage, we have explored the utilization of motor imagery as an alternative solution. Here, we will briefly introduce our efforts made on motor imagery-based "mind control" of our rat-robots system [50].

5.4.1 Configuration of the BBI System

The BBI system is comprised of an electroencephalography (EEG)-based noninvasive BMI device, a host computer, the rat-robot navigation control system, and a video camera-based visual feedback system. The EEG-based BMI device is applied for motor imagery signal acquisition from human operators. The signals are sent to the host computer through Bluetooth, where motor intentions are recognized and

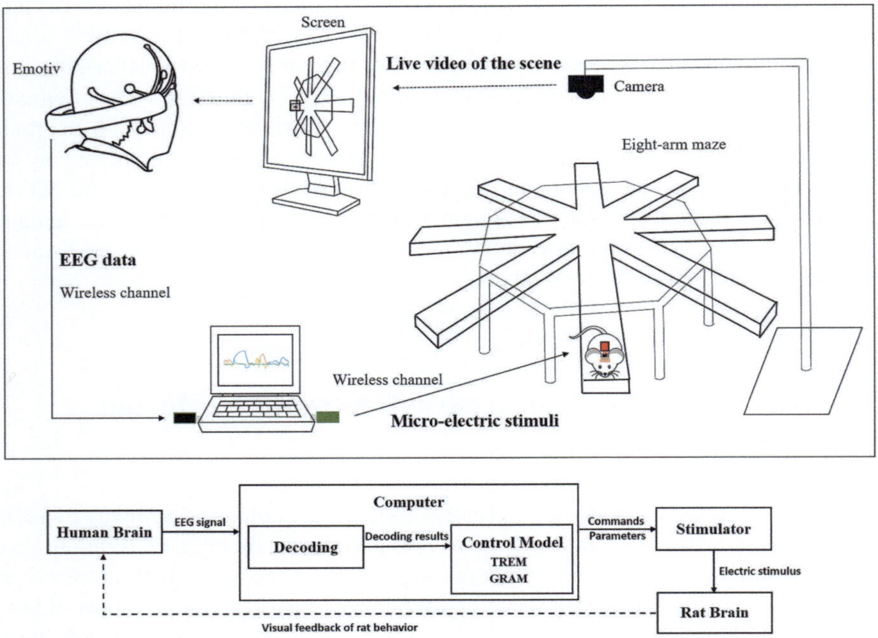

Fig. 5.8 The framework of mind-controlled rat-robot system. (From [50], Fig.1)

classified. The decoding results of motor imagery signals are then translated into locomotion control commands of the rat-robots. With the help of the remote control system, the navigation commands from the operators can finally be applied on rats through stimulator backpack.

To form a closed-loop control system, we have chosen video information as feedback signals. The visual feedback information is provided by a bird's-eye camera, with which the entire experimental field is supervised. The recorded images are sent back to an LCD screen and presented in front of the operator; thus the rat's location can be traced by an operator in real time. Figure 5.8 shows the framework of a complete BBI system.

5.4.2 Implementation of Motor Imagery-Based BMI

Different from typical manual control model, the mind control signals need to be detected and deciphered before controlling stimulations delivered to rat-robot. The operator's "thoughts" are recorded by noninvasive EEG devices. In our current BBI system, the Emotiv EPOC Neuroheadset (Emotiv Company) [51] is employed for EEG signal recording, providing up to 14 recording channels and wireless information transmission in real time. The neural signal obtained is transmitted to the host computer for further processing. As the SSVEP signals are liable to distract the operator's attention from the visual feedback of rat-robot's locomotion states, we tried a combination of nictation signals and motor imagery signals as mind control signal sources, where the nictation signal is used for reward control and motor imagery as turning cues. The eye blinks are detected by amplitude changes from raw data recorded with the frontopolar channel of the Emotiv device. And the turning cues are decoded from motor intent of left and/or right arm movement of the operator. This imagination signal requires the operators to modulate their oscillation rhythms of sensorimotor cortex into the upper mu frequency band (10–14 Hz). The power spectrum of the imagination raw data is calculated as the main character for motor intent decoding. After several sessions of signal decoding training, motor intents can be detected and evaluated by classification algorithms.

To eliminate the noises from real motor intent signals, thresholds are set as criteria for corresponding control stimulation delivering. Decoding results are detected as real intentions only when the resulting values are above the thresholds, while subliminal ones are considered as noises. However, the threshold needs to be determined carefully, since higher thresholds can efficiently reduce false alarm rate while increasing the miss rate at the same time. During rat-robot navigation control, either the increasing of fake control stimulations or the missing of key turning commands may cause failures in rat's locomotion control, resulting in lowered control efficiency. One of the feasible alternatives is to use changing gradients as the evaluation criteria. Recorded data shows that compared to background noises, the real intention signals display a larger gradient at the beginning. By calculating the changing gradients of the values decoded, the control efficiency can be significantly improved while ensuring a low false alarm rate as well.

Open-field tasks show that the system performs a mean delay of 158 ms between the detection of motor intents and the delivering of electrical stimulations, which is sufficient for stable control of rat's locomotion. Further eight-armed maze tasks also show an overall success rate above 80%. These control results suggest that our BBI system realizes a high-performance mind-controlled rat-robot navigation model.

5.5 Optimized Solutions for Control Strategy

In previous sections we have demonstrated a navigation system manipulated by human operators with hand or even by the mind. However, for mass preparations, human training seems to be inefficient, since experiences and concentrations are required for the training sessions. Besides, lack of information feedback is another main drawback of current rat-robot model. These restrictions hold the rat-robot technique back in demonstration stage rather than real practical applications. One of the solutions is to employ artificial intelligence as controlling assistant. In this section, we will demonstrate our attempts made on automatic control systems and other artificial equipment assisted control strategies.

5.5.1 Automatic Control System Designed for Rat Navigation

Machine learning gives computers artificial intelligence in many aspects. With the help of this technology, robots can be controlled automatically by computers. Thus, it is natural to expect an automatic controlled rat-robot system. However, unlike traditional mechanical robots, there are several challenges of controlling these living creatures. Self-consciousness and biological intelligence are the major differences between bio-robots and traditional robots. To rat-robots, stimulations are more like "cues" to induce proper movements rather than force driven, resulting in uncertainties of reactions to the same command. It makes the decision logic hard to be modeled with explicitness and preciseness. In addition, the body of rat-robot is variable among individuals and not rigid while moving, making it difficult to extract locomotion information of the animal. Without precise locomotion parameters, it is impossible to evaluate the feedback of control command and thus difficult to make the following decisions.

To solve the problem of locomotion parameters, an extra bird's-eye camera is employed to supervise the entire experimental scene for real-time movement capture of the rat. The rat's locomotion information including body position and head orientation is tracked and extracted with imaging processing techniques [52]. A decision algorithm based on state machines can then be used to choose proper command to induce rat-robot to accomplish a preset navigation task. The schematics of the navigation system are illustrated in Fig. 5.9.

As to individual variations, we borrow experiences from human navigating operations and propose a new method to learn and imitate this process [53]. Different from the traditional methods that try to describe controlling logic explicitly, the new model regards the navigation as a whole procedure and learns to make controlling decision based on the operations made by human operators. In this model, a General Regression Neural Network (GRNN) is applied as the mathematical model for control decision learning. A total of two steps are included in this model. During the first training stage, the control commands made by human operators and the corresponding locomotion information of rat-robot (including body position and head orientation) are extracted and synchronized as the labels and inputs to the model. During this short training session, the GRNN configures the network with the instances. After the training sessions, the GRNN model is able to generate controlling instructions automatically and thereby takes place of human operators to navigate the rat-robot. This newly proposed method displays comparable performance to human operators with high accuracy and reasonable stimulation instructions quantities, indicating that the basic locomotion as position and

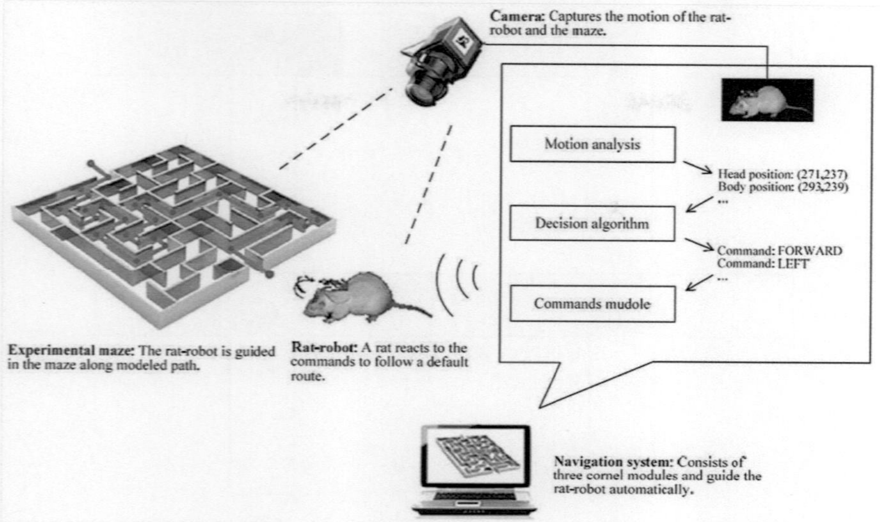

Fig. 5.9 The framework of automatic navigation system for rat-robot. (From [54], Fig. 5)

orientation can provide enough evidences for controlling decision-making [54]. This conclusion is essential for further studies of bio-robots.

5.5.2 A Reward-Driven Control System

The control strategy mentioned above mainly focuses on simplifying the decision logic by mimic the controlling principles conducted by human operators as a whole. In this process, three directional commands are needed to be classified from locomotion information. In fact, thanks to its biological intelligence, the rat-robots have shown the capabilities of reward seeking during daily training. It is well known that rats have remarkable ability in navigation [55]. In spatial learning studies, the rats are trained to seek for rewards and correct their behaviors whenever the rewards are deprived [56]. Therefore, it is feasible to use single reward command to guide the rat-robot, further reducing the decision-making procedure of the automatic controlling strategy.

According to this idea, we have designed a new automatic controlling strategy by applying reward-only commands [57]. This method simplifies the automatic navigation algorithm largely by taking full advantage of the rats' learning capabilities. A schematic diagram of control strategy is illustrated in Fig. 5.10. Before the navigation test, the destination of the task is preset, and the corresponding navigation route is yielded by path planning algorithm. During navigation task, the rat-robot receives rewards frequently when it performs ideal behaviors as walking along the right route. If it walks into incorrect direction, the rewards are cut off as a cue. The rat-robot

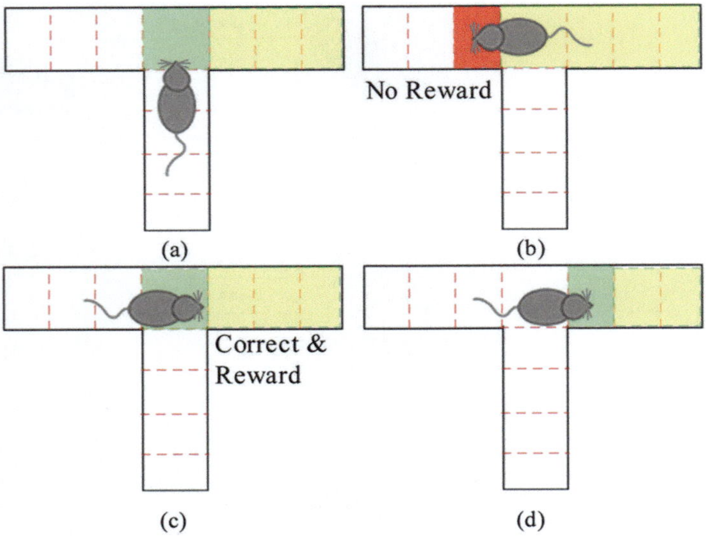

Fig. 5.10 Schematic of the reward-only automatic control model for rat-robot navigation. The green grids indicate current correct locations of the rat, the yellow grids denote the ideal route toward destinations, and the red ones represent the wrong route. (**a**) The rat-robot enters the crossing area of the maze following reward stimulation without turning cues. (**b**) The rat turns into one of the arms at its own will. As entering a wrong grid, no reward is given, and the ideal route is re-computed. (**c**) The rat turns back seeking for reward and enters the correct direction. As the route is reset, the rat gets reward again. (**d**) The rat moves forward following the reward stimulation toward the targeted end. (From [57], Fig. 1)

would realize its error and correct to proper direction for rewards. After several days of training with the automatic control system, the rat-robots show high performances in simple T-maze tasks, with their turning directional preferences dropping along with the training session. These results indicate a successful attempt at combining biological intelligence with artificial intelligence, providing new ideas for further studies on bio-robot controlling and intelligence hybrid.

Furthermore, to enhance the capability of exploring un-configured environment, we propose another reinforcement learning process-involved graded reward-generating algorithm [58]. Q-learning procedure, one of the classical reinforcement learning methods, is introduced in this method, to generate an incremental sequence of electrical reward according to the distance from the preset destination. During navigation task, the rat-robot receives higher reward stimulations when it approaches closer to the given destination. After the initial trial-and-error test, the animal learns to seek the maximum reward spots in an unknown environment. Since the rat has excellent spatial recognition, the rat-robot and the reinforcement learning algorithm can convergent to an optimal route by co-learning. This work has significant inspiration for the practical development of bio-robot navigation with hybrid intelligence.

5.5.3 *Cognition Enhancement of Rat-Robotics with Visual/Audio Cue*

Another advantage of hybrid intelligence is that we can balance biological intelligence with artificial intelligence. Utilizing of animals instead of mechanics overcomes difficulties faced by traditional robotics such as motion flexibility and stability, and meanwhile, external devices can be a compensation for rat's own limitations. For example, it is broadly known that rats communicate with ultrasound band (higher than 20 kHz) [55], which is far beyond human speech wavelength range (300–3400 Hz). It seems impossible for rats to understand or even hear voice commands from human beings. Nevertheless, with the help of speech recognition module, computers can directly translate vocal navigation commands into electrical stimulations without hand operations [59].

In addition, when the rats are equipped with sensory devices, environmental information can be acquired and analyzed by computers remotely. This equipment upgrade may encourage the rat-robot to move further into larger areas. Considering the poor vision of rats, one of the most intuitive approaches is taking images. With miniature cameras mounted on backpack, environmental imaging information can be captured and transferred back to a host computer. Operators can then guide the rat-robot from far beyond visual range. Furthermore, with the technology of image recognition, objects of interest can be detected efficiently by algorithms. The object detection results can in turn be taken as guidance on adjusting the moving directions of rat-robot toward the object. Incorporated with object detection methods, this closed-loop model can accomplish beyond visual range control of rat robotics automatically, which is of great significance for remote searching tasks [60].

5.6 Conclusions and Future Perspectives

For decades, great progress has been made in locomotion control of rat-robots. Commands "written" into the rat's brain have become more and more efficient. The main idea of locomotion control is to use electrical stimulation to mimic natural perceptions and feelings that result in corresponding behaviors of rats. By carefully selecting stimulation sites and parameters, rats can be guided in complex environments. Integrated with artificial intelligence, rats can even be controlled automatically by computers to reach a preset destination. On the contrary, there are also many limitations to what has so far been accomplished with rat navigation system. The major problem is the lack of feedback information of the environment surroundings. Without efficient information feedback, operators can hardly control the rats to avoid obstacles or adjust heading direction towards targets. Video cameras with image stabilization technique mentioned in Sect. 5.2 are one of the solutions. However, the angle of view restricts the ranges of information obtainment. Besides, weights of bear load also limit quantity and dimension of equipment that can be carried by rats.

In actual situations, odor information is always of great importance. In fact, rodents have excellent sense of smells. Given their small sizes and light body weights, rodents are more suitable for demining than traditional canine. APOPO (Anti-Personnel Landmines Detection Product Development), a non-governmental organization, has been developing technologies for landmine detection using African Giant Pouched rats (*Cricetomys gambianus*) since 1996. Although the trained rats showed a relatively high average indication rate of 76%, the score was varied among individuals while taking a long period of training as well. When taking advantage of BMI technology, the training paradigms are simplified. The electrical stimulation can provide more stable control of rat's locomotion, and the virtual reward stimulation may accelerate combination learning between target odor and reward. Moreover, with a bidirectional BMI, odor information might be read out straightly from the rat's brain, and more information could then be decoded including odorant components and concentrations. These readouts might largely enrich operator's knowledge of environmental information, leading to more precise navigation control of rat-robots.

Clinically, BMIs are designed for movement assistance and rehabilitation therapy of paralyzed patients. An ideal BMI system requires bidirectional, closed-loop information communication. Besides reading out motion intentions through brain-to-machine pathway to drive prosthetics, sensory information feedback is of equal importance. Compared to neural signal decoding, this recoding process through machine-to-brain pathway still needs a long way to go deep into. To write in sensory information properly, stimulation modes, target sites, and stimulation parameters are all needed to be taken into consideration carefully. As our rat-robot model is based on virtual sensation generation, studies on locomotion control also shed lights on stimulation targets and parameter determinations. Through behavioral and imaging tests, stimulation parameters can be optimized for more natural sensation regeneration. Inspiringly, researchers from Pittsburgh have recently succeeded in restoring tactile sensation through intracortical micro-stimulation upon human being [13]. By delivering electrical micro-stimulation into corresponding representation locations in somatosensory cortex, stable sensations can be evoked naturally. Similar to the relationship between somatosensory cortex and thalamus in rodents, some study suggested more stable encoding in thalamus than in cortex with nonhuman primates [61]. As to stimulation methods, for clinical consideration, researchers are making great efforts on seeking for possible noninvasive stimulation approaches. Transcranial direct current stimulation (tDCS) and transcranial magnetic stimulation (TMS) are two of the most commonly used noninvasive stimulation methods for neural activity modulation. Both tDCS and TMS are considered safe and painless for human use; nonetheless, lacking spatial specificity, these two stimulation methods are not suitable for localized stimulations. Ultrasound has recently been shown to be capable of inducing reversible changes of neural activity noninvasively. Unlike tDCS and TMS, focused ultrasound can be applied to deep brain structures and hit the target specifically with greater spatial precision [62]. Though it still needs to be assessed before clinical trials, this technique and other new ones will have impacts on neural modulations.

References

1. Fritsch G, Hitzig E (1870) Uber die elektrische Erregbarkeit des Grossihirns. Arch Anat Physiol Wissen Med 37:300–332
2. Gildenberg PL (2005) Evolution of neuromodulation. Stereotact Funct Neurosurg 83 (2–3):71–79
3. Owen SL, Green AL, Stein JF et al (2006) Deep brain stimulation for the alleviation of post-stroke neuropathic pain. Pain 120(1–2):202–206
4. Marchand S, Kupers RC, Bushnell MC et al (2003) Analgesic and placebo effects of thalamic stimulation. Pain 105(3):481–488
5. Krack P, Batir A, Van Blercom N et al (2003) Five-year follow-up of bilateral stimulation of the subthalamic nucleus in advanced Parkinson's disease. N Engl J Med 349(20):1925–1934
6. Bittar RG, Burn SC, Bain PG et al (2005) Deep brain stimulation for movement disorders and pain. J Clin Neurosci 12(4):457–463
7. Wilson BS, Dorman MF (2008) Cochlear implants: a remarkable past and a brilliant future. Hear Res 242(1–2):3–21
8. Shannon RV (2012) Advances in auditory prostheses. Curr Opin Neurol 25(1):61–66
9. Chow AY, Chow VY, Packo KH et al (2004) The artificial silicon retina microchip for the treatment of vision loss from retinitis pigmentosa. Arch Ophthalmol 122(4):460–469
10. Yanai D, Weiland JD, Mahadevappa M et al (2007) Visual performance using a retinal prosthesis in three subjects with retinitis pigmentosa. Am J Ophthalmol 143(5):820–827
11. Romo R, Hernandez A, Zainos A et al (2000) Sensing without touching: psychophysical performance based on cortical microstimulation. Neuron 26(1):273–278
12. Houweling AR, Brecht M (2007) Behavioural report of single neuron stimulation in somato-sensory cortex. Nature 451(7174):65–68
13. Flesher SN, Collinger JL, Foldes ST et al (2016) Intracortical microstimulation of human somatosensory cortex. Sci Transl Med 8(361):141r–361r
14. Holzer R, Shimoyama I (1997) Locomotion control of a bio-robotic system via electric stimulation. In: Intelligent robots and systems, 1997. IROS'97., Proceedings of the 1997 IEEE/RSJ international conference on IEEE, pp 1514–1519
15. Bozkurt A, Gilmour R, Stern D et al (2008) MEMS based bioelectronic neuromuscular interfaces for insect cyborg flight control. In: Proceedings IEEE micro electro mechanical systems IEEE, p 160
16. Bozkurt A, Gilmour RF Jr, Lal A (2009) Balloon-assisted flight of radio-controlled insect biobots. IEEE Trans Biomed Eng 56(9):2304–2307
17. Sato H, Berry CW, Peeri Y et al (2009) Remote radio control of insect flight. Front Integr Neurosci 3:24
18. Bao L, Zheng N, Zhao H et al (2011) Flight control of tethered honeybees using neural electrical stimulation. In: International IEEE EMBS conference on neural engineering IEEE, pp 558–561
19. Talwar SK, Xu S, Hawley ES et al (2002) Direct brain stimulation enables training of complex behaviors in freely roaming animals. In: Proceedings of annual international conference of the IEEE engineering in medicine and biology society IEEE, pp 2076–2077
20. Gomes W, Perez D, Catipovic J (2006) Autonomous shark tag with neural reading and stimulation capability for open-ocean experiments. EOS Trans Am Geophys Union 87:36
21. Brown S (2006) Stealth sharks to patrol the high seas. New Sci 189(2541):30–31
22. Wang W, Guo C, Sun J et al (2009) Locomotion elicited by electrical stimulation in the midbrain of the Lizard Gekko gecko. In: Budiyono A, Riyanto B, Joelianto E (eds) Studies in computational intelligence. Springer, Berlin, p 145
23. Talwar SK, Xu S, Hawley ES et al (2002) Rat navigation guided by remote control. Nature 417 (6884):37–38
24. Hermer-Vazquez L, Hermer-Vazquez R, Rybinnik I et al (2005) Rapid learning and flexible memory in "habit" tasks in rats trained with brain stimulation reward. Physiol Behav 84 (5):753–759

25. Doty RW (1969) Electrical stimulation of the brain in behavioral context. Annu Rev Psychol 20:289–320
26. Feng Z, Chen W, Ye X et al (2007) A remote control training system for rat navigation in complicated environment. J Zheijang Univ Sci A 8(2):323–330
27. Francis JT, Xu S, Chapin JK (2008) Proprioceptive and cutaneous representations in the rat ventral Posterolateral thalamus. J Neurophysiol 99(5):2291–2304
28. Xu K, Zhang J, Zhou H et al (2016) A novel turning behavior control method for rat-robot through the stimulation of ventral posteromedial thalamic nucleus. Behav Brain Res 298 (Pt B):150–157
29. Zhang J, Zhang S, Yu C et al (2018) Intrinsic optical imaging study on cortical responses to electrical stimulation in ventral posterior medial nucleus of thalamus. Brain Res 1684:40–49
30. Zhu Y, Xu K, Xu C et al (2016) PET mapping for brain-computer Interface stimulation of the Ventroposterior medial nucleus of the thalamus in rats with implanted electrodes. J Nucl Med 57(7):1141–1145
31. Keay KA, Bandler R (2001) Parallel circuits mediating distinct emotional coping reactions to different types of stress. Neurosci Biobehav Rev 25(7):669–678
32. Brandao ML, Troncoso AC, Silva M et al (2003) The relevance of neuronal substrates of defense in the midbrain tectum to anxiety and stress: empirical and conceptual considerations. Eur J Pharmacol 463(1–3):225–233
33. Lin J, Yu C, Jia J et al (2010) Using dlPAG-evoked immobile behavior in animal-robotics navigation. 2010 5th international conference on computer science & education, (iccse 2010) - hefei, china, 1295–1298
34. Huai R, Yang J and Wang H et al (2009) A new robo-animals navigation method guided by the remote control. In: Shi R, Fu WJ, Wang YQ, Wang HB(eds) International conference on biomedical engineering and informatics. IEEE, pp 1170–1173
35. Koo B, Koh CS, Park HY et al (2017) Manipulation of rat movement via Nigrostriatal stimulation controlled by human visually evoked potentials. Sci Rep 7(1):2340
36. Ye X, Wang P, Liu J et al (2008) A portable telemetry system for brain stimulation and neuronal activity recording in freely behaving small animals. J Neurosci Methods 174(2):186–193
37. Guo S, Chen S, Zhang Q et al (2014) Optogenetic activation of the excitatory neurons expressing CaMKIIFC; in the ventral tegmental area upregulates the locomotor activity of free behaving rats. Biomed Res Int 2014:687469
38. Guo S, Zhou H, Zhang J et al (2013) A multi-electrode array coupled with fiberoptic for deep-brain optical neuromodulation and electrical recording. In: Engineering in medicine and biology society (EMBC), 2013 35th annual international conference of the IEEE, pp 2752–2755
39. Aravanis AM, Wang LP, Zhang F et al (2007) An optical neural interface: in vivo control of rodent motor cortex with integrated fiberoptic and optogenetic technology. J Neural Eng 4(3): S143–S156
40. Wang J, Wagner F, Borton DA et al (2012) Integrated device for combined optical neuromodulation and electrical recording for chronic in vivo applications. J Neural Eng 9 (1):16001
41. Chen S, Zhou H, Guo S et al (2015) Optogenetics based rat-robot control: optical stimulation encodes "stop" and "escape" commands. Ann Biomed Eng 43(8):1851–1864
42. Guo S, Zhou H, Wang Y et al (2014) A rat-robot control system based on optogenetics. In: Ren L, Wang H, Dai Z (eds) Applied mechanics and materials, pp 848–852
43. Min BK, Marzelli MJ, Yoo SS (2010) Neuroimaging-based approaches in the brain-computer interface. Trends Biotechnol 28(11):552–560
44. Pais-Vieira M, Lebedev M, Kunicki C et al (2013) A brain-to-brain interface for real-time sharing of sensorimotor information. Sci Rep 3:1319
45. Deadwyler SA, Berger TW, Sweatt AJ et al (2013) Donor/recipient enhancement of memory in rat hippocampus. Front Syst Neurosci 7:120
46. Pais-Vieira M, Chiuffa G, Lebedev M et al (2015) Building an organic computing device with multiple interconnected brains. Sci Rep 5:11869

47. Grau C, Ginhoux R, Riera A et al (2014) Conscious brain-to-brain communication in humans using non-invasive technologies. PLoS One 9(8):e105225
48. Rao RP, Stocco A, Bryan M et al (2014) A direct brain-to-brain interface in humans. PLoS One 9(11):e111332
49. Yoo SS, Kim H, Filandrianos E et al (2013) Non-invasive brain-to-brain interface (BBI): establishing functional links between two brains. PLoS One 8(4):e60410
50. Zhang S, Yuan S, Huang L, Zheng X, Wu W, Xu K, Pan G (2019) Human mind control of rat cyborg's continuous locomotion with wireless brain-to-brain interface. Sci Rep 9:1321. https://doi.org/10.1038/s41598-018-36885-0
51. Martinez-Leon J, Cano-Izquierdo J, Ibarrola J (2016) Are low cost brain computer Interface headsets ready for motor imagery applications? Expert Syst Appl 49:136–144
52. Zhang Y, Sun C, Zheng N, et al (2010) An Automatic Control System for Ratbot Navigation [C]// 2010 IEEE/ACM International Conference on Green Computing and Communications & 2010 IEEE/ACM International Conference on Cyber, Physical and Social Computing. IEEE Computer Society, 2010, Hangzhou, China
53. Sun C, Zheng N, Zhang X et al (2011) An automatic control model for rat-robot. In: Engineering in medicine and biology society, EMBC, 2011 Annual international conference of the IEEE, Boston, USA
54. Sun C, Zheng N, Zhang X et al (2013) Automatic navigation for rat-robots with Modeling of the human guidance. J Bionic Eng 10(1):46–56
55. Whishaw IQ, Kolb B (2004) The behavior of the laboratory rat: a handbook with tests. Oxford University Press, Oxford
56. Hull CL, Spence KW (1938) "Correction" vs "non-correction" method of trial-and-error learning in rats. J Comp Psychol 25(1):127–145
57. Sun C, Zhang X, Zheng N et al (2012) Bio-robots automatic navigation with electrical reward stimulation. Engineering in medicine and biology society (EMBC), 2012 annual international conference of the IEEE,San Diego, California, USA
58. Zhang C, Sun C, Gao L et al (2013) Bio-robots automatic navigation with graded electric reward stimulation based on reinforcement learning. Conf Proc IEEE Eng Med Biol Soc 2013:6901–6904
59. Wu Z, Yang Y, Xia B et al (2014) Speech interaction with a rat. Chin Sci Bull 59 (28):3579–3584
60. Wang Y, Lu M, Wu Z et al (2015) Visual Cue-guided rat cyborg for automatic navigation. IEEE Comput Intel Mag 10(2):42–52
61. Song W, Semework M (2015) Tactile representation in somatosensory thalamus (VPL) and cortex (S1) of awake primate and the plasticity induced by VPL neuroprosthetic stimulation. Brain Res 1625:301–313
62. Yoo SS, Bystritsky A, Lee JH et al (2011) Focused ultrasound modulates region-specific brain activity. NeuroImage 56(3):1267–1275

Chapter 6
Realizing Efficient EMG-Based Prosthetic Control Strategy

Guanglin Li, Oluwarotimi Williams Samuel, Chuang Lin, Mojisola Grace Asogbon, Peng Fang, and Paul Oluwagbengba Idowu

Abstract As an integral part of the body, the limb poses dexterous and fine motor grasping and sensing capabilities that enable humans to effectively communicate with their environment during activities of daily living (ADL). Hence, limb loss severely limits individuals' ability especially when they need to perform tasks requiring their limb functions during ADL, thus leading to decreased quality of life. To effectively restore limb functions in amputees, the advanced prostheses that are controlled by electromyography (EMG) signal have been widely investigated and used. Since EMG signals reflect neural activity, they would contain information on the muscle activation related to limb motions. Pattern recognition-based myoelectric control is an important branch of the EMG-based prosthetic control. And the EMG-based prosthetic control theoretically supports multiple degrees of freedom movements that allows amputees to intuitively manipulate the device. This chapter focuses on EMG-based prosthetic control strategy that involves utilizing intelligent computational technique to decode upper limb movement intentions from which control commands are derived. Additionally, different techniques/methods for improving the overall performance of EMG-based prostheses control strategy were introduced and discussed in this chapter.

G. Li (✉) · O. W. Samuel · C. Lin · P. Fang
CAS Key Laboratory of Human-Machine Intelligence-Synergy Systems, Shenzhen Institutes of Advanced Technology (SIAT), Chinese Academy of Science (CAS), Shenzhen, China

Institute of Advanced Integration Technology, SIAT, Shenzhen, China
e-mail: gl.li@siat.ac.cn

M. G. Asogbon · P. O. Idowu
CAS Key Laboratory of Human-Machine Intelligence-Synergy Systems, Shenzhen Institutes of Advanced Technology (SIAT), Chinese Academy of Science (CAS), Shenzhen, China

Institute of Advanced Integration Technology, SIAT, Shenzhen, China

Shenzhen College of Advanced Technology, University of Chinese Academy of Sciences, Shenzhen, China

X. Zheng (ed.), *Neural Interface: Frontiers and Applications*,
Advances in Experimental Medicine and Biology 1101,
https://doi.org/10.1007/978-981-13-2050-7_6

Keywords Myoelectric signal · Pattern recognition · Limb motion · Upper limb amputee · Prosthesis · Rehabilitation robot

6.1 Introduction

Human upper limb represent an integral part of the body that provides incredible array of functionalities during activities of daily living (ADL). In addition to possessing dexterous and fine motor grasping/sensing capabilities, the upper limb enables individuals to freely and effectively communicate with their natural environments through artistic expressions, sign languages, and greetings (handshake/hand gesture), among others. Therefore, the loss of one or both arms severely limits individuals' ability especially when they need to perform tasks requiring their arm functions during ADL, thus leading to decreased quality of life [1]. Limb loss also has profound negative impact on amputees including stump pain or phantom limb pain, or perhaps even both, depression, changes in body image, and psychological burden [1–5]. Under certain circumstance, upper limb loss may also affect amputees' stability (dis-balance) during mobility, thus making them prone to falls or collisions with other individuals/objects. To this end, limb loss causes a devastating experience necessitating the need for proper psychological and physical rehabilitation strategy for arm amputees.

A large number of the commercially available upper limb prostheses are either mechanically (body-powered) or electrically (motorized) driven. Meanwhile, body-powered upper limb prosthetic devices were primarily developed as alternative to aid amputees in carrying out ADL requiring their arm functions [6]. Fundamentally, these prostheses operate by using cables to link the movement of the body to the device and thus actuating simple hand tasks such as opening or closing of the hook/gripper. To achieve simple hand grasp tasks, the amputee would need to possess a significant amount control over his/her shoulder, chest, and even the residual limb which should be long enough to support the motion. Although body-powered prostheses have fine cosmetic appearances that are comparable to the human natural arm [7], they are nonintuitive and require significant amount of efforts (burdensome) to perform simple hand tasks and as well support only single degree of freedom (DOF) movement per time [7]. With these limitations, motorized upper limb prostheses are considered as viable alternative for intuitive restoration of multiple DOF arm functions in amputees. At the forefront of this technology are the prostheses that utilize electromyography (EMG) recordings to characterize upper limb movement intentions of the amputees [8–12]. Thus, the properties of the EMG signals are encoded in form of control commands to the prosthetic controller which then actuates the device accordingly. It is noteworthy that in the last two to three decades, the motorized upper limb prosthetic technology has progressively advanced leading to the development of better control methods as well as designs. Hence, this article focuses on the EMG-based prosthetic control with emphases on methods/technologies that could help improve the overall performance and acceptability of the prostheses system.

The remaining part of the chapter is organized as follows: Section 6.2 discusses different EMG-based prosthetic control methods; Section 6.3 introduces some potential techniques for improving the performance of EMG-based prosthetic control; Section 6.4 presents the concluding remarks.

6.2 EMG-Based Prosthetic Control Methods

6.2.1 Background

Fundamentally, EMG signals represent electrical manifestation of neuromuscular activation associated with a contracting muscle. Such signals have been proven to contain rich set of neural information from which different types of upper limb movements could be accurately decoded, thus making them useful control input to prosthetic devices as well as other forms of rehabilitation robots [8, 13]. Additionally, EMG signals carry significant information for diagnosis of neuromuscular disorders and other muscle-related diseases. It has been established that EMG signal patterns for a specific limb movement are repeatable over time and distinct from the patterns of other limb movements. By exploring this phenomenon, researchers have been able to decode a range of arm movements from EMG recordings using intelligent computational methods. To this end, dozens of previous studies have reiterated the potential of EMG-based method in the control of multifunctional prostheses [14–16]. Thus, the following section will focus on discussing the currently applied EMG-based prosthetic control schemes including: pattern recognition (PR) control method and the simultaneous and proportional control (SPC) method.

6.2.2 Pattern Recognition-Based Prosthetic Control

Although the concept of EMG pattern recognition (EMG-PR) is by no means new, it has been proven over time to be promising especially for improving the dexterity of control in upper limb prosthetic devices in comparison with the conventional amplitude-based methods. It is noteworthy that the first generation of pattern recognition-driven prosthetic controllers was built in the late 1960s/early 1970s [17–19]. Since then, advancement in signal processing, multichannel recording, and microprocessor technology has expedited implementation of EMG-PR method in embedded control systems. Also, with a lot of synergistic efforts from the academia and industries, EMG-PR-based prosthetic arms have been recently made available for commercial use.

EMG-PR prosthetic control method is primarily based on the assumption that EMG signal patterns corresponding to a particular class of limb movement are consistently repeatable across trials and as well distinct from EMG signal patterns of the other classes. Due to the promising nature of this control method, it has been

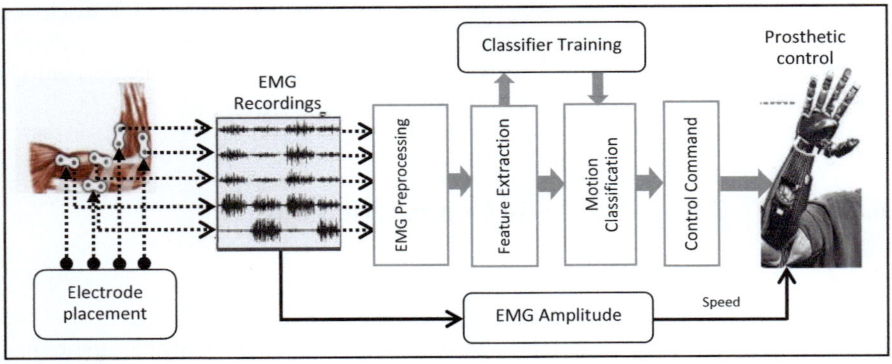

Fig. 6.1 Schematic diagram of EMG-PR-based prosthetic control system

widely explored in the last two to three decades with significant progress reported in terms of its robustness. The control schemes as a whole consist of a sequence of steps beginning with: EMG signal acquisition, preprocessing of the recorded signals, extraction of feature set, classification of the extracted features, and issuing of control commands to drive prosthesis. Figure 6.1 shows the schematic diagram of a typical EMG-PR-based prosthetic control system.

In the above represented control scheme, myoelectric signals are collected via electrodes placed on the skin surface underlying the arm muscles. Often times, the electrodes are integrated with miniaturized amplifier to improve the quality of the recorded signals through standard amplification and filtering procedures. Afterwards, the acquired EMG recordings are subjected to preprocessing procedure, so as to attenuate inherent interferences. In this regard, the 50/60 Hz notch filter is usually applied to the raw acquired signals to minimize interferences resulting from power line noise. In addition to the notch filter, a band-pass filter with cut of frequency between 10/20 Hz and 500 Hz is often applied to the EMG signals to obtain the most useful spectrum with high motor information content.

EMG signals recorded during targeted limb movements are continuous in nature, thus for a specific length of recording, a decision corresponding to a subset of the observed limb movements may be required to actuate the prosthesis. Therefore, the use of data segmentation technique in handling the preprocessed data prior to feature extraction is necessary. This step helps increase the probability of the prosthetic control system achieving high accuracy as well as quick response time. Ideally in real life applications, a response time of no more than 300 ms is required [20]. Note that the response time (which is a function of the *segment length* and the *processing time*) is the time required to produce the control commands that drive the prostheses. In a study conducted by Englehart and Hudgins [21], it was shown that by adopting continuous segmentation on a *steady-state* EMG signal, the segment length can be reduced to about 128 ms without a significant reduction in classification accuracy.

Also, it has been shown previously that segmented *steady-state* EMG signals achieved better accuracy compared to segmented *transient* EMG signals, and less degradation were recorded with shorter segments of the *steady-state* EMG data. After considering the segment length and the state of the data, the third important data segmentation factor is the windowing technique adopted. Data segmentation is often achieved via either of the following windowing techniques: *sliding analyses windowing* and *adjacent windowing* method [20]. The adjacent windowing method uses disjoint segments of predefined length for its feature extraction; and the intended limb movement is classified after a certain processing delay. Because the processing time often denotes a small fraction of segment length, the processor is found to be mostly idle during the remaining time of the segment length. Meanwhile, in the sliding windowing method, the new window segment slides over the current window segment, with an increment time less than the segment length. In comparison, this sliding windowing method takes advantage of the processor's idle time in adjacent windowing method to produce more classified outputs. Thus, the sliding windowing technique has been widely applied to *steady-state* EMG signals for enhanced feature extraction process in most recent studies on EMG-PR-based prosthetic control.

Aside from the feature extraction stage, the choice of classification algorithm has been reported to be another important factor in EMG-PR-based prosthetic control systems. Thus, choosing an efficient classifier becomes necessary to ensure accurate and consistent classification performance. In this regard, a number of classification schemes have been examined with respect to their performance in identifying different upper limb movements. For instance, classification schemes based on Bayesian statistics [22], adaptive neural networks (ANN) [23], fuzzy logic [24], support vector machine [25], and linear discriminant analysis (LDA) [11–26] were previously studied. The ANN-based classifier and the LDA classifier were found to be more accurate for the prediction of limb movement intention. However, the LDA has been widely applied because it is relatively less complex in terms of implementation and performs well in comparison with the ANN and other complex classification Schemes [26].

After reaching a decision on the classification scheme to adopt, a classifier is developed by utilizing part of extracted EMG feature vector. Afterwards, the performance of the built classifier in identifying multiple classes of limb movement intent is evaluated using the testing set as obtained from the remaining part of the feature vector. The performance of the EMG-PR-driven prostheses in terms of its controllability is mostly examined in both off-line and real-time metrics. In general, metrics including recognition rate (classification accuracy) of limb movements/subjects, the time taken to complete a motion (motion completion time), number of successfully completed motion (motion completion rate), and the time taken to select the target motion (motion selection time) are often considered. Also, the performance of the device could be assessed based on the number success recording reaching and grasping tasks.

6.2.3 Simultaneous and Proportional EMG-Based Prosthetic Control

Multiple DOF pattern recognition-based prostheses control schemes have recently been implemented to enable upper limb amputees perform the usual arm function tasks during ADL. Even with this remarkable achievement, the available EMG-PR-based controlled prostheses do not support simultaneous control of multiple DOFs [27, 28]. This limitation prevents upper limb prostheses users from having the natural feel of the coordinated joint control often associated with an intact limb. Although advanced prosthetic arms [29, 30], including multiple DOF wrists, offer the mechanical means to restore arm movements, there is still a significant need for systems that enable simultaneous control of such devices.

To develop a simultaneous prosthetic control scheme which offers multiple DOF movements, different approaches have been proposed in the recent years. These approaches include the use of artificial neural networks for joint kinematics predictions during upper limb movements [31], analysis of the muscle synergies underlying a range of upper limb movements [32], and the utilization of joint kinetics [33]. While these methods have been reported to be promising toward clinical realization of upper limb prostheses, they however mostly focus on providing control for limited classes of limb movements such as wrist without hand or have imposed identical velocities on all active degrees of freedom movements and thus do not provide independent proportional control of individual degrees of freedom.

Additionally, toward achieving simultaneous and proportional control scheme, some investigators developed a model of muscle synergy driven by nonnegative matrix factorization (NMF) method to resolve myoelectric signal factorization issue that is capable of hindering continuous control [32, 34]. The NMF technique often utilizes a DOF-wise training approach, based on which simultaneous control of multi-DOF is achieved via linearly combined single degree of freedom. This kind of strategy supports simultaneous and proportional myoelectric-based control [32, 34]. Despite the promising results of this scheme, there are still some shortcomings that have hindered its real life implementation. The most critical disadvantage of the NMF-driven EMG-based prostheses is that it needs a DOF-wise calibration during which requires a user to activate precisely single DOF in a specific manner, e.g., firstly, trigger DOF1 for several trials followed by DOF2. Besides the inconvenience and the time needed for the execution of this phase, more notably, if the prostheses users are unable to trigger an individual DOF (i.e., the activation patterns during the training session are not exactly of single DOF), the DOF-wise-driven NMF approach will produce inadequate muscle synergy matrix that may result in poor control outcome. This behavior could be attributed to the fact that the classical NMF-based scheme lacks the ability to produce a factorization that is sparse enough for efficient control of multiple DOF [35, 36]. Hence, we proposed an alternative approach to the classical NMF that identifies the factorization which separates the generated basis functions robustly and concurrently in a quasi-unsupervised way to simultaneously and proportionally control the prostheses.

In the proposed approach, the subject does not need to follow a predefined sequence to activate single DOF in the calibration phase, and can even activate more than one DOF simultaneously in the training stage, which is not allowed in the classical NMF method. Nevertheless, the idea of the synergy matrix could be realized in one step, which is contrary to the conventional NMF approach that attempts to sequentially extract the bases for each DOF. For this reason, constraints were included in the factorization method and the method is at maximizing the sparseness of the resulting control outcome. The constraints associated with the sparseness limits the space of possible NMF solutions. Precisely, the solution with bases functions associated to single DOF, which is the target solution, represents the sparsest of the other solutions. In this regard, the factorization with constraints usually does not need prior calibrations/activations of single DOF and could be utilized to any set of target tasks carried out by the user with no any form of strict labeling. The proposed sparse nonnegative matrix factorization (SNMF)-based simultaneous and proportional myo-control scheme is presented mathematically as follows.

In myoelectric control, multiple DOF limb movements could be re-represented as a linearly combined individual DOFs, and the synergy basis and control input are nonnegative [34]. Suppose we represent the root mean square (RMS) values of an N-channel with T-length surface myoelectric signal as $\mathbf{Z}$, in which the t-th column is the myoelectric signal at a given time t and m represents the number of DOFs. Thus, the multichannel observation can be approximated by the product of a $N \times 2m$ latent nonnegative synergy matrix $\mathbf{W}$ and a $2m \times T$ latent nonnegative control signal matrix. $\mathbf{F}$ is expressed in the following equation:

$$\mathbf{Z}_{N\times T} \approx \mathbf{W}_{N\times 2m}\mathbf{F}_{2m\times T} \tag{6.1}$$

While the sparseness of the solution is obtained by imposing a specific single DOF activation sequence for calibration, the above factorization problem can be solved in a semi-supervised way with the summation property of DOFs in the muscle signal domain.

A mathematical way to generate a sparse solution without the need for the specific set of calibration data for individual DOF activation is utilized. Meanwhile, this is achieved by imposing a sparseness constraint on the control signals, which leads to the proposed SNMF Scheme [37].

Mathematically, the extent (degree) of sparseness could be controlled by utilizing either l1-norm or l0-norm. To build a computationally efficient system, we chose the l1-norm-based sparse NMF scheme, SNMF [37]. Meanwhile, the objective function of the SNMF method is expressed mathematically as follows:

$$\min_{\mathbf{W},\mathbf{F}} \frac{1}{2}\|\mathbf{Z}-\mathbf{W}\mathbf{F}\|_{Fro}^2 + \lambda\sum_{t=1}^{T}\|\mathbf{F}(:,t)\|_1^2 \quad s.t.\ \mathbf{W},\mathbf{F} \geq \mathbf{0},$$

$$\Leftrightarrow \min_{\mathbf{W},\mathbf{F}} \frac{1}{2}\left\|\mathbf{Z}-\left[W_1^+, W_1^-, \cdots W_m^+, W_m^-\right]\begin{bmatrix} F_1^+ \\ F_1^- \\ \vdots \\ F_m^+ \\ F_m^- \end{bmatrix}\right\|_{Fro}^2 + \lambda\sum_{t=1}^{T}\left\|\begin{bmatrix} F(:,t)_1^+ \\ F(:,t)_1^- \\ \vdots \\ F(:,t)_m^+ \\ F(:,t)_m^- \end{bmatrix}\right\|_1^2$$
$$s.t.\ W_{ij}, F_{ij} \geq \mathbf{0} \tag{6.2}$$

where **F** (:, *t*) is the *t*-th column vector of **F** [Eq. (6.1)], *Fro* is the Frobenius norm, and $\lambda > 0$ is a regularization parameter to balance the accuracy of the factorization and the degree of sparseness of **F**. In the experiment, we choose the optimalλ through cross-validation with the value of $m = 2$. The superscripts "+" and "–" denote the positive and negative direction of each DOF, respectively. The above formulation can be rewritten as:

$$\min_{\mathbf{W},\mathbf{F}} \left\|\begin{pmatrix} \mathbf{W} \\ \sqrt{\lambda}\mathbf{e}_{1\times 2m} \end{pmatrix}\mathbf{F} - \begin{pmatrix} \mathbf{Z} \\ \mathbf{0}_{1\times T} \end{pmatrix}\right\|_{Fro}^2 \quad s.t.\ \mathbf{W},\mathbf{F} \geq \mathbf{0}, \tag{6.3}$$

where $\mathbf{e}_{1\times 2m}$represent a row vector that contains values equal to 1 and $\mathbf{0}_{1\times T}$represent entries with values equal to 0. Equation (6.3) could be adequately solved via the alternating nonnegative least squares (ANLS) approach. Meanwhile, for the ANLS method, **F** and **W** are iteratively updated by adjusting the other one:

$$\mathbf{F}^{(k+1)} = \underset{\mathbf{F}}{\operatorname{argmin}} \left\|\begin{pmatrix} \mathbf{W}^{(k)} \\ \sqrt{\lambda}\mathbf{e}_{1\times 2m} \end{pmatrix}\mathbf{F} - \begin{pmatrix} \mathbf{Z} \\ \mathbf{0}_{1\times T} \end{pmatrix}\right\|_{Fro}^2 \tag{6.4}$$

$$\mathbf{W}^{(k+1)} = \underset{\mathbf{F}}{\operatorname{argmin}} \left\|\mathbf{W}\mathbf{F}^{(k+1)} - \mathbf{Z}\right\|_{Fro}^2 \tag{6.5}$$

It could be seen that Eqs. (6.4) and (6.5) represent the classical least square problem, and either of the equation has a closed-form solution. Meanwhile, the derivation in [37] reveals that the SNMF method could converge to a stationary point. The SNMF method can equally extract all the basis functions in a single step from the EMG recordings produced via arbitrary combinations of DOFs per user. Only the ordering of the control input signals (basis functions) would be undetermined using this method, but this issue could be simply addressed. In practice, the labels of the first two single DOFs need to be recorded in order to decide the sequence of the basis in the matrix (**F**) from which control signal is obtained, thus making the method quasi-unsupervised instead of completely unsupervised.

Having computed the synergy basis matrix **W** from the recorded myoelectric signals, it is assumed to be a constant (at least over the period of the test experiments). To estimate the control input signal associated with the intended DOF activation level, a pseudo inverse of W (represented as W+) is applied, and it was multiplied with the newly obtained surface myoelectric signals $\left(\widetilde{Z}\right)$. Thus, the estimated control input signal $(\tilde{\mathbf{F}})$ is defined as follows:

$$\tilde{\mathbf{F}} = \mathbf{W}^{+}\tilde{\mathbf{Z}} \tag{6.6}$$

where

$$\tilde{\mathbf{F}} = \left[\tilde{F}_1^{+}; \tilde{F}_1^{-}; \tilde{F}_2^{+}; \tilde{F}_2^{-};\right] \tag{6.7}$$

Toward ensuring that no components are surpassed by the others, the individual component of $\tilde{\mathbf{F}}$ in (6.7) is subject to a normalization process based on its maximum value. An estimate of the control input signal in (6.7) is thus scaled by the scalar correction factorsτ_{ij} that are utilized to account for the indetermination of the signal power (range of control signals) during the factorization process:

$$\begin{cases} \tilde{F}_1 = \tau_{11}\tilde{F}_1^{+} - \tau_{12}\tilde{F}_1^{-} \\ \tilde{F}_2 = \tau_{21}\tilde{F}_2^{+} - \tau_{22}\tilde{F}_2^{-} \end{cases} \tag{6.8}$$

where we name $\tilde{F}_1$ and $\tilde{F}_2$the control signals for DOF1 and DOF2, respectively. The multiplicative factors $\tau_{ij}$are determined for each subject so that the final control signals,$\tilde{\mathbf{F}} = \left[\tilde{F}_1; \tilde{F}_2\right]$ match the range of joint angles in the respective DOFs. The tau-s parameters are manually set before operating online tasks according to different subjects. The obtained control signals, low-pass filtered at 6 Hz (kinematic bandwidth), are then used by the subjects for controlling the DOFs.

It is noteworthy that the proposed SNMF scheme is aimed at robust and concurrent extraction of basis functions for myo-control in upper limb prostheses with a quasi-unsupervised scheme. This factorization scheme shown in our study has merits compared to the classical NMF because it selects the sparsest solution among infinite feasible solutions. To that end, there is no need for pre-calibration of the constraint tasks by users but rather the constraint on the solution. The method has been compared in similar conditions to the NMF and linear regression with results proving it to be superior over these methods even when a single DOF activation was used for the factorization which is an ideal condition for the NMF. Furthermore, the approach could be used by factorizing the myoelectric signals from arbitrary tasks by combining individual DOFs. The proposed factorization algorithm allows robust simultaneous and proportional control and is superior to previous supervised algorithms, and its minimal supervision characteristic paves the way to online adaptation in myoelectric control.

6.3 Neural Interfaces for Improvement of EMG-Based Prosthetic Control

6.3.1 Background

The previously described sections show that EMG-based prostheses control method could potentially allow amputees to intuitively manipulate the prosthetic arm with high level of dexterity and as well achieve multiple DOF controls. Developing EMG-based prostheses with these characteristics mainly depends on whether the users have enough residual muscles which act as source of EMG control signals. For instance, individuals with below the elbow amputation (transradial amputees) generally have enough residual muscles for the control of wrist functions such as wrist flexion/extension, wrist pronation/supination, and so on. Also, these category of amputees could provide enough EMG signals from their residual forearm muscles to control hand and even finger movements as reported in a number of previous studies [7]. On the other hand, individuals with high-level amputations (transhumeral or shoulder disarticulation) usually do not have enough residual muscles to provide myoelectric signals for the successful control of the wrist and hand movements although they can produce enough EMG signals for the control of elbow functions (elbow flexion/extension). Thus, intuitively controlling multiple DOF hand/wrist movements becomes a major challenge for these categories of amputees. One possible means of addressing the issue of insufficient generation of myoelectric control signals for these users would be to consider neural machine interfaces approaches such as targeted muscle reinnervation (TMR) and targeted nerve function replacement (TNFR). The theoretical background is that muscles can be considered as the bio-amplifier of the neural signals.

6.3.2 Targeted Muscle Reinnervation Technology

Toward enabling high-level upper limb amputees producing sufficient myoelectric control signals again, a neural machine interface technology known as TMR was proposed by Todd Kuiken et al. [28]. TMR is a surgical procedure that involves the transfer of residual nerves from an amputated limb unto alternative muscle group usually located around the chest region. During the surgery, brachial plexus nerves that provide motor control and sensory feedback prior to amputation are transferred to the arm or chest muscles that are biomechanically nonfunctional. Afterward, a functional link is established between the implanted nerves and the targeted muscles, thereby inducing the contractility of targeted muscles and enabling the generation of additional myoelectric signals. Importantly, this surgical technique is aimed at restoring the peripheral nerve functions in amputees thus enabling the provision of sufficient myoelectric control signals again. A schematic diagram of the TMR surgical procedure is conceptualized in Fig. 6.2a. For the subject shown in

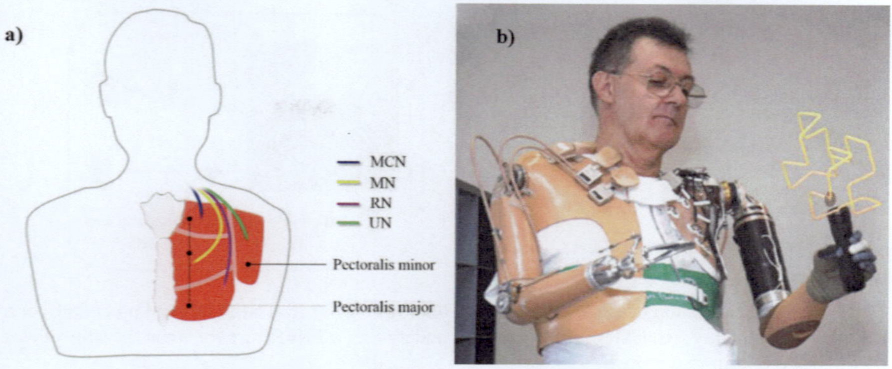

Fig. 6.2 The schematic representation of the TMR technology. (**a**) TMR surgical procedure for shoulder disarticulation amputee. (**b**) A bilateral amputee with successful TMR surgery controlling a prosthetic arm [28]

Fig. 6.2b, the surgery was performed on his right chest region, and it involves transferring the median, ulnar, radial, and musculocutaneous nerves to the subject's pectoralis major muscle segmented into four parts, respectively.

The above described neural machine interface technology for upper limb prostheses control has been implemented and tested on a number of human subjects. The test sessions conducted with a bilateral amputee (Fig. 6.2b) shows that the high-level amputee could efficiently perform additional hand function tasks such as drinking and eating which could not be achieved prior to the TMR surgery. Importantly, the success of performing additional tasks was objectively linked to the realization of more EMG control signals after the TMR surgery.

6.3.3 *Targeted Nerve Function Reconstruction Technology*

Despite the success recorded by the TMR technology, it still has some limitations which have slowed down its widespread adoption. For instance, TMR surgery involves establishing functional links between the nerves and targeted muscles leading to two major challenges: First, the targeted muscle would atrophy with time because it has been denied of the required nutrients provided by the original nerve which has been replaced with the implanted nerve. Second, intramuscular nerve distribution (a key indicator of the motor function recovery) will also degenerate after a while due to the lack of nutrients from the surrounding cell body.

Peripheral nerves which play an important role after TMR surgery have been reported to have a slow growth rate. That is, it takes about 3~6 months or even more after TMR for proper redistribution of the implanted nerve into the targeted muscle, thus leading to atrophy of the targeted muscle. Aside the slow growth rate, the absence of nerve supplements is another factor responsible for atrophy of the

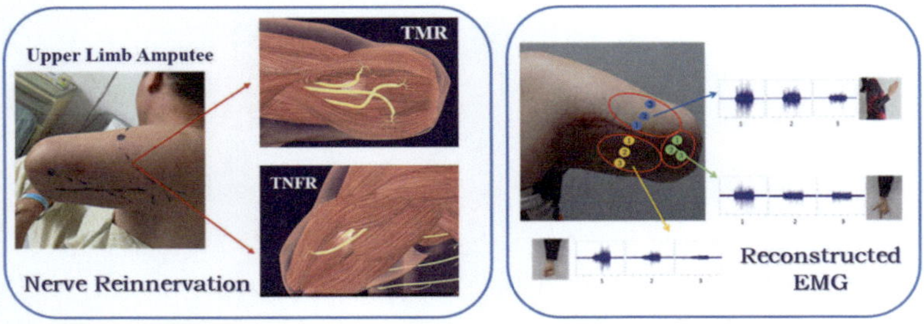

Fig. 6.3 A representation of the TMR/TNFR technology. (**a**) TMR/TNFR surgical procedure for a transhumeral amputee, (**b**) EMG recordings after undergoing a TNFR surgery when the subject was doing different arm movements

targeted muscle [38–40]. In such a situation, high-level upper limb amputees would need training and treatment over an extended period for new functional connections to be established between the targeted muscle and the implanted nerves. It is worth noting that end-to-end anastomosis has been proven to be the best approach for restoring the functions of injured nerves in patients [41]. Thus based on this principle, targeted nerve function reconstruction (TNFR) technology is considered to address the abovementioned problems associated with TMR. This is because TNFR has the potential to better restore the motor functions associated with multiple degrees of freedom upper limb movements. Aiming at improving intramuscular nerve distribution as well as preventing muscle atrophy, TNFR establishes functional links between the nerves of the targeted muscles and nerves from the amputated region/arm of the individual. A schematic representation of the TNFR technology is shown in Fig. 6.3.

Shown in Fig. 6.3 is a male transhumeral amputee that underwent a TNFR surgery. He actually got his left limb amputated in 2012 as a result of injury sustained from a rolling machine. At the time of the surgery (2015), the amputee was approximately 48 years old. Prior to the TNFR surgery, his residual limb was subjected to a number of assessments to ensure that he meets up with the requirement. From the assessments, we found that the residual limb has a length of 22 cm and has not been infected. In addition, the results of an MRI scan show that the musculus biceps brachii and musculus triceps brachii of the residual limb were intact with neuroma tissue formed at median nerve, ulnar nerve, and musculospiral nerve terminals. Afterward, the TNFR surgery was performed on the subject by transecting the MCN, MN, RN, and UN from the residual limb and implanting them into the corresponding MCN, MN, RN, and UN on the amputee's chest region.

About 7 days after the surgery, the operation regions were observed twice a week for 8 consecutive weeks. Indexes, such as status of the incision (healing rate), skin color and temperature, and limb peripheral circumference, were closely monitored. The pre-surgery and post-surgery experiences of the subject were also utilized to assess the limb status. These experiences include the subject's feelings of missing fingers, hand, and arm as well as his experience during observing different movements (finger, wrist, hand, and elbow). Subsequently, a TNFR testing environment was setup to demonstrate the feasibility of our proposed technology. In the testing environment shown in Fig. 6.3b, the amputee was able to control the virtual prostheses reliably across a range of upper limb movements compared to his pre-TNFR experience. It is noteworthy that the TNFR technology is proposed by our research group and we hope it will revolutionize prostheses control technology and other related EMG-based rehabilitation device technologies in the future.

We further built a number of animal models recently to compare the performance of proposed TNFR technology with that of TMR and found the TNFR model obviously demonstrates better intramuscular nerve distribution compared to the TMR model. Additionally, we found out that bicep brachii muscles on the operational sides of the TNFR model achieved better wet mass compared to that of the TMR model.

6.3.4 Restoration of Sensory Feedback

Tactile sensation is an essential element for a dexterous hand manipulation. In an ideal bidirectional hand prostheses, the user's intentions are first decoded through the efferent pathway, and then a nearly "natural" sensory feedback is delivered through the remnant afferent pathways, simultaneously in real time, as illustrated in Fig. 6.4.

The lack of proper sensory feedback mechanism limits prostheses users in grasping and manipulating objects, which makes the use of prostheses cumbersome and uncomfortable, leading to prosthetic abandonment. In fact, lack of sensory feedback schemes in the currently available multiple-DOF upper-limb prostheses have slowed down its clinical and commercial success [42]. Improper feedback of tactile perception may also lead to "over-grasp" causing damage to an object or "under-grasp" causing slip of object. Report from previous survey indicated that about 95% of male amputees desire to have force feedback in their prostheses and 88% placed more emphasis on grip force and proprioceptive information of prosthesis position and movement. With the rejection rate of upper limb prostheses reaching 39% and the increasing rate of people living with the loss of a limb projected to be doubled by the year 2050 [43], the need for integrating efficient sensory feedback mechanism into the prostheses becomes more necessary.

To restore the lost sensory feedback, Raspopovic et al. [44] were able to stimulate the median and ulnar nerve fascicles using transversal multi-channel intrafascicular electrodes (TIMEs) connected to the artificial hand sensors. The subject was able to control the prosthesis through information provided by the artificial sensors. This feedback enabled users to effectively control the grasping force of the prosthesis with no visual or auditory feedback. To restore a sense of touch that is close to the natural perception, the artificial receptors known as sensors need to first register the characteristics of the sensed object/environment [45], which works in a similar way to the receptor organs in humans. These sensory

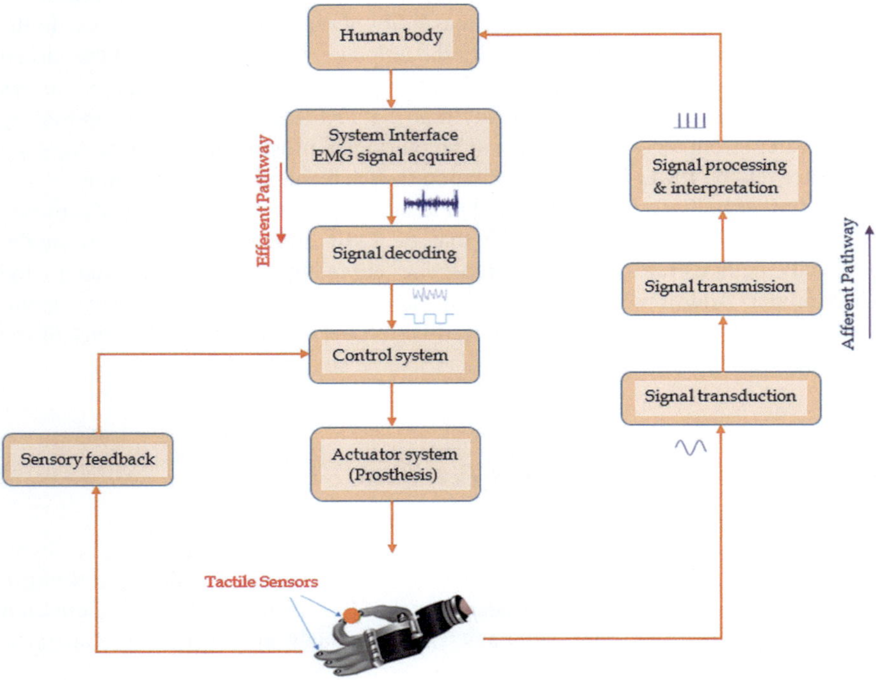

Fig. 6.4 Closed loop architecture for restoring sensory feedback

receptors are broadly classified into pain receptors, cold receptors, warm receptors, and four mechanoreceptors. Then, the receptors encode information into trains of action potentials and also possess characteristic innocuous temperature variation evoked by different afferents between ~5 and 48 °C [46]. The mechanoreceptors are grouped into slow-adapting receptors (SA-I and SA-II) and fast-adapting receptors (FA-I and FA-II) having different receptive fields to which they respond to mechanical stimulus, as well as different rates of adaptation. Also, tactile stimuli are transferred from the artificial receptors to the intact skin through the actuators. At this stage, signals are transported from receptors through nerve fibers to the brain where complex information are interpreted. The time taken for the signal to get to the brain is less than tens of milliseconds [47]. And lastly, a learning process is initiated by the central nervous system to adapt to the new type of afferent signals. A closed loop architecture for restoring a sensory feedback is described in Fig. 6.4. This figure shows the transduction and interfacing system which transmit the electronic signals and the control system which performs signal processing and interpretation.

To provide a sensory feedback mechanism, data are firstly acquired through the use of tactile sensors that come in contact with the target object (Fig. 6.4). Secondly, the analog-to-digital converter digitizes the signal at the signal transduction stage prior to further analysis. Before further signal processing, unwanted signal disturbances called artifact are removed by applying a wavelet transform and empirical mode decomposition strategies to reduce the noise in the signal [48]. At the last stage, the transmitted signal which mimics the real signal (action potential) is then sent to the brain for interpretation.

With the invasive mechanism, action potentials traveling in the form of electrical signals through the nervous systems could possibly regain all somatic sensory information with modality matching by implanting neural electrodes in the peripheral nerves [49]. This approach is able to restore the sense of touch by means of implantable nerve interfaces but always associated with risk of surgical procedure, and their long-term usability needs to be fully investigated [50]. There are three major approaches for invasive sensory feedback which are peripheral nervous system (PNS) stimulation, targeted sensory reinnervation (TSR), and central nervous system (CNS) stimulation. On the other hand, the promising noninvasive mechanism could be realized through techniques such as mechanotactile, vibrotactile, or electrotactile stimulations. To achieve any of these stimulation techniques, the characteristics of the actuator (weight, size, and power consumption) would need to be considered. Importantly, more research focus has been shifted toward the noninvasive sensory feedback mechanism in the recent years with a lot of work currently ongoing in this area.

In summary, integrating an efficient noninvasive sensory feedback mechanism into the currently available multifunctional upper limb prostheses will not only improve the control performance of the prostheses but also give users a feel of their true natural environment during ADL.

6.4 Conclusion

The human upper limb is an important part of the body that enables the accomplishment of a range of tasks in workplace, at home, and in social gathering during ADL. The loss of this part of the body severely limits the capability of amputees while undergoing ADL. In this regard, prosthetic device driven by different control methods have been built to restore the lost limb functions in amputees. At the forefront of these prostheses, control method is the EMG-based control which has attracted considerable attention in both the academia and industry in the recent years. Hence, this chapter focuses on EMG-based prostheses control methods with emphases on pattern recognition-based control as well as simultaneous and proportional control techniques, with details on the fundamental procedure of decoding limb movement intents. Also, a number of techniques for improving the overall performance of EMG-based prosthetic system were discussed with emphases on the advantages and limitations.

Acknowledgments This work was supported in part by the National Natural Science Foundation of China (#U1613222, #81850410557), the Shenzhen Basic Research Grant (#JCYJ20160331185848286), and the Outstanding Youth Innovation Research Fund of Shenzhen Institutes of Advanced Technology, Chinese Academy of Sciences (#Y7G016).

References

1. Holzer LA, Sevelda F, Fraberger G, Bluder O, Kickinger W, Holzer G (2014) Body image and self-esteem in lower-limb amputees. PLoS One 9(3):e92943
2. Geertzen J et al (2015) Dutch evidence-based guidelines for amputation and prosthetics of the lower extremity: amputation surgery and postoperative management Part 1. Prosthetics Orthot Int 39(5):351–360
3. Mckechnie PS, John A (2014) Anxiety and depression following traumatic limb amputation: a systematic review. Injury 45(12):1859–1866
4. Physical and Emotional effects of limb amputation: http://www.seriousinjurylaw.co.uk/other-serious-claims/amputation/effects-of-amputation/Date. Last accessed 26 Sept 2017
5. Đurović A, Ilić D, Brdareski Z, Plavšić A, Đurđević S (2007) Pain, functional status, social function and conditions of habitation in elderly unilaterally lower limb amputees. Vojnosanit Pregl 64(12):837–843
6. Muilenburg AL, LeBlanc MA (1989) Body-powered upper-limb components. In: Comprehensive management of the upper-limb amputee. Springer, New York, NY, pp 28–38
7. Li G (2011) Chapter 6: Electromyography pattern-recognition-based control of powered multifunctional upper-limb prostheses. In: Mizrahi J (ed) Advances in applied electromyography, IntechOpen (IntechOpen Limited), London, pp 99–117. ISBN: 978-953-307-382-8
8. Samuel OW, Li X, Geng Y, Asogbon MG, Fang P, Huang Z, Li G (2017) Resolving the adverse impact of mobility on myoelectric pattern recognition in upper-limb multifunctional prostheses. Comput Biol Med 90:76–87
9. Scheme E, Englehart K (2011) Electromyogram pattern recognition for control of powered upper-limb prostheses: state of the art and challenges for clinical use. J Rehabil Res Dev 48:643–660
10. Smith LH, Hargrove LJ, Lock BA, Kuiken TA (2011) Determining the optimal window length for pattern recognition-based myoelectric control: balancing the competing effects of classification error and controller delay. IEEE Trans Neural Syst Rehabil Eng 19(2):186–192
11. Li G, Schultz AE, Kuiken TA (2010) Quantifying pattern recognition-based myoelectric control of multifunctional transradial prostheses. IEEE Trans Neural Syst Rehabil Eng 18(2):185–192
12. Li G, Li Y, Yu L, Geng Y (2011) Conditioning and sampling issues of EMG signals in motion recognition of multifunctional myoelectric prostheses. Ann Biomed Eng 39(6):1779–1787
13. Samuel OW, Zhou H, Li X, Wang H, Zhang H, Sangaiah AK, Li G (2017) Pattern recognition of electromyography signals based on novel time domain features for amputees' limb motion classification. Comput Electr Eng 2017:1–10
14. Adewuyi AA, Hargrove LJ, Kuiken TA (2016) An analysis of intrinsic and extrinsic hand muscle EMG for improved pattern recognition control. IEEE Trans Neural Syst Rehabil Eng 24 (4):485–494
15. Li X, Samuel OW, Zhang X, Wang H, Fang P, Li G (2017) A motion-classification strategy based on sEMG-EEG signal combination for upper-limb amputees. J Neuroeng Rehabil 14(1):2
16. Geng Y, Samuel OW, Wei Y, Li G (2017) Improving the robustness of real-time myoelectric pattern recognition against arm position changes in Transradial amputees. Biomed Res Int 2017:5090454
17. Finley FR, Wirta RW (1967) Myocoder studies of multiple myopotential response. Arch Phys Med Rehabil 48(11):598–601. [PMID: 6060789]

18. Lawrence P, Herberts P, Kadefors R (1973) Experiences with a multifunctional hand prosthesis controlled by myoelectric patterns. In: Gav rilovic MM, Wilson AB Jr (eds) Advances in external control of human extremities. Etan, Belgrade, pp 47–65
19. Lyman JH, Freedy A, Prior R (1976) Fundamental and applied research related to the design and development of upper-limb externally powered prostheses. Bull Prosthet Res 13:184–195
20. Oskoei MA, Hu H (2007) Myoelectric control systems—a survey. Biomed Signal Process Control 2(4):275–294
21. Englehart K, Hudgins B (2003) A robust, real-time control scheme for multifunction myoelectric control. IEEE Trans Biomed Eng 50(7):848–854
21. Khushaba RN, Al-Timemy A, Kodagoda S, Nazarpour K (2016) Combined influence of forearm orientation and muscular contraction on EMG pattern recognition. Expert Syst Appl 61:154–161
22. Huang YH, Englehart K, Hudgins B, Chan AD (2005) A Gaussian mixture model based classification scheme for myoelectric control of powered upper limb prostheses. IEEE Trans Biomed Eng 52:1801–1811
23. Hudgins B, Parker P, Scott RN (1993) A new strategy for multifunction myoelectric control. IEEE Trans Biomed Eng 40(1):82–94
24. Ajiboye AB, Weir RF (2005) A heuristic fuzzy logic approach to EMG pattern recognition for multifunctional prosthesis control. IEEE Trans Neural Syst Rehabil Eng 13:280–291
25. Oskoei MA, Hu H (2008) Support vector machine-based classification scheme for myoelectric control applied to upper limb. IEEE Trans Biomed Eng 55(8):1956–1965
26. Hargrove L, Englehart K, Hudgins B (2007) A comparison of surface and intramuscular myoelectric signal classification. IEEE Trans Biomed Eng 54:847–853
27. Coapt (2013) Available: http://coaptengineering.com/
28. Kuiken TA, Li G, Lock BA, Lipschutz RD, Miller LA, Stubblefield KA, Englehart KB (2009) Targeted muscle reinnervation for real-time myoelectric control of multifunction artificial arms. JAMA 301:619–628
29. Harris A, Katyal K, Para M, Thomas J (2011) Revolutionizing prosthetics software technology. In: Systems, man, and cybernetics (SMC), 2011 IEEE international conference on, October, IEEE, pp 2877–2884
30. Samuel OW, Asogbon MG, Geng Y, Chen S, Fang P, Lin C, Wang L, Li G (2018) A novel time-domain descriptor for improved prediction of upper limb movement intent in EMG-PR system. In: Engineering in medicine and biology society (EMBC), 40th annual international conference of the IEEE, 2018, July 17–21, 2018, Honolulu, Hawaii, USA
31. Muceli S, Farina D (2012) Simultaneous and proportional estimation of hand kinematics from EMG during mirrored movements at multiple degrees-of-freedom. IEEE Trans Neural Syst Rehabil Eng 20:371–378
32. Jiang N, Englehart KB, Parker PA (2009) Extracting simultaneous and proportional neural control information for multiple-DOF prostheses from the surface Electromyographic signal. IEEE Trans Biomed Eng 56:1070–1080
33. Samuel OW et al (2016) Examining the effect of subjects' mobility on upper-limb motion identification based on EMG-pattern recognition. In: Intelligent robot systems (ACIRS), Asia-Pacific conference, IEEE, 137–14
34. Jiang N, Rehbaum H, Vujaklija I, Graimann B, Farina D (2014) Intuitive, online, simultaneous, and proportional myoelectric control over two degrees-of-freedom in upper limb amputees. IEEE Trans Neural Syst Rehabil Eng 22(3):501–510
35. Lee DD, Seung HS (2001) Algorithms for non-negative matrix factorization. In: Advances in neural information processing systems. MIT Press, Cambridge, MA, pp 556–562
36. Lee DD, Seung HS (1999) Learning the parts of objects by non-negative matrix factorization. Nature 401(6755):788–791
37. Kim H, Park H (2007) Sparse non-negative matrix factorizations via alternating non-negativity-constrained least squares for microarray data analysis. Bioinformatics 23(12):1495–1502
38. Kuiken TA, Childress DS, Rymer WZ (1995) The hyper-reinnervation of rat skeletal muscle. Brain Res 676:113–123

39. Williams HB (1996) The value of continuous electrical muscle stimulation using a completely implantable system in the preservation of muscle function following motor nerve injury and repair: an experimental study. Microsurgery 17:589–596
40. Nicolaidis SC, Williams HB (2001) Muscle preservation using an implantable electrical system after nerve injury and repair. Microsurgery 21:241–247
41. Grinsell D, Keating CP (2014) Peripheral nerve reconstruction after injury: a review of clinical and experimental therapies. J Biomed Res Int 2014:698256
42. Samuel OW, Asogbon MG, Geng Y, Al-Timemy AH, Pirbhulal S, Ji N et al (2019) Intelligent EMG pattern recognition control method for upper-limb multifunctional prostheses: advances, current challenges, and future prospects. IEEE Access 7:10150–10165
43. Ziegler-Graham K, MacKenzie J, Ephraim P, Travison T, Brookmeyer R (2008) Estimating the prevalence of limb loss in the United States: 2005 to 2050. Arch Phys Med Rehabil 89(3):422–429
44. Raspopovic S, Capogrosso M, Petrini FM, Bonizzato M, Rigosa J, DiPino G, Carpaneto J, Controzzi M, Boretius T, Fernandez E, Granata G, Oddo CM, Citi L, Ciancio AL, Cipriani C, Carrozza MC, Jensen W, Guglielmelli E, Stieglitz T, Rossini PM, Micera S (2014) Restoring natural sensory feedback in real-time bidirectional hand prostheses. Sci Transl Med 6:222ra19
45. Christian A, Anders B, Sven-Olof F, Fredik S, Goran L, Birgitta R (2012) Sensory feedback from a prosthetic hand based on air-mediated pressure from the hand to the forearm skin. J Rehabil Med 44:702–707
46. Chortos A, Liu J, Bao Z (2016) Pursuing prosthetic electronic skin. Nat Mater 15(9):937–950
47. Zou L, Ge C, Wang ZJ, Cretu E, Li X (2017) Novel tactile sensor technology and smart tactile sensing systems: a review. Sensors 17(11):2653
48. Saal HP, Bensmaia SJ (2015) Biomimetic approaches to bionic touch through a peripheral nerve interface. Neuropsychologia 79:344–353
49. D'anna E, Petrini FM, Artoni F, Popovic I, Simanić I, Raspopovic S, Micera S (2017) A somatotopic bidirectional hand prosthesis with transcutaneous electrical nerve stimulation based sensory feedback. Sci Rep 7(1):10930
50. Micera S (2016) Staying in touch: toward the restoration of sensory feedback in hand prostheses using peripheral neural stimulation. IEEE Pulse 7(3):16–19

Chapter 7
Neural Interface: Frontiers and Applications

Cochlear Implants

Xiaoan Sun, Sui Huang, and Ningyuan Wang

Abstract The theory and implementation of modern cochlear implant are presented in this chapter. Major signal processing strategies of cochlear implants are discussed in detail. Hardware implementation including wireless signal transmission circuit, integrated circuit design of implant circuit, and neural response measurement circuit are provided in the latter part of the chapter. Finally, new technologies that are likely to improve the performance of current cochlear implants are introduced.

Keywords Cochlear implant · Signal processing strategy · Neural response measurement

7.1 Introduction of Cochlear Implant

A cochlear implant is a surgically implanted electronic device that provides functional hearing to a person with severe to profound sensorineural hearing loss. It bypasses a non-functional inner ear and stimulates the auditory nerve directly with patterns of electrical currents derived from incoming sounds. A cochlear implant system usually consists of an external sound processor and an implanted current stimulator. Cochlear implant is widely accepted as the only effective way to restore functional hearing for deaf people in daily clinical practice and is the most successful neural prosthesis in current world. Up to date, about 450,000 people worldwide have received cochlear implants; half of them are pediatric users who were able to develop near normal language skills and merged seamlessly into mainstream schools and society. As the technology of cochlear implant improves, and the risk associated with the device and the surgery decrease over the years, more and more people

---

X. Sun (✉) · S. Huang · N. Wang
Nurotron Biotechnology Inc., Irvine, CA, USA
e-mail: xsun@nurotron.com

X. Zheng (ed.), *Neural Interface: Frontiers and Applications*,
Advances in Experimental Medicine and Biology 1101,
https://doi.org/10.1007/978-981-13-2050-7_7

benefit from the device each year. The clinical threshold to be considered as a cochlear implant candidate has been dropped from 100 dB hearing loss in 1980s and 1990s to 85 dB hearing loss in current years. Therefore, the boundary between hearing aids user and cochlear implant candidate becomes obscure and interleaved.

According to census by World Health Organization in 2010, there are about 360 million hearing-impaired people worldwide, representing the second largest group of handicapped people. This number grows annually as a result of 0.1–0.2% rate of congenital deaf among newborns. In addition, consequential deaf from disease, medicine intoxication, trauma, environmental noise, and genetic disorders also contribute to the increase of deaf population. Traditionally, large amount of resource is required to set up rehabilitation facilities for deaf people, yet the effectiveness of these rehabilitations is very limited. They still live within the deaf society and find it hard to communicate with the outside world. The appearance of cochlear implants changes the situation fundamentally. Most modern cochlear implant recipients can carry a normal conversation in a quite environment, although they still find it challenging to communicate in a noisy environment.

The development of modern cochlear implant is a long and interesting journey, including joint efforts from physiologists, physicians, and engineers. In 1957, French physician Djourno and his colleagues induced successful hearing with electrical stimulation in two totally deaf patients. The news spurred an intensive level of research activities worldwide to restore hearing to deaf people. In 1961, William House in Los Angeles, California, implanted a single-channel cochlear implant in the scalar tympani of two deaf people. Both subjects report useful hearing from electrical stimulation although the initial devices only lasted 2 weeks. In 1978, Graeme Clark in Australia developed a multi-channel cochlear implant and implanted in two deaf people. Although single-channel cochlear implant won the first round as the first FDA-approved device for clinical use, multi-channel device is the standard for modern cochlear implants because of its superior performance.

A cochlear implant usually includes an internal implant device and an external sound processor, as shown in Fig. 7.1. The external sound processor (1) powered by battery pack (2) picks up sound through microphone and processes and encodes sound signal, which is then transmitted wirelessly by transmission coil (3) to the receiver coil (4) of internal device. The current stimulator (5) of internal device converts the received signals into electrical impulses and sends the impulses through electrode body (6) to an electrode array (7). The electrical impulses from the electrode array (7) stimulate the auditory nerves (8), allowing the brain to perceive sound.

Now there are four major cochlear implant providers in the world: Cochlear Corporation in Australia (www.cochlear.com), Med-El Corporation in Austria (www.medel.com), Advanced Bionics Corporation in the United States (www.advancedbionics.com), and Nurotron Biotechnology Inc. in China (www.nurotron.com). There are a couple of smaller cochlear implant companies with no significant market shares. They all provide multi-channel devices with 12–24 electrodes, while Nurotron device has the most intra-cochlear electrodes of 24. Functions like virtual channel, electrode impedance measurement, and neural response telemetry also become standard for modern cochlear implant devices. The recognition rate of short sentences in quiet is around 80% for the products of all four major manufactures.

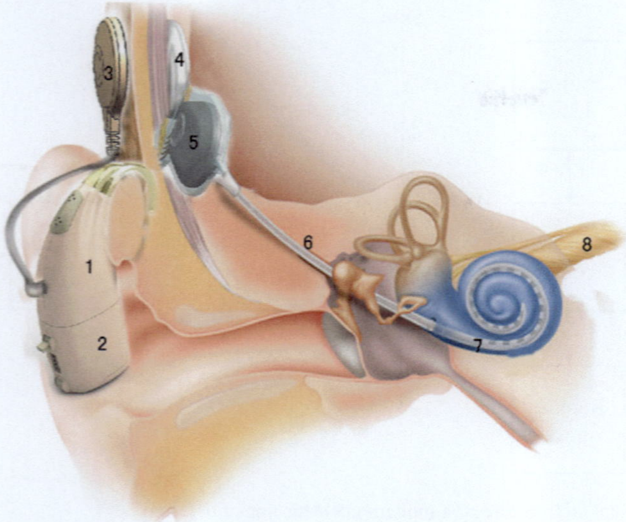

Fig. 7.1 External device of cochlear implant system includes (1) sound processor, (2) battery pack, (3) transmission coil. Internal device of cochlear implant system includes (4) receiver coil, (5) current stimulator, (6) body of electrode, (7) electrode array, which stimulator auditory nerves (8) with current pulses

7.2 System Specification

The ultimate goal of cochlear implant is to restore effective hearing for severe to profound hearing-impaired people through safe electrical stimulation of auditory nerve. To this end, the design and manufacture of a cochlear implant system should satisfy requirements in three aspects. First, the material of the implant device should be safe for long-term implantation. Second, the circuit of the implant should be able to deliver safe, reliable, and accurate current stimulation. And last, the structure of the implant should consider surgical requirement and provide enough protection for electronic components. The architecture of a modern cochlear implant system is shown in Fig. 7.2, consisting of a speech processor, a RF unit, an internal unit, an electrode array, a remote controller, and a fitting unit [76].

7.2.1 Speech Processor

The speech processor shown in Fig. 7.3 is the core of the external part of the cochlear implant system. The entire speech processor is powered by a battery block. The clock block provides clock signal to all blocks except the power amplifier and the reverse telemetry decoder. The audio signal from natural environment or other

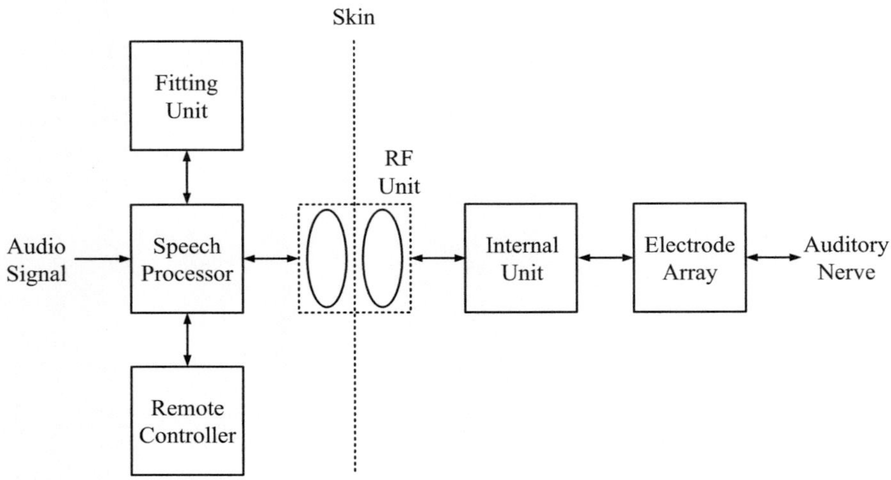

Fig. 7.2 Typical architecture of a modern cochlear implant system

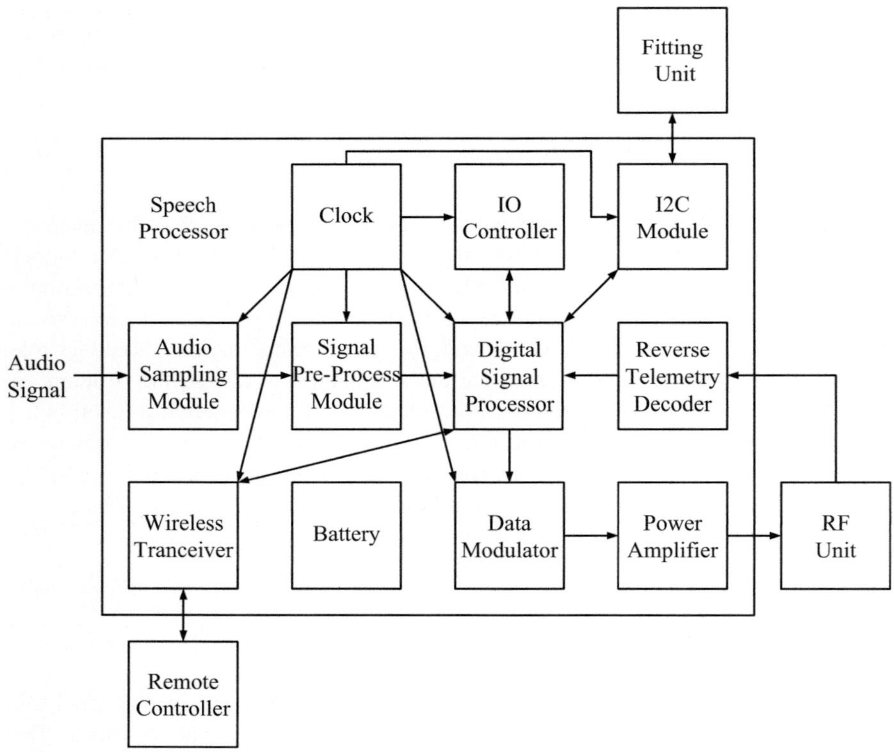

Fig. 7.3 Block diagram of speech processor

electrical devices (TV, radio, tele-coil, etc.) is first sampled, digitized, and synchronized in audio sampling module. Processed by signal pre-processing module and digital signal processor (DSP) based on certain encoding strategies, the output of audio sampling module is converted to data that contains stimulation information. Then the data are modulated at data modulation block by radio frequency (RF) carrier signal and sent to power amplifier (PA). The amplified signal is eventually sent to RF unit. The reverse telemetry decoder detects and modulates the signal from the RF unit when the system is in the status of reverse telemetry. The decoded signal goes into the DSP for further operation. The I2C module and wireless transceiver, which are both controlled by the DSP, are used to communicate with the fitting unit and remote controller, respectively. The IO controller controls indication lights and buttons on the speech processor.

7.2.2 RF Unit

The RF unit is composed of a transmitter coil and a receiver coil connected with their resonant circuits, which are on the outside and inside of human body, respectively. Through this inductive link, the stimulating information as well as power is transmitted into the internal unit to control the unit to stimulate the auditory nerve. More interestingly, using the reverse telemetry technique, the data from the implant can be sent back to the speech processor with the same link.

7.2.3 Remote Controller

The remote controller is used to execute some regular operation, for example, turning on and off the system, adjusting the sound volume, changing fitting maps, and so on. In some applications, it could also be used to check the status of the speech processor and the internal unit. Most of the recent cochlear implant products favor this wireless control unit because it is much more convenient for users or guardians of users to manipulate the device instead of pushing buttons on the speech processors.

7.2.4 Fitting Unit

The fitting unit, which is not worn by users, is only used by clinicians for clinical uses, e.g., impedance test, neural response measurement, fitting program, etc. The fitting unit can bi-directionally communicate with the speech processor. On one hand, when clinicians want to optimize performance of devices, different stimulating information can be sent to speech processors from fitting programs. On the other hand, when they need to detect the condition of implants, the internal data from implants can also be sent back to the unit.

7.2.5 Internal Unit

The block diagram of the internal unit is shown in Fig. 7.4. The RF signal from the RF unit is divided into two paths. One goes into the power generator to provide a high voltage for the stimulating circuit and power supply for all other internal circuits. The other path goes to RF signal demodulator. If the RF signal is a command signal, the demodulator sends the signal to the command decoder, which stores the decoded information in the setting register. If the RF signal is a data signal, the data decoder is activated, and the stimulating data is sent to the data distributor. The output of the demodulator is also the source of the clock generator, which is basically a clock recovery circuit to provide a stable clock signal for other blocks. The stimulating circuit is controlled by data distributor that decides current amplitude and electrode address and by the timing controller that controls the stimulating time. The stimulating current finally flows from the stimulating circuit to the electrode array. The recording circuit is used to record the voltage signal in the electrode, which is converted to the digital signal and transmitted back to RF unit through the reverse telemetry modulator.

7.2.6 Electrode Array

The electrode array is a major part of the implant, which is inserted into the cochlea. The stimulating current pulse goes out from the stimulating electrode to activate the neurons and flows back to the return electrode in the end. To avoid the DC voltage shift accumulation, normally the current pulse has two phases with positive and negative current as shown in Fig. 7.5.

7.3 Signal Processing Strategies

7.3.1 Introduction

Speech processor can be described as the brain of a cochlear implant. Acoustic signals are processed and encoded into specific parameters of electric pulses using particular signal processing methods, termed as "signal coding strategy".

In the CI system, the signal coding strategy plays an extremely important role in generating auditory perception of CI recipients (e.g., [72, 74]), which determines

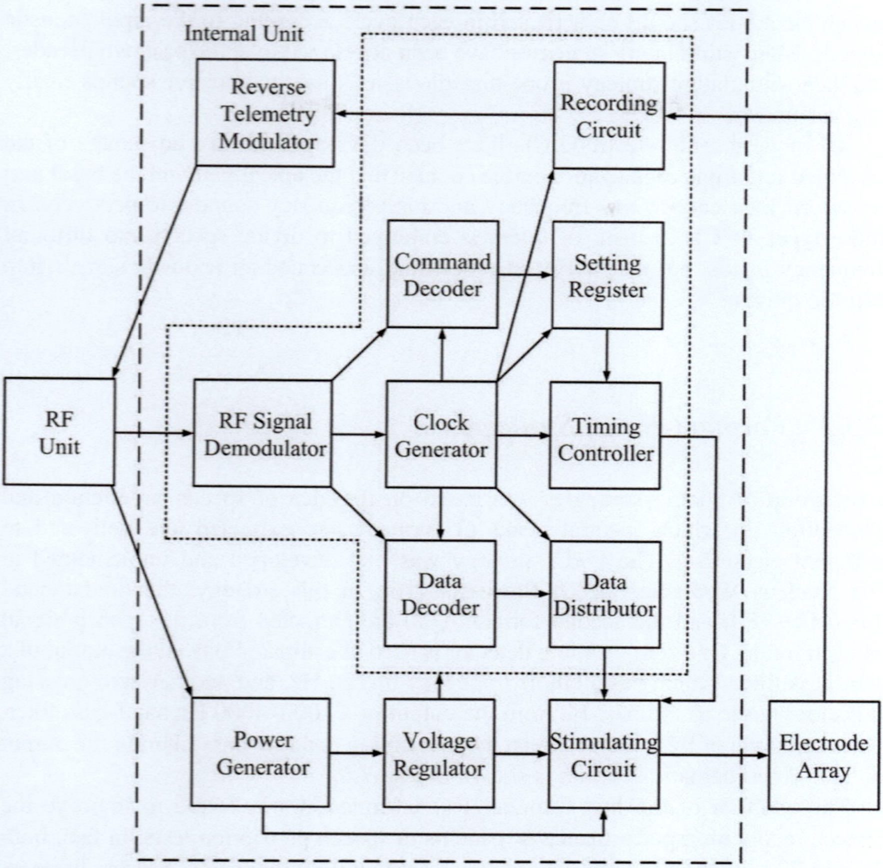

Fig. 7.4 Block diagram of internal unit

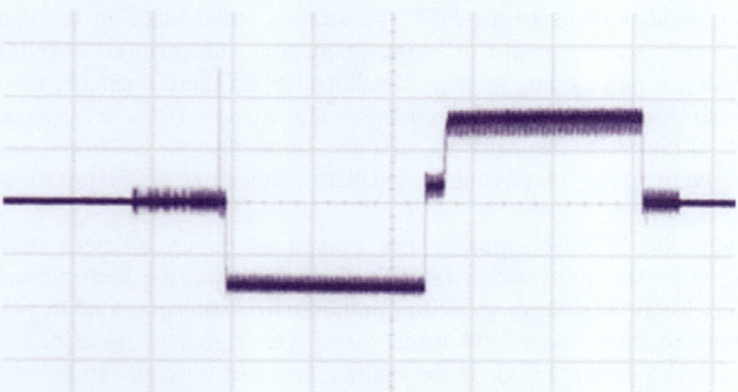

Fig. 7.5 A typical biphasic stimulating pulse using the return electrode as reference

which electrodes should be activated in each cycle according to the input acoustic signals. Many stimulating strategies have been developed over the past two decades. An ideal stimulating strategy is one that allows a CI user to perceive sounds clearly and accurately.

All modern multi-electrode CIs have been developed to take advantage of the so-called tonotopic organization in the cochlea that the apical part and the basal part of the cochlea encode low-frequency and high-frequency sounds, respectively. In these types of CIs, a bank of filters is employed to divide speech into different frequency bands, but they differ in processing and encoding acoustic signals into electric pulses.

7.3.2 Formant-Based Strategies

One group of coding strategies was based on the idea of speech production and perception, in which spectral peaks or formants are extracted and delivered to different electrodes. The F0/F2 strategy was first developed and implemented in the Nucleus Wearable Speech Processor [16]. In this strategy, the fundamental frequency (F0) and the second formant (F2) are extracted from the speech signal in each frame. One zero-crossing detector is used to estimate F0 from the output of a low-pass filter, which has a cutoff frequency of 270 Hz, and another zero-crossing detector is used to estimate F2 from the output of a 1000–4000 Hz band-pass filter. The amplitude of F2 is then estimated by rectifying and low-pass filtering the output signal. The schematic diagram is shown in Fig. 7.6.

The addition of the first formant (F1) information was found to improve the speech recognition performance of patients in speech perception tests. In fact, both F1 and F2 were reported important in vowel recognition for normal-hearing listeners [24]. Dowell et al. [24] found that the average scores on word recognition increased from 30% to 63% after switching from F0/F2 to F0/F1/F2 strategy.

Further improvements to the F0/F1/F2 strategy were made in the late 1980s, including the involvement of high-frequency information. MULTIPEAK (or MPEAK), a new coding strategy developed upon these changes, was used to extract both formants and high-frequency information from the speech signal (as shown in Fig. 7.7). The MPEAK strategy stimulates four electrodes at a rate of F0 pulses/sec (pps) for voiced sounds, and at an average rate of 250 pps for unvoiced sounds [57].

Although the MPEAK strategy has proved to be an efficient strategy for extracting important information from the speech signal, the performance is still limited. The MPEAK strategy and F0/F2 and F0/F1/F2 strategies tend to make errors in formant extraction, especially when the speech signal is embedded in noise. However, all the attempts made in the improvements of formant extraction methods motivated the development of the next generation coding strategies for CIs.

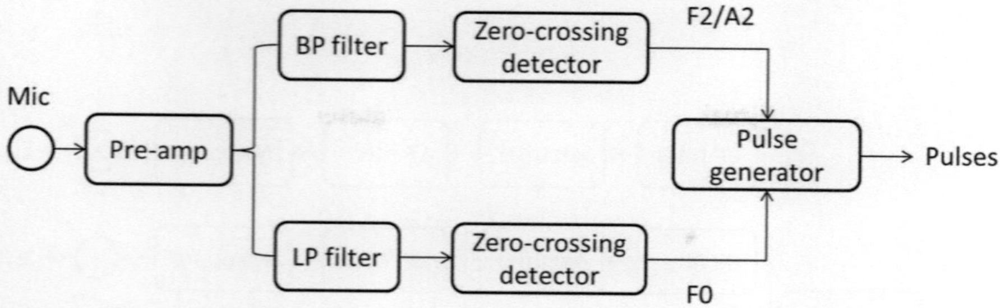

Fig. 7.6 Block diagram of F0/F2 strategy

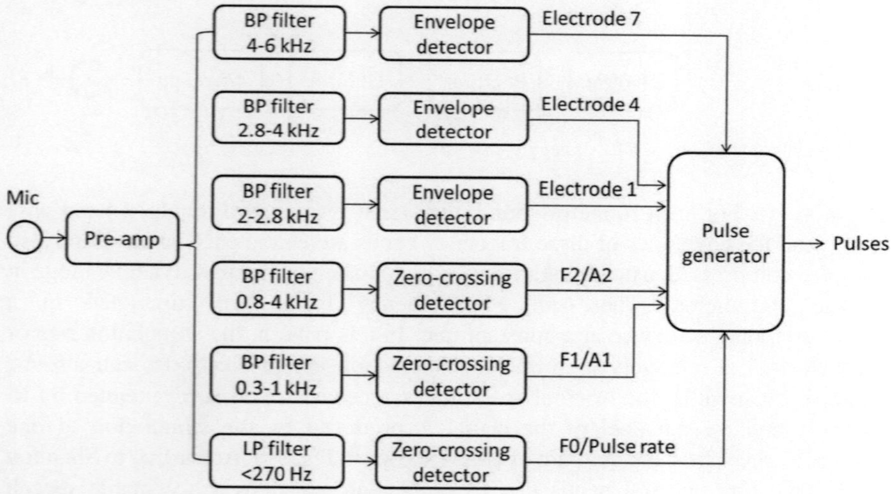

Fig. 7.7 Block diagram of MPEAK strategy

7.3.3 Temporal Envelope-Based Strategies

Another thought of spectrum regeneration with electric stimulation was based on the implicit encoding of temporal envelope cues and could be traced back to earlier work in speech vocoder [25]. Most of algorithms implemented in current commercial devices origin from this idea.

7.3.3.1 Continuous Interleaved Sampling (CIS) Strategy

For modern multi-channel cochlear implants, which have M (e.g., 12–24) implanted electrodes, CIS is a strategy used in the speech processors of all major cochlear implant manufacturers [72]. CIS is a multi-band coding strategy, in which the temporal envelope of signal in each band is extracted at a cutoff frequency of

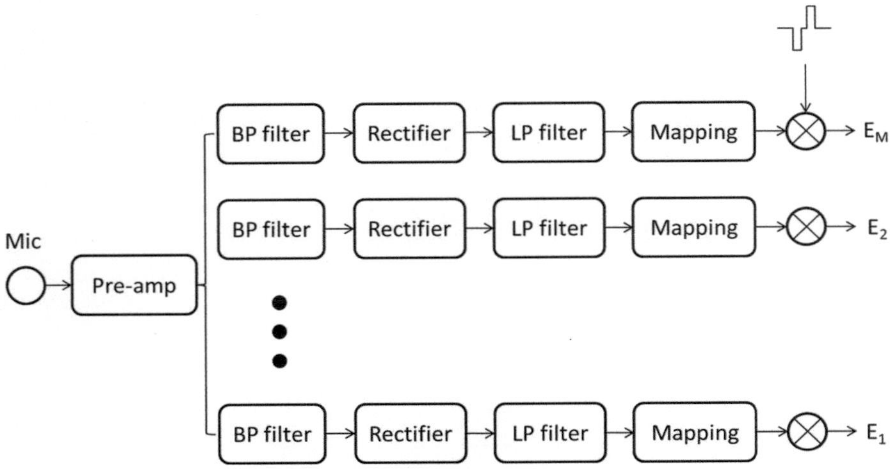

Fig. 7.8 Block diagram of CIS strategy

160–400 Hz. For each frame of sound, the signal is applied through M band-pass filters, and the envelopes of these frequency bands are extracted. The envelope also has to be compressed, usually logarithmically, to match the narrow dynamic range in electric stimulation. Then, all M frequency bands are stimulated in a non-overlapping sequence at a constant rate. In this pattern, the stimulation rate of each channel is relatively high and the interaction across electrodes can also be reduced. Meanwhile, the original spectrum of received sound is regenerated by M channels and one channel of the signal is produced by the stimulation of one electrode. The schematic diagram of CIS is shown in Fig. 7.8. According to Shannon et al. [56], four spectral bands in CIS processing can deliver acceptable speech intelligibility in quiet.

7.3.3.2 n-of-m Strategies

A group of coding strategies based on channel selection or "spectral peak picking" scheme is represented by "n-of-m strategy" (Fig. 7.9), which is designed in part to reduce the density of stimulation while still reserving the most important aspects of speech signals [38]. In each cycle, only N channels with largest amplitudes are selected from all M channels and stimulated. The deletion of stimulation from low-energy channels for each frame of sound may reduce the interference across electrodes and restrict the area of stimulated region in the cochlea. Taking the fact into consideration that low-level channels usually do not contain significant information, such "unmasking" may improve the perception of sounds by patients. In addition, selection of the channels containing higher energy in each frame may enhance the signal-to-noise ratio in signal domain situations.

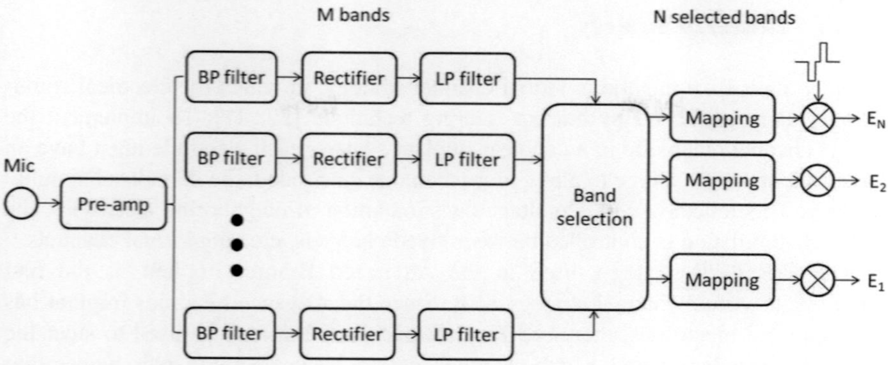

Fig. 7.9 Schematic diagram of n-of-m strategy

The ACE and SPEAK strategy (differ in stimulation rate), used in the Nucleus implant, consists of 22 fixed channels. The input signal is divided into 22 frequency bands for each frame just as CIS strategy. After the envelope information for each frequency band is extracted by rectifying and low-pass filtering, 6–10 frequency bands with the largest amplitudes are stimulated. The number of stimulated channels could be set by audiologists. Electrodes corresponding to the selected channels are then activated, and the original spectrum is reproduced by 6–10 fixed channels [49].

Advanced peak selection (APS) is a coding strategy employed by the Nurotron device [76]. The APS strategy can also be defined as an "n-of-m" strategy, with the number of frequency bands being 24. The number of channels being stimulated in each frame is typically set to 6–8, dependent on the number of available electrodes or the feedback from patients. In the APS strategy, stimulation rate can be adjusted to improve the representation of temporal fine structure. The typical stimulating rate of each selected electrode is normally set to 680 pps or 890 pps.

7.3.4 Coding Strategies with Advanced Cues

As described above, both n-of-m and CIS strategies only employ a certain number of channels to reproduce the original sound spectrum, which means only 12–24 electrodes can currently be implanted into a CI user's ear to generate 12–24 fixed channels. However, the population of auditory nerve fibers in a human cochlear is around 30,000. Due to the limitation of the electrode number and design, the spectral resolution of a CI user is normally worse than that of a normal-hearing listener, which severely weakens their pitch and tone perception. Therefore, a school of new coding strategies designed to improve pitch or tone perception of CI users has been proposed in recent years.

7.3.4.1 HiRes120 Strategy

Another alternative method is virtual channel strategy, in which the electrical stimulation pattern is shaped by current steering technique [22, 41]. To implement the virtual channel technique in a cochlear implant system, each electrode must have an independent power source to allow more than one electrode to be stimulated simultaneously. Theoretically, with simultaneous stimulation of neighboring electrodes, the locus of stimulation is controlled between fixed channels, creating virtual channels.

The HiRes120 strategy, used in the Advanced Bionics implant, is the first commercial virtual channel strategy [40]. Since the Advanced Bionics implant has 16 implanted electrodes, there are 15 electrode pairs that can be used to steer the electrical stimulation and 8 sub-channels within each electrode pair, generating 120 channels in total. Figure 7.10 is a diagrammatic representation of the HiRes120 strategy. Similar to CIS strategy, in each frame of sound, the signal is divided into 15 frequency bands and the envelope of each band is extracted. Next, 15 spectral peaks are also derived, which indicate the most important frequency component within each channel. These spectral peaks are then steered by corresponding electrode pairs based on the virtual channel technique. However, according to results from clinical tests, HiRes120 strategy did not achieve expectation of researchers.

7.3.4.2 C-Tone Strategy

CI signal processing significantly affects lexical tone recognition due to the restricted spectral resolution of CI users, which is to the disadvantage of overall speech understanding of tonal languages (e.g., [30]). One way to improve tone recognition rate is to deliver information other than temporal envelope. Since F0 has been found to carry most of the lexically meaningful information for tonal languages, Luo and Fu [47] evaluated a novel method to improve Chinese tone recognition in which amplitude contour was modified to better resemble the F0 contour. In this study, results showed significantly better tone recognition with the amplitude contour modification. An amplitude contour manipulation algorithm (C-tone) that originate from this idea was implemented in Nurotron implant system to challenge the difficulties in lexical tone recognition of CI users [51].

C-tone was implemented as a tone-enhanced version of APS strategy as shown in Fig. 7.11. A voice activity detector (VAD) is applied at the frontend of the algorithm process. In each frame, the first step of C-tone is to estimate the F0 with the average magnitude difference function (AMDF) algorithm [54]. Second, the input signal is passed through a similar process as APS strategy. Finally, the amplitude envelope of each selected stimulation channel is adjusted based on the relative change of F0 comparing to that estimated from the previous frame. An increment in F0 results in an enhancement in the amplitude envelope and vice versa. The preliminary data suggest that real-time enhancement of the amplitude envelope contour in C-tone may benefit Chinese CI users' tone recognition.

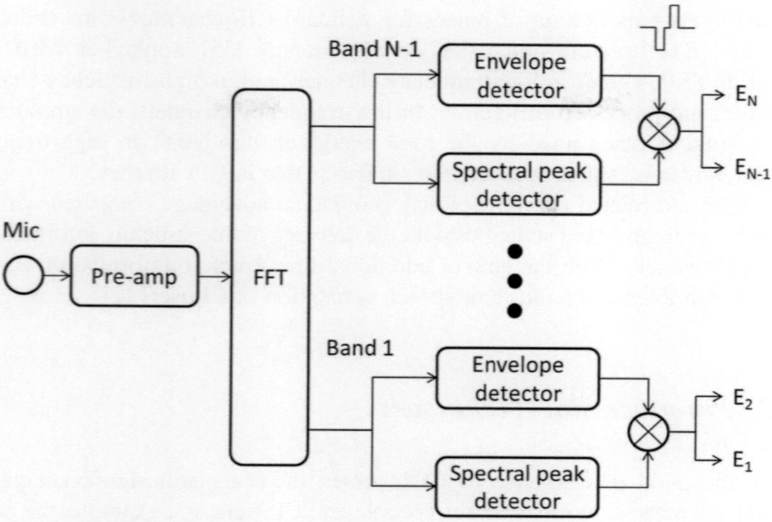

Fig. 7.10 Block diagram of HiRes120 strategy

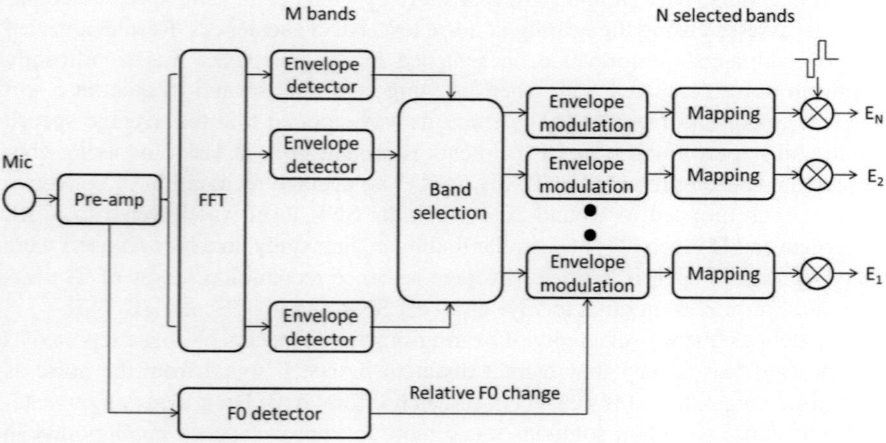

Fig. 7.11 Block diagram of C-tone

7.3.4.3 Fine Structure Processing (FSP) Strategy

New processing options have also been introduced to increase the transmission of fine structure information compared with the CIS and other n-of-m strategies in current use. In one approach, the timing of positive zero crossings in the output of the four band-pass filters with the lowest center frequency dominates the presentation of a short group of pulses for the corresponding channels of stimulation as opposed to the continuous presentation of pulses for standard CIS channels. This strategy is called the "fine structure processing" (FSP) strategy [55], applied in MED-EL's implant. In FSP, signals in low-frequency channels and in high-frequency channels are treated and processed differently. In low-frequency channels, the amplitude of output signal is determined by the total energy in this band. In high-frequency channels, however, signal is processed similar as that in CIS strategy.

The FSP and related approaches may provide an advantage compared with CIS and other envelope-based strategies with the delivery of fine structure information in the lower channels. With the help of additional fine structure information, FSP has been reported to enhance music or speech perception of CI users [2].

7.3.5 Pre-processing Techniques

Despite the good performance of CI in quiet, there are still significant gaps in performance between normal-hearing people and CI users. For example, the performance of CI users in speech perception is extremely poor with additive noise. At least a 15-dB loss in functional signal-to-noise ratio (SNR) could be produced in a steady-state noise background [75]. In a study by Firszt et al. [28], speech recognition was assessed using the hearing in noise test (HINT) sentences. Results revealed that CI recipients' performance on sentence recognition tasks was significantly poorer in noise compared with when listening at speech stimuli in quiet at a soft level. In Spahr and Dorman [61]'s study, it was reported that the average speech intelligibility performance of CI recipients is around 40% in tasks involving conversational speech presented at 10 dB SNR. The average recognition of conversational speech dropped to around 20% when the SNR level was lowered to 5 dB. Fetterman and Domico [27] had similar finding in their study, in which CI users were asked to identify speech in noise. Average sentence recognition scores of CI users decreased from 82% in quiet to 73% at 10 dB SNR and to 47% at 5 dB SNR.

The poor frequency selectivity of hearing-impaired listeners has been reported as a significant reason why they cannot distinguish speech signal from the noise as normal-hearing listeners [64]. Recent research efforts have been focusing on state-of-the-art noise reduction solutions much more to improve speech intelligibility in noisy environments. Since CIs deliver electrical pulses to stimulate auditory nerves to help restore hearing sensation, multiple signal processing algorithms have been applied to convert acoustic signals into electrical stimuli (e.g., [46, 74]). As we mentioned above, the majority of CI users can score high in open-set speech recognition tasks regardless of the device or speech coding strategy used (e.g., [58, 61]). However, few of them could overcome the problems of additive noise.

Many of current noise reduction algorithms can improve output SNR, but few of them can really enhance the speech intelligibility (e.g., [44]). The generated speech distortion has been considered as one of the reasons. A traditional method to reduce the effect of noise is to apply a gain calculated from estimated SNR level to suppress the noise. However, since the power of noise cannot be accurately estimated, the speech signal can be either amplified or attenuated due to the underestimation or overestimation to the noise power, respectively. Considerable speech distortion is then introduced [39], resulting in no or even negative benefits in speech intelligibility [33].

In general, noise reduction algorithms designed for CIs could be divided into two classes: single-microphone and multi-microphone methods. Single-microphone approaches rely mostly on statistical models of speech and noise, and therefore can only remove noise with different temporal and spectral features as speech signals. In recent years, there has been a growing tendency toward the use of noise reduction methods with multiple microphones in CI devices (e.g., [42, 73]). Large improvements in SNR and therefore considerable benefits in speech intelligibility can be obtained, only if the target speech and the noise sources are located at different locations in space. In the following session, both single-channel and multi-channel noise reduction techniques are reviewed.

7.3.5.1 Single-Channel Noise Reduction Methods

One of the first proposed single-channel noise reduction algorithms is spectral subtraction, based on a maximum likelihood (ML) estimator, and has been implemented in numerous applications. Its efficiency and low computational complexity have resulted in its widespread use. The general idea of spectral subtraction is to estimate the spectrum of noise during gaps in the speech, and then to remove it from the noisy signal [66]. The performance of this class of algorithms depends critically on the accuracy of noise estimation. A conventional method to estimate the noise is to update the noise power cumulatively during speech gaps and hold it unchanged during speech frames. To determine when speech is present or absent, a voice activity detector (VAD) is required [4].

Despite its attractive simplicity, spectral subtraction has a number of shortcomings, which limit its use in applications such as hearing aids and CIs. One commonly audible artifact produced by the technique is termed "musical noise" and is generated primarily by inaccurate estimation of the noise spectrum. Since only the average power of the noise can be estimated over time, short-term, potentially important details, such as momentary spectral peaks and valleys, are ignored by the algorithm. As a result, after the estimated average noise is removed, those residual components can produce annoying tonal sounds, which can worsen, rather than improve, speech intelligibility and perceived quality.

Over the past three decades, much effort has gone into developing methods that remove or reduce musical noise [17, 52]. A common technique is to set up a noise floor, that when the signal level is under a certain threshold, it will be left unprocessed [4]. A factor could be used to control the strength of this technique, that how much the signal should be removed when its level is under the threshold. Moreover, it has been noticed that the influence of noise on speech signal may not be unified; it is rational to apply different control factors at different frequencies.

Wiener filtering is another well-studied technique in speech enhancement, based on optimal minimum mean square error (MMSE) estimator of each speech spectral component [13, 45]. Martin [48] developed an algorithm based on the combination of Wiener filtering and spectral subtraction to overcome the limitation of VAD, so that in a speech frame, noise information could still be updated by a Wiener filter. Extended Wiener filters have been proposed to further enhance the performance of noise reduction algorithms. Multi-channel Wiener filtering, for example, has been tested [68, 69]. In addition, the speech distortion weighted Wiener filter was developed to reduce speech distortion, in order to maintain speech intelligibility while cleaning speech [21, 62].

Most commercial CI devices employ single-channel noise reduction algorithm due to its simplicity and efficiency. A recent proposed technique named as eVoice applied in Nurotron speech processor, which can adapt to the external SNR conditions so as to generate stable auditory perception, was evaluated with speech perception test of CI users. Results are exhibited in Fig. 7.12. The percent of correct rates of subjects using APS strategy were compared with the performance of the

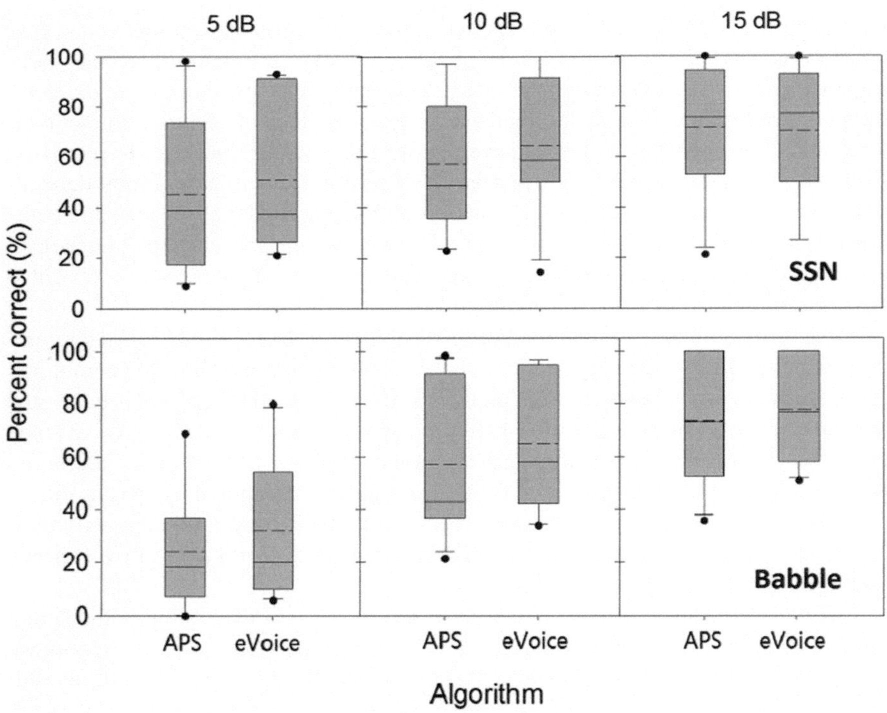

Fig. 7.12 Results of speech perception test in evaluating the effects of eVoice. *Notes*: APS represents using APS strategy along and eVoice indicates using eVoice as the pre-processing technique prior to the application of APS strategy. SSN and Babble represent speech-shaped noise and babble noise, respectively

same group of subjects when using APS and eVoice together. In summary, eVoice improves speech intelligibility of CI users in 5- and 10-dB SNR conditions in the presence of both speech-shaped noise and babble noise. Meanwhile, most subjects reported the preference of using eVoice in noisy environments.

7.3.5.2 Multi-Channel Noise Reduction Methods

Cocktail party problem [14] as one of the most challenging tasks for hearing-impaired listeners has been attempted to be solved via multi-channel algorithms. The crucial ability of completing this task is to separate the mixture of target speech and interfering noise or speech. One group of methods is blind source separation (BSS), which does not require the prior knowledge of microphone array and spatial information of target speech (e.g., [70, 71]). Information-based criteria are usually employed in such methods. A representative work was done by Douglas and Sun [23], using convolutive blind source separation with natural gradient algorithm to automatically resolve mixed recordings from the real environment. In this study, it was found that the modified BSS technique with linear prediction filters was able to complete the task of separating the mixture of two-channel speech signals.

Beamforming is another group of multi-microphone spatial enhancement technique via source separation and works by exploiting phase differences in the signals arriving at two spatially separated microphones from a particular direction. Beamformers can maximally attenuate noise at a null point whereas reserve the target speech in front (Fig. 7.13). This pattern of directionality can either be fixed to generate a static pattern of directionality or be adaptively updated to suppress moving noise sources. In order for beamforming to work, a number of assumptions typically have to be achieved: the target should always be located within a narrow range on angles in front of the listener and the noise source should be spatially separated from the target, meanwhile, the two microphones have to be well matched and calibrated.

Over the past two decades, a few beamforming techniques have been used to enhance speech perception of hearing-impaired listeners in noise. In a CI or hearing aids, the combination of two microphones can detect the differences in amplitude and phase of received signals and then generate the directionality of beam pattern to block noise coming from directions other than that of the target.

In 2005, BEAM strategy, a modified dual-microphone technique was implemented in the behind the ear (BTE) speech processor in Cochlear's Nucleus Freedom CI system. In BEAM, the first step is implemented through a spatially adaptive system that integrates the front directional microphone (which is changed to an omnidirectional microphone later in Nucleus 5 processor) and a rear omnidirectional microphone to separate speech from noise. The output from the rear omnidirectional microphone is firstly filtered through a fixed finite impulse response (FIR) filter and then subtracted from the delayed output from the front directional

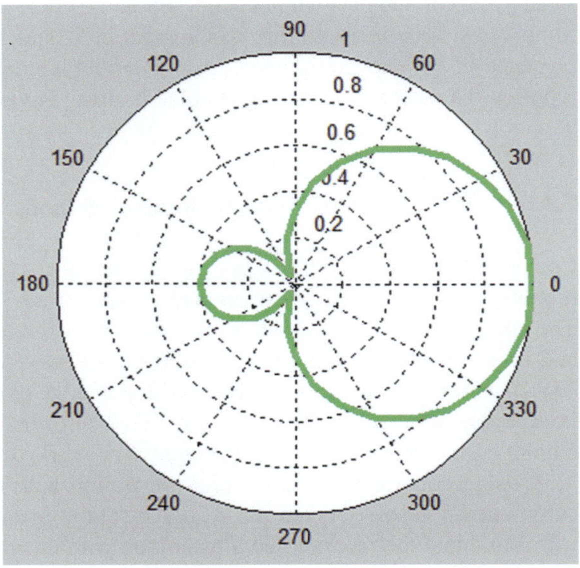

Fig. 7.13 A typical beam pattern of beamforming technique. *Note*: 0°and 180°represent front and back, respectively

microphone to create the noise reference [73]. The filtered signal from the omnidirectional microphone is then added to the delayed signal from the directional microphone to create the speech reference. The reference signals are then delivered to an adaptive filter to remove additional noise, which can be updated with a least-mean-squares (LMS) or a normalized least-mean-squares (NLMS) method. This bilateral noise reduction system was tested and showed an average SRT improvement of 11 dB over a directional microphone.

7.4 Wireless Signal Transmission

In this section, we mainly focus on the analog circuit of cochlear implant system, and several critical analog blocks of wireless signal and power transmission are discussed in detail in the following parts.

As mentioned in the previous section, the internal unit of the cochlear implant system is powered by the inductive link for several considerations [76]. First, the skin thickness of the implanting location is about 3–10 mm, where the wireless transfer efficiency is still acceptable. Second, the expected implanting duration for the cochlear implant is more than 10 years, and it is difficult to use the implanted battery to sustain such a long period, because unlike pacemakers, the cochlear implant has to stimulate the nerve all the time so that its power consumption is relatively large. Third, both surgeon and patients request a small implanted device for easy operation in the surgery and gentle appearance from outside to see.

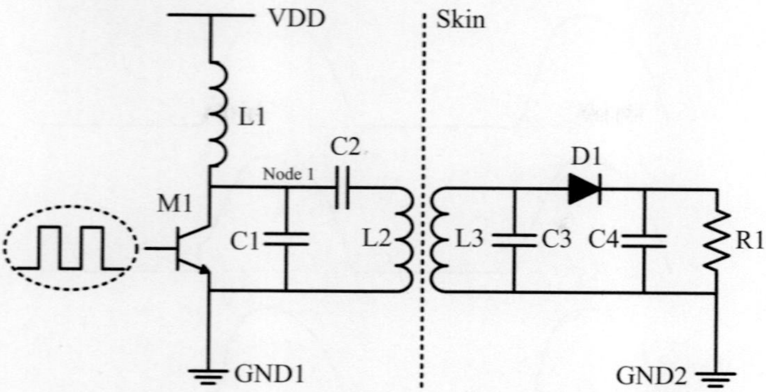

Fig. 7.14 A typical class E power amplifier with inductive link

Figure 7.14 shows the basic schematic of a power amplifier with an inductive link. The class E power amplifier drives the primary RF coil L2 that sends power inductively to the secondary RF coil L3. C2 and C3 are used to be resonant in series and in parallel with L2 and L3 at the working frequency, respectively. The RF signal on L3-C3 tank then is rectified to provide the power for the internal unit and stimulating electrodes. M1 is the main power transistor for the class E amplifier, which is usually controlled by the modulated signal. L1 is the choke inductor to provide supply current, and C1 is used to adjust the waveform of node 1 to make sure that M1 is on right at the moment when the voltage of node 1 is close to ground shown in Fig. 7.15a [60].

However, due to the variability of patient skin-flap-thickness and stimulating current, the equivalent load of the power amplifier varies. Therefore, in many cases as shown in Fig. 7.15b and c, the class E amplifier does not operate perfectly and the entire power transfer efficiency reduced. To solve this issue, many researches have been done. In Fig. 7.16a, the peak detector is used to find the zero crossing of current going through L2, and M1 is trigger by adding a delay t_{delay}, which can be expressed as:

$$t_{\mathrm{delay}} = \frac{T}{2\pi}\left(\pi - \sin^{-1}\left(\frac{i_{\mathrm{dd}}}{i_{\mathrm{L2,P}}}\right)\right) \tag{7.1}$$

where T is the period of the input signal; i_{dd} is the supply current from L1; $i_{L2,\ P}$ is the peak current going through L2 [37]. Figure 7.16b shows another method that tries to achieve zero-voltage switching condition [3]. In this technique, when Vd is lower than the reference voltage, M1 is on and the switch losses are minimized. Both techniques show good results with different distances between the primary and secondary coils. However, these proposed circuits could consume additional power, especially because they are running at a high speed. In some cases, the saved power could be less than the power these circuits dissipate.

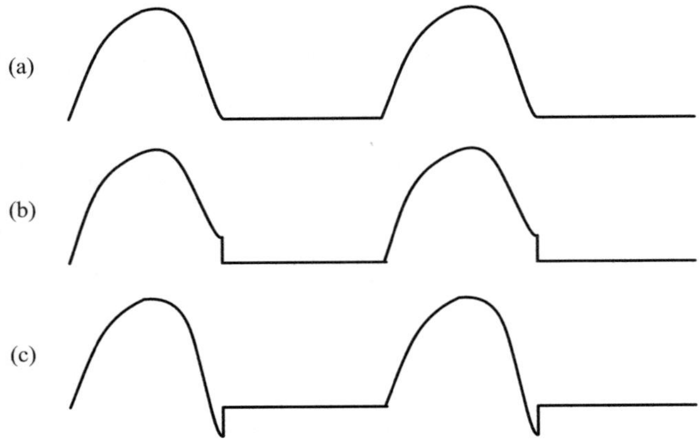

Fig. 7.15 Waveforms of node 1 when the coupling factor of the inductive link varies

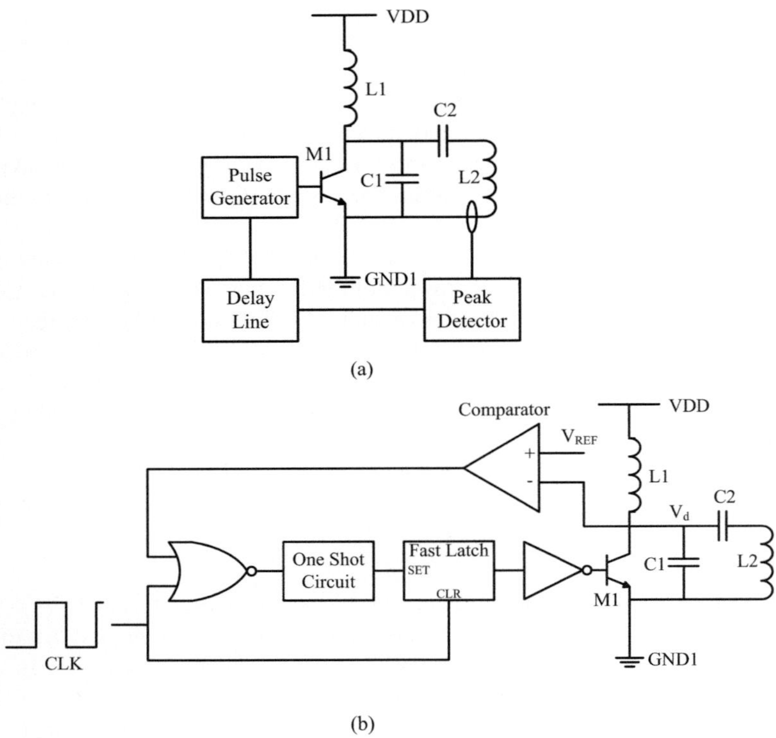

Fig. 7.16 (**a**) Using peak detector to find the zero crossing of current going through L2. (**b**) Using comparator to guarantee V_d is near ground when M1 is on

The efficiency of the inductive link is determined by the coupling coefficient k, the quality factor of L2 Q_2, the quality factor of L3 Q_3, and the effective quality factor of load network Q_L and can be simplified as:

$$\eta_{Link} = \left(\frac{k^2 Q_2 Q_3'}{1 + k^2 Q_2 Q_3'}\right)\left(\frac{Q_3}{Q_3 + Q_L}\right) \tag{7.2}$$

where $Q_3' = Q_3 Q_L/(Q_3 + Q_L)$. From (7.2), we can clearly see that with a higher coupling coefficient, the transfer efficiency of the inductive link increases. It means that the efficiency increases with thinner skin thickness. Also, Q_2 should be as large as possible to increase the efficiency. Normally k and Q_L varies in different circumstances so that it is difficult to find a determined Q_3 to maximize the efficiency. Moreover, the system usually uses on-off keying modulation, which means that the input signal of the amplifier is not a continuous signal. The complexity of the design needs to be more carefully considered.

7.5 ASIC Design

Stimulating auditory nerves and recording neural response are two main functions of the internal unit in a cochlear system. To reduce dimension and power consumption of implant, the internal circuit is implemented with application specific integrated circuit (ASIC). Usually, the function blocks inside the dashed border of Fig. 7.4 is achieved by ASIC.

7.5.1 Current Source

In Fig. 7.17, a current mode stimulating circuit is shown [77]. Twenty-four stimulating electrodes are connected to node E1, E2, . . . E24, respectively. MP1 and MP2 are two reference electrodes. Two types of current sources, P-source and N-source, can be selected by controlling switch S5 and S6. The reason why current mode is chosen for stimulating circuit is because the charge injected into tissue could be precisely controlled. To clearly describe the mechanism of this circuit, choose electrode 1 as the stimulating electrode and MP1 as the reference electrode. Meanwhile, make S5 closed and S6 open to select P-type current source. In the cathodic phase, S2[1] and S3[1] are closed, and the current flows from MP1 to E1. In the anodic phase, S1[1] and S4[1] are closed and the current flows from E1 to MP1.

Stimulating charge balance is one of the most important factors to be considered in the circuit design. The tissue damage does not occur in chronic stimulation of cochlear implants if residual DC current is lower than 100 nA [34]. In this biphasic stimulation as explained above, the positive and negative charge going through the

Fig. 7.17 Current mode stimulating circuit

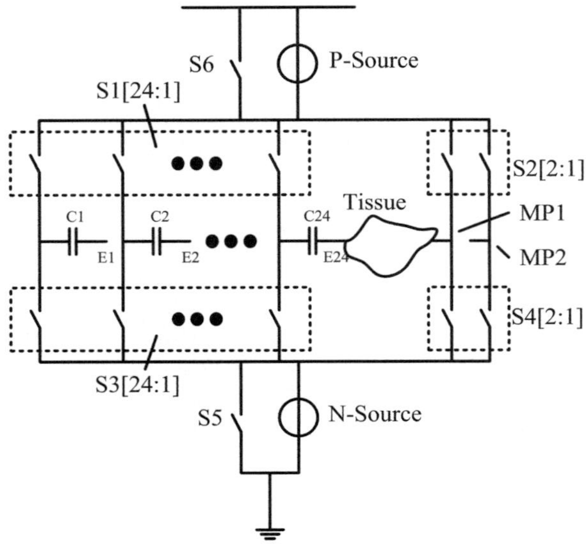

tissue should be balanced ideally, since both phases utilize the same current source. To further suppress imbalanced DC charge, a DC blocking capacitor, C_{block}, in series with each electrode is usually inserted as shown in Fig. 7.18. Moreover, when stimulation is completed, switches S7 and S8 are closed in Fig. 7.18 to guarantee that the stimulating and reference electrodes are on the same electrical potential [15].

7.5.2 *Recording Circuit*

A robust recording circuit is the key to successfully capture the electrically evoked compound action potential (ECAP) signal. As shown in Fig. 7.19, the recording circuit normally is composed of four blocks [12]. The targeted signal is received by the sensing electrode, and then transmitted to the band-pass filter to remove undesired noise. The variable gain amplifier (VGA), whose gain is set according to different circumstances, is used to amplify this signal, because the amplitude of this type of signals is only several millivolt or even smaller. At last, an analog-to-digital converter (ADC) is employed to measure this signal and send the data out.

Mismatch compensation is a critical technique for VGA design, since the amplitude of the input signal is comparable with the input offset of the operational amplifier. In Fig. 7.20, a typical VGA circuit with mismatch compensation is shown [53]. In this circuit, the first stage amplifier consists of Gm1, Gm2, and R, and A1 is the second stage amplifier. Switches S7, S8, S9, and S10 are used to change the gain of amplifier A1. The rest of switches are used for mismatch compensation of the first stage. In detail, S1, S2, S5, and S6 are turned on first. At

Fig. 7.18 Stimulating circuit with charge balancing techniques

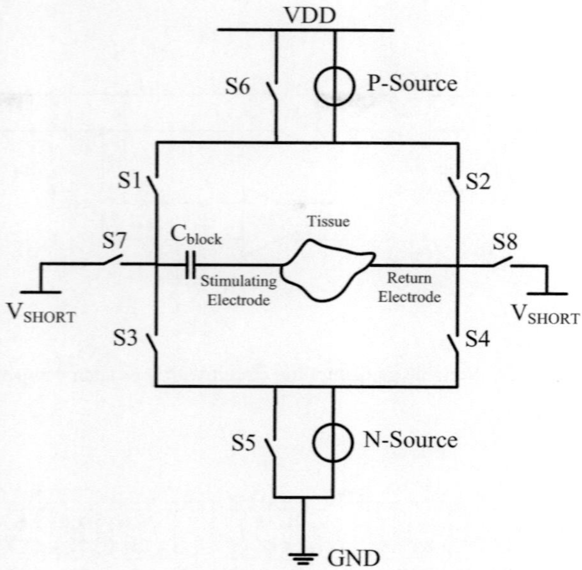

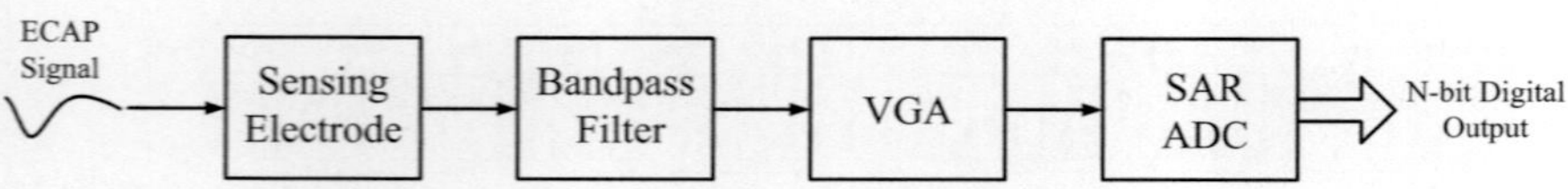

Fig. 7.19 A block diagram of neural response recording circuit

this moment, only the input offset V_{OS} is amplified by the first stage. Then S3 and S4 are also turned on, and the output of the first stage V_{OUT} becomes:

$$V_{OUT} = \frac{g_{m1}}{g_{m2}} V_{OS} \tag{7.3}$$

where g_{m1} and g_{m2} are the trans-conductance of Gm1 and Gm2, respectively. The input offset information is stored in C3 and C4 by turning S1, S2, S3, S4, S5, and S6 off. Therefore, the offset of the first stage is cancelled by the output of Gm2.

Low power, small area, and high acquisition precision are three major factors for ADC design in the cochlear implant, which should be considered carefully. Since the neural response is relatively slow (~KHz), the speed of this circuit is not needed to be very high. Based on the discussion above, the successive approximation (SAR) ADC is one of the most suitable candidates for ADC design in the internal unit [12]. Figure 7.21a shows a 12-bit fully differential SAR ADC circuit. The basic theory is using the output of a 12-bit DAC to compare with the input signal [19]. After comparing 12 times, the ADC finally gets all the digital output. Unlike flash ADC and pipeline ADC, SAR ADC only needs one comparator so that both

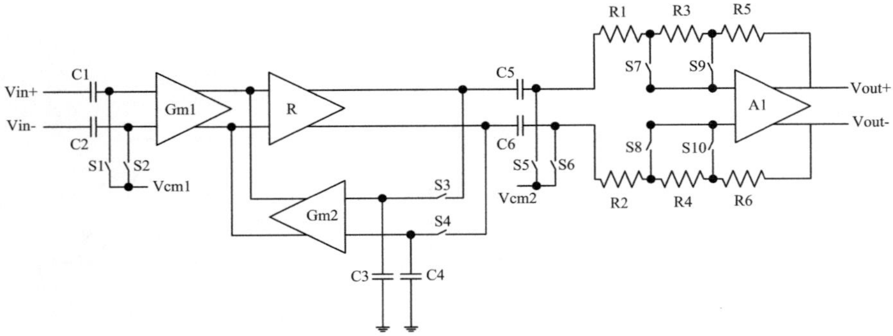

Fig. 7.20 Variable gain amplifier circuit with mismatch compensation

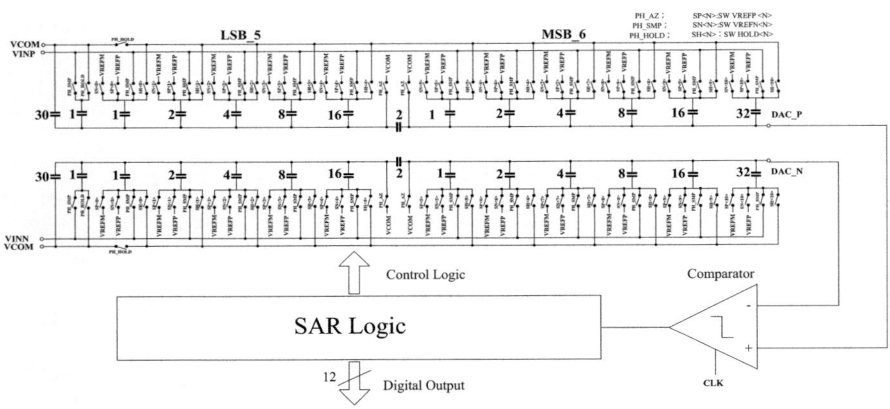

(a)

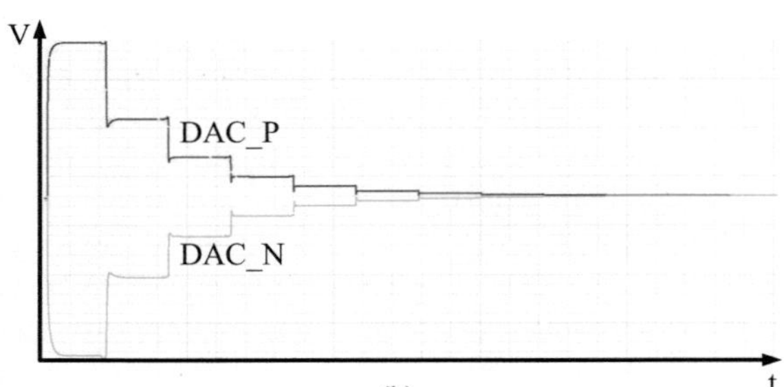

(b)

Fig. 7.21 (**a**) A 12-bit SAR ADC circuit. (**b**) Simulation waveforms of DAC_P and DAC_N

power and area are saved. Figure 7.21b shows the simulation results of DAC_P and DAC_N. We can clearly see that the comparator compares two input signals one stage by one stage in sequential way and the input of next stage is the subtraction of the output of this stage and the output of the DAC.

7.5.3 Reverse Telemetry Circuit

In a cochlear system, the implant in some cases needs to reversely transmit data to the speech processor, for example, neural impedance test data, neural response measurement data, or indication of operation status. Load-shift keying (LSK), which was referred as "reflectance modulation," is a commonly used technique for reverse telemetry [65]. The essence of LSK is the variation of the equivalent impedance in the primary coil, which could be influenced by changing the load of secondary coil. Therefore, this telemetry can be achieved using the same inductive link with forward transmission. Fig. 7.22a shows a modulation circuit that is based on a circuit configuration modulator [65]. When switch S1 is open, the circuit on the implant side is a full-wave rectifier. When S1 is closed, the circuit becomes a voltage clamp circuit. Using this method, the load of the secondary coil is successfully modulated by controlling S1. In Fig. 7.22b, LSK is realized by changing the resonant capacitor of the secondary coil [77]. This technique is especially effective for half-wave rectifier circuit.

The demodulator circuit is used to sense and digitize the variation of the signal amplitude on the external coil, which is modulated by the implant. Unlike other regular amplitude-shift keying (ASK) demodulator, this circuit is suffered from variable signal amplitudes, a shifting DC level, and an undetermined digitizing threshold due to the variation of patient's skin depth and the misalignment of the coils. Figure 7.23 shows that a demodulation architecture, which is based on a compact CMOS integrated circuit solution, is proposed to greatly enhance the transmission robustness, increase the maximum bit rate, reduce the bit error rate, and save the board area [35].

The entire architecture is composed of four blocks: sensor, rectifier, amplifier, and digitizer. An additional small coil (Fig. 7.24a) is introduced to pick up the modulated signal to separate the forward and reverse telemetry and reduce the interaction between each other. A half-wave rectification (Fig. 7.24 b) is implemented to filter high-frequency components and remove the negative part of the signal. In this block, the RC filter should be carefully designed to minimize the load of the additional coil to alleviate its effect on the main coil and save power consumption. The amplifier (Fig. 7.24c) is used to amplify the output of the rectifier. It has two modes: a normal mode and a sleep mode. The amplifier is in the sleep mode while the processor is transmitting regular data containing sound information to the implant. In this mode, a switch that is connected between the input of the amplifier and a constant DC voltage source is on and the DC level of the input is determined. Once the amplifier goes into the normal mode, the switch becomes off, and the common mode voltage of the input is still well defined. A high-pass feedback network is employed in the

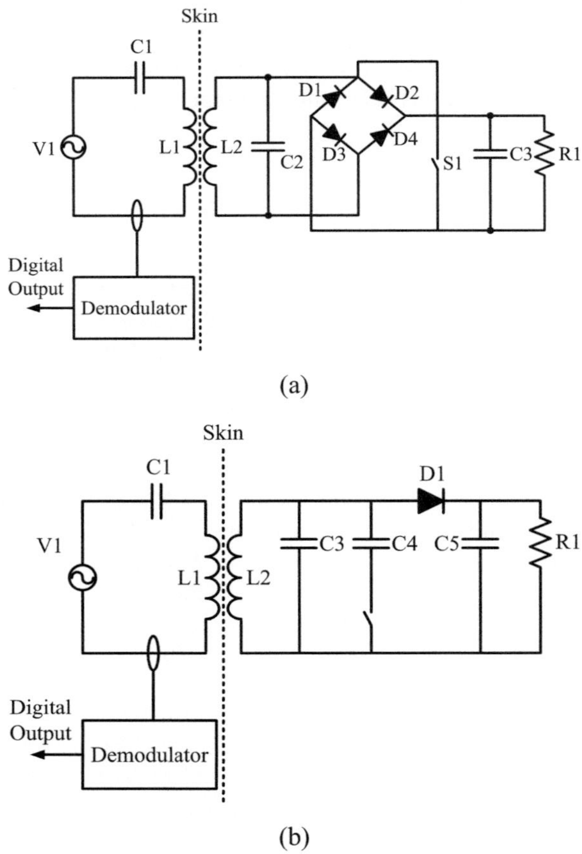

Fig. 7.22 Reverse telemetry circuit based on (**a**) half-wave rectifier and (**b**) full-wave rectifier

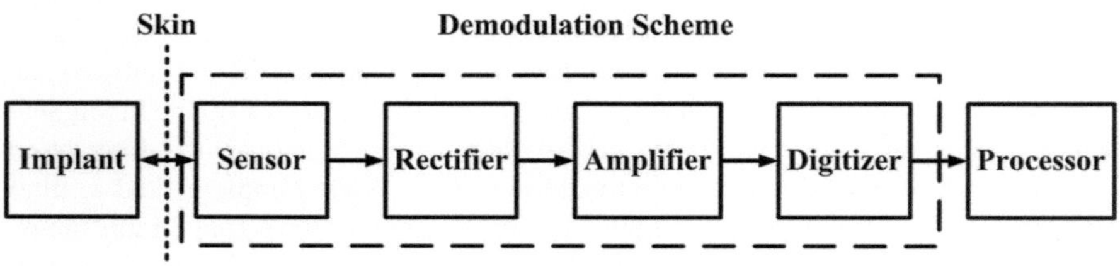

Fig. 7.23 A block diagram of the demodulator

amplifier so that the voltage on the capacitor in the network is the integral of the output signal. This voltage is used as a reference for the digitization block to differentiate digital “1” and “0”. A Schmitt trigger (Fig. 7.24d) is implemented to convert the amplified analog signal to a clean digital signal. To further suppress random and deterministic noise, a hysteresis characteristic is utilized to avoid

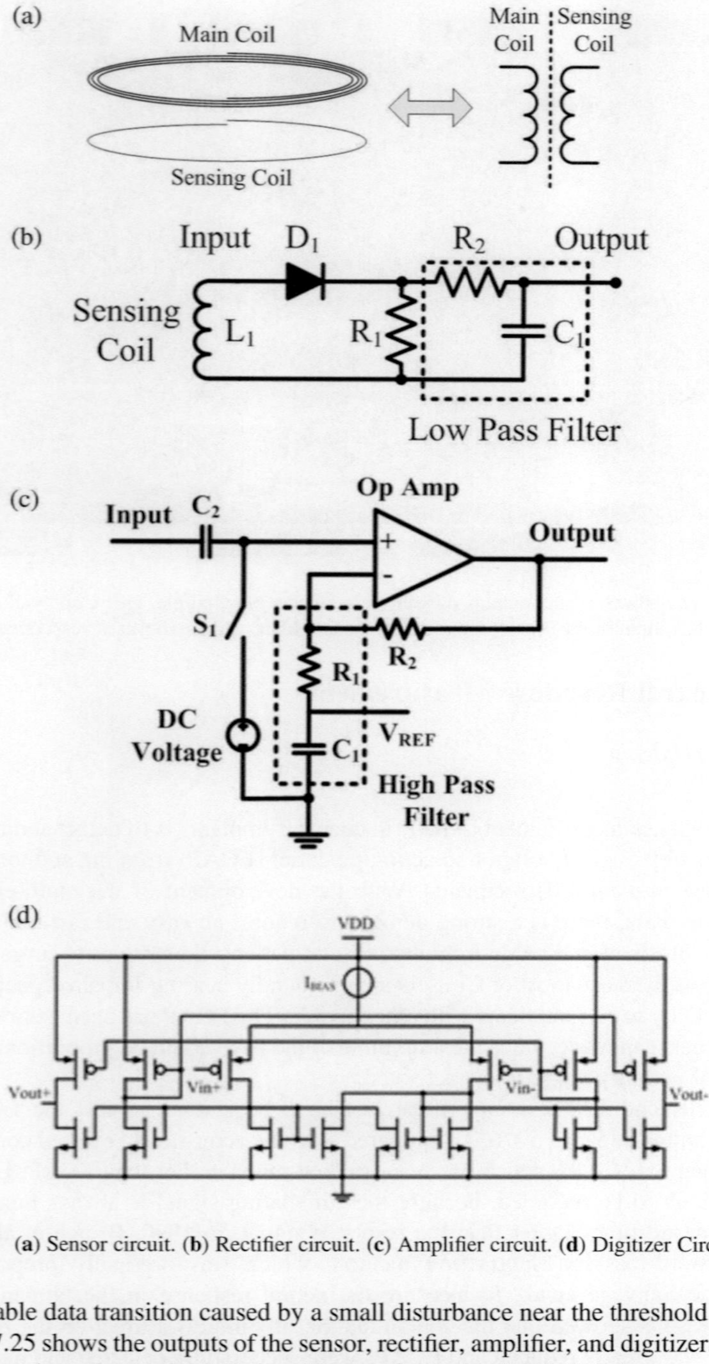

Fig. 7.24 (**a**) Sensor circuit. (**b**) Rectifier circuit. (**c**) Amplifier circuit. (**d**) Digitizer Circuit

undesirable data transition caused by a small disturbance near the threshold voltage. Figure 7.25 shows the outputs of the sensor, rectifier, amplifier, and digitizer. We can clearly see how the modulated signal is transformed to be digital data.

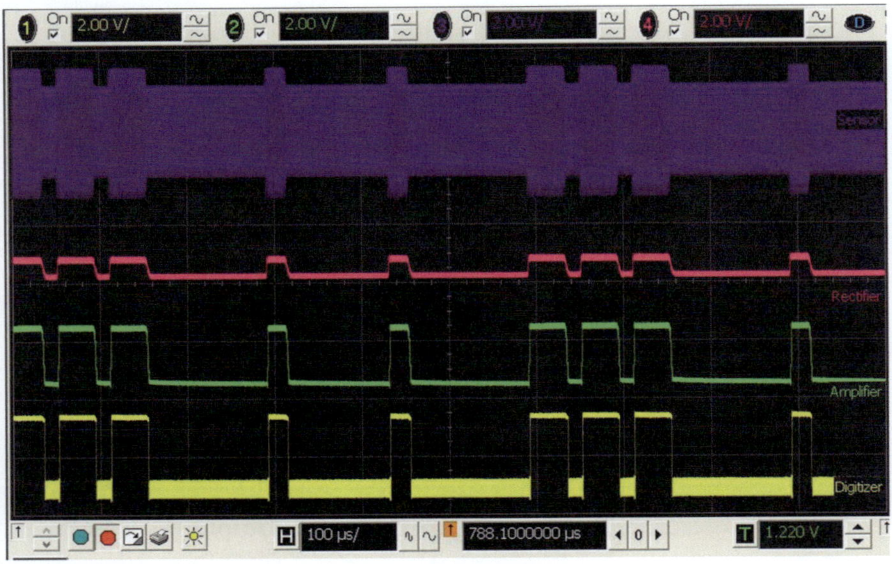

Fig. 7.25 Waveforms of the outputs of four main blocks: blue, purple, green, and yellow waveforms are the output of the sensor, the rectifier, the amplifier, and the digitizer, respectively

7.6 Neural Response Measurement

7.6.1 History

Neural response measurement (NRM) in cochlear implants is to detect and measure the electrically evoked compound action potential (ECAP) from the auditory nerve in cochlear implant (CI) recipients. With the development of the multi-electrode cochlear implant, there is a strong demand to request an easy and fast tool to give objective information directly from devices and patients themselves to surgeons and audiologists, because most of CI users are profoundly hearing-impaired people who have difficulty to communicate with other people. The ECAP has been demonstrated to be the best candidate, since the amplitude of the EAP could be proportional to the number of auditory nerve fibers activated.

In 1970s and 1980s, many researches have been done to test the ECAP in different animals [5, 11, 50, 63]. Compared with the acoustically evoked compound action potential (CAP), which has been studied much earlier than ECAP, ECAP is very difficult to be recorded, because the stimulating signal is always hundred or even thousand times larger than the response signal. In 1990, Brown et al. firstly used forward-masking subtraction method, which was originally proposed by Charlet de Sauvage et al., to measure the neural response in the human beings [6]. The results showed that the contamination of stimulus artifact on the recorded response was greatly reduced and ECAPs were successfully detected and measured.

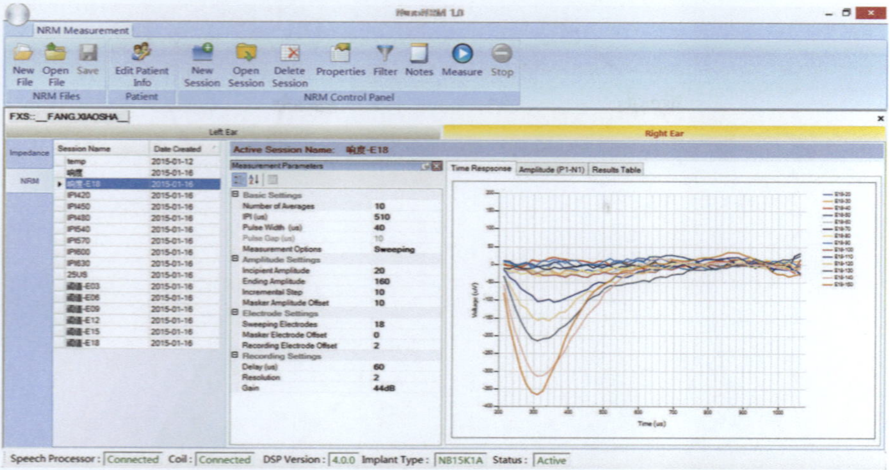

Fig. 7.26 Neural response measurement (NRM) tool

Based on this technique, the first commercial ECAP measurement tool, whose name was called "neural response telemetry" (NRT), was available in 1998. The initial results showed that based on a carefully designed protocol, the ECAP can be reliably recorded in most patients and most electrodes. After that, the similar tools with different names ("neural response imaging," "auditory nerve response telemetry," "neural response measurement") were issued by different brands of CI producers. Nowadays, ECAP measurement tools (Fig. 7.26) become one of the most useful tools to provide objective information of CI patients for clinical evaluation from surgeons and audiologists.

7.6.2 *Electrically Evoked Compound Action Potential*

Essentially, the ECAP is the Wave I of the auditory brainstem response (ABR). Once the auditory nerve is stimulated by an electrical signal, which is larger than a specific threshold, the ECAP is generated. Figure 7.27 shows two types of ECAPs, which were recorded with intra-cochlear electrodes in Nurotron Venus users. In most cases, the ECAP, as shown in Fig. 7.27a, is composed of a negative peak (N1) occurring at approximately 0.2–0.4 ms following stimulus onset and a much smaller positive peak (P2) at approximately 0.6–0.8 ms [1, 6, 7, 18]. But two-peak ECAPs as shown in Fig. 7.27b can also be occasionally recorded, whose incidence is around 10–20% [43, 67]. In this case, the N1 occurs at similar time as the one-peak case, but the P1 and the P2 typically occur at around 0.4–0.5 ms and 0.6–0.7 ms, respectively [43, 67].

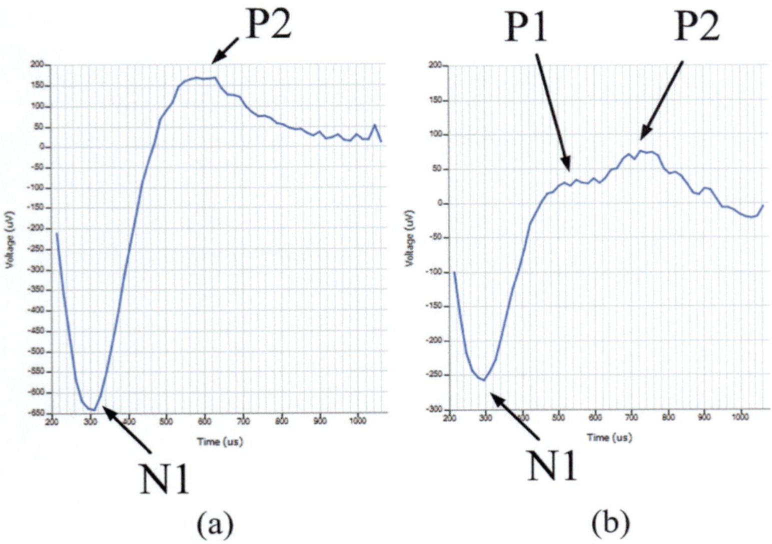

Fig. 7.27 The recorded ECAPs with (**a**) one positive peak and (**b**) two positive peaks

The amplitude of the ECAP varies from about 20 μV to 2 mV, and its maximum magnitude usually is larger than the electrically evoked ABR [8]. Because of its large amplitude, myogenic activity has little influence on the ECAP [32]. Moreover, maturation of the central auditory system also is not a main factor that will cause any changes in the ECAP. Therefore, the difference of ECAPs between adult and pediatric CI users is very small [6, 26, 31], and the recorded waveforms do not change with the increase of the duration of CI use. However, many factors also lead to the variation of the latency and amplitude of the ECAP recorded, for example, the stimulation level, the polarity of the stimulation, the surviving neural population, the intra-cochlear test electrode location, and so on. The amplitude of the ECAP is the voltage difference between N1 and P2 peaks. As shown in Fig. 7.28, ECAP amplitude increases as the stimulation level increases.

The ECAP threshold and slope of the ECAP amplitude growth, as shown in Fig. 7.29, are two major points in the neural response measurement, in which a number of researchers are interested. The ECAP threshold is the stimulation current level, at which the neural response measurement is able to capture the ECAP waveform. Many studies indicate that the ECAP threshold falls between the behavioral threshold (T-level) and comfort levels (C-level) of CI users [7, 9, 10, 18, 29, 59]. The slope of the ECAP amplitude growth is also widely used to objectively predict the fitting program of CI users [10, 29].

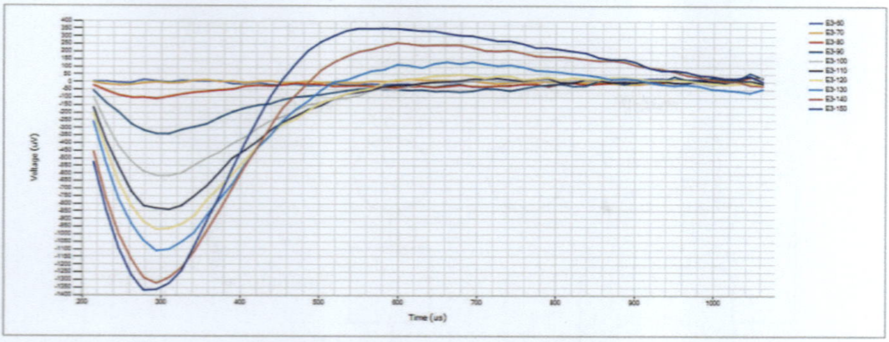

Fig. 7.28 Recorded ECAP waveform with different amplitudes of stimulation current

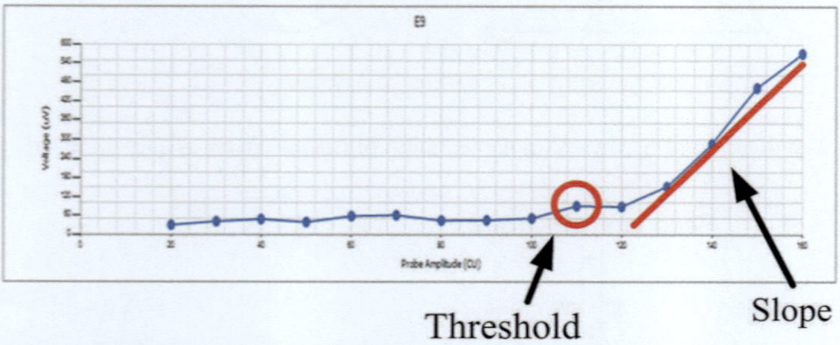

Fig. 7.29 The relationship between the stimulation current level and the amplitude of ECAP

7.6.3 Hardware Implementation

Hardware implementation is one of the key factors to successfully record the neural response. Figure 7.30 shows how the cochlear implant system realizes the NRM function. Generally, at the beginning the computer, where the NRM tool is installed and all the parameters are set, controls the speech processor to send a command to the implant with the information about the electrode and stimulation current level. The modulated signal, which is encoded by the speech processor according to this command, is transmitted through the inductive link between the external coil and the internal coil. The stimulator decodes the signal from the internal coil and sends a biphasic current pulse to the stimulating electrode (e.g., electrode 3 in Fig. 7.31). Once the electrode finishes the stimulation, the forward transmission is completed. After a determined delay, the recording electrode (e.g., electrode 5 in Fig. 7.31) starts to record the voltage signal it receives. This voltage signal firstly is amplified by an amplifier with a very large gain (usually more than 60 dB). Then the signal is digitized by a high-resolution ADC (usually more than 10 bits). Later, all the data are reversely transmitted to the speech processor by the

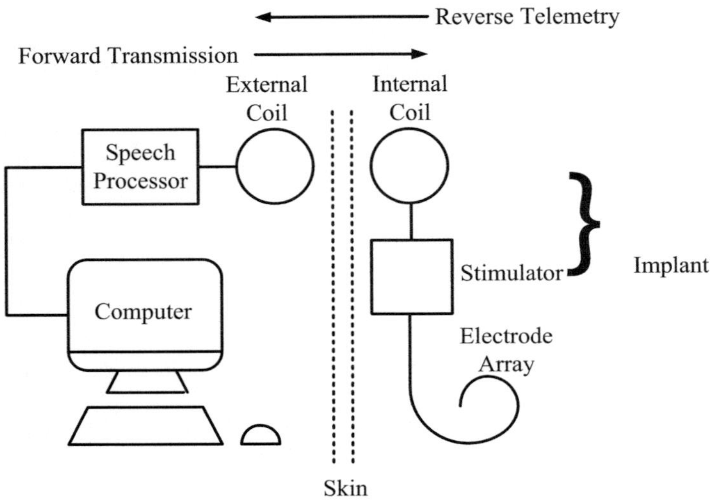

Fig. 7.30 The diagram of hardware implementation of neural response measurement

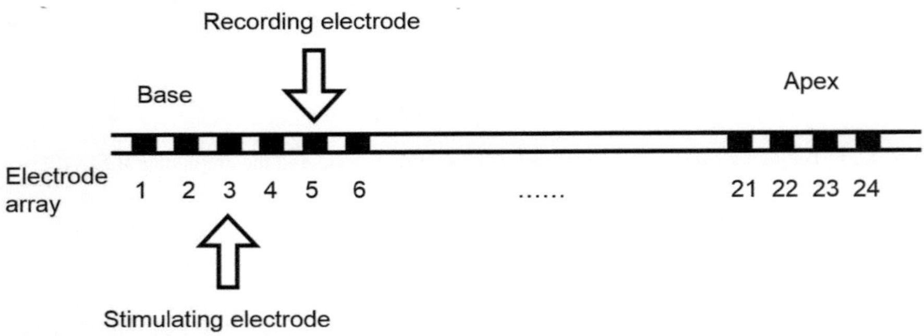

Fig. 7.31 Electrode array. *Note*: In this case, electrode 3 is the stimulating electrode, and electrode 5 is the recording electrode

reverse telemetry technique. In the end, the computer completes the analysis and provides the report based on the data which are sent from the speech processor.

The entire system is very flexible so that the tool could change a number of parameters. For example, the gain of the amplifier in the stimulator should be variable. In most cases, a large gain (e.g., 60 dB) is required to amplify very weak neural signal. By contrast, in some other cases, a relatively small gain (e.g., 40 dB) is necessary to avoid a severe saturation problem. In addition, one electrode should be able to consecutively generate two stimulating current pulses within a period, which varies from 50 μs to several milliseconds. Moreover, the turn-on time of the analog-to-digital converter (ADC) in the stimulator should also be well controlled, and the requirement of the ADC's turn-on duration time is at least 1 ms. Furthermore, the roles between recording and stimulating electrodes can be easily changed or switched. In the real application, every electrode could be either a stimulating electrode or a recording electrode.

7.6.4 Measurement Techniques

The forward-masking subtraction method, which is briefly introduced above, now becomes the routine technique for most commercial neural response measurement tools [1, 6, 7, 9, 20]. The main purpose of this tool is to cancel the artifact caused by the probe stimulus, which is substantially larger than the neural response. The details of this method shown in Fig. 7.32 are described below.

Four types of stimulation are used each time in the measurement. For type A, the probe is stimulated alone at t_1. For type B, the masker is stimulated in the same electrode at t_0, and then the probe is stimulated at t_1. For type C, the masker alone is stimulated at t_0. For type D, zero stimuli for both masker and probe are presented. Each type of stimulation is measured separately. The main theory behind this technique is that the neurons in the auditory nerve will be in the refractory state for a short while after electrical stimulation. Therefore, the measurement results of type A, type B, type C, and type D, A, B, C, and D, are presented below, respectively:

$$\mathrm{A} = \mathrm{SA} + \mathrm{PA} + \mathrm{PN} \tag{7.4}$$

$$\mathrm{B} = \mathrm{SA} + \mathrm{MA} + \mathrm{MN} + \mathrm{PA} \tag{7.5}$$

$$\mathrm{C} = \mathrm{SA} + \mathrm{MA} + \mathrm{MN} \tag{7.6}$$

$$\mathrm{D} = \mathrm{SA} \tag{7.7}$$

where SA is the system artifact; PA is the probe artifact; PN is the neural response caused by the probe; MA is the masker artifact; and MN is the neural response caused by the masker. In (7.5), the reason why PN is missed is because in this case, the auditory nerve is in the refractory state and has no response for the probe stimulus. Combining (7.4), (7.5), (7.6) and (7.7), the final measurement result of the ECAP (FMRE) is:

$$\mathrm{FMRE} = \mathrm{A} - \mathrm{B} + \mathrm{C} - \mathrm{D} = \mathrm{PN}. \tag{7.8}$$

Not surprisingly, the recorded signal is equal to the neural response caused by the probe. According to (7.8), it is known that the amplitude of the probe stimulus is variable and directly related to the amplitude of the ECAP. For the masker, the stimulating current should be at a high level to make sure enough neurons are stimulated by the masker and go into the refractory state [1].

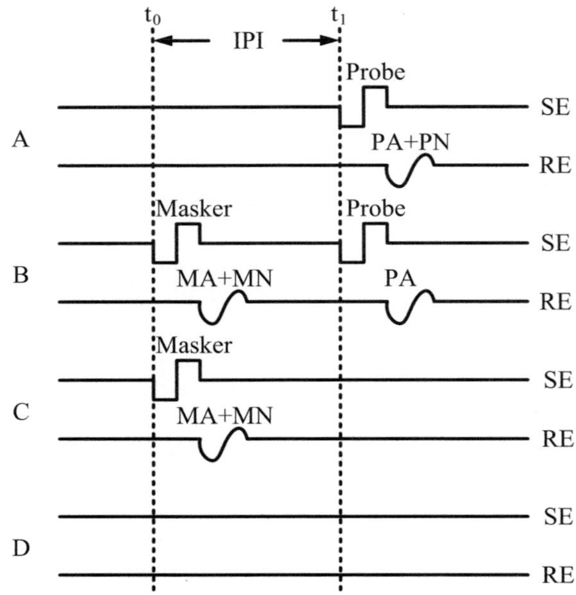

Fig. 7.32 The forward-masking subtraction method. *Note*: For each case, the upper one is the electrical pulse generated by the stimulating electrode (SE), and the lower one is the measured signal recorded by the recording electrode (RE)

In practice, the results of this method are affected by several non-ideal factors. The saturation issue is one of the most serious factors, which could lead to the failure of neural response measurement. In Fig. 7.32, it shows that the signal the recording circuit received is comprised of the stimulus artifact and the neural response. However, the amplitude of the stimulus artifact is much larger than the neural response. Accordingly, if the amplitude of the stimulus artifact is too large so that the amplifier is saturated at its largest output, the ECAP could not be captured after measurement. In this case, reducing amplifier's gain and increasing the recording delay after stimulation could improve the final results of ECAP measurement.

In addition, changing the inter-pulse interval (IPI), which is the time difference between t_0 and t_1 in type B stimulation, could also change the amplitude of recorded ECAP. On one hand, the number of neurons in the refractory state after the masker stimulus in type B stimulation gradually decreases with increasing the IPI. Thus more and more neurons in the auditory nerve have responses to the probe stimulus in type B stimulation, which may reduce the amplitude of the ECAP based on (7.8). On the other hand, if the IPI is too short, the integration of probe and masker stimulation current may reduce the amplitude of the ECAP as well [1]. Figure 7.33 shows a group of ECAP signals, which were captured by NRM tool. It clearly shows that the amplitude of ECAP of this patient reaches the maximum value when the IPI is 510 μs.

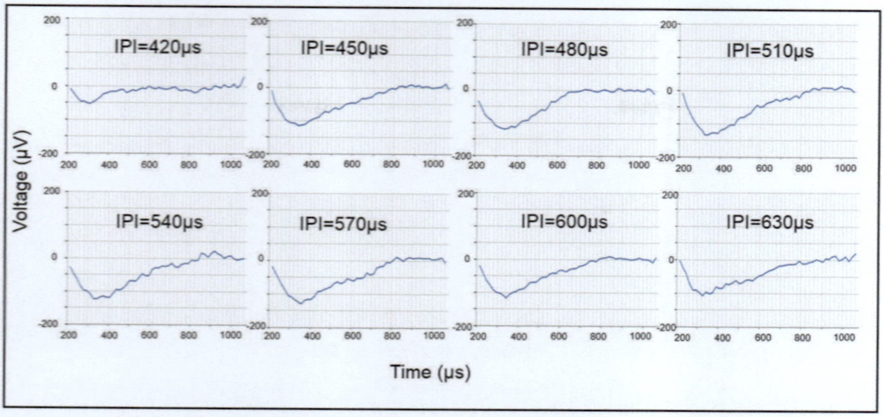

Fig. 7.33 The ECAP signal of patient P1 using NRM tool (stimulation of 130 CL was applied at electrode 12)

7.6.5 Clinical Uses

As mentioned in the previous section, the invention of the reliable ECAP measurement meets great demands from surgeons and audiologists for their clinical work. The neural response measurement now is clinically used in the following applications [36].

First, the ECAP measurement is used to evaluate and verify the neural response from the auditory nerve. In many cases, this objective information is very important for surgeons and audiologists. For example, surgeons can use it to monitor whether the electrode arrays are well placed in the cochlea in surgery. After surgery, the recovering condition of CI recipients varies significantly, and the audiologists can use this information to judge the effectiveness of the electrical stimulation in the auditory nerve fibers.

Second, the electrode and device can be functionally verified. Even though the impedance test also can offer useful information about the device in some degree, the capture of the ECAP signal is the direct proof that the device works properly inside the cochlea.

Third, the ECAP measurement creates a great convenience for programming the speech processor. As we all know, the first and very critical step to use CIs is to find a map with corresponding T and C levels, which is fitted with every user's own circumstance. Many studies prove that the NRM threshold fall between T- and C-levels [7, 9, 10, 18, 29, 59]. In Figs. 7.34, 21 adult subjects are tested using NRM tool, and the preliminary results also support this conclusion. Thus the ECAP threshold can guide audiologists how to start to test T and C levels individually. More importantly, for the people who poorly communicate with audiologists, for example, very young children, the results greatly assist audiologists to build relatively accurate maps for those people.

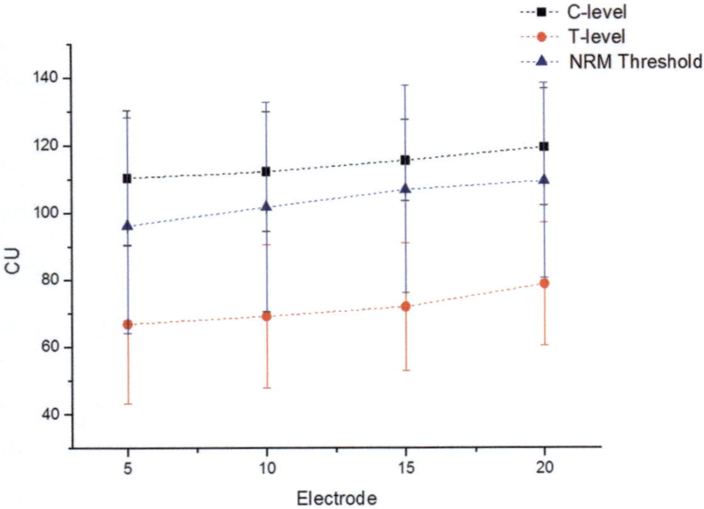

Fig. 7.34 Preliminary results of the relationship between NRM threshold and T- and C- levels using NRM tool

Compared with other measurement techniques, e.g., electrically auditory brainstem response (EABR), ECAP measurement is easy and convenient to operate, because it does not require special equipment, extra recording electrodes, and entirely different software. All the measurement is achieved based on the standard CI product itself and the tool is embedded into the programming software. Additionally, there is no strict requirement to limit patient's status so that it is easy to cooperate with patient to complete test, which is essentially useful for pediatric CI users. Moreover, the ECAP does not have an obvious difference between adults and children, a typical protocol with determined parameters can be used for most patients. Thus the operation of audiologists is much simplified. Because of the advantages discussed above, neural response measurement becomes a very promising technique for more clinical applications.

7.7 Emerging Technologies

Three pioneers of cochlear implant Graeme Clark, Ingeborg Hochmair, and Blake Wilson received the 2013 Lasker~DeBakey Clinical Medical Research Award for their outstanding contributions to the development of the modern cochlear implant. Meanwhile, active and diverse research is currently taking place in the area of cochlear implant. Compared with normal-hearing people, cochlear implant recipients find it hard to listen in noisy environment or to listen to music or tonal languages. Studies on conveying fine structure information, combined acoustic and electric hearing, and bilateral implantation have been proposed to improve the performance of cochlear implant listeners in above mentioned difficult tasks. Recently, optogenetic stimulation of auditory nerve has been proposed to develop optical cochlear implant, which can significantly reduce the channel interference of electrical stimulation. Cochlear implant is on its way to confer more benefit to mankind.

References

1. Abbas PJ, Brown CJ, Shallop JK, Firszt JB, Hughes ML, Hong SH, Staller SJ (1999) Summary of results using the nucleus CI24M implant to record the electrically evoked compound action potential. Ear Hear 20:45–59
2. Arnoldner C, Riss D, Brunner M, Durisin M, Baumgartner WD, Hamzavi JS (2007) Speech and music perception with the new fine structure speech coding strategy: preliminary results. Acta Otolaryngol 127:1298–1303
3. Baker M, Sarpeshkar R (2007) Feedback analysis and design of RF power links for low-power bionic systems. IEEE Trans Biomed Circuits Syst 1:28–38
4. Boll SF (1979) Suppression of acoustic noise in speech using spectral subtraction. IEEE Trans Acoust Speech Signal Process 27:113–120
5. Brown CJ, Abbas PJ (1990) Electrically evoked whole-nerve action potentials: parametric data from the cat. J Acoust Soc Am 88:2205–2210
6. Brown CJ, Abbas PJ, Gantz B (1990) Electrically evoked whole-nerve action potentials: data from human cochlear implant user. J Acoust Soc Am 88(3):1385–1391
7. Brown CJ, Abbas PJ, Gantz B (1998) Preliminary experience with neural response telemetry in the nucleus CI24M cochlear implant. Am J Otol 19:320–327
8. Brown CJ, Hughes ML, Lopez SM, Abbas PJ (1999) Relationship between EABR thresholds and levels used to program the Clarion speech processor. Ann Otol Rhinol Laryngol 108(Suppl. 177):50–57
9. Brown CJ, Hughes ML, Luk B, Abbas PJ, Wolaver A, Gervais J (2000) The relationship between EAP and EABR thresholds and levels used to program the nucleus 24 speech processor: data from adults. Ear Hear 21:151–163
10. Cafarelli Dees D, Dillier N, Lai WK et al (2005) Normative findings of electrically evoked compound action potential measurements using the neural response telemetry of the nucleus CI24M cochlear implant system. Audiol Neurotol 10:105–116
11. Charlet de Sauvage R, Cazals Y, Erre JP, Aran JM (1983) Acoustically derived auditory nerve action potential evoked by electrical stimulation: an estimation of the waveform of single unit contribution. J Acoust Soc Am 73:616–627
12. Chaturvedi V, Anand T, Amrutur B (2013) An 8-to-1 bit 1-MS/s SAR ADC with VGA and integrated data compression for neural recording. IEEE Trans Very Large Scale Integr VLSI Syst 21(11):2034–2044
13. Chen JD, Benesty J, Huang Y, Doclo S (2006) New insights into the noise reduction Wiener filter. IEEE Trans Audio Speech Lang Process 14:1218–1234
14. Cherry C (1953) Some experiments on the recognition of speech with one and two ears. J Acoust Soc Am 25:975–981
15. Chun H, Yang Y, Lehmann T (2014) Safety ensuring retinal prosthesis with precise charge balance and low power consumption. IEEE Trans Biomed Circuits Syst 8(1):108–118
16. Clark GM, Tong YC, Dowell RC (1984) Comparison of two cochlear implant speech-processing strategies. Ann Otol Rhinol Laryngol 93:127–131
17. Crozier PM, Cheetham BMG, Holt C, Munday E (1993) Speech enhancement employing spectral subtraction and linear predictive analysis. Electron Lett 29:1094–1095
18. Cullington H (2000) Preliminary neural response telemetry results. Br J Audiol 34:131–140

19. Culurcielllo E, Andreou A (2003) An 8-bit, 1mW successive approximation ADC in SOI CMOS. In: Proceedings of IEEE international symposium on circuit and system, IEEE, Bangkok, Thailand pp 301–304
20. Dillier N, Lai WK, Almqvist B, Frohne C, Müller-Deile J, Stecker M, von Wallenberg E (2002) Measurement of the electrically evoked compound action potential via a neural response telemetry system. Ann Otol Rhinol Laryngol 111(5):407–414
21. Doclo S, Spriet A, Wouters J, Moonen M (2007) Frequency-domain criterion for the speech distortion weighted multichannel Wiener filter for robust noise reduction. Speech Comm 49:636–656
22. Donaldson GS, Kreft HA, Litvak L (2005) Place-pitch discrimination of single- versus dual-electrode stimuli by cochlear implant users (L). J Acoust Soc Am 118(2):623–626
23. Douglas SC, Sun XA (2003) Convolutive blind separation of speech mixtures using the natural gradient. Speech Comm 39:65–78
24. Dowell R, Seligman P, Blamey P, Clark G (1987) Evaluation of a two-formant-estimating speech processor for a multiple-channel cochlear prosthesis. Ann Otol Rhinol Laryngol 96 (Suppl. 128):132–134
25. Dudley H (1939) The vocoder. Bell labs record, 18(4), pp 122–126
26. Eisen MD, Franck KH (2004) Electrically evoked compound action potential amplitude growth functions and HiResolution programming levels in pediatric CII implant users. Ear Hear 25:528–538
27. Fetterman BL, Domico EH (2002) Speech recognition in background noise of cochlear implant patients. Otolaryngol Head Neck Surg 126:257–263
28. Firszt JB, Holden LK, Skinner MW, Tobey EA, Peterson A, Gaggl W, Runge-Samuelson CL, Wackym PA (2004) Recognition of speech presented at soft to loud levels by adult cochlear implant recipients of three cochlear implant systems. Ear Hear 25:375–387
29. Franck KH, Norton SJ (2001) Estimation of psychophyical levels using the electrically evoked compound action potential measured with the neural response telemetry capabilities of Cochlear Corporation's CI24M device. Ear Hear 22:289–299
30. Fu QJ, Zeng FG, Shannon RV, Soli SD (1998) Importance of tonal envelope cues in Chinese speech recognition. J Acoust Soc Am 104:505–510
31. Gordon KA, Papsin BC, Harrison RV (2004) Toward a battery of behavioral and objective measures to achieve optimal cochlear implant stimulation levels in children. Ear Hear 25:447–463
32. He S, Teagle HFB, Buchman CA (2017) The electrically evoked compound action potential: from laboratory to clinic. Front Neurosci 11(Article 339):1–20
33. Hu Y, Loizou PC (2007) A comparative intelligibility study of single-microphone noise reduction algorithms. J Acoust Soc Am 122:1777
34. Huang C, Shepherd R, Center P, Seligman P, Tabor B (1999) Electrical stimulation of the auditory nerve: direct current measurement in vivo. IEEE Trans Biomed Eng 46(4):461–469
35. Huang S, Xia B, Wang S, Sun X (2015) A novel demodulation technique in reverse telemetry for cochlear device. In: Conference on implantable auditory prostheses, p 119
36. Hughes ML (2010) Fundamentals of clinical ECAP measures in cochlear implants, Part 1: use of the ECAP in speech processor programming, 2nd edn, Audiology Online, November 8, 2010, Article 2347
37. Kendir GA, Liu W, Bashirullah R, Wang G, Humayun MS, Weiland J (2005) An optimal design methodology for inductive power link with class-E amplifier. IEEE Trans Circuits Syst Regul Pap 52(5):857–866
38. Kiefer J, Hohl S, Stürzebecher E, Pfennigdorff T, Gstöett-ner W (2001) Comparison of speech recognition with different speech coding strategies (SPEAK, CIS, and ACE) and their relationship to telemetric measures of compound action potentials in the nucleus CI 24M cochlear implant system. Audiology 40(1):32–42
39. Kim G, Loizou PC (2011) Gain-induced speech distortions and the absence of intelligibility benefit with existing noise-reduction algorithms. J Acoust Soc Am 130:1581–1596

40. Koch DB, Osberger MJ, Segal P, Kessler D (2004) HiResolution and conventional sound processing in the HiResolution bionic ear: using appropriate outcome measures to assess speech recognition ability. Audiol Neurootol 9(4):214–223
41. Koch DB, Downing M, Osberger MJ, Litvak L (2007) Using current steering to increase spectral resolution in CII and HiRes 90K users. Ear Hear 28(2 Suppl):39S–41S
42. Kokkinakis K, Loizou PC (2008) Using blind source separation techniques to improve speech recognition in bilateral cochlear implant patients. J Acoust Soc Am 123:2379–2390
43. Lai WK, Dillier N (2000) A simple two-component model of the electrically evoked compound action potential in the human cochlea. Audiol Neurotol 5:333–345
44. Li N, Loizou PC (2008) Factors influencing intelligibility of ideal binary-masked speech: implications for noise reduction. J Acoust Soc Am 123:1673–1682
45. Lim JS, Oppenheim AV (1979) Enhancement and bandwidth compression of Noisy speech. Proc IEEE 67:1586–1604
46. Loizou PC, Poroy O, Dorman M (2000) The effect of parametric variations of cochlear implant processors on speech understanding. J Acoust Soc Am 108:790–802
47. Luo X, Fu QJ (2004) Enhancing Chinese tone recognition by manipulating amplitude envelope: implications for cochlear implants. J Acoust Soc Am 116:3659–3667
48. Martin R (1994) Spectral subtraction based on minimum statistics. In: Proceedings European signal process, Edingburgh, Scotland, U.K. pp 1182–1185
49. McDermott HJ, McKay CM, Vandali AE (1992) A new portable sound processor for the University of Melbourne/Nucleus Limited multielectrode cochlear implant. J Acoust Soc Am 91(3367–3371):1992
50. Nagal D (1974) Compound action potential of the cochlear nerve evoked electrically. Arch Otorhinolaryngol 206:293–298
51. Ping L, Wang N, Tang G, Lu T, Yin L, Tu W, Fu QJ (2017) Implementation and preliminary evaluation of 'C-tone': a novel algorithm to improve lexical tone recognition in Mandarin-speaking cochlear implant users. Cochlear Implants Int 18(5):240–249
52. Plapous C, Marro C, Scalart P (2006) Improved signal-to-noise ratio estimation for speech enhancement. IEEE Trans Audio Speech Lang Process 14:2098–2108
53. Razavi B (2001) Design of analog CMOS integrated circuits. McGraw-Hill, New York, pp 471–479
54. Ross M, Shaffer H, Cohen A, Freudberg R, Manley H (1974) Average magnitude difference function pitch extractor. IEEE Trans Acoust Speech Signal Process 22:10
55. Schatzer R, Krenmayr A, Au DK, Zierhofer C (2010) Temporal fine structure in Cochlear implants: preliminary speech perception results in Cantonese speaking implant users. Acta Otolaryngol 130:1031–1039
56. Shannon RV, Zeng FG, Kamath V (1995) Speech recognition with primarily temporal cues. Science 270:303–304
57. Skinner MW, Holden LK, Holden TA (1991) Performance of postlinguistically deaf adults with the wearable speech processor (WSP III) and mini speech processor (MSP) of the nucleus multi-electrode Cochlear implant. Ear Hear 12:3–22
58. Skinner MW, Holden LK, Whitford LA, Plant KL, Psarros C, Holden TA (2002) Speech recognition with the nucleus 24 SPEAK, ACE, and CIS speech coding strategies in newly implanted adults. Ear Hear 23:207–223
59. Smoorenburg GF, Willeboer C, van Dijk JE (2002) Speech perception in nucleus CI24M cochlear implant users with processor settings based on electrically evoked compound action potential thresholds. Audiol Neurotol 7:335–347
60. Sokal NO, Sokal AD (1975) Class-E- a new class of high-efficiency tuned single-ended switching power amplifiers. IEEE J Solid State Circuits 10(3):168–176
61. Spahr AJ, Dorman MF (2004) Performance of subjects fit with the advanced bionics CII and nucleus 3G cochlear implant devices. Arch Otolaryngol Head Neck Surg 130:624–628
62. Spriet A, Moonen M, Wouters J (2004) Spatially pre-processed speech distortion weighted multi-channel Wiener filtering for noise reduction. Signal Process 84:2367–2387

63. Stypulkowski PH, van den Honert C (1984) Physiological properties of the electrically stimulated auditory nerve. I. Compound action potential recordings. Hear Res 14:205–223
64. Summers IR, Al-Dabbagh AD (1982) Simulated loss of frequency selectivity and its effects on speech perception. Acoust Lett 5:129–132
65. Tang Z, Smith B, Schild JH, Peckham PH (1995) Data transmission from an implantable biotelemeter by load-shift keying using circuit configuration modulator. IEEE Trans Biomed Eng 42(5):524–528
66. Vary P (1985) Noise suppression by spectral magnitude estimation – mechanism and theoretical limits. Signal Process 8:387–400
67. van de Heyning P, Arauz SL, Atlas M, Baumgartner WD, Caversaccio M, Chester-Browne R et al (2016) Electrically evoked compound action potentials are different depending on the site of cochlear stimulation. Cochlear Implants Int 17:251–262
68. Van den Bogaert T, Doclo S, Wouters J, Moonen M (2009) Speech enhancement with multichannel wiener filter techniques in multimicrophone binaural hearing aids. J Acoust Soc Am 125:360–371
69. Van Dun B, Wouters J, Moonen M (2007) Multi-channel wiener filtering based auditory steady-state response detection. In: IEEE international conference on acoustics, speech and signal processing, Honolulu, Hawaii, USA pp 929–932
70. Van Gerven S, Van Compernolle D (1995) Signal separation by symmetric adaptive decorrelation: stability, convergence, and uniqueness. IEEE Trans Signal Process 43:1602–1612
71. Weinstein E, Feder M, Oppenheim AV (1993) Multi-channel signal separation by decorrelation. IEEE Trans Signal Process 1:405–413
72. Wilson BS, Finley CC, Lawson DT, Wolford RD, Eddington DK, Rabinowitz WM (1991) Better speech recognition with cochlear implants. Nature 352:236–238
73. Wouters J, Vanden Berghe J (2001) Speech recognition in noise for cochlear implantees with a two-microphone monaural adaptive noise reduction system. Ear Hear 22:420–430
74. Zeng FG (2004) Trends in cochlear implants. Trends Amplif 8:1–34
75. Zeng FG, Nie K, Stickney GS, Kong YY, Vongphoe M, Bhargave A, Wei C, Cao K (2005) Speech recognition with amplitude and frequency modulations. Proc Natl Acad Sci U S A 102:2293–2298
76. Zeng FG, Rebscher SJ, Fu QJ, Chen H, Sun X, Yin L, Ping L, Feng H, Yang S, Gong S, Yang B, Kang HY, Gao N, Chi F (2015) Development and evaluation of the Nurotron 26-electrode cochlear implant system. Hear Res 322:188–199
77. Zeng F, Rebscher S, Harrison W, Sun X, Feng H (2008) Cochlear implants: system design, integration, and evaluation. IEEE Rev Biomed Eng 1:115–142

Chapter 8
Neuromodulation for Pain Management

Jing Wang and Zhe Chen

Abstract Pain is a salient and complex sensory experience with important affective and cognitive dimensions. The current definition of pain relies on subjective reports in both humans and experimental animals. Such definition lacks basic mechanistic insights and can lead to a high degree of variability. Research on biomarkers for pain has previously focused on genetic analysis. However, recent advances in human neuroimaging and research in animal models have begun to show the promise of a circuit-based neural signature for pain. At the treatment level, pharmacological therapy for pain remains limited. Neuromodulation has emerged as a specific form of treatment without the systemic side effects of pharmacotherapies. In this review, we will discuss some of the current neuromodulatory modalities for pain, research on newer targets, as well as emerging possibility for an integrated brain-computer interface approach for pain management.

Keywords Pain management · Neuromodulation · Brain-computer interfaces · Circuit-based neural signature

8.1 Introduction

Pain is a complex human experience which contains sensory, affective, and cognitive dimensions. The sensory component of pain provides information about the exact location, timing, and character of a noxious stimulus – the stimulus that provokes pain. The affective component includes feelings of annoyance, anger,

J. Wang (✉)
Department of Anesthesiology, Perioperative Care and Pain Medicine, Department of Neuroscience and Physiology, New York University School of Medicine, New York, NY, USA
e-mail: Jing.Wang2@nyumc.org

Z. Chen
Department of Psychiatry, Department of Neuroscience and Physiology, New York University School of Medicine, New York, NY, USA

X. Zheng (ed.), *Neural Interface: Frontiers and Applications*, Advances in Experimental Medicine and Biology 1101,
https://doi.org/10.1007/978-981-13-2050-7_8

sadness, anxiety, and depression in response to that noxious stimulus. Finally, the cognitive component of pain dictates behavioral responses to this stimulus. During a painful episode, these three components are integrated and contribute to the full experience of pain.

The Webster dictionary definition of pain is "physical suffering or discomfort caused by illness or injury." However, this definition does not provide any anatomical or mechanistic dimensions. Nor does it necessarily delineate the contribution of the three components of pain. Compatible with this vague definition, the clinical diagnosis of pain is established purely by verbal report and thus remains highly subjective. Therefore, an objective diagnosis of pain can greatly improve both our scientific understanding of pain and its clinical management.

At the treatment level, pharmacological options are limited. Opioids remain the most efficacious analgesics, and these drugs have been utilized for pain management for over 1000 years. Opioids, however, carry significant side effects, including constipation, sedation, respiratory depression, immunosuppression, neuro-hormonal dysregulation, dependence, and addiction. There is currently a great debate regarding the use of opioids in the management of chronic non-malignant pain. Despite enormous research over the last century, most of which focuses on spinal and peripheral nervous systems, there have been relatively few new discoveries of pharmacological therapy for pain. Neuromodulation, on the other hand, has emerged as a potentially more specific form of treatment without the systemic side effects of pharmacotherapies. In this review, we will discuss some of the current neuromodulatory treatment modalities for pain, as well as ongoing research to identify newer targets. We will also discuss the possibility for an integrated brain-computer interface approach for pain management that involves both decoding pain signals and neuromodulation of pain.

8.2 Spinal Neuromodulation for Pain

8.2.1 Theory for Spinal Cord Modulation

One of the key concepts in the scientific understanding of pain comes from a classic paper by Melzack and Wall [1]. In this paper, Melzack and Wall argued that there are possible gates of control for nociceptive and non-nociceptive signals at the spinal cord level. More specifically, pyramidal or excitatory neurons in the spinal dorsal horn (SDH) receive sensory inputs – both nociceptive and non-nociceptive – from the peripheral primary sensory neurons, and then in turn transmit this sensory information by projection to the thalamus and brain stem. The SDH, however, also contains interneurons, which are GABAergic and inhibitory. These interneurons also receive peripheral sensory information, but instead of projecting to the brain, they project to other pyramidal neurons in the SDH to inhibit these neurons from providing ascending signals. Thus, non-nociceptive sensory inputs to the SDH can activate interneurons to inhibit the pain signal – in a sense shutting the gate for pain.

Based on this theory for gate control, spinal cord stimulation (SCS) was developed as a way to treat pain. It was thought that external electric stimulator can simulate the firing of interneurons to inhibit ascending pyramidal neurons in the SDH. As a result, SCS can inhibit the transmission of the pain signal from the periphery to the brain. Surgically, SCS includes a trial phase and a permanent implantation phase. During the trial phase, a trial stimulator is implanted into the epidural space under fluoroscopy guidance. The stimulator is then turned on, and the frequency of the stimulation and the location of the stimulator are adjusted over a period of hours to ensure patient comfort and analgesic effects. As the goal of the SCS therapy is to replace the actual pain sensation with paresthesia, the SCS leads need to cover the body area where the patient's pain is generated [2]. If the trial is successful and the patient is able to tolerate the side effects, a permanent implant is then done in the operating room. The permanent implantation is procedurally similar to the trial lead placement, but the leads and battery are implanted in a pocket inside the patient. The battery can be remotely charged. In some cases, permanent lead implantation is done by laminectomy and the leads are sutured to the dura to ensure stable placement [2].

8.2.2 *Current Clinical Practice*

The frequency of spinal cord stimulation is typically 40–60 Hz [2]. Newer advances, however, include the use of very high-frequency stimulation (10,000 Hz) and burst stimulation. Through mechanisms that are not entirely understood, such higher frequency stimulations have been shown in a limited number of preliminary studies to confer superior pain relief compared with traditional stimulation protocols [3–7].

Numerous studies have confirmed the analgesic efficacy for SCS. In two recent meta-analyses of SCS treatment, it was found that >50% of the patients experienced pain relief with an average follow-up of 24 months [8, 9]. The pain syndromes most responsive to therapy include lumbosacral radiculitis that is refractory to pharmacological and interventional treatments, failed back surgery syndrome, diabetic peripheral neuropathy, and other distal peripheral neuropathic syndromes [2]. In Europe, SCS has also been utilized for ischemic chest pain [2].

8.2.3 *Limitations of Spinal Modulation for Pain*

Despite its efficacy in pain treatment, there are several distinct disadvantages for SCS. First, the SCS is implanted in the epidural space, and with a constant movement of the spine, there is a high risk for lead displacement, with rates reported to be as high as 30% in some studies [10]. Another key disadvantage of SCS is that it requires constant stimulation. This is because while SCS was developed as a neuromodulation technique, it was not developed as a demand-based neuromodulation technique. As a result, many patients develop tolerance to SCS, and in a significant subset of patients, SCS loses its analgesic efficacy over time. In addition, continuous stimulation makes side effects more problematic. The main side effect of SCS is paresthesia. Over time, many patients elect to turn off the SCS due to the continuous bothersome paresthesia.

Theoretically, a demand-based approach to neuromodulation can alleviate many of the side effects associated with SCS. SCS would be turned on only when pain is sensed. This way, the patient will not experience a long period of paresthesia when there is no pain signal. The design for a demand-based neuromodulation approach, however, requires real-time decoding of pain signals, which has not been achieved at the level of SDH and DRG neurons.

8.3 Brain Neuromodulation for Pain

After it first emerged 70 years ago, brain neuromodulation has been studied to be a treatment modality for refractory pain conditions. Some of the earlier studies are in rodents. A number of brain targets have been sought, with the overall goal of reducing the activities in these pain-producing or processing regions with high-frequency stimulations in a protocol that is similar to deep brain stimulation (DBS) utilized for Parkinson's disease [11]. In some cases, direct stimulation with lower frequency has been attempted to activate the region of interest. These targets include the anterior cingulate cortex (ACC), motor cortex, sensory thalamus, intralaminar parafascicular complex (CMP), periacqueductal gray (PAG), and nucleus accumbens (NAc). A number of studies targeting these areas have shown a varying degree of success in pain relief.

8.3.1 Current Practice of Brain Modulation for Pain

Cortical Stimulations

The ACC is well known to be a critical hub for the processing for the affective component of pain [12–22]. Not surprisingly, it was one of the first regions to be targeted for pain relief. The first case series of electrical stimulation of the ACC as a treatment for chronic pain was published in 1960 [23]. A large number of these cases involved pain from spinal cord injuries [24, 25]. In these studies, bilateral stimulation using a DBS protocol achieved temporary pain relief as measured by the patient's pain reporting and medication dosages. In particular, patients who received ACC stimulation reported a decrease in the affective component of pain. It was reported that these patients still felt pain, but "it didn't bother" them as much. However, pain relief was not found to be enduring. In a more recent study of 15 patients with chronic pain, chronic DBS of the ACC using implanted electrodes produced an analgesic effect in five patients [26]. The side effects have in general been mild in most studies.

Although not a region known for pain processing, the motor cortex has also been targeted for pain relief. Motor cortex stimulation (MCS) has been evaluated particularly in central pain syndromes such as post-stroke pain or refractory trigeminal neuralgia. The rationale for MCS comes from the observation that motor cortex stimulation achieved sensory responses, and treatment as early as in 1971 has shown partial analgesic effects [27]. Modern approach to MCS began with a study in 1991, where eight of 12 patients were able to experience some degree of pain relief after 1 year of MCS with implanted electrodes [28]. A number of studies since then have demonstrated efficacy for MCS, particularly for post-stroke pain syndrome, where up to 70% of patients responded positively, and facial pain, with up to 80% efficacy [29–34]. In these studies, a chronic implant with stimulation of 5–130 Hz has been investigated to be efficacious.

Subcortical Stimulations

The role of the sensory thalamus in pain regulation is well established, as studies have shown that thalamic injury such as in the case of stroke can lead to persistent central pain or post-stroke pain. The thalamus is a highly heterogeneous structure, and the regions involved in pain processing include the ventroposterior lateral (VPL) and ventroposterior medial (VPM) nuclei. Whereas the lateral thalamus is thought to project to the somatosensory cortex (S1) and provide accurate sensory information related to pain, the VPM has been shown to project to a variety of cortical and subcortical structures and is thought to be involved in the affective processing of pain. Hyperactivity of the thalamus has been observed in neuropathic pain conditions [35–37]. In an early and well-cited study, four of five facial pain patients who received chronic VPM stimulation demonstrated pain relief [38]. Subsequently, stimulation of both VPM and VPL regions have been carried in a number of studies. Studies have shown that contralateral VPM stimulation can result in persistent analgesia as well as paresthesia, whereas study with neuropathic pain patients showed that thalamic stimulation can result in improvement in pain scores [39–41]. The technical difficulty associated with thalamic stimulation, however, is considerable, due to its position in the brain and its numerous subdivisions with highly divergent functions. Perhaps as a result of these anatomic and structural concerns, more recent studies have shown more conflicting results [42, 43].

The PAG projects to the rostroventral medial medulla (RVM), which in turn projects to the spinal dorsal horn neurons, and depending on the exact neuronal subgroups and circuit dynamics, this projection can provide either descending inhibition or facilitation of ascending pain signals [44]. Given its well-established involvement in descending pain regulation, the PAG has been pursued as a promising target in a number of studies [45–47]. Recent data, however, did not demonstrate consistent and enduring pain relief [24, 43, 48–51]. Some of these studies showed that PAG stimulation results in pain relief, whereas other studies did not demonstrate persistent analgesic effect. These divergent results may be due to the fact that the PAG can project to the RVM, but it projects to two distinct populations of RVM cells. Whereas the off cells in the RVM produce pain relief, the on cells actually give rise to pain intensification. As a result, a constant stimulation approach without differentiating on- or off-cells in the PAG may not be highly effective. Furthermore, PAG stimulation appears to cause significant side effects including nausea, nystagmus, vertigo, and nausea [45].

Additional targets in the brain include the NAc, CMP, hypothalamus (HT), and the internal capsule (IC). The NAc forms the ventral striatum and is known for its role in reward-based behaviors [52]. However, the NAc has also been shown in recent studies to play a crucial role in pain and analgesia [53–55]. Specifically, studies have shown that the projection from the medial prefrontal cortex (PFC) to the NAc can produce relief of both sensory and affective components of pain in acute as well as chronic pain conditions [56]. There are in fact case studies of post-stroke (central) pain where stimulation of the NAc when combined with PAG stimulation can achieve significant pain relief [57]. Meanwhile, the CMP has high levels of expression of opioid receptors and is anatomically connected to the thalamus, ACC, and NAc, and thus it has also been postulated to function as a target for pain treatment [58]. Surgical stimulation of the CMP, however, has achieved variable results. Some studies have shown CMP stimulation to have modest analgesic effects in central or neuropathic pain syndromes, whereas other studies have failed to reproduce such results with any significant time frame [59–61]. Finally, a recent number of studies have shown that DBS of the posterior HT demonstrated relief for cluster headache as well as for cephalgias [62–64]. The rationale for this approach is that cluster headache has a circadian component, which in turn is regulated in large part by the posterior HT [65]. These studies, however, are relatively small in sample size. Meanwhile, the earliest study on IC stimulation in 1974 showed analgesic efficacy [66, 67]. Subsequent studies in this area have also shown that DBS in this area may have a role in reducing neuropathic pain [68–70]. The mechanisms for the analgesic effects of IC stimulation, however, are not well understood.

8.3.2 Technical Considerations for Brain Modulation for Pain

Electrical leads for MCS or DBS are typically two side by side four-contact electrodes. MCS is achieved with an amplitude of 0.5–10 volts, with frequency ranging from 5 to 130 Hz, with 40–100 Hz being the most common, and pulse widths spanning from 60 to 450 ms. Meanwhile, standard DBS electrodes can be implanted for pain similar to implants for Parkinson's patients. Unfortunately, due to the small number of studies, the difficulty in enrollment, and the nature of the implantation surgery, MCS and DBS protocols have not been well investigated in studies using large sample sizes. Important factors for consideration for cortical and subcortical stimulation include the duration of stimulation, frequency of stimulation, and the possibility of low-frequency stimulation to activate neurons vs high-frequency stimulation to inhibit neuronal activities. The desired outcome would be for the device to perform stimulation or inhibition of the neuronal networks during the occurrence of pain episodes. The ultimate goal, meanwhile, is to achieve demand-based automated stimulation. The accomplishment of this goal, however, requires accurate decoding of the pain signals in the brain.

8.4 Decoding Pain in the Brain

8.4.1 *Electrophysiological Recordings in Rodents and Non-human Primates*

From a neural coding perspective, the goal of pain encoding is to understand how neurons represent and transmit the information of pain. Current decoding of pain relies on neuroimaging techniques such as fMRI or PET. These neuroimaging studies have a few critical limitations. First, at the anatomical level, fMRI imaging experiments cannot provide desirable single-cell resolution. Second, fMRI measurements reflect the blood-flow changes, and such changes indirectly indicate brain activity. Therefore, it only measures the correlation between blood flow and neural activity instead of direct neurophysiological measurements. Nociceptive information is distributed so widely that blood-flow changes sometimes cannot accurately depict primary and secondary neural responses. Third, imaging studies cannot identify the causal relationship between circuit change and pain behaviors. This lack of causality limits the mechanistic insight that neuronal codes can provide. Fourth, the limited temporal resolution of imaging studies precludes the analysis of ultrafast circuit and synaptic events. It is the last constraint that makes various imaging techniques less ideal tools for real-time decoding in the brain. Due to these limitations, circuit mechanisms of pain, particularly the affective and cognitive dimensions of pain, remain poorly defined. On the other hand, in vivo electrophysiological recordings in animal studies have provided insight on pain mechanism at cellular and circuit levels and at a millisecond timescale. Below we will review a few important electrophysiological studies in rodents and non-human primates. We focus on the primary neural circuits that modulate the pain stimuli, including thermal, mechanical, inflammatory, and neuropathic pain.

Recent animal research on pain has focused on spinal and peripheral systems. Spinal cord and peripheral systems mostly inform nociception. Pain, however, is a multi-dimensional experience and includes sensory, affective, and cognitive dimensions. Therefore, attempts to decode pain should ideally include circuitry in the brain. Meanwhile, there is no "pain cortex," as pain is perceived and processed in a distributed network [71]. Among the many brain areas, the S1 and ACC are two of the most studied areas for pain-related responses. The S1 has long been thought to encode the sensory component of pain, whereas the ACC has been known as a key anatomic region for understanding the aversive and other affective manifestations of pain [72, 73]. Other brain regions that are engaged in pain processing include the medial prefrontal cortex (mPFC) [74], insular cortex [75],

NAc [76], thalamus including the thalamic reticular nucleus (TRN), medial dorsal (MD) nucleus, and other thalamic nuclei [77, 78].

In rodent experiments, neurophysiological recordings in pain studies were reported in both anesthetized and freely behaving animals. For instance, it was found that neurons in the TRN and ventroposterior lateral nucleus responded to CO_2 laser in anesthetized rats [79]. These neurons have shown specificity in their responses and hence have been thought to be nociceptive thalamic neurons. However, in another independent report, it was found that none of the mechanical nociceptive neurons showed a clear relationship between discharge and thermal pain temperature [80]. The differences in these studies may be attributed to the fact that animals were under anesthesia, which can make the magnitude and duration of the sensory information transfer variable.

Physiological recordings in conscious, behaving rats have been collected in S1 and ACC [81–83], as well as in VPL and MD [83]. Specifically, S1 and VPL are within the lateral sensory pain pathway, whereas ACC and MD are within the medial pathway more involved in the affective and cognitive processing of pain. Kuo and Yen [81] compared noxious responses in S1 and ACC and observed that ACC neurons were less frequently activated and had a longer latency of onset in response to thermal pain induced by a CO_2 laser [81]. Additional experiments showed that neuronal responses in both the S1 and ACC increased with the intensity of the laser heat; and the stimulus-response curve of S1 ensemble activity had a steeper slope than that of the ACC. Furthermore, S1 neurons were better at coding laser-heat intensity than ACC neurons, whereas more ACC neurons were involved in conditioned fear associated with an electric shock than S1 neurons. These results suggest differential contributions of the S1 and ACC to sensory and affective dimensions of pain. With the simultaneous recording of multiple brain areas in lateral and medial pain pathways, Zhang et al. [83] reported significant correlations between laser intensity and the number of responsive neurons, the firing rates, as well as the mass spike counts. In addition, their results also showed a comparable coding capacity of medial and lateral pain pathway neurons [83]. Furthermore, Zhang et al. used a linear discriminant analysis for pain intensities based on the mass spike counts of neuronal recordings from multiple brain areas. Although several electrophysiological investigations in rats similar to the above studies have established that activities in the S1 and ACC are altered by pain stimuli, very little effort has been devoted specifically to decoding analysis, especially for the onset detection of pain signals.

In non-human primate experiments, Chudler et al. [84] have found that nociceptive neurons in the S1 of anesthetized monkeys were responsive to noxious thermal stimulation applied to the face [84]. One type of nociceptive S1 neurons showed increased firing rate with increasing stimulus intensity, whereas another type of nociceptive S1 neurons showed saturated response once reaching a plateau above a specific temperature. At the behavioral level, the intensity of pain sensation or pain rating in some ways parallels the neuronal responses of nociceptive S1 neurons. In another study, Iwata et al. [85] recorded ACC neurons from awake behaving monkeys [85]. Positively and negatively modulated ACC neurons were reported in response to a heat stimulus on the face. Their results further suggested that ACC neurons are involved in the attention to pain and escape from pain but not in the sensory-discriminative aspect of pain.

8.5 Potential for a Demand-Based Neuromodulation Approach

From a treatment perspective, a neuromodulation is a potential option for refractory pain. As discussed above, both spinal and brain stimulation has the potential for modulating the pain experience [86]. In addition to therapeutic benefits, the use of brain stimulation also allows mechanistic understanding of brain circuits that underlie the pain experience. However, current techniques such as deep brain stimulation or transcranial magnetic stimulation (TMS) lack optimal targets in the brain and require constant stimulation resulting in unacceptable side effects.

Demand-based neuromodulation, or closed-loop control for pain, provides an appealing alternative strategy. This strategy can be combined with either spinal or brain modulation. The realization of a demand-based neuromodulatory approach depends critically on two factors: first, accurate and timely identification of the pain signal, and second, low-latency neuromodulatory control. In other words, timing matters in closed-loop neuromodulation therapy [87].

8.5.1 Challenges for Demanded-Based Pain Modulation

In neuroscience and clinical experiments, a closed-loop brain-machine interface (BMI) offers a powerful platform to investigate the causal link between brain circuits and behavior [88]. A basic closed-loop BMI system for pain modulation (Fig. 8.1) would consist of a detection arm (which is aimed for detecting the onset of pain signals) and a treatment arm (which is aimed for delivering neuromodulation).

For the detection arm, the central challenge in change-point detection is the speed and accuracy of detection. In real-time processing, the objective of *quickest detection* is to detect changes as quickly as possible in an online fashion, and the requirement of time-delay for closed-loop feedback usually lies within tens or hundreds of milliseconds.

For the treatment arm, the first step is to identify effective neural circuits for modulating pain. Some candidate targets have been discussed above, including the nucleus accumbens [89], ACC [90], and thalamus [79, 91]. The second step is to apply neuromodulation methods. There are also a few strategies based on invasive or noninvasive stimulations. In animal studies, the emergence of optogenetics [56, 92–96] and chemogenetics [96] offer intriguing options. While not yet applicable in humans due to the nature of viral transfer, these techniques allow quick and

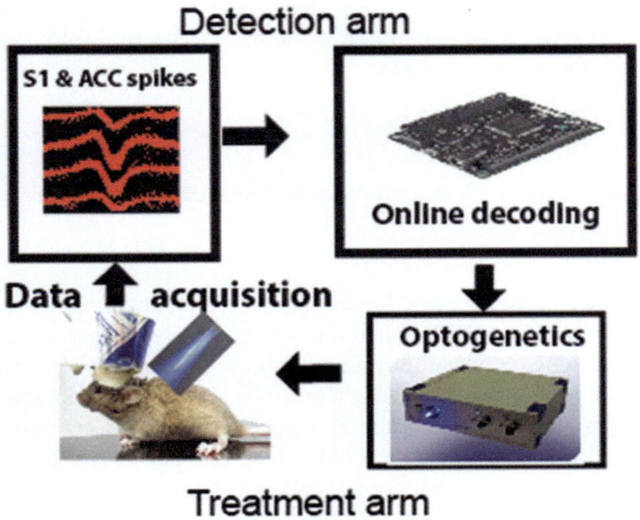

Fig. 8.1 A schematic of the rodent closed-loop BMI system. A closed-loop system consists of the detection arm (detection of the onset of pain signals via online neural decoding) and treatment arm (neuromodulation via electrical or optogenetic stimulation)

targeted methods to alter brain circuits. Such techniques can be used to perform feasibility studies for therapeutics as well as to probe circuits in basic mechanistic investigations.

8.5.2 Pilot Rodent Studies

In our investigations, we have used a rat model to study mechanisms of acute pain, including acute thermal pain and acute mechanical pain. In thermal pain experiments, we used a blue diode-pumped solid-state laser applied 1 mm from the plantar surface of the hind paw contralateral to the brain recording site to generate pain in freely moving rats. The laser was turned on by a transistor-to-transistor pulse generator until paw withdrawal was observed. The fiber output power was calibrated by compact power and energy at the beginning of every recording day [90, 97]. We used custom-designed microdrives (32 channels) to simultaneously record neuronal ensemble spike activity from the rat S1 and ACC areas. Both spike (>300 Hz) and local field potential (<300 Hz) data were obtained.

For detection of the onset of pain signals, we have designed model-based and model-free algorithms based on the input of (off-line or on-line) sorted neuronal ensemble spike activity [97–99].

A Model-Based Approach

Notations: Given C neuronal recordings, let $\boldsymbol{y}_k = [y_{1,k}, \ldots, y_{C,k}]$ denote a C-dimensional population vector, with each element consisting of the neuronal spike count at the kth time bin (bin size Δ); let exp.($\boldsymbol{\eta}_k$) denote the Poisson firing rate vector for C neurons. We assumed that the data follows a latent-state Poisson linear dynamical system (PLDS), where the latent univariate variable z_k denotes an unobserved common input that drives neuronal ensemble spiking activity $\boldsymbol{y}_k$ as follows [97, 98]:

$$z_k = az_{k-1} + \epsilon_k \tag{8.1}$$

$$\boldsymbol{\eta}_k = \mathbf{c}z_k + \mathbf{d}, \tag{8.2}$$

$$\boldsymbol{y}_k \tilde{\ } Poisson(\exp(\boldsymbol{\eta}_k)\Delta), \tag{8.3}$$

where the state Eq. (8.1) is a first-order autoregressive (AR) model ($0 < |a| < 1$) driven by a Gaussian noise process with zero mean and variance $N(0, \sigma^2)$. We used an iterative expectation-maximization (EM) algorithm to estimate the latent state (E-step) and the unknown parameters $\{a, \mathbf{c}, \mathbf{d}, \sigma^2\}$ (M-step). Alternatively, we transformed the Poisson population vector (nonnegative) into a Gaussian random vector and obtained a transformed linear dynamical system (TLDS). In this case, the inference procedure was simpler due to the approximation methods. In on-line applications, once the model parameters are identified, we used a recursive (forward) filter to estimate the latent state variable [97, 99].

From the estimated latent state $\hat{z}_k$, we further computed the Z-score related to the baseline: $Z = \frac{z - \text{mean}(z_{\text{baseline}})}{\text{SD}(z_{\text{baseline}})}$. Under the assumption that the Z-score is standard normally distributed, we converted it to the one-tailed P-value. The criterion of Z-score change was determined by a critical threshold for reaching statistical significance.

A Model-Free Approach

We designed a greedy CUSUM (cumulative sum) algorithm. The CUSUM method is aimed to detect a change point based on the instantaneous log-likelihood ratio (LLR):

$$\text{LLR}_{f_1 \| f_0} = y\log\frac{\lambda_1}{\lambda_0} - (\lambda_1 - \lambda_0) \tag{8.4}$$

where λ_0 and $\lambda_1$are the rate parameters of two Poisson distributions $f_0$and f_1. We assume that the baseline rate λ_0 is known or can be estimated directly from data. We define an empirical criterion for a significant change in λ_1 such that $\lambda_1^{sig} = \lambda_0 \pm 2\sqrt{\lambda_0}$ in either a positive or negative direction. The CUSUM algorithm computes the instantaneous LLR $s_{c,k}$ for the cth neuron and update the cumulative sum as follows:

$$S_k = \max_c\{S_{c,k}\} = \max_c\{\max\{0, S_{c,k-1} + s_{c,k}\}\} \tag{8.5}$$

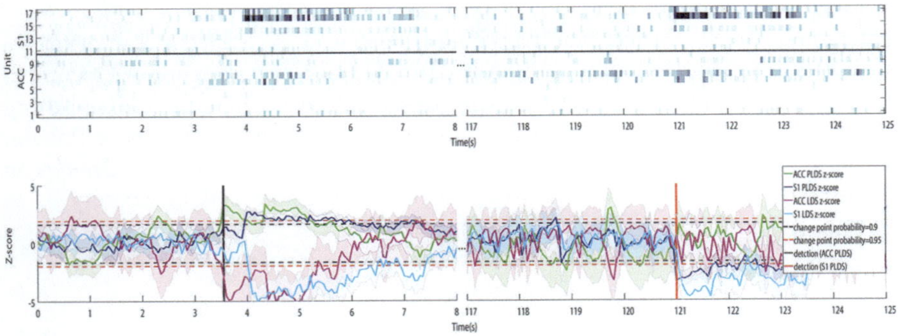

Fig. 8.2 Online detection of the onset of thermal acute pain. *Top*: Spike counts of simultaneously recorded, online-sorted 17 units (seven S1 units and ten ACC units) from a freely behaving rat. *Bottom*: Continuous tracking and detection of pain signals derived from the rat S1 and ACC activity. Shaded area denotes the CI of Z-score. Vertical lines denote the moments of change-point detection. In this example, the detection results of the PLDS were better than those derived from the TLDS and model-free method (not shown) [100]

where $S_0 = 0$. The technical details are referred to [99]. Once the cumulative sum S_k is above a predetermined threshold θ_0 and the trend continuous more than two consecutive steps, then a decision is made and S_k is reset. The threshold θ_0 in the CUSUM algorithm controls the false alarm rate. An empirical choice is to use the test statistic (twofold log-likelihood) being a chi-square distribution with 1 degree of freedom: $\chi^2_{1,(1-\alpha)}$. If we set $\alpha = 0.01$, we have $\theta_0 = 0.5 \times 6.64 = 3.38$.

To illustrate these two methods, we apply these two methods to simultaneous recordings of rat S1 and ACC neuronal populations. Two representative examples of online detection of pain signals are shown in Fig. 8.2.

Therefore, as these pilot studies demonstrate, the S1 and ACC neuronal ensemble activity encode important information about pain signals. Even more importantly, these experiments demonstrate that integrating neural signals from multiple brain areas will further improve the decoding accuracy for the onset of pain. Developing online statistical methods for detecting the onset of pain signals in closed-loop neural interfaces can open new research opportunities for both acute and chronic pain. Such methods, when combined with spinal or brain modulation, can have the potential to achieve pain control with limited side effects.

8.6 Summary

Pain is a complex experience. In a subset of patients who suffer from chronic pain, it is a debilitating experience. The current state-of-the-art neuromodulation therapy is spinal cord stimulation. While this treatment modality is well validated, it continues to have side effects. Neuromodulation in the brain, meanwhile, has also been explored, but optimal targets remain lacking. Recent advances in neuro-engineering and computational neuroscience, however, when combined with an improved understanding of neurocircuitry of pain, allow for the possibility of demand-based neuromodulation of pain. The possibility of a brain-computer interface approach can allow better mechanistic understanding of basic pain mechanisms, and at the same time, it can provide much-needed treatment for refractory pain syndromes.

References

1. Melzack R, Wall PD (1965) Pain mechanisms: a new theory. Science 150(3699):971–979
2. Deer TR et al (2014) The appropriate use of neurostimulation of the spinal cord and peripheral nervous system for the treatment of chronic pain and ischemic diseases: the Neuromodulation Appropriateness Consensus Committee. Neuromodulation 17(6):515–550. discussion 550
3. Liem L et al (2015) One-year outcomes of spinal cord stimulation of the dorsal root ganglion in the treatment of chronic neuropathic pain. Neuromodulation 18(1):41–48. discussion 48-9
4. De Ridder D et al (2013) Burst spinal cord stimulation for limb and back pain. World Neurosurg 80(5):642–649 e1
5. De Ridder D et al (2010) Burst spinal cord stimulation: toward paresthesia-free pain suppression. Neurosurgery 66(5):986–990
6. Van Buyten JP et al (2013) High-frequency spinal cord stimulation for the treatment of chronic back pain patients: results of a prospective multicenter European clinical study. Neuromodulation 16(1):59–65. discussion 65-6
7. Tiede J et al (2013) Novel spinal cord stimulation parameters in patients with predominant back pain. Neuromodulation 16(4):370–375
8. Taylor RS et al (2014) Predictors of pain relief following spinal cord stimulation in chronic back and leg pain and failed back surgery syndrome: a systematic review and meta-regression analysis. Pain Pract 14(6):489–505
9. Grider JS et al (2016) Effectiveness of spinal cord stimulation in chronic spinal pain: a systematic review. Pain Physician 19(1):E33–E54
10. Deer TR et al (2014) The appropriate use of neurostimulation: avoidance and treatment of complications of neurostimulation therapies for the treatment of chronic pain. Neuromodulation 17(6):571–598
11. Levy RM, Lamb S, Adams JE (1987) Treatment of chronic pain by deep brain stimulation: long term follow-up and review of the literature. Neurosurgery 21(6):885–893
12. Turnbull IM (1972) Bilateral cingulumotomy combined with thalamotomy or mesencephalic tractotomy for pain. Surg Gynecol Obstet 134(6):958–962
13. Craig AD et al (1996) Functional imaging of an illusion of pain. Nature 384(6606):258–260
14. Rainville P et al (1997) Pain affect encoded in human anterior cingulate but not somatosensory cortex. Science 277(5328):968–971
15. Foltz EL, White LE (1968) The role of rostral cingulumotomy in "pain" relief. Int J Neurol 6(3–4):353–373
16. Talbot JD et al (1995) Evaluation of pain perception after anterior capsulotomy: a case report. Somatosens Mot Res 12(2):115–126
17. Koyama T, Kato K, Mikami A (2000) During pain-avoidance neurons activated in the macaque anterior cingulate and caudate. Neurosci Lett 283(1):17–20
18. Koyama T et al (2001) Anterior cingulate activity during pain-avoidance and reward tasks in monkeys. Neurosci Res 39(4):421–430
19. Qu C et al (2011) Lesion of the rostral anterior cingulate cortex eliminates the aversiveness of spontaneous neuropathic pain following partial or complete axotomy. Pain 152(7):1641–1648

20. Johansen JP, Fields HL, Manning BH (2001) The affective component of pain in rodents: direct evidence for a contribution of the anterior cingulate cortex. Proc Natl Acad Sci U S A 98(14):8077–8082
21. LaGraize SC et al (2006) Selective regulation of pain affect following activation of the opioid anterior cingulate cortex system. Exp Neurol 197(1):22–30
22. Lubar JF (1964) Effect of medial cortical lesions on the avoidance behavior of the cat. J Comp Physiol Psychol 58:38–46
23. Lewin W, Whitty CW (1960) Effects of anterior cingulate stimulation in conscious human subjects. J Neurophysiol 23:445–447
24. Boccard SG et al (2013) Long-term outcomes of deep brain stimulation for neuropathic pain. Neurosurgery 72(2):221–230. discussion 231
25. Spooner J et al (2007) Neuromodulation of the cingulum for neuropathic pain after spinal cord injury. Case report. J Neurosurg 107(1):169–172
26. Boccard SG et al (2014) Deep brain stimulation of the anterior cingulate cortex: targeting the affective component of chronic pain. Neuroreport 25(2):83–88
27. Lende RA, Kirsch WM, Druckman R (1971) Relief of facial pain after combined removal of precentral and postcentral cortex. J Neurosurg 34(4):537–543
28. Tsubokawa T et al (1991) Chronic motor cortex stimulation for the treatment of central pain. Acta Neurochir Suppl (Wien) 52:137–139
29. Katayama Y, Tsubokawa T, Yamamoto T (1994) Chronic motor cortex stimulation for central deafferentation pain: experience with bulbar pain secondary to Wallenberg syndrome. Stereotact Funct Neurosurg 62(1–4):295–299
30. Katayama Y, Fukaya C, Yamamoto T (1998) Poststroke pain control by chronic motor cortex stimulation: neurological characteristics predicting a favorable response. J Neurosurg 89(4):585–591
31. Herregodts P et al (1995) Cortical stimulation for central neuropathic pain: 3-D surface MRI for easy determination of the motor cortex. Acta Neurochir Suppl 64:132–135
32. Ebel H et al (1996) Chronic precentral stimulation in trigeminal neuropathic pain. Acta Neurochir 138(11):1300–1306
33. Nguyen JP et al (2000) Treatment of central and neuropathic facial pain by chronic stimulation of the motor cortex: value of neuronavigation guidance systems for the localization of the motor cortex. Neurochirurgie 46(5):483–491
34. Garcia-Larrea L et al (1999) Electrical stimulation of motor cortex for pain control: a combined PET-scan and electrophysiological study. Pain 83(2):259–273
35. Lenz FA et al (1989) Characteristics of the bursting pattern of action potentials that occurs in the thalamus of patients with central pain. Brain Res 496(1–2):357–360
36. Tasker RR et al (1987) Thalamic microelectrode recording and microstimulation in central and deafferentation pain. Appl Neurophysiol 50(1–6):414–417
37. Lenz FA et al (1987) Abnormal single-unit activity recorded in the somatosensory thalamus of a quadriplegic patient with central pain. Pain 31(2):225–236
38. Hosobuchi Y, Adams JE, Rutkin B (1973) Chronic thalamic stimulation for the control of facial anesthesia dolorosa. Arch Neurol 29(3):158–161
39. Mazars G, Merienne L, Cioloca C (1974) Treatment of certain types of pain with implantable thalamic stimulators. Neurochirurgie 20(2):117–124
40. Mazars G, Merienne L, Ciolocca C (1973) Intermittent analgesic thalamic stimulation. Preliminary note. Rev Neurol (Paris) 128(4):273–279
41. Turnbull IM, Shulman R, Woodhurst WB (1980) Thalamic stimulation for neuropathic pain. J Neurosurg 52(4):486–493
42. Marchand S et al (2003) Analgesic and placebo effects of thalamic stimulation. Pain 105(3):481–488
43. Coffey RJ (2001) Deep brain stimulation for chronic pain: results of two multicenter trials and a structured review. Pain Med 2(3):183–192

44. Heinricher MM, Cheng ZF, Fields HL (1987) Evidence for two classes of nociceptive modulating neurons in the periaqueductal gray. J Neurosci 7(1):271–278
45. Richardson DE, Akil H (1977) Pain reduction by electrical brain stimulation in man. Part 1: acute administration in periaqueductal and periventricular sites. J Neurosurg 47(2):178–183
46. Richardson DE, Akil H (1977) Pain reduction by electrical brain stimulation in man. Part 2: chronic self-administration in the periventricular gray matter. J Neurosurg 47(2):184–194
47. Hosobuchi Y, Adams JE, Linchitz R (1977) Pain relief by electrical stimulation of the central gray matter in humans and its reversal by naloxone. Science 197(4299):183–186
48. Nandi D et al (2003) Thalamic field potentials in chronic central pain treated by periventricular gray stimulation -- a series of eight cases. Pain 101(1–2):97–107
49. Owen SL et al (2007) Deep brain stimulation for neuropathic pain. Acta Neurochir Suppl 97(Pt 2):111–116
50. Green AL et al (2004) N-of-1 trials for assessing the efficacy of deep brain stimulation in neuropathic pain. Neuromodulation 7(2):76–81
51. Bittar RG et al (2005) Deep brain stimulation for pain relief: a meta-analysis. J Clin Neurosci 12(5):515–519
52. Nestler EJ, Carlezon WA Jr (2006) The mesolimbic dopamine reward circuit in depression. Biol Psychiatry 59(12):1151–1159
53. Goffer Y et al (2013) Calcium-permeable AMPA receptors in the nucleus accumbens regulate depression-like behaviors in the chronic neuropathic pain state. J Neurosci 33(48):19034–19044
54. Su C et al (2015) Persistent pain alters AMPA receptor subunit levels in the nucleus accumbens. Mol Brain 8(1):46
55. Baliki MN et al (2010) Predicting value of pain and analgesia: nucleus accumbens response to noxious stimuli changes in the presence of chronic pain. Neuron 66(1):149–160
56. Lee M et al (2015) Activation of corticostriatal circuitry relieves chronic neuropathic pain. J Neurosci 35(13):5247–5259
57. Mallory GW, Abulseoud O, Hwang SC, Gorman DA, Stead SM, Klassen BT, Sandroni P, Watson JC, Lee KH (2012) The nucleus accumbens as a potential target for central poststroke pain. Mayo Clin Proc 87(10):1025–1031
58. Weigel R, Krauss JK (2004) Center median-parafascicular complex and pain control. Review from a neurosurgical perspective. Stereotact Funct Neurosurg 82(2–3):115–126
59. Andy OJ (1980) Parafascicular-center median nuclei stimulation for intractable pain and dyskinesia (painful-dyskinesia). Appl Neurophysiol 43(3–5):133–144
60. Krauss JK et al (2002) Deep brain stimulation of the centre median-parafascicular complex in patients with movement disorders. J Neurol Neurosurg Psychiatry 72(4):546–548
61. Ray CD, Burton CV (1980) Deep brain stimulation for severe, chronic pain. Acta Neurochir Suppl (Wien) 30:289–293
62. Franzini A et al (2003) Stimulation of the posterior hypothalamus for treatment of chronic intractable cluster headaches: first reported series. Neurosurgery 52(5):1095–1099. discussion 1099-101
63. Schoenen J et al (2005) Hypothalamic stimulation in chronic cluster headache: a pilot study of efficacy and mode of action. Brain 128.(Pt 4:940–947
64. Green AL et al (2006) Deep brain stimulation for neuropathic cephalalgia. Cephalalgia 26(5):561–567
65. Magis D, Schoenen J (2012) Advances and challenges in neurostimulation for headaches. Lancet Neurol 11(8):708–719
66. Fields HL, Adams JE (1974) Pain after cortical injury relieved by electrical stimulation of the internal capsule. Brain 97(1):169–178
67. Adams JE, Hosobuchi Y, Fields HL (1974) Stimulation of internal capsule for relief of chronic pain. J Neurosurg 41(6):740–744
68. Hosobuchi Y, Adams JE, Rutkin B (1975) Chronic thalamic and internal capsule stimulation for the control of central pain. Surg Neurol 4(1):91–92

69. Namba S et al (1984) Electrical stimulation of the posterior limb of the internal capsule for treatment of thalamic pain. Appl Neurophysiol 47(3):137–148
70. Namba S et al (1985) Sensory and motor responses to deep brain stimulation. Correlation with anatomical structures. J Neurosurg 63(2):224–234
71. Mano H, Seymour B (2015) Pain: a distributed brain information network? PLoS Biol 13(1): e1002037
72. Sewards TV, Sewards MA (2002) The medial pain system: neural representations of the motivational aspect of pain. Brain Res Bull 59(3):163–180
73. Bushnell MC, Ceko M, Low LA (2013) Cognitive and emotional control of pain and its disruption in chronic pain. Nat Rev Neurosci 14(7):502–511
74. Apkarian AV et al (2004) Chronic back pain is associated with decreased prefrontal and thalamic gray matter density. J Neurosci 24(46):10410–10415
75. Segerdahl AR et al (2015) The dorsal posterior insula subserves a fundamental role in human pain. Nat Neurosci 18(4):499–500
76. Baliki MN et al (2012) Corticostriatal functional connectivity predicts transition to chronic back pain. Nat Neurosci 15(8):1117–1119
77. Lenz FA et al (2004) The role of the thalamus in pain. Suppl Clin Neurophysiol 57:50–61
78. Dostrovsky JO (2000) Role of thalamus in pain. Prog Brain Res 129:245–257
79. Yen CT, Shaw FZ (2003) Reticular thalamic responses to nociceptive inputs in anesthetized rats. Brain Res 968(2):179–191
80. Peschanski M, Guilbaud G, Gautron M (1981) Posterior intralaminar region in rat: neuronal responses to noxious and nonnoxious cutaneous stimuli. Exp Neurol 72(1):226–238
81. Kuo CC, Yen CT (2005) Comparison of anterior cingulate and primary somatosensory neuronal responses to noxious laser-heat stimuli in conscious, behaving rats. J Neurophysiol 94(3):1825–1836
82. Kuo CC et al (2009) Differential involvement of the anterior cingulate and primary sensorimotor cortices in sensory and affective functions of pain. J Neurophysiol 101(3):1201–1210
83. Zhang Y et al (2011) Ensemble encoding of nociceptive stimulus intensity in the rat medial and lateral pain systems. Mol Pain 7:64
84. Chudler EH et al (1990) Responses of nociceptive SI neurons in monkeys and pain sensation in humans elicited by noxious thermal stimulation: effect of interstimulus interval. J Neurophysiol 63(3):559–569
85. Iwata K et al (2005) Anterior cingulate cortical neuronal activity during perception of noxious thermal stimuli in monkeys. J Neurophysiol 94(3):1980–1991
86. Riva-Posse P et al (2013) Practical considerations in the development and refinement of subcallosal cingulate white matter deep brain stimulation for treatment-resistant depression. World Neurosurg 80(3–4):S27.e25–S27.e34
87. Sun FT, Morrell MJ (2014) Closed-loop neurostimulation: the clinical experience. Neurotherapeutics 11(3):553–563
88. Thakor NV (2013) Translating the brain-machine interface. Sci Transl Med 5(210):210ps17
89. Rauschecker JP et al (2015) Frontostriatal gating of tinnitus and chronic pain. Trends Cogn Sci 19(10):567–578
90. Zhang Q et al (2017) Chronic pain induces generalized enhancement of aversion. elife 6: e25302
91. Wilson HD, Uhelski ML, Fuchs PN (2008) Examining the role of the medial thalamus in modulating the affective dimension of pain. Brain Res 1229:90–99
92. Daou I et al (2013) Remote optogenetic activation and sensitization of pain pathways in freely moving mice. J Neurosci 33(47):18631–18640
93. Iyer SM et al (2014) Virally mediated optogenetic excitation and inhibition of pain in freely moving nontransgenic mice. Nat Biotechnol 32(3):274–278
94. Gu L et al (2015) Pain inhibition by optogenetic activation of specific anterior cingulate cortical neurons. PLoS One 10(2):e0117746

95. Zhang Z et al (2015) Role of Prelimbic GABAergic circuits in sensory and emotional aspects of neuropathic pain. Cell Rep 12(5):752–759
96. Iyer SM et al (2016) Optogenetic and chemogenetic strategies for sustained inhibition of pain. Sci Rep 6:30570
97. Chen Z et al (2017) Deciphering neuronal population codes for acute thermal pain. J Neural Eng 14(3):036023
98. Chen Z, Wang J (2016) Statistical analysis of neuronal population codes for encoding acute pain. In: Proceedings of IEEE ICASSP. IEEE Press, New York, pp 829–833
99. Chen Z, Hu S, Zhang Q, Wang J (2017) Quickest detection for abrupt changes in neuronal ensemble spiking activity using model-based and model-free approaches. In: Proceedings of the IEEE conference neural engineering. IEEE Press, New York, pp 481–484
100. Hu S, Zhang Q, Wang J, Chen Z (2017) Real-time particle filtering and smoothing algorithms for detecting abrupt changes in neural ensemble spike activity. J Neurophys 119(4):1394–1410

Chapter 9
Future of Neural Interfaces

Farah Laiwalla and Arto Nurmikko

Abstract The technological ability to capture electrophysiological activity of populations of cortical neurons through chronic implantable devices has led to significant advancements in the field of brain-computer interfaces. Recent progress in the field has been driven by developments in integrated microelectronics, wireless communications, materials science, and computational neuroscience. Here, we review major device development landmarks in the arena of neural interfaces from FDA-approved clinical systems to prototype head-mounted and fully implantable wireless systems for multi-channel neural recording. Additionally, we provide an outlook toward next-generation, highly miniaturized technologies for minimally invasive, vastly parallel neural interfaces for naturalistic, closed-loop neuroprostheses.

Keywords Brain-computer interfaces · Wireless neural interfaces · Spatially distributed neural sensors · Neuroprostheses

9.1 Overview: Where We Are Now

This volume contains many illustrative examples of the early development and specific applications, across a spectrum of contemporary techniques, whereby direct electronic access to brain circuits, primarily from/to the neocortex, offers opportunities for basic brain science, emergence of medical devices for assisting and/or correcting neurological deficits, as well as discovering potential device-based therapies as an alternative to pharmaceuticals. The emergence of electronic neural interfaces, in particular, reflects outcomes of worldwide multidisciplinary work over the past two decades whereby device engineering, neuroscience, and clinical needs have met at the intersection of fundamental and applied brain science. Here,

---

F. Laiwalla · A. Nurmikko (✉)
School of Engineering, Brown University, Providence, RI, USA
e-mail: arto_nurmikko@brown.edu

X. Zheng (ed.), *Neural Interface: Frontiers and Applications*,
Advances in Experimental Medicine and Biology 1101,
https://doi.org/10.1007/978-981-13-2050-7_9

we consider progress—and challenges ahead—for the development of brain-computer interfaces for neural prostheses. We narrow the focus to cortical brain-computer interfaces (BCI) in offering a subjective view of future assistive technologies, particularly those applicable to severe neurological injury to the motor system such as in case of paralysis from neuromotor diseases, stroke, or spinal cord injury. To gain a broader perspective, the reader is referred elsewhere for reviews of neural interfaces including those which engage deeper brain structures such as deep brain stimulation (DBS) which has reached medical device maturity in widespread treatment of, e.g., Parkinsonian disorder [33]. The DBS technology is one example of neuromodulation application of electrical stimulation; many other approved electrical stimulation-based medical devices are also in widespread clinical use, such as spinal cord treatment for chronic pain.

When compared to maturity of the DBS-like neuromodulation technologies, and leaving questions about their therapeutic efficacy to clinicians, cortical BCIs are still at relative infancy, limited at this writing to perhaps up to a dozen human pilot trials worldwide. The fundamental challenges are, in our opinion, considerably more difficult than devices which deploy electrical stimulation from a handful of electrodes near/at the anatomical central or peripheral nervous system target. In particular, for BCIs, one must be able to record cortical circuit activity at sufficient level of detail and then decode typically many parallel channels of electrophysiological activity (action potentials, field potentials, etc.). The decoding, in turn, for essentially reconstructing cortical network activity in a predictive context requires sophisticated computational models and tools to decipher, e.g., a subject's intention to move a limb.

One set of illustrative examples of progress is from research where intracortical microelectrode arrays (MEA) are used to record population dynamics from (<1 cm^2 area) cortical patches at single neuron-level and subsequent application of stochastic state-space dynamical models to decode, e.g., movement planning intention of a subject. The genesis and progress have been well documented in a number of review articles and book chapters. Suffice it to say here that from early discoveries in non-human primates [9, 21, 30, 51], the adaptation of these methods has enabled tetraplegic subjects to control communication devices and electromechanical devices such as robotic arms and hands [3, 13, 22, 23, 46, 52]. The cortical locations where such recordings are made have typically involved primary and/or premotor and parietal areas where, e.g., direction and velocity tuning is distinctly encoded in subsets of neurons. While specific tuning properties of neurons can be helpful across a population which is part of an interconnected network with manifolds of internal state representation (e.g., motor and visual cortices), new decoding techniques are emerging using, e.g., deep learning methods, where presence of such tuning properties is not necessary for successful decoding.

While it follows from biophysical fundamentals that intracortical probes with their inherently single-neuron space-time resolution offer the most direct means to communicate with neuronal populations and their associate networks, they are by no means the only, and not always necessary, requirement for building electronic neural interfaces for prosthetic and other assistive use. As shown elsewhere in this volume, whether scalp-based (EEG) electrode arrays [36], epidural or subdural electrocorticography (ECoG) [34, 50], peripheral nerves, or elsewhere, there are multiple means to acquire meaningful control signals for assistive use. Broadly, for any assistive medical technology, the overall cost-to-benefit calculations must weigh in many factors: engineering and technological complexity, required medical procedures (such as surgery), safety and reliability—and above all, the level of required performance of the neural interface from the user's point of view.

This chapter examines electrophysiological BCIs only. While the brain is an electrical (electrochemical) biological machine with remarkable level of performance (e.g., per Watt of metabolic power) and lends itself readily to electrophysiology, active research is being pursued to search for neural interfaces relying on recording/stimulation modalities which range from optical to acoustic (ultrasound), to magnetic to chemical modalities, and combinations thereof. Time will tell if these alternative biophysical agents/modalities can compete or augment the direct electrical/electronic approaches.

We summarize in Sect. 2 some recent progress where innovation of *wireless* neural interfaces can now be envisioned to enable mobile BCIs, for untethered movement of subjects. In Sect. 3, we peek into the early ambitious efforts to develop very large-scale neural recording and microstimulation systems where scaling the electrophysiological access to, say, thousands of cortical points requires entirely new ways of approaching the problem, from developing new types of neural probes to re-examining systems level neuroengineering concepts.

9.2 Examples of State-of-the-Art Electronic Brain Interfaces

From a systems level perspective, a brain-computer interface may be broadly partitioned into (i) the physical neural probe (electrodes), (ii) an electronic custom integrated circuit (IC) core for signal acquistion, (iii) the (generally noisy) signal conditioning and telemetry, and (iv) the computing modules with dedicated back-end hardware embedding computational algorithms specific to the neural decoding task. The last decade has seen significant advances in each of these areas, even if a challenging road still lies ahead to meet the overarching goal of developing versatile BCI systems that can adaptively engage with the underlying non-stationary neural circuits in a closed-loop fashion, while compatible with long-term chronic implantation of neural probes in a (future) mobile subject without eliciting significant immune response.

The fundamental challenges of chronic intracortical or deep-brain neural probe design are manyfold. First, it is arguably desirable that the spatial scale and density of microscale probes reach close to single neuron resolution (say, ~20–100 μm electrode proximity for a layer V pyramidal neural cell body). Second, the probe material in direct electrical contact with tissue must enable high signal to noise ratio (SNR) recordings and high-charge-density biphasic microstimulation (the latter being critical for eventual fully closed-loop BCIs) while maintaining designed form factor (rigidity, etc.) and microscale surface characteristics that hopefully limit unavoidable immunologic response. There have been multiple recent reports of successful refinement of electrode microfabrication methodologies to yield, for example, high-density surface ECoG electrodes [26, 49], as well as penetrating 3-D probes [40] approaching an electrode pitch of tens of microns for high-density neural sampling. In contrast to traditional epi and intracortical approaches, Oxley et al. [37] have described a minimally invasive intravascular approach for the placement of a "stentrode" in cortical veins to access neural information at high spatial resolution (where the latter approach is currently being commercialized by Synchron med). Meanwhile, material science research has provided several promising candidates for the integration of next-generation BCI probes in addition to established work horses of Pt and IrOx, from silicon-carbide [15, 16], carbon nanotubes [12], nanoelectronic threads [32] to hybrids of conductive polymers like poly(3,4-ethylenedioxythiophene) (PEDOT) [10, 11, 27]. Success of optical techniques such as optogenetics has led to the development of hybrid probes integrating, for example, optical stimulation and electronic recording onto the same intracortical platform [22, 28, 45], where the latter target closed-loop hybrid optoelectronic BCIs in rodents and non-human primates. Successful transitioning of these novel experimental probe technologies for a chronic implantable BCI is the next major challenge and goal.

The electronic core of a BCI system invariably includes signal amplifiers, conditioning (filters) and dedicated analog-to-digital neural signal conversion. A number of groups worldwide have been developing custom IC solutions for low-power and low-noise neural signal acquisition, particularly focusing on channel scalability (from a handful to >100) and ease of physical integration with the typically monolithic multi-element neural probe. While it is not possible to review these developments exhaustively here, we highlight one recent work, namely, the "Neuropixel" probe system where up to 1356 recording channels were microfabricated within one silicon shank with integrated CMOS signal preconditioning [25, 39]. Systems such as these which allow high-resolution, sub-acute (up to a few months) of robust, wired access to the brain are instrumental in driving neuroscientific knowledge forward, yet are limited for translational chronic use by the complex percutaneous connectors in terms of full subcutaneous implantation. In general, adding a wireless telecommunication capability to neural probes defines one crucial need to extend the utility of these methods to freely moving animals and clinical applications.

9.2.1 BCI Going Wireless: General Considerations

In designing wireless neural interfaces, it is important to consider the requirements and constraints in the full ecosystem context. Relevant questions to ask up front might include, e.g., how a given intracortical MEA or subdural ECoG array may be interfaced with the electronic payload, the latter now located within the hermetically sealed envelope of the wireless device system. A key question is where within or on the body will the electronics package locate. We start with the design where the system electronics are configured as a head-mounted compact package whereby percutaneous/transcranial connectorization from, e.g., an MEA, is required. One rationale for this configuration as a starting choice on the wireless path is that silicon-based commercial intracortical MEAs are now widely available with a variety of channel counts compatible with small-animal (rodent) as well as non-human and human primate models, including reasonable chronic robustness demonstrated in clinical trials.

Given that contemporary intracortical (or subdual ECoG) sensor arrays typically sample on the order of 100 nodes (single neurons for MEAs, summated field potentials for ECoGs), it is remarkable how such pronounced under-sampling of neuronal population can be successful in enabling subjects to control, e.g., a robotic hand/arm with multiple degrees of freedom of motion. The good news is that this limited amount of neural data sets not unreasonable requirements for wireless telemetry. As also noted elsewhere in this volume, much credit to BCI development must be assigned to (scalable) sophisticated statistical models of neuronal dynamics which have produced powerful algorithms for effective decoding of, e.g., patient's movement intention, from this under-sampled pool of "noisy" brain signals, for real-time assistive device use. Since the "noise" typically hides valuable neural information, it is important that the signal fidelity of a wireless link be maintained (use of data compression techniques must thus be approached judiciously). The wireless neural population data provides direct inputs to the neuro-computational decoding models. There are numerous approaches to neural decoding; one representative example is state-space dynamical modeling, which distils multidimensional neural data (spikes and field potentials) to lower dimensional cortical state representation, suitable for interpretation both by visualization (graphics algorithms), as well as for prosthetic use through forwarding the outputs to assistive electronic devices (e.g., direct cortically operated laptops). The role of machine learning applied to neural decoding and encoding is a rapidly evolving new area of computational neuroscience, with very promising preliminary results suggestive of significant performance and efficiency benefits for real-time BCIs. We bypass the field of neural decoding/encoding in this article and refer the reader to the rich literature, which by now exists for this topic.

From a generic system level view, a fully wired BCI instrumentation collects the raw neural (multi-channel) data via multi-wire cables into electronic neural signal processors (composed of combinations of analog and digital electronics). The digitized data is fed to computers where even today's desktop machines have enough on-board processing power to carry out much of neural decoding to run a simple set of assistive devices. Figure 9.1 shows a block diagram of the basic electronic ecosystem which is more or less common to most (non-human and human) primate researchers in the field.

The next level of significance in advancing this type of electronic brain interface neurotechnology (even if still in early days) is to "untether the patient" by engineering direct wireless access to neural probes such as depicted in Fig. 9.2. While for external body wearable biosensors (whether EEG, EKG, etc.), the transition to a

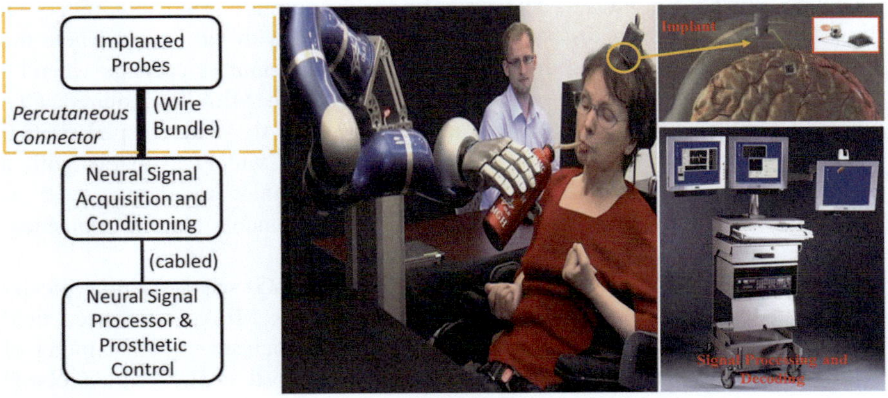

Fig. 9.1 A block diagram view of a multi-channel neural recording system for brain-computer interface applications. Multi-channel implantable microelectrode arrays are connectorized using wire bundles, and external cabled electronics and computers are used for signal acquisition and processing

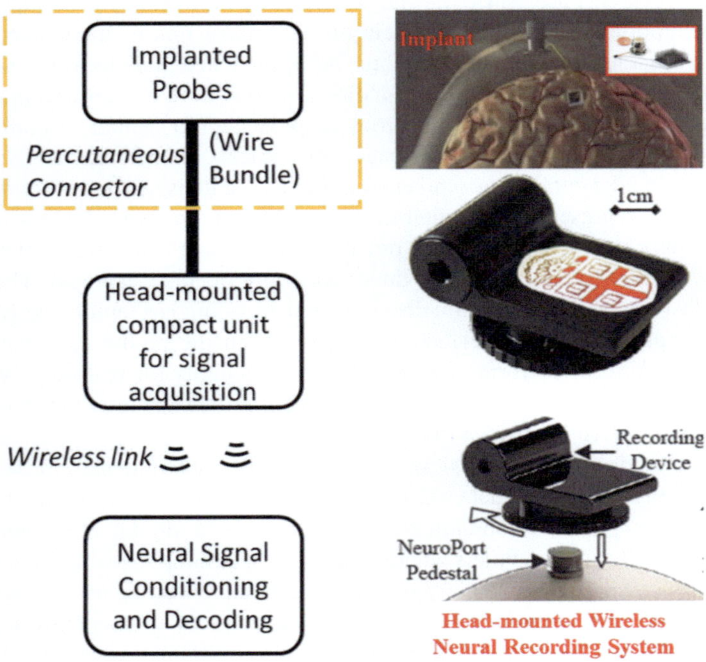

Fig. 9.2 Block diagram and representative device view of a head-mounted wireless neural recording system [55]. Integrated microelectronics are placed within a compact package powered by a replaceable Li-ion battery

wireless system for enhancing subject's mobility is already possible by adapting any number of wireless electronics developed in the past decade for consumer and industrial use, the situation for implanted wireless BCIs poses a number of challenges. These include the following which must be designed and ultimately integrated seamlessly into one viable medical grade implant device system:

- Very low-power and high-fidelity analog and digital (custom) application specific integrated circuits (ASICs) for broadband (spikes, field potentials) neural signal acquisition and electrical microstimulation.
- Very low-power radio frequency transmitter capable of high data rate transmission (~100 MBits/s and beyond), custom designed as an integrated circuit ("RF ASIC").
- Connectorization of the multi-element neural probe front-end (such as one or more MEAs) to and integration with the on-board ASICs.
- Packaging the electronics within a compact housing which is hermetically viable for long-term chronic use.
- Designing external RF receivers which capture the wirelessly emitted, digitally-encoded signals for inputs to computational devices for decoding.
- Strategy for powering the implant electronics (internal battery or RF inductive coupling by external coils).

These engineering challenges are further bound by the first priority in any medical device candidate—safety—which overrides even the main functional claim of the device, here the neurotechnological performance. In the wireless device system which we use here as an example (from our laboratories), safety has multiple components such as (and revisited further below):

- Electrical safety: the electrical energy required by the active ASICs in the implant must have zero accidental probability of discharging into the body/brain tissue.
- Mechanical safety: if the implant unit locates, for example, in the subject's head, the packaging of the electronics must be correspondingly impact-proof.
- Chemical safety: the hermetic sealing of the electronics package must ensure that there is no leakage of toxic or other chemically harmful materials into tissue.
- Thermal safety: ensure that maximum temperature, e.g., in the subcutaneous vicinity of the implant during operation (or recharging batteries), does not exceed $\Delta T < 2.0\ ^\circ\text{C}$ and that corresponding temperature rise in the cortical space are kept at $\Delta T < 0.1\ ^\circ\text{C}$.
- Electromagnetic (EM) safety: RF (or similar wireless) signal and power transmission of the device are secure, and are not susceptible to external EM interference by ambient RF traffic and vice versa.

Most of these requirements follow the footsteps of established protocols, such as those for pacemakers or cochlear implants for human use, thereby mirroring established regulatory guidelines determined by FDA in the USA. We note, however, that at the frontiers of neurotechnology such as discussed in this volume, the device systems involve considerably complex electronics and require orders of magnitude larger wireless data rates so that early discussions by the technology developers with regulatory agencies is very important.

A number of neurotechnology research groups have developed compact, externally head-mounted wireless neural recording systems as the first step, whereby this approach has allowed leveraging commercially available microelectronics for the development of proof-of-concept wireless BCI systems. Among some of the earliest innovation in this arena, a set of contributions came from the Shenoy group at Stanford. This team has built a series of incrementally sophisticated "Hermes" series of wireless neural recording platforms [12, 19, 30, 38] beginning from a two-channel Hermes-B [41] to a 96-channel battery-operated Hermes-E [17]. In the laboratories of the authors of this chapter, we have similarly developed an external battery-operated, 96-channel broadband neural recording system, featuring custom ICs for neural signal capture and conditioning as well as a dedicated RF IC for 3.5 GHz short-range (~4 m), very-low power wireless transmission with a net power consumption of 61.2 mW [55]. The latter is now a licensed technology which is commercially available, allowing animal researchers to conduct studies on freely moving animals, for example in the context of foraging and naturalistic locomotion.

In addition to the systems level achievements highlighted above, new research is particularly focused at pushing the limits of microelectronic technology to attain improved performance for wireless neural recording with ever-shrinking power budgets. This research extends from subsystems utilizing commercial technologies such as UWB radios, WiFi, and Zigbee to highly optimized low-power integrated approaches. For example, a 4096 channel multiplexed ECoG recording chip with 64-channel amplifiers and 5.12 Mbps data rata, transmitted at 7.8 GHz using UWB radio has been described by Ando et al. [2] with a net power consumption of 1400 mW. Schwarz et al. [42] similarly describe a 128-channel head-mounted system working with microwire array implants, but now integrating the capability for on-board neural signal processing and bidirectional communications with a power consumption of 264 mW, or ~2 mW per channel. Several other groups have described multi-channel ASICs with sub-milliwatt power consumptions per channel while representing various trade-offs between signal compression and power budgets [5, 8]. Finally, multi-modal sensing and stimulation, for example, integrating dopamine sensing and electrical stimulation alongside electrical neural recording [7, 22], and approaches incorporating optogenetic approaches, are increasingly building a niche in the BCI component level landscape which is anticipated to grow significantly in future.

9.2.2 Fully Implanted Wireless Devices—A Case Example

The development of head-mounted and other outside-body wireless neural recording systems has provided a new BCI tool for neuroscience research by enabling more complex and naturalistic animal experiments. However, the need for a percutaneous connection still poses a limitation for BCI use of these systems, in particular, limiting their translation into the human/clinical BCI domain. In considering the bridge from a head-mounted to a fully implanted wireless neural interface device system, we reiterate key challenges which have to be addressed, typically through extensive animal testing and performance validation:

- Implantation of the system's electronic components within the body in a hermetically encapsulated unit, and choice for the anatomical location of this payload relative to the actual neural probes, where the electronic package has a form factor compatible with subcutaneous placement, for example, in the epicranial space.
- ASIC design and optimization for ultra-low power, consistent with thermal safety guidelines defined by FDA for active brain implants (maximum of 0.5 °C increment in tissue temperature). Extensive thermal simulations and metrology of implants in animal models are typically required.
- Implant power management (rechargeable batteries or continuous inductive coupling) and transcutaneous wireless communication, mitigating the impact of tissue RF absorption, and keeping within the specific absorption rate (SAR) limits prescribed by *IEEE Std C95.1–2005*.

The basic microelectronic building blocks of a fully implantable neural interface system are nearly identical to those described in previous sections; the caveat is that now the spatial scale and performance metrics and requirements (such as those listed above) for a fully implantable device are far more stringent. As the case example for this chapter, the authors' labs have developed a 100-channel hermetically-sealed, fully implantable broadband wireless neural recording system utilizing a 100-channel neural amplifier ASIC, 12-bit SAR ADCs, and a custom RF transmitter IC. The components mirror those developed for the head-mounted system described previously, but now focus on a system architecture with a titanium (Ti)-based hermetic enclosure. The Ti-enclosure has been equipped with a sapphire window to provide full electromagnetic transparency for wireless charging and telecommunications. Further, a custom planar interconnect interface has been built to feed the 100 microwires from the intracortical MEA to the active electronics via multi-channel custom high-density hermetic feedthroughs as shown in Fig. 9.3 [6]. In keeping with the need for further

Fig. 9.3 A fully implantable wireless device for neural recording using titanium hermetic enclosure, ceramic feedthroughs, and a sapphire window for broadband wireless communication. Device is powered using a rechargeable Li-ion battery, with multi-hour operation between recharges

electronic integration, there have been several recent efforts to develop system on chip electronics (SoC), where a single chip solution provides neural recording in conjunction with wireless charging, compressive sensing or spike detection, electrical stimulation and bidirectional telecommunications [4, 31, 36]. These advancements are important elements for the further development of fully-implantable systems, but the most critical current bottle-neck for these devices is the availability of scalable hermetic packaging solutions. Hermetic sealing approaches being further studied include ceramic, metallic, or thin-film materials, among others. In addition to the materials choice, a major engineering challenge is to microfabricate the large number of electrical "feedthroughs" without compromising the hermeticity of the implant.

9.3 New Horizons: Large-Scale Neural Interfaces to 10,000 Nodes and Beyond

There has been significant progress in enhancing and maintaining neural access through dense-electrode arrays and biocompatible surface modifications as described in previous sections, yet the scalability of practical neural interfaces to orders of magnitude higher numbers of recording and/or stimulating nodes (thousands and tens of thousands) remains a persistent challenge. As we move from the realm of highly controlled, experimental BCIs to more naturalistic, deployable systems, neural task complexity (NTC) is expected to grow. To gauge the expected number of microelectrode sites with increasing NTC, Gao et al. [18] have postulated that neural decoding would be critically dependent on access to significantly higher numbers of neurons at high densities, ideally from an anatomically diverse neural population. Figure 9.4 represents the outcome from a high-dimensionality neural state-space dynamical theoretical model. Current monolithic multi-electrode constructs are, in our view, inherently incompatible with large-scale implementations of BCIs due to a variety of anatomical and fabrication constraints. For example,

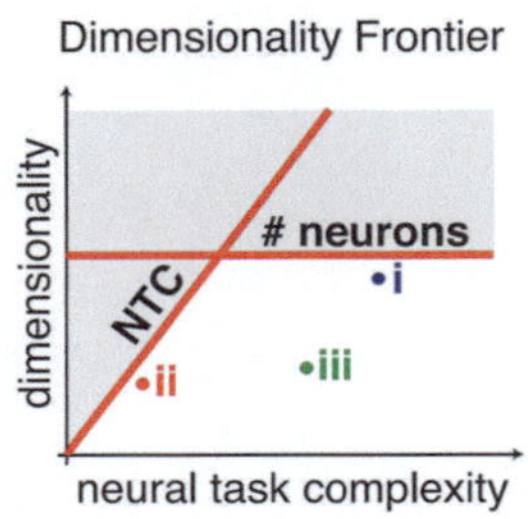

Fig. 9.4 A dimensionality frontier in motor cortical data. Allowed possibilities of dimensionality D and neural task complexity, NTC, exhibit three distinct regimes: (i) the number of recorded neurons M but not NTC restricts dimensionality, (ii) NTC but not M restricts D, and (iii) D is far less than both M and NT C, reflecting an unexplained circuit constraint beyond smoothness and task simplicity. (Courtesy of [18])

traditional hermetic sealing approaches require bulky Ti or ceramic cases, which add significant volume overhead to device size. The intricacy of interconnect design, particularly in the case of a multi-areal implant, is an added complication. While the short tethers between intracortical arrays and the wireless electronics described in the previous section represent a significant improvement over percutaneous tethers, they are still relying on the traditional monolithic intra- or epicortical electrode arrays. Looking forward, one needs to consider paradigm shifts in approaching future BCI-compatible neural interface designs to mitigate these issues in the design of vastly scalable systems. One such approach is to split individual recording and/or neurostimulation electrodes into separate autonomous microscale active electronic chiplets, each with its own internal electronics and (wireless) means to communicate with a central information processing hub. The idea takes advantage of the remarkable progress made by silicon microelectronics in the past 20 years which has pushed component and transistor sizes deep into the sub-100 nm regime. For neural and BCI applications, having access to such CMOS technologies is very attractive provided that commensurate expertise is available for custom design of “mixed-signal” CMOS chips (analog-digital-RF).

Among contemporary work is the concept of free-floating individual “neural dust” sensors in conjunction with a sub-cranial interrogator which was proposed by a number of neuroengineering researchers, most notably Seo et al. [43] (we show concrete examples from our own work below to illustrate a related approach for wireless microsensors and stimulators). The proposed system by Seo et al. [43] involves ultrasonic power and telemetry to circumvent the challenges of efficient electromagnetic energy coupling at millimeter-scale and has been demonstrated at the level of a single 1 mm^3 recording node in the peripheral nervous system [44]. While the concept is presented as an ultra-miniaturizable sensor system, the authors identify several fabrication and materials challenges requiring further advances in CMOS die post-processing completion of microscale implantable microdevices. Elsewhere, a purely electromagnetic-based distributed neural sensor has been demonstrated by Yeon et al. [53], where a three-coil resonant near field inductively coupled system is used to improve the efficiency of wireless power transfer to a millimeter-sized free-floating wireless implantable neural recording system (FF-WINeR). The approach to assembling the FF-WINeR sensor node by manual techniques is also described by Yeon et al. [54] including integration of microwire electrodes, ASICs with through-silicon vias (TSVs), discrete microcoils, and surface mount components. Finally, hermetic sealing with polymeric film deposition (parylene in this case) was applied to form a single “push-pin” recording node measuring $1.05 \times 1.05 \times 0.3\ mm^3$. Here too, the authors outline a number of challenges ahead in scaling to many devices given the number of heterogeneous materials and the complexity plus fabrication requirements comprising each sensor node.

With these snapshots of recent early work, we support the view that spatially distributed sensor systems, once developed as full-fledged medical devices can contribute to future breakthroughs in the field of BCIs while providing new tools for brain science. There are, however, several caveats for the development of a truly scalable distributed sensor system. First, it is imperative that it be possible to pursue the fabrication of sensors in a uniform, high throughput batch process. Second, sophisticated and ultra-miniaturized sensor node electronics are needed to support semi-autonomous operation including critical functions such as memory and networking. Third, the full neural interface system needs to be defined in the context of a highly-efficient multi-node telecom network. This includes the design and implementation of a compact wearable telecommunications hub that would drive and tune the performance specifications of individual sensor nodes in an adaptive fashion, depending on the specific neural interface application. Fourth, the next frontier in BCI development beyond neural recording for decoding must include neural encoding implemented as patterned cortical microstimulation by a spatially distributed ensemble. This capability needs to be integrated into the individual nodes to form an independent, bidirectional neural interfacing element. Fifth, for chronic implants, it is critical to develop the means for (thin film) hermetic seals which provide a chemically impermeable envelope for the microdevices over decades in the body environment. Finally, the implementation of any neural interface system for use in a human subject requires a real-time neural encoding/decoding using a wearable neurocomputational processor with wireless telemetry to supporting computational platforms, including cloud computing. The concept image of Fig. 9.5 shows schematically the general approach (with some specifics to the approach adopted in the authors' laboratories).

Designing a system to meet the above-mentioned caveats is a challenging proposition on multiple levels. Recent work in our laboratories has focused on developing a spatially distributed implantable wireless "cortical communication" system toward these aspirations, which is built around 500 × 500 μm individually addressable custom microelectronic "neurograins" chiplets (see Fig. 9.6, which also

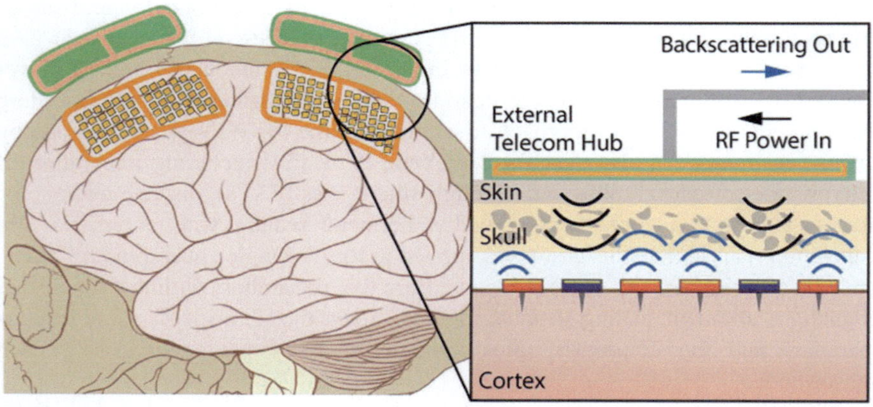

Fig. 9.5 Overview diagram of the spatially distributed implanted wireless neurograin read-out-write-in network—in full deployable system context. Here, two cortical areas are targeted. The external telecom hub in this representation is synonymous with neural computation engine

		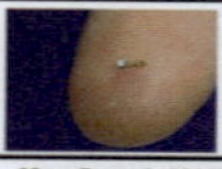			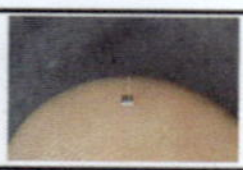
Reference	**[Muller R et al 2018**	**[Seo D et al, 2016]**	**[Johnson B et al 2018]**	**[Yeon P et al [2018]**	**Brown/UCSD work**
Applications	Recording (CNS)	Recording (PNS)	Stimulation (PNS)	Recording (CNS)	**Record./Stim. (CNS)**
Number of Ch	64	1	1	1	**Up to 800**
IC Area [mm^2]/Ch	**0.09 (Note A)**	**0.032 (Note B)**	1	1.1	0.25
Total Volume [mm^3]	-	2.4	2.2	<1	**<0.0075 (note C)**
Power Consumption [uW]/Ch	3.4	120	9.7	<300	**<30**
Input Ref.Noise [uVrms]	1.3	180	-	3.78	1.6
Wireless Powering	300MHz RF	1.85MHz *Ultrasonic*	1.85MHz *Ultrasonic*	131 MHz RF	915 MHz RF
Telemetry	LSK	Backscattering	Backscattering	IR-UWB	**BPSK (TDMA)**
Data rate (Mbps)	1	0.5	-	0.8	**10**
Power supply (V)	0.5	-	3 V	1.8	0.7
ADC	VCO	-	-	VCO	SAR
ADC Resolution	15	8	-	10	8
Technology	65 nm CMOS	Discrete	65 nm CMOS	0.35 um CMOS	65 nm CMOS
Hermetic Sealing	?	?	?	?	**ALD > 10 years**

Note A: A multichannel system that shares many of the components across the channels
Note B: The ASIC an incomplete system – many of area savings come from using external components such as piezoelectric transducers
Note C: Volume savings include thin substrate but includes hermetic seals.

Fig. 9.6 Comparison of some mm-scale neural microdevices using electrical and/or ultrasound

compares several other recent microdevices). Each chiplet integrates RF-energy harvesting, broadband neural recording or cortical microstimulation. Ensembles of neurograins operate as a synchronized, networked bidirectional RF telecommunication network, scalable in principle across populations of nodes up to tens of thousands. We have chosen to utilize near-field electromagnetic coupling at ~1 GHz to mitigate RF tissue absorption for SAR compliance and designed a single RF-channel simultaneous power-data link for network level simplicity. An external RF telecommunications hub, implemented as a wearable module, wirelessly manages implant performance, and is envisioned also to integrate real-time data processing for neural decoding [19, 20]. Further, we have addressed the microscale packaging challenge by utilizing a batch-process stacked multi-layer thin-film deposition process which is able to yield packaged individual devices that are <0.01 mm^3 in physical volume. This thin film hermetic sealing technique using atomic layer epitaxy, has been demonstrated to provide implant impermeability in accelerated lifetime tests exceeding 10 years [24]. These features have been validated in the context of ex vivo rodent brain slices as well as acute in vivo experiments [29], and now mature for transition to a specific application context.

The above-described system from our labs is one example of an approach for a completely untethered neural interface system that is minimally invasive from the perspective of implant volume burden. Alternative approaches involve flexible, floating epicortical probes with very high channel counts [48] versus integration of genetic approaches to modify neural responsiveness (optogenetics, etc.) toward the realization of next-generation highly efficient, integrated neural interface systems.

9.4 Summary: Outlook for High-Performance BCIs

In this chapter, we have described approaches to high data rate neural interfaces via examples of technical solutions to intracortical/intracranial recording systems. While early clinical trials are under way, and mainly use well-established, but bulky cabled systems, there is a need and opportunity to invest resources to pursue innovative approaches across the entire ecosystem from micro- and optoelectronic probe arrays to chipscale integration of tailored on-chip signal processing and storage functions, to name two areas.

The microminiaturization approaches reviewed here offer a particular opportunity to enhance the functionality of the wireless interfaces by adding neuromodulation/stimulation capabilities. Among such functions is the implementation of patterned electrical (e.g., [14, 47]) or/and optical microstimulation for enabling truly bidirectional communication opportunities with the brain. Efforts to develop such multi-node targeted microstimulation tools are being pursued in several academic and commercial laboratories and are likely to reach primates in the near term. These efforts must link closely with neural decoding/encoding models using combination of theoretical neuroscience and machine learning tools.

We re-emphasize the importance of close synergy between closed-loop sense/stimulate-based algorithms which have recently made progress in deep brain stimulation [1]; however, the challenges encountered herein, for bidirectional interfaces comprising hundreds of channels (nodes) and beyond require that large amounts of real-time data, recorded from the nervous system, must be processed and decoded (e.g., [38, 52]). Among the many challenges to this aspiration are the inherent variability and statistical entropy-driven fluctuations in neural circuits, the latter requiring approaches which can adapt to such "non-stationarities."

Acknowledgments The authors are very grateful to many members, past and present, in their laboratory. These include Jihun Lee, Joonsoo Jeong, David Borton, Ming Yin, Y.-K. Song, William R. Patterson III, Naubahar Agha, and Chris Heelan, among others. At Brown University, the authors are part of a synergistic brain science and neurotechnology effort with many colleagues including John Donoghue, Leigh Hochberg, David Rosler, John Simeral, and Carlos Vargas-Irwin whom we thank for their continuing input and expertise. Special thanks are also extended to Krishna Shenoy and his colleagues at Stanford for providing early leadership in the field.

All animal procedures referred in this chapter were conducted according to protocols approved by Institutional Animal Care and Use Committee (IACUC) at each institution. Research in the authors' laboratory was supported by US National Institutes of Health, Defense Advanced Projects Agency and the National Science Foundation.

References

1. Afshar P, Khambhati A, Stanslaski S, Carlson D, Jensen R, Dani S, ..., Denison T (2013) A translational platform for prototyping closed-loop neuromodulation systems. Front Neural Circuits 6:117

2. Ando H, Takizawa K, Yoshida T, Matsushita K, Hirata M, Suzuki T (2016) Wireless multichannel neural recording with a 128-mbps UWB transmitter for an implantable brain-machine interfaces. IEEE Trans Biomed Circuits Syst 10(6):1068–1078
3. Aflalo T, Kellis S, Klaes C, Lee B, Shi Y, Pejsa K, …, Liu C (2015) Decoding motor imagery from the posterior parietal cortex of a tetraplegic human. Science 348(6237):906–910
4. Biederman W, Yeager DJ, Narevsky N, Leverett J, Neely R, Carmena JM, …, Rabaey JM (2015) A 4.78 mm 2 fully-integrated neuromodulation SoC combining 64 acquisition channels with digital compression and simultaneous dual stimulation. IEEE J Solid State Circuits 50(4):1038–1047
5. Bonfanti A, Ceravolo M, Zambra G, Gusmeroli R, Borghi T, Spinelli AS, Lacaita AL (2010) A multi-channel low-power IC for neural spike recording with data compression and narrowband 400-MHz MC-FSK wireless transmission. In: 2010 Proceedings of ESSCIRC, September. IEEE, pp 330–333
6. Borton DA, Yin M, Aceros J, Nurmikko A (2013) An implantable wireless neural interface for recording cortical circuit dynamics in moving primates. J Neural Eng 10(2):026010
7. Bozorgzadeh B, Schuweiler DR, Bobak MJ, Garris PA, Mohseni P (2016) Neurochemostat: a neural interface SoC with integrated chemometrics for closed-loop regulation of brain dopamine. IEEE Trans Biomed Circuits Syst 10(3):654–667
8. Brenna S, Padovan F, Neviani A, Bevilacqua A, Bonfanti A, Lacaita AL (2016) A 64-channel 965-$\mu\text {W} $ neural recording SoC with UWB wireless transmission in 130-nm CMOS. IEEE Trans Circuits Syst Express Briefs 63(6):528–532
9. Carmena JM, Lebedev MA, Crist RE, O'Doherty JE, Santucci DM, Dimitrov DF, …, Nicolelis MA (2003) Learning to control a brain–machine interface for reaching and grasping by primates. PLoS Biol 1(2):e42
10. Castagnola E, Maiolo L, Maggiolini E, Minotti A, Marrani M, Maita F, …, Fadiga L (2015a) PEDOT-CNT-coated low-impedance, ultra-flexible, and brain-conformable micro-ECoG arrays. IEEE Trans Neural Syst Rehabil Eng 23(3):342–350
11. Castagnola V, Descamps E, Lecestre A, Dahan L, Remaud J, Nowak LG, Bergaud C (2015b) Parylene-based flexible neural probes with PEDOT coated surface for brain stimulation and recording. Biosens Bioelectron 67:450–457
12. Chen G, Dodson B, Hedges DM, Steffensen SC, Harb JN, Puleo C, …, Davis RC (2018) Fabrication of high aspect ratio millimeter-tall free-standing carbon nanotube-based microelectrode arrays. ACS Biomater Sci Eng 4(5):1900–1907
13. Collinger JL, Wodlinger B, Downey JE, Wang W, Tyler-Kabara EC, Weber DJ, …, Schwartz AB (2013) High-performance neuroprosthetic control by an individual with tetraplegia. Lancet 381(9866):557–564
14. Dadarlat MC, O'doherty JE, Sabes PN (2015) A learning-based approach to artificial sensory feedback leads to optimal integration. Nat Neurosci 18(1):138
15. Deku F, Cohen Y, Joshi-Imre A, Kanneganti A, Gardner TJ, Cogan SF (2018) Amorphous silicon carbide ultramicroelectrode arrays for neural stimulation and recording. J Neural Eng 15(1):016007
16. Diaz-Botia CA, Luna LE, Neely RM, Chamanzar M, Carraro C, Carmena JM, …, Maharbiz MM (2017) A silicon carbide array for electrocorticography and peripheral nerve recording. J Neural Eng 14(5):056006
17. Gao H, Walker RM, Nuyujukian P, Makinwa KA, Shenoy KV, Murmann B, Meng TH (2012) HermesE: a 96-channel full data rate direct neural Interface in 0.13μm CMOS. IEEE J Solid State Circuits 47(4):1043–1055
18. Gao P, Trautmann E, Byron MY, Santhanam G, Ryu S, Shenoy K, Ganguli S (2017) A theory of multineuronal dimensionality, dynamics and measurement. bioRxiv 1:214262
19. Heelan C, Komar J, Vargas-Irwin CE, Simeral JD, Nurmikko AV (2015) A mobile embedded platform for high performance neural signal computation and communication. In: 2015 IEEE biomedical circuits and systems conference (BioCAS), October. IEEE, pp 1–4

20. Heelan C, Nurmikko AV, Truccolo W (2018) FPGA implementation of deep-learning recurrent neural networks with sub-millisecond real-time latency for BCI-decoding of large-scale neural sensors (104 nodes). In: 2018 40th annual international conference of the ieee engineering in medicine and biology society (EMBC), July. IEEE, pp 1070–1073
21. Hochberg LR, Serruya MD, Friehs GM, Mukand JA, Saleh M, Caplan AH, . . ., Donoghue JP (2006) Neuronal ensemble control of prosthetic devices by a human with tetraplegia. Nature 442(7099):164
22. Hochberg LR, Bacher D, Jarosiewicz B, Masse NY, Simeral JD, Vogel J, . . ., Donoghue JP (2012) Reach and grasp by people with tetraplegia using a neurally controlled robotic arm. Nature 485(7398):372
23. Jarosiewicz B, Sarma AA, Bacher D, Masse NY, Simeral JD, Sorice B, . . ., Cash SS (2015) Virtual typing by people with tetraplegia using a self-calibrating intracortical brain-computer interface. Sci Trans Med 7(313):313ra179
24. Jeong J, Laiwalla F, Lee J, Ritasalo R, Pudas M, Larson L, . . ., Nurmikko A (2019) Conformal hermetic sealing of wireless microelectronic implantable Chiplets by multilayered atomic layer deposition (ALD). Adv Funct Mater 29(5):1806440
25. Jun JJ, Steinmetz NA, Siegle JH, Denman DJ, Bauza M, Barbarits B et al (2017) Fully integrated silicon probes for high-density recording of neural activity. Nature 551(7679):232
26. Khodagholy D, Gelinas JN, Thesen T, Doyle W, Devinsky O, Malliaras GG, Buzsáki G (2015) NeuroGrid: recording action potentials from the surface of the brain. Nat Neurosci 18(2):310
27. Kozai TD, Catt K, Du Z, Na K, Srivannavit O, Razi-ul MH, . . ., Cui XT (2016) Chronic in vivo evaluation of PEDOT/CNT for stable neural recordings. IEEE Trans Biomed Eng 63(1):111–119
28. Kwon KY, Lee HM, Ghovanloo M, Weber A, Li W (2015) Design, fabrication, and packaging of an integrated, wirelessly-powered optrode array for optogenetics application. Front Syst Neurosci 9:69
29. Lee J, Mok E, Huang J, Cui L, Lee AH, Leung VW, Mercier P, Shellhammer S, Larson L, Asbeck A, Song YK, Nurmikko A, Laiwalla F (2019) An implantable wireless network of distributed microscale sensors for neural applications. IEEE EMBS conference on neural engineering 2019
30. Leuthardt EC, Schalk G, Wolpaw JR, Ojemann JG, Moran DW (2004) A brain–computer interface using electrocorticographic signals in humans. J Neural Eng 1(2):63
31. Liu X, Zhang M, Xiong T, Richardson AG, Lucas TH, Chin PS, . . ., Van der Spiegel J (2016) A fully integrated wireless compressed sensing neural signal acquisition system for chronic recording and brain machine interface. IEEE Trans Biomed Circuits Syst 10(4):874–883
32. Luan L, Wei X, Zhao Z, Siegel JJ, Potnis O, Tuppen CA, . . ., Dunn AK (2017) Ultraflexible nanoelectronic probes form reliable, glial scar–free neural integration. Sci Adv 3(2):e1601966
33. McDermott H (2016) Neurobionics: treatments for disorders of the central nervous system. In: Neurobionics: the biomedical engineering of neural prostheses. Wiley, Hoboken, pp 213–230
34. Mestais CS, Charvet G, Sauter-Starace F, Foerster M, Ratel D, Benabid AL (2015) WIMAGINE: wireless 64-channel ECoG recording implant for long term clinical applications. IEEE Trans Neural Syst Rehabil Eng 23(1):10–21
35. Mirbozorgi SA, Bahrami H, Sawan M, Rusch LA, Gosselin B (2016) A single-chip full-duplex high speed transceiver for multi-site stimulating and recording neural implants. IEEE Trans Biomed Circuits Syst 10(3):643–653
36. Müller-Putz GR, Schwarz A, Pereira J, Ofner P (2016) From classic motor imagery to complex movement intention decoding: the noninvasive Graz-BCI approach. In: Progress in brain research, vol 228. Elsevier, pp 39–70
37. Oxley TJ, Opie NL, John SE, Rind GS, Ronayne SM, Wheeler TL, . . ., Steward C (2016) Minimally invasive endovascular stent-electrode array for high-fidelity, chronic recordings of cortical neural activity. Nat Biotechnol 34(3):320

38. Park YS, Hochberg LR, Eskandar EN, Cash SS, Truccolo W (2013) Early detection of human epileptic seizures based on intracortical local field potentials. In: 2013 6th international IEEE/EMBS conference on neural engineering (NER), November. IEEE, pp 323–326
39. Raducanu BC, Yazicioglu RF, Lopez CM, Ballini M, Putzeys J, Wang S et al (2016) Time multiplexed active neural probe with 678 parallel recording sites. In: 2016 46th European solid-state device research conference (ESSDERC). IEEE, pp 385–388
40. Rios G, Lubenov EV, Chi D, Roukes ML, Siapas AG (2016) Nanofabricated neural probes for dense 3-D recordings of brain activity. Nano Lett 16(11):6857–6862
41. Santhanam G, Linderman MD, Gilja V, Afshar A, Ryu SI, Meng TH, Shenoy KV (2007) HermesB: a continuous neural recording system for freely behaving primates. IEEE Trans Biomed Eng 54(11):2037–2050
42. Schwarz DA, Lebedev MA, Hanson TL, Dimitrov DF, Lehew G, Meloy J, . . ., Ramakrishnan A (2014) Chronic, wireless recordings of large-scale brain activity in freely moving rhesus monkeys. Nat Methods 11(6):670
43. Seo D, Carmena JM, Rabaey JM, Alon E, Maharbiz MM (2013) Neural dust: an ultrasonic, low power solution for chronic brain-machine interfaces. arXiv preprint arXiv:1307.2196
44. Seo D, Neely RM, Shen K, Singhal U, Alon E, Rabaey JM, . . ., Maharbiz MM (2016) Wireless recording in the peripheral nervous system with ultrasonic neural dust. Neuron 91(3):529–539
45. Seymour EÇ, Freedman DS, Gökkavas M, Özbay E, Sahin M, Ünlü MS (2014) Improved selectivity from a wavelength addressable device for wireless stimulation of neural tissue. Front Neuroeng 7:5
46. Simeral JD, Kim SP, Black MJ, Donoghue JP, Hochberg LR (2011) Neural control of cursor trajectory and click by a human with tetraplegia 1000 days after implant of an intracortical microelectrode array. J Neural Eng 8(2):025027
47. Tabot GA, Dammann JF, Berg JA, Tenore FV, Boback JL, Vogelstein RJ, Bensmaia SJ (2013) Restoring the sense of touch with a prosthetic hand through a brain interface. Proc Natl Acad Sci 110(45):18279–18284
48. Tsai D, Sawyer D, Bradd A, Yuste R, Shepard KL (2017) A very large-scale microelectrode array for cellular-resolution electrophysiology. Nat Commun 8(1):1802
49. Viventi J, Kim DH, Vigeland L, Frechette ES, Blanco JA, Kim YS, . . ., Wulsin DF (2011) Flexible, foldable, actively multiplexed, high-density electrode array for mapping brain activity in vivo. Nat Neurosci 14(12):1599
50. Wang W, Collinger JL, Degenhart AD, Tyler-Kabara EC, Schwartz AB, Moran DW, . . ., Kelly JW (2013) An electrocorticographic brain interface in an individual with tetraplegia. PLoS One 8(2):e55344
51. Wessberg J, Stambaugh CR, Kralik JD, Beck PD, Laubach M, Chapin JK, . . ., Nicolelis MA (2000) Real-time prediction of hand trajectory by ensembles of cortical neurons in primates. Nature 408(6810):361
52. Wodlinger B, Downey JE, Tyler-Kabara EC, Schwartz AB, Boninger ML, Collinger JL (2014) Ten-dimensional anthropomorphic arm control in a human brain− machine interface: difficulties, solutions, and limitations. J Neural Eng 12(1):016011
53. Yeon P, Mirbozorgi S, Ash B, Eckhardt H, Ghovanloo M (2016) Fabrication and microassembly of a mm-sized floating probe for a distributed wireless neural interface. Micromachines 7(9):154
54. Yeon P, Gonzalez JL, Zia M, Rajan SK, May GS, Bakir MS, Ghovanloo M (2017) Microfabrication, assembly, and hermetic packaging of mm-sized free-floating neural probes. In 2017 IEEE biomedical circuits and systems conference (BioCAS), October. IEEE, pp 1–4
55. Yin M, Borton DA, Komar J, Agha N, Lu Y, Li H, . . ., Larson L (2014) Wireless neurosensor for full-spectrum electrophysiology recordings during free behavior. Neuron 84(6):1170–1182

MIX
Papier aus verantwortungsvollen Quellen
Paper from responsible sources
FSC® C105338

If you have any concerns about our products,
you can contact us on
ProductSafety@springernature.com

In case Publisher is established outside the EU,
the EU authorized representative is:
Springer Nature Customer Service Center GmbH
Europaplatz 3, 69115 Heidelberg, Germany

Printed by Libri Plureos GmbH
in Hamburg, Germany

The Story of Indian Manufacturing

Vijay K. Seth

The Story of Indian Manufacturing

Encounters with the Mughal and British Empires (1498–1947)

palgrave
macmillan

Ane Books
Pvt. Ltd.

Vijay K. Seth
Faculty of Management Studies
University of Delhi
New Delhi, India

International Management Institute
New Delhi, India

ISBN 978-981-10-5573-7 ISBN 978-981-10-5574-4 (eBook)
DOI 10.1007/978-981-10-5574-4

Library of Congress Control Number: 2017951616

Cover Credit: PjrStudio / Alamy Stock Photo and Mohammed Anwarul Kabir Choudhury / Alamy Stock Photo

Printed on acid-free paper

This Palgrave Macmillan imprint is published by Springer Nature
The registered company is Springer Nature Singapore Pte Ltd.
The registered company address is: 152 Beach Road, #21-01/04 Gateway East, Singapore 189721, Singapore

To the princess of the story of my life, Arpita

Preface

History informs us that most of the economies of the world that were able to sustain long-term and consistent higher rates of growth, and were able to transform themselves from the status of underdeveloped to developed economies, have experienced revolution in their manufacturing sector. The revolutionary changes that occurred took place during what is known as the Industrial Revolution. Development in the manufacturing sector provides conditions, when economies develop capability to create unlimited amounts of wealth. The manufacturing sector also facilitates the transfer of manpower from farming to manufacturing. The transfer of manpower results in higher standards of living. It is now being increasingly realized that the nature of development and efficiency of other sectors also depends on the level of growth of the manufacturing sector of the economy. It is due to these important historical facts, that in the literature on economics of development, industrialization and development have become coterminous.

The process of this kind of transformation initially occurred in Britain, and then it spread to different economies of Europe and America. After World War I, it spread to the U.S.S.R. In the recent past, this kind of transformation has been experienced by the economies of Japan, South Korea and China. However, the Indian economy has yet to receive the gains from this kind of transformation.

It is quite an important historical fact that even before the birth of the Industrial Revolution, India produced around twenty-five per cent of globally traded manufactured goods. This too with the help of small household workshops, where craftsmen worked with hand tools. India experienced de-industrialization after 1750. Since then, India has never been able to achieve the status of an important manufacturing nation of the world. This has happened, despite the fact, that it has a well-diversified manufacturing sector. The Indian economy has in fact deviated from the empirically accepted path of development, which has been established by several scholars. This historical paradox motivated me to explore the puzzle of how a country that produced high quality handicrafts, which were traded in exchange of bullion, lost the race of becoming an important manufacturing nation of the world. It was quite obvious to me that the roots of this paradox lie in the past.

The present work is an outcome of my intellectual journey, which I undertook to seek an answer to the problem. My journey was made quite comfortable by the path cleared by the scholars, who have made contribution in the areas of history, history of business, economic history, history of technology, sociology and economic theory. This scholar provided me with a well-laid path, with enough light to guide me to my destination.

However, so many times I felt I was walking alone in darkness. The isolation sometimes created doubt in my mind as to whether I was following a correct path or not. In such situations, to know the appropriateness of my path, I shared my ideas with other scholars by writing about my understanding about different issues in professional journals. In the course of writing the present work, I published three papers, which were based on certain portions of the chapters of the present book. These papers appeared in the *Journal of Management Research* and *Global Business Review*. The feedback that I received from fellow travellers gave me confidence that I am following the correct course.

I do not claim that I have created an original account of the story of Indian manufacturing. What I have attempted here is to place the facts about Indian manufacturing, which were spread across different disciplines of the social sciences, into one place, in a framework to develop the story of Indian manufacturing. Now the book is in your hands, and I hope you will enjoy reading it.

New Delhi, India Vijay K. Seth

Acknowledgements

I have been working on this book for almost seven years. In the course of writing the book, I have benefited immensely from the writings of some of the great scholars. Their writings have both stimulated new thoughts and understanding, and have provided a new direction to my work. Here, I would like to mention the writings of Professors E. J. Hobsbawn, David Landes, Maxine Berg, William Lazonick, C. F. Sabel, M. J. Priore and Andre Gunder Frank.

I will also like to place on record the contribution of my teachers at the Delhi School of Economics, Professors Dharma Kumar, J. Krishnamurthy and Om Prakash, whose writings and lectures on economic history have provided initial advantage to my exploration on the subject. I have also learnt a great deal from the writings of economic and business historians, who have made significant contributions to the literature on India. Writings of some of these scholars have impacted my study significantly. These scholars are Professors Burton Stein, C. A. Bayly, Morris David Morris, Irfan Habib, Sanjay Sumbramanyan, Rajat Kanta Ray, Trithankar Ray, Prasannan Parthasarthi and Dwizendra Tripathi.

The quality of the work improved substantially from my interactions with Professors K. L. Krishna, J. P. S. Ubroi, B. N. Goldar and Raghav Gaiha.

I would also like to acknowledge the benefits of valuable comments that I received from my colleagues at the Faculty of Management Studies of the University of Delhi, where I completed a major part of my study. Here I would like to mention the names of Professors Joy Mitra and Vivek Suneja, Drs Harsh Vardhan Verma, Amit Bandhan and Pankaj Sinha. I would also like to place on record the continuous encouragement and motivation that I received from Padma Shri recipient, Prof Bakul Dholakia, Director General of IMI, where I am working at present.

I am also grateful to the librarians of the Ratan Tata Library of Delhi School of Economics; the Institute of Economic Growth, Delhi; Central Library of the University of Delhi; Jawaharlal Nehru University; Faculty of Management Studies, University of Delhi; and International Management Institute, New Delhi.

I am also indebted to my brother-in law, the late Prof B. B. L. Sharma who kept on motivating me to finish the book as soon as possible. Maybe he knew that due to delays in writing of this book, he might not have been able to read it before he went on his heavenly journey. I would also like to mention the contribution of my nephew Mr Pankaj Vianayak Sharma, who quietly listened to various facts that I have mentioned in the book regarding various aspects of the economy, society and polity of India.

I will fail in my duty if I do not mention the name of my doctoral students, Bharat Singh, Ritu Solanki, Lubina Zaidi and Prushottam Arya, who helped me in several ways.

Finally, I would like to mention that despite the availability of all aforementioned resources , this book would not have been completed without the quiet contribution made by my wife, Meera Seth. I am also equally grateful to my daughter, Arpita Seth, who not only brought immense happiness in my life, but also gave purpose to my existence. I am dedicating this book to her because it is only recently that I have begun to realize her importance in my life when she left my household to live with her husband Vishal Saxena.

Vijay K. Seth

Contents

CHAPTER 1

Introduction

Economists have divided all economic activities that take place in an economy into three broad categories. In the first category are included all those economic activities that human beings have been performing since the beginning of civilization. These activities are where persons work with different elements of nature viz. agriculture, animal husbandry, fishing, etc. and are called 'primary activities'. In the second category are included those economic activities where human beings work with manmade elements, such as manufacturing and processing. Through these activities, workers add value to natural resources by processing them or by transforming intermediate inputs and components into finished goods. These activities are described as 'secondary activities'. Along with these economic activities, there are others in which human beings make improvements to the state of products and persons through a variety of intangibles. These activities are categorized as 'tertiary activities' or 'services'. In economic analysis, these broad categories of economic activities are described as the three sectors of the economy.

The relative importance of each sector or group of economic activities for an economy can be estimated by measuring its contribution to the national income or employment. Development economists have evolved certain stylized facts, based on their empirical research, regarding the relationship between inter-sectional trends in income and employment and the level of development of an economy.[1] These stylized facts are expressed in terms of relative importance of different sectors of an economy at different stages of development. These stages are also described as periods of structural

V.K. Seth, *The Story of Indian Manufacturing*,
DOI 10.1007/978-981-10-5574-4_1

changes. Initially, structural change in terms of relative contribution to national income and employment occurs from primary to secondary sector and later to services. These inter-sectoral shifts have been described as three distinct phases of development experienced by economies during three distinct historical periods.

All developed economies were initially dominated by primary economic activities, when a substantial proportion of their working population was engaged in primary economic activities. The second phase began between 1760 and 1880, described as the post-agricultural period, during which these developed economies experienced (at varying dates) that the proportion of working population engaged in primary economic activity registered a marked decline and shifted to work in the manufacturing sector (i.e., secondary sector). This phase emerges as a consequence when an economy achieves industrial revolution. The third phase began after the 1960s, which is described as the post-industrial phase, when these economies experienced (at varying dates) decline in employment in the manufacturing sector and rise in the proportion of working population in services. The post-industrial phase of development is also described as the emergence of the 'service economy[2]'. These inter-sectoral trends in employment are accompanied by similar trends in the contribution of different sectors to national income.

Contrary to these trends, in the case of Indian economy, it has been observed that services are emerging as an important contributor to its growth process. Since the 1980s, the service sector has been contributing more than 50 per cent to its domestic product. This fact has led to the belief, amongst observers, that the Indian economy has evolved its own new path of development in which the economy leapfrogs from being an economy dominated by primary economic activities directly to a stage of development where services occupy the dominant position. In this process of leapfrogging, the Indian economy has skipped the stage in its process of development where the manufacturing sector generally dominates. This has happened because the Indian economy has missed industrial revolution. This historical fact is quite puzzling, because according to my own understanding of history, India was once an important manufacturing nation, much before the advent of the Industrial Revolution in England. I have also read in historical writings that the manufactured products of India were globally traded in exchange of bullion, because at that time other economies had nothing to offer in exchange for them except bullion.

Skipping the stage in the process of development in which the manufacturing sector emerges as a dominant sector of the economy is quite a significant loss. Historically speaking, in the literature on economic development, dominance of the manufacturing sector in the economy and economic development are considered synonymous. Only those economies that have undergone a manufacturing revolution have been able to achieve a long-term sustained process of development. It is the emergence of the manufacturing revolution that provided them with the possibility of continuous growth in the material wealth of the economy. The possibility of growth in the agrarian economies is constrained by the availability of fertile land, and even fertility of land is subject to diminishing returns.

Even Aristotle, while writing about sectoral composition of economic activities of an economy, suggested that primary activities are necessary for livelihood, but cannot become the source of unlimited wealth. He divided economic activities into two broad categories. The first category of economic activities consisted of activities that he termed as 'natural art of acquisition', and the other set of economic activities he defined as 'man-made art of exchange'. The first category of economic activities included those activities that helped man to obtain livelihood from nature. Aristotle believed that Mother Nature has the capacity to provide for only bare subsistence, in the same way milk is provided for the young infant by the mother. Thus, activities that depend on nature, nature herself sets limitation.[3]

It is also an important historical fact that the efficiency and quality of products produced by other sectors of the economy largely depends on the existence of a well-developed manufacturing sector. The production of agricultural capital goods and inputs by the manufacturing sector, for example, tractor combine harvesters, fertilizers and water pumps, has transformed the agricultural sector. Similarly, production of varying means of transportation, communication and information by the manufacturing sector have altered the way business is conducted in modern economies, across all sectors. One cannot ignore the importance of modern equipment such as CT-scan, X-ray and ECG machines, and several pharmaceutical products produced by the manufacturing sector, which have improved efficiency, reliability and quality of the healthcare sector.

In view of these important contributions of the manufacturing sector in the process of development of an economy, it becomes a matter of great concern that an economy, which was once upon a time an important manufacturing nation of the world, is now known for its service sector. This historical puzzle needs an appropriate answer that will tell us how a major manufacturing nation has to skip that stage of development in which the manufacturing sector makes significant contribution to income as well as employment. The historical puzzle motivated me to explore the story underlying this paradox in greater details. *A Story of Indian Manufacturing* is the outcome of my intellectual journey of exploration for finding an answer to this puzzle. I have followed a particular route for my intellectual journey. I do not claim that it was the best route to follow. I will welcome if scholars follow different routes to solve the puzzle from various perspectives to enrich our understanding. There are possibilities that different routes may lead to different destinations or may converge at the same point. It is due to this reason that the title of this book is *A Story of Indian Manufacturing* rather than ***The*** *Story of Indian Manufacturing*. My intellectual journey led me to the study of several writings of historians. This exposure has happened as a consequence of the nature of my exploration. I hope my historian friends will pardon me for intruding into their domain, which is almost like the entry of a mosquito into one's sacred space.

1.1 Traditional Flexible Manufacturing: Beginning the Story of Indian Manufacturing

The roots of the word 'manufacture' are two Latin words', *manu* and *facers*. Here, *manu* means 'hand' and *facer* means 'to make'. This signifies that manufacturing in earlier times was associated with handicrafts. This early form of manufacturing in India, in its embryonic stage, was carried out by artisans in the villages since ages. The process of social and economic division of labour created separation between craftsmen and farmers. Despite the separation of functions, for a considerable period of history, craftsmen remained an integral part of the agrarian economy. It is also an important fact that the separation of functions between craftsmen and the rest of the village economy was organized on caste lines. The social organization of production in the village economy did not facilitate the origin of market-mediated processes of exchange between craftsmen and other economic activities that were part of the village economy. This was happening because of the existence of a unique type of symbiotic relationship between artisans and other occupational groups in the villages. This symbiotic relationship was created by a distinct Indian social institution known as the '*jajmani system*'.[4] Through this institution, society had evolved a 'gift exchange' relationship amongst different occupational groups to provide the requisite supply of different kinds of manpower needed to run a village economy.

However, most of the economies of the world, especially of Europe and South and Southeast Asia, experienced the emergence of a peculiar form of manufacturing during the sixteenth, and in some cases during the seventeenth century, which has been variously described as the protean stage of industrialization,[5] industry before industrialization,[6] proto-factory,[7] nascent capitalism,[8] commercial manufacturing[9] and proto-industry.[10] Of all the terms used to describe this phenomenon, the term 'proto-industry' caught the attention of a large number of scholars. Initially, this term was used by Freudenberger and Redlick.[11] However, it received the attention that it deserved from the scholars after it was described in the writings of Mendles.[12] Proto-industry was used as one of the themes for discussion at the International Economic History Congress held in Budapest. The term *proto*, used as a prefix to the term proto-industry, belongs to the Greek language. It means embryonic, original, first time, earliest and primitive. In this sense, proto-industry stands for a stage of manufacturing that contained preconditions (i.e., seeds) for the emergence of modern manufacturing.

The proto-industry emerged as a culmination of a long, drawn-out process of evolution that began with Neolithic revolution.[13] The Neolithic revolution began with the domestication of animals and plants.[14] It brought together, at a higher level, two of the oldest economic activities of human societies: food gathering and hunting. The domestication of plants and animals created a process whereby human beings came to control, increase and improve the availability of food. During this stage of human existence, the easily available elements obtained from the exploitation of earth and nature were integrated into one operating system. The birth of these new economic activities led to the migration of human beings down from highlands and through valleys, on to vast alluvial plains. Along with the pattern of their migration, the nature of farming also underwent changes. It was, earlier, in the form of slash and burn, then bush fallowing, and subsequently short fallowing. When human beings began to occupy vast alluvial fertile river valleys, farming became a more organized annual single crop or multi-crop operation.[15] The increasing importance of fertile land led to increasing attachment of human beings to land, which made them more and more immobile. This attachment to land made it inevitable for human beings to give up their nomadic existence and forced them to embrace settled life.

The settled life set in motion several kinds of forces, which completely transformed the life of human beings. It gave birth to social organizations, collective learnings and beliefs. These provided conditions that facilitated the discovery of new crops, methods of cultivation and improved quality of animals. The expanding knowledge about plants and animals created possibilities for the production of food far exceeding the producers' own requirement. This remarkable development entails the emergence of economic surplus, which could sustain several economic activities other than the production of food, like crafts, trade, art, music, etc. The availability of surplus of foods also resulted in the formation of towns and urban culture and handicrafts. The crafts provided preconditions for the origin of the stage that has been described here as proto-industry.

Proto-industrialization has been accepted by scholars of economic history as a stage that existed during the period between the emergence of the putting-out system and consolidation of factory form of production organization in manufacturing. Some of the scholars view proto-industrialization as a transitional stage from *Kaufsystem*, in which production was dominated by petty commodity production, to *verlag system*, which was based on a capital-intensive form of manufacturing.[16] Proto-industry as a form of organization of production was characterized by dominance of a crafts-based method of production carried out in family-owned workshops. These household workshops were integrated into an operating system through the putting-out system organized by the merchants.

This form of organization of production required '*circulating capital*' rather than '*fixed capital*'. This distinction has emerged as a consequence of the use of simple tools and equipment by artisans, which they owned themselves. Therefore, there was no need to make large investments in fixed capital. The process of production was based on division of labour, where the production process was sub-divided into several specialized craft functions. These characteristics of proto-manufacturing provided it several kinds of flexibilities. These flexibilities got embedded into the organization of production. These same flexibilities are now built into the modern flexible production system.[17] These flexibilities establish similarities between proto-industry and modern flexible manufacturing; the only thing that separates them from each other is the level of technology being used.

Most of the criticism against the concept of proto-industry pertains to describing it as a stage in the evolutionary process of manufacturing.[18] I arrived at the term 'primitive flexible manufacturing' by combining the dominant characteristic of proto-industry (i.e., flexibility), with one of the synonyms of the word proto (i.e., primitive).[19] However, realizing that the word primitive connotes with primitive societies, I am using here the term 'traditional flexible manufacturing'. This is being done not to describe it as a stage in the evolution of modern manufacturing, but as a method of organization of traditional crafts into a system of production. This system of organization of production in manufacturing developed the capacity to satisfy the needs of mass markets.

Traditional flexible manufacturing emerged as a consequence of the loss of independence of craftsmen to merchants who supplied them raw materials or working capital along with their subsistence needs to control the marketing of their final output for mass markets. The emerging bondage of independent craftsmen to merchants was part of market expansion, resulting from centralized authority of state, monetization of the economy and emergence of urban agglomerations, which were integrated with a global network of long-distance trade.

1.2 Traditional Flexible Manufacturing and the Mughal Empire

Looking at the preconditions that were necessary for the successful growth of traditional flexible manufacturing, I am beginning the story of Indian manufacturing from c.1500 of the Christian Era (all subsequent dates are in Christian Era). The reason being that around this time two important historical events occurred that significantly influenced the course of Indian manufacturing. One was the entry of the European merchants through a sea route after the discovery of the sea route to India by Vasco da Gama in c.1498 CE, and the other was the beginning of the Mughal Empire in c.1526 CE with the entry of Mughals by land route. These two concurrent events provided necessary and sufficient conditions for the emergence of traditional flexible manufacturing in India. These were also two intertwined episodes of globalization, which performed a vital role in determining the pattern of growth of Indian manufacturing. The Mughal Empire created an integrated economy of continental size, and European trading companies expanded the commercial connectivity of the Indian economy to Europe and South and Southeast Asia.

Mughals were perhaps the first rulers of India who ruled over the land mass, which has been subsequently ruled by the East India Company (EIC), the British Raj, and the governments of Independent India, respectively. The continuity of events associated with Indian manufacturing occurring in the same geographical space provides geographical continuity to the story. This makes it possible to show that the whole drama of Indian manufacturing has been played out on the same stage. Story without geography becomes mythology, and when a story is narrated in the context of geography, it becomes history. I read somewhere that 'history without geography is a picture without a frame'. Moreover, India as we visualize it today took its shape during the Mughal Empire. Therefore, beginning the story of Indian manufacturing from c.1500 CE will help us to understand how it was conditioned by different political regimes that ruled the same geographical space at different points in time. This will make the story more interesting because it will be narrated in the context of the affairs of Indian manufacturing with different political regimes. Successful emulation of growth of manufacturing depends on the compatibility between technology, organization of production, and political environment emanating from the distribution of power amongst different strata of society. Political regimes have the capability to redistribute benefits in favour of the economic agent that can bring about higher rates of growth of manufacturing. The redistribution of power in favour of transformational economic agents is necessary because history tell us that, most frequently, political regimes are monopolized by those vested interests that obtain maximum gains from status quo.

Each society works according to its structure, which consists of institutions, rules, conventions and their enforcement. Therefore, it is the structure of the society that determines the rules and regulations that create the economic environment in which economic agents make decisions. The structure of the society also determines the kind of belief system that will be dominant for making decisions. Institutions constitute the structure that different political regimes impose on a geographical space, which create formal structure of incentives and disincentives. The created incentive structure influences the choices made by economic agents. Societies prosper when political regimes provide institutional support to those economic agents that are engaged in the economic activities that take the society to the next higher level of economic development. Societies may also experience decline and stagnation if the incentive structure promotes unproductive activities.[20] Keeping in my mind these lessons provided by institutional economies, and historical experience of social change, as recorded by economic historians, I am beginning to narrate the story of

Indian manufacturing as it has been conditioned by different political regimes that ruled in India since c.1500 CE. However, in the present book, I am focusing on the impact of only those political regimes that ruled India before its independence. Therefore, I am narrating the study of Indian manufacturing until c.1947 CE. I am writing a separate book to analyse the impact of those political regimes that Indian manufacturing experienced after independence.

In Chap. 2 of the present study, I describe the relationship that emerged between the economic environment created by the Mughal Empire and traditional flexible manufacturing. This is because during this period, traditional flexible manufacturing was emerging as an important form of organization of production in manufacturing. In this chapter, attempts have also been made to understand the impact of the socio-economic environment prevalent during the Mughal Empire and the composition of Indian manufacturing. The chapter also traces the sources of competitive advantage of traditional flexible manufacturing in supplying manufactured products to mass markets. It was due to the competitive advantage of this form of manufacturing that India emerged as an important producer as well as exporter of manufactured products during the Mughal Empire.

The Mughal Empire established a symbiotic relationship with traditional flexible manufacturing, known to the people living during this period as the *Dadni system*. The symbiosis between traditional flexible manufacturing and the Mughal Empire was so intense that the dawn of the latter determined the rise of the former. [The slow erosion in the authority of the Mughal state, after the death of Aurangzeb in c.1707 CE, until the Battle of Plassey (c.1757 CE, i.e., the beginning of the rule of the EIC) until the Mughal Empire lasted, fragmentation occurred in the polity of India.] There emerged different political entities, which redrew the political map of India. Some of these were large political entities like Mysore, Hyderabad, Bengal, Marathas and Sikhs, whereas some of them were small entities like Awadh, Ruhelas and Rajputana. The EIC remained primarily a trading company during this period. These five decades of political uncertainty were a mixed experience for Indian manufacturing. Handicrafts either registered a decline or reorganized themselves into different organizational structures at different locations depending on the size of the new political entity and attitude of the rulers towards merchants and craftsmen. This is because traditional flexible manufacturing was just one of the several modes in which handicraft manufacturing could be organized to supply products in the mass market. Crafts-based production can be embedded into several kinds of organizational structures.

1.3 Transformation of Traditional Flexible Manufacturing into Modern Manufacturing

It is a well-known historical fact that while traditional flexible manufacturing was experiencing slow and steady decline after the 1750s in India, the traditional flexible manufacturing of Britain was preparing to transform into modern manufacturing. It was in Britain, where in a short span of time, a series of inventions related to spinning and weaving of cotton sprang up simultaneously. These inventions, along with the invention of the steam engine, amongst many other technological changes, transformed the nature of manufacturing in Britain.[21] Despite the existence of traditional flexible manufacturing simultaneously in several economies of the world, including India, transformation of traditional flexible manufacturing into modern manufacturing first occurred in Britain. This process has been described by historians as the Industrial Revolution.[22]

Modern manufacturing is characterized by the replacement of human muscles by machines that are run with the help of inanimate powers. The integration of tools and inanimate powers gave rise to immense possibilities for the ever-changing unit of the manufacturing firm. However, to take advantage of these immense possibilities, merchants have to transform their circulating capital into fixed capital. The moment merchants did this they became slaves of their own investments. The use of machines also led to the concentration of various stages of production under one roof and, for the first time, craftsmen left their household workshops to work as workers in the *factory*. This new form of manufacturing organized into a centralized unit of production came to be known as '*factory*'. These factories were generally owner managed, which led to convergence between ownership and control. These changes altered the meaning of manufacturing from manu-facer (i.e., made of hand) to machino-facer (i.e., made of machines). This transformation explains how new techniques of production necessitated the emergence of a new form of organization of production in the form of factory or mill where owners of tools and equipment became industrialists and enjoyed control over production processes and management of the production unit.

Modern manufacturing crated an economy that was a great departure from earlier agrarian economies. In predominantly agrarian economies, even if they had traditional flexible manufacturing, process of development was conditioned by constraints imposed by nature. These economies depended on organic materials for food, clothing, housing and fuel. The supply of these materials was fixed in the long run, due to the fixed factor–land. This also applied to the modes of transportation that were being used in these economies based on animal power. In such economies, once land available with them was exhausted, diminishing returns to land were inevitable; if one views the diminishing returns on land along with Malthusian understanding regarding trends in population, escape from poverty becomes extremely difficult. The only alternative available for them to escape from poverty was to practice '*Smithian division of labour*'.[23]

When modern manufacturing emerged after the Industrial Revolution in Britain, it led to substitution of organic materials by mineral-based energy sources (i.e., the use of inanimate sources of energy to run machines). This new development resulted in the creation of a new world that no longer depended entirely on the vagaries of nature; it created a world in which possibilities of development did not depend entirely on land but on how the society itself managed its economy.[24] This departure signified a structural break from the past, by shifting its growth trajectory based on '*Smithian growth process*' to '*Promethian* growth process'[25] or '*Schumpeterian growth* process'.[26] It is also an important historical fact that most economies that have been able to internalize modern manufacturing, have also become developed economies. Due to this reason, since the emergence of industrial revolution, dominance of the manufacturing industry in an economy and economic development have become coterminous. It has now become an important stylized fact that only those economies that have achieved manufacturing revolution have been able to experience long-term durable economic development.

In Chap. 3 of the book, the entire process of transformation of traditional flexible manufacturing into modern manufacturing, as it occurred in Britain, is described in detail. This chapter also analyses how British manufacturing, which had emerged from the transformation of traditional flexible manufacturing, retained some of its characteristics. As a consequence, British manufacturing remained primarily consisting of family-owned enterprises, generally small in size and engaged in specialized production processes. Due to these characteristics, it also retained built-in flexibilities that were attributes of traditional flexible manufacturing. The most important reason for including this chapter in the book is to show how British industrial revolution conditioned the growth of manufacturing in other countries, especially in those countries that Britain had integrated as its colonies.

1.4 Traditional Flexible Manufacturing and the East India Company

As the British economy was experiencing transition from being an agrarian economy to an industrialized economy after industrial revolution, the East India Company was slowly becoming the ruler of the Indian sub-continent, after the Battle of Plassey in c.1757. The reincarnation of the East India Company into the ruler of the country altered the economic environment of the Indian economy as experienced by its traditional flexible manufacturing. The East India Company was established in the year c.1600, when Queen Elizabeth I of Britain singed the charter that granted to the company monopoly to trade with the East. The company started its first factory (i.e., a warehouse) at Surat in c.1608.[27,28] In c.1640, it established its next factory at Madras, and in c.1667 at Calcutta. It also obtained territorial rights over Bombay, when the king of England married the princess of Spain, because Bombay was given to the king as dower. In c.1717, the emperor Farrukhsiyar gave several concessions to the East India Company; the most important being the exemption from all levies on goods passing from one district to the other in Bengal. This concession was the major bone of contention between local rulers and traders with the East India Company.

The gradual decline in the authority of the Mughal emperors, after the death of Aurangzeb in c.1707, had increased the frequency of conflict between different regional powers. The increasing political risk forced the East India Company to protect its ships and offices of factors (agents). This necessity facilitated the process of transformation of factories into forts (like Fort William and Fort George) and raising its own private army. The emerging gaps in the structure of political power provided opportunities to the East India Company to develop territorial interests in the Indian sub-continent. In c.1757, with the Battle of Plassey, the process of territorial occupation of India began. In 1765, the East India Company obtained *Diwani rights* (i.e., right to collect land revenue) over Bengal from the Mughal emperor. After consolidating its power in Bengal, it slowly expanded its political control on other important centres of traditional flexible manufacturing, viz. Bihar (1765), Eastern Uttar Pradesh (1775), Malabar Coast and Mysore (1792–1801), and Surat and Ahmedabad (1800–1803).[29]

In new circumstances, the nature of interaction between the East India Company and the Indian economy qualitatively changed. Prior to c.1765, the East India Company had to pay in bullion for the purchases that it made for Indian manufactured items, which it traded globally. The company also depended on Indian merchants to procure these manufactured products from artisans. In the new dispensation, purchases of Indian manufactured products could be financed by the land revenue that the company collected from the Indian farmers; the other remaining part could be financed by the profits that it made from the selling of British machine-made goods, which it sold in India. The increasing political dominance of the East India Company provided support to the British merchants who slowly reduced the role of Indian merchants in different markets. They also started directly establishing relationships with Indian craftsmen, and in certain cases, they directly employed craftsmen in factories owned by the East India Company.[30] In these factories, Indian craftsmen were employed at very low wages, which the craftsmen had to accept because the company, from merely being a trading company, had now become the ruler. Through its policies, the East India Company marginalized the role of Indian merchants from the organization of production of traditional flexible manufacturing. This led to a disconnect between the Indian merchants and craftsmen, which provided precondition for the decline of traditional flexible manufacturing.

This period, which witnessed the rising dominance of the East India Company in the economic and political affairs of the Indian economy, was unfolding certain fundamental changes in the economy and polity of Britain—from being run by feudal lords and shopkeepers to an economy where manufacturers were dominating. This transformation created a new class of people known as industrialists, whose interests were quite at odds with the interests of the old order of merchants and landlords. The ideological premises of the new social order were articulated by Adam Smith in his famous books *The Wealth of Nations*, which appeared c. 1776. The new ideology systematically attacked the interests of merchants. Since the East India Company was a trading company and represented the interests of British traders, it also became an object of criticism in the new ideology. The most fundamental criticism of the East India Company is given by Adam Smith in his book.[31] However, in the conflict of interest between merchants and industrialists, the interest of industrialists prevailed.

The emerging dominance of the interest of industrialists in the economic and political life of Britain slowly created circumstances that led to erosion in the autonomy of the East India Company in managing the affairs in India. To protect the interest of the infant manufacturing sector in Britain, heavy tariffs and penalties were imposed on imports of Indian manufactured products. Eventually, c.1813, the monopoly of the East India Company to trade with the East ended. Finally, after the first war of independence (which is described in British historical writings as the Sepoy Mutiny) in 1857, the rule of the company also ended. After c.1857, the Indian economy became an integral part of the British Empire. This was the beginning of the '*British Raj*'.

The impact of the rule of the East India Company, which lasted for a century in India (1757–1857), on Indian traditional flexible manufacturing is discussed in detail in Chap. 4 of the book. The detailed analysis of the period amply demonstrates how the conflict between the interest of merchants and industrialists, which was becoming intense in Britain, negatively affected the traditional flexible manufacturing. This conflict, in fact, transformed the Indian economy from being an exporter of manufactured products to an exporter of raw materials for the British manufacturing industries. The traditional craftsmen and artisans lost their independence and their livelihood and had no choice but to go back to villages in search of their subsistence. Naturally, they were compelled to give up their traditional craft skills and embrace works related to agricultural activities. This process has been described by the nationalist and Marxian historians and economists of India as the process of '*de-industrialization*'.[32] The issues related to the debates on de-industrialization are also discussed in Chap. 4 of the book. To understand the consequences of the rule of the East India Company on the traditional flexible manufacturing of the Indian sub-continent, it is essential to know how the company was organized. The East India Company was the first joint-stock company, and being so it experienced the consequences of separation between '*ownership*' and '*control*'. This separation provides opportunities, to the people who controlled the day-to-day functioning of the company, for opportunistic behaviour. To study the consequences of opportunistic behaviour of the managements of corporate entities, and methods to regulate their behaviour, in order to harmonize the interests of different stakeholders of the company, a new discipline has emerged known as 'corporate governance'.

It has been observed that the organizational structure of the company was such that it provided ample opportunities for its management to indulge in opportunistic behaviour. The situation was more serious with respect to the managers who were working in the Indian sub-continent. Due to geographical distancing between the head office of the company and its operation in the Indian sub-continent, the chances of opportunistic behaviour on the part of the officials of the company working in India were much greater. Therefore, I discuss these problems in greater detail in Chap. 5. This chapter studies the East India Company purely as a corporate entity to illustrate how different the behaviour of the first joint-stock company was from the behaviour of other contemporary corporations.

1.5 Spatial Spread of Modern Manufacturing

Once modern manufacturing was firmly established in Britain, this powerful force began to spread to other European economies, and simultaneously to other distant economies where European settlers were dominating the polity and economy as well. In this sense spatial spread of modern manufacturing to the economies of America, Canada, Australia and India, was in a way an extension of modern manufacturing in the periphery of the European core region. Indian economy experienced the birth of modern manufacturing c.1854. The manufacturing enterprises that emerged at that time in India were in those lines of manufacturing where Indian manufacturers had a distinct cost-related competitive advantage over British manufacturing. These cost advantages in production were so substantial that they could also overcome the cost disadvantages created by discriminating tariff structures. Therefore, modern manufacturing began in India with the opening of enterprises in two textile industries—cotton and jute textiles. Both industries had a long tradition of being organized into handicrafts production and also had local availability of raw materials. Cowasjee Davar established the first cotton textile mill in July 1854 at Tardeo in Bombay,[33] and in the same year, George Aukland established a jute textile mill, Auckland Company, at Rishra in Seerampur in Bengal.[34]

It is also an important historical fact that while modern manufacturing was finding its roots in India, the nature of modern manufacturing was changing globally. Due to these global changes, British manufacturing was slowly losing its status of being the 'manufacturer of the world'. By the 1860s, a new industrializing economy, America, began to give stiff competition to British manufacturers. Its manufacturing firms were emerging as global players. The competitive advantage of US manufacturing largely depended on the availability of raw materials and natural resources, accompanied by increasing inflow of an 'immigrant' workforce. The competitive advantage of US manufacturing was also enormously facilitated by the existence of railways, telegraph and telephone. US manufacturing could leverage the advantage of these innovations, because the United States started growing around the same time as these innovations were commercially exploited on a large scale. In fact, some of the development economists have established that the railway acted as '*leading sector*' of the economy.[35] In the case of Britain, these innovations came into being after a considerable lapse of time (around seven to eight decades) from the time modern manufacturing emerged there in the 1760s.[36] For instance, railways were developed in Britain around the 1830s, the telegraph around the 1840s,[37] and the global connectivity of the telegraph was established in the 1870s.[38] The US manufacturing enterprises received enormous benefits from these innovations because these innovations provided them logistic support to cater to large markets. These benefits were the same as provided by the introduction of information technology to contemporary manufacturers. This advantage was manifested in the emergence of large-scale enterprises. Early entrepreneurs in US manufacturing like Rockefeller, Morgan and Carnegie established large enterprises, which was also facilitated by the introduction of '*methods of mass production*'. The process began with the introduction of the assembly line in the Ford Motor Car Company c.1911, which lead to the birth of Fordism. The existence of large-scale enterprises also provided objective conditions for the introduction of methods of 'scientific management', which were evolved by Frederick Winslow Taylor and are now described as *Taylorism.*[39]

Firms in America experienced rapid expansion in their scale of operations, from being single-plant, single-product and single-location to multi-product, multi-plant and multi-national firms. Even increasing size of the enterprise necessitated the dependence of firms on savings of persons other than the promoters of the firm. This led to the birth of joint-stock companies, making a divorce between 'ownership' and 'control' inevitable. In new circumstances, ownership rested with the stockholders, whereas control was in the hands of professional managers. This phenomenon has been described as '*managerial revolution*'.[40] These three interrelated innovations in the method of production and its organization provided distinct identity to the American manufacturing enterprises. These three innovations affected the nature of manufacturing very significantly. The use of the assembly line increased the size of the enterprises due to economies of scale and bulk production. The methods of Taylorism pushed the extent of division of labour at the shop floor to the minutest conceivable task. The separation between ownership and control gave birth to managerial hierarchies, which removed the managerial constraint, which kept the size of family firms very small in Britain.

These three unique characteristic of US manufacturing have given support to the view that technological changes are generally path-determined, and they follow a path determined by socio-economic and political environment of the economy, collectively reflected in the institutional setting in which manufacturing enterprises grow. Since the emergence of these two, and subsequently several distinct patterns of growth of manufacturing, scholars are trying to find the answer to a most pertinent question regarding global spread of modern manufacturing. The question is whether manufacturing will retain its structural uniformity in all economies where it is internalized, or will assume different forms or characteristics emanating from the temporal and spatial variations in the history, culture, factor endowments and institutions of different economies.

The former alternative view is based on the premise that temporal and spatial variations in the forces that have bearing on the process of development of modern manufacturing do not really matter, because it is an autonomous scaler process independent of socio-political and economic systems. The alternative strand of thought highlights the differences in the nature of innovations and variety of forms that manufacturing can assume in the organization of production over time and space. Therefore, this view emphasizes the uniqueness in the pattern of growth of manufacturing at different points in time and space. The uniqueness emerges because the growth of manufacturing is embedded in different institutional settings in different economies at a point in time, and in the same economy over a period of time. Hence, according to this strand of thought, one of the important tasks of the scholar, under these circumstances, is to identify the unique attributes of country-specific manufacturing and explain how these attributes have determined the unique path or trajectory of development of modern manufacturing.

After accepting these historical facts about modern manufacturing enterprises, economists have evolved the term '*long waves*' rather than stages to describe paradigm shifts in manufacturing activities. This term was initially used by a Russian economist, Nikolai Kondratieff[41] and further evolved by Joseph Schumpeter.[42] The term 'long wave' rests on the premise that each long wave is associated with certain activities, which provide it distinguishing characteristics. Schumpeter identified that underlying each long wave is a '*clustering of innovations*', which provides each wave a distinct identity. This suggests that innovations in manufacturing are not isolated events but rather part of an interrelated technological system. The wave may consist of new products, new processes or new organizational structures because the ones already in use by leading enterprises get diffused very fast and become almost like a paradigm. Each wave in manufacturing embodies a distinct combination of products, processes and organizational forms, which also collectively influence the other parts of economy.[43] Therefore, each wave may result in deep-rooted qualitative changes in the economy also. These qualitative changes are necessitated by a need for realignment of policy regime with new economy, which is being evolved as a consequence of a new wave. Sometimes these waves become all-pervasive and alter the fundamental relationship between technological change, political change and social change.[44] The new waves may emerge in the same economy at different points in time or in different economies at a single point in time.

The important facts regarding the characteristics of modern manufacturing that are country-specific or unique facilitated the birth of the '*national system of innovation*'. The historical origins of this concept can be traced back to the writings of Friedrich List (1841). However, he uses the term '*National System of Production*'. The National System of Production includes manufacturing, national institutions (including institutions that create and diffuse knowledge and skills), infrastructure and networks of transportation and communication for transporting people, ideas and goods. List gave overriding importance to building national infrastructure and institutions that are engaged in the creation and accumulation of '*mental capital*' (human capital). He also gave equal importance to processes that facilitate proper utilization of mental capital in initiating the process of economic development through state intervention. He had no confidence in the '*invisible hand of the market*' to do the magic as envisaged by Adam Smith.[45,46]

In its cotemporary form, the concept of a national system of innovation emerged in an unpublished paper written by Christopher Freeman, which he had prepared for the Organisation for Economic Co-operation and Development (OECD) expert group on science, technology and competitiveness. In this paper, he argues that short-term strategies that are used by economies, by manipulating factor prices and exchange rates, play only a marginal role in increasing the competitiveness of the manufacturing sector of an economy. He emphasized that competitive advantage of the manufacturing sector of any economy depends on how efficiently it assimilates new technological systems and how far the new technological systems match or mismatch with the institutions prevalent in the economy. An economy that might have been growing faster with one form of dominant technological paradigm may suddenly experience stagnation, or down turn, if its institutions are not able to respond to the new emerging technological paradigm.

Several scholars subsequently contributed to the expansion and refinement of the concept of a national system of innovation.[47] The process of enrichment of the concept led to numerous variants of the concept viz. '*social system of innovation*',[48] or '*national business system*'.[49] Despite the fact that there exist several competing formulations regarding this concept, contemporaneously, there are certain basic or core characteristics of all the variants of the concepts that are common. The first common characteristic, which is similar in all the formulations of the concept of a national system of innovation, is that each national system differs in terms of the area of its specialization, which is determined by its knowledge base.[50] The second common characteristic is that the knowledge base of the economy is a more important determinant of the performance of the national system. The third common characteristic is that the knowledge base of the national system of innovation is more localized and it does not move from one location to another rapidly. This is because knowledge is embedded in the institutions and routines in the firms. The knowledge base is localized because it is different from information.

All formulations regarding the global spread of manufacturing activities amply suggest that the initial evolution of modern manufacturing might be a spontaneous historical event, but its subsequent spread was not possible without the active role of the state, both in terms of nurturing modern manufacturing and of creating national systems of innovation. Another lesson we have learnt from the experience of global spread of manufacturing is that when modern manufacturing spreads into economies that had well-developed traditional manufacturing, the result was almost always to create a hybrid structure. One part of the hybrid structure consisted of manufacturing working on capitalist lines, using modern technology, with large-scale enterprises, and another part consisted of working on non-capitalist social formations using traditional technology on a small scale. The structure of manufacturing that existed in India, at the time of independence, manifested this phenomenon of dualistic structure.

1.6 Alternative Paths to Develop Modern Manufacturing

The theoretical constructs, mentioned previously, regarding global spread of manufacturing, amply suggest that the story of Indian manufacturing ought to be different from the story of manufacturing of other economies. The difference emanates from the fact that these formulations are based on the premise that the economy is governed by the national state that acts as a facilitator of modern manufacturing. This basic premise was missing in the case of India during the British Raj. The British Raj was a unique experience for Indian manufacturing because unlike the earlier rulers (the Mughals), who had accepted India as the land of their destiny, the Raj never thought of India as their promised land. In fact, they believed that they were on a civilizing mission to the land of barbarians. Therefore, at that stage of history, when the country needed the existence of a nation state, it was ruled by an imperialist state. For the first time in Indian history, the processes regarding policy and decision making remained outside its territories, and decision makers remained strangers and alien to Indians.

During the rule of the East India Company, the power elites of Britain had transient interest in the affairs of the Indian economy. The beginning of the Raj brought India to centre stage because it emerged as an important jewel in the British crown. The centrality of India in the British Empire was conceived differently by different constituents of the power elite of British power structure. Marx has very appropriately described the vested interests of these various segments of British power structure in his famous article, which he penned for the *New York Daily Tribune* dated 8 August c.1853. He writes—'British aristocracy wanted to conquer it, the moneyocracy to plunder it, and milliocracy to under sell it'.[51] Amongst all these vested interests, the interests of British aristocracy and moneyocracy were already fulfilled during the rule of the East India Company. During British Raj, there emerged paramountcy of the interests of British milliocracy. The hegemony of the milliocracy began with the removal of monopoly power of the East India Company to trade with India c.1813, and was completed with the end of the rule of the East India Company c.1857.

British aristocrats and merchants used protectionist policies to protect their vested interest. In the early days of the birth of British modern manufacturing, it also helped the British milliocracy to protect their infant manufacturing sector. To lay the foundation of modern manufacturing in Britain, tariffs were continuously imposed since c.1813. However, by the

1850s, Britain had emerged as an undisputed 'Workshop of the World' and its manufacturers did not require protection. Protected trade essentially protected the interests of landlords and merchants and harmed the interest of industrialists. By imposing tariffs on imported raw materials from other countries, including India, costs of raw materials increased for industrialists. Similarly, by taxing imports of food items, government increased the cost of living, which necessitated the payment of higher wages to the working classes. Milliocracy of Britain had absolute advantage in the production of machine-made goods. These new factors helped in articulating the policy of free trade by the British manufacturers. Keeping in mind the interests of British manufacturers, British government strictly followed the policy of free trade. It was at this stage, when there was need to protect infant modern manufacturing in India, that Indian manufacturers were experiencing a political regime that was religiously following the policy of free trade articulated by Adam Smith and David Ricardo.[52] The political regime was totally indifferent to the needs and inspirations of the Indian manufacturers.

It is obvious that during the same period, all other economies that were governed by nation states imposed tariffs to protect and promote their infant modern manufacturing. Bismarck raised tariffs in Germany in c.1875, and tariffs were raised in France in c.1881, Russia in c.1882, America in c.1890 (McKinley Tariffs) and Canada in c.1897 (Dingley Tariffs).[53] By following the policy of free trade under the guise of Smithian and Ricardian arguments in support of free trade, British government served the interest of British milliocracy. This policy had very obvious consequences for the Indian manufacturers. It led to decline of traditional crafts-based manufacturing in India because these small household workshops could not compete with British machine-made goods. This policy also helped 'simultaneously' in postponing the emergence of modern manufacturing because it needed protection at the stage of infancy. Hence, these policies, on the one hand, created conditions for the decline of traditional handicrafts, and, on the other hand, arrested the growth of modern manufacturing. In this way, the free trade policy transformed a manufacturing nation into a supplier of raw materials.

In order to increase the availability of Indian raw materials to British manufacturing industries, the British Raj invested lots of resources for the construction of railways in India. The process of expansion of railways began in India in c.1853 when railway connectivity was created between Bombay and Thane. The objective for constructing railways was essentially to connect centres of production of raw materials with ports. Due to the very objective of its creation, railways did not emerge as a leading sector of the Indian economy, as it emerged in the case of the United States.[54] However, it shifted the cropping pattern, from subsistence food crops, in favour of cash crops. This shift further increased the supply of raw materials.

Manufacturing enterprises of most economies also received support from the government in terms of government purchases. Governments are very large buyers of manufactured products. By buying these products locally, governments create additional demand for manufacturers. However, in India, since its inception, the British Raj followed the policy of buying stores from England. This policy of government purchases, on the one hand, provided support to British manufactures, and, on the other hand, led to a substantial increase in sterling payment. Despite several suggestions made by persons from the establishment of the Raj, Indian manufacturers had to wait for a substantial period to take the advantage of government purchases. Buying of stores from India became inevitable during World War I and World War II when trade between Britain and India was suspended due to lack of safety of sea routes. Wars provided unintended protection to the Indian manufacturing sector and also compelled the British government to source its requirements from Indian manufacturers.

One of the unique innovations in the organization of production, which was predominantly used by the Indian manufacturing enterprises, was the managing agency system. This system of organization of business emerged in response to the prevailing circumstances experienced by the promoters of manufacturing enterprises during the British Raj. The managing agency system grew after the formulation of the Companies Act 1850, which permitted the formation of joint-stock companies in India. The promotion of modern manufacturing, in joint-stock form, needed the availability of managerial skills and large volumes of capital. In the absence of a well-developed capital market and shortage of managerial skills, British agency houses, which were conducting business in India as agents of British companies, provided managerial skills as well as capital. These agency houses promoted joint-stock companies to enter into manufacturing businesses and appointed themselves as managing agents of the companies, in place of an elected board of directors. Following this form of business organization, they created a situation where private agency houses began to manage public capital. Later on, the same business model was used by Indian promoters of modern manufacturing like Tata Sons, Birla Brother and Nauroji Wadia & Co.

It is significant to note that, despite having to cope with a hostile and generally indifferent political regime during the British Raj, due to its certain inherent competitive advantages in some of the manufacturing products, modern manufacturing enterprises began to spring up in India. The historical circumstances that emerged after the birth of the British Raj ensured the availability of all important prerequisites that are necessary for the growth of modern manufacturing. The Indian sub-continent always had a substantial supply of entrepreneurs by way of several communities of merchants operating in different parts of the economy. These merchants had accumulated financial resources from long-distance trade in commodities like cotton, jute and opium, and several manufactured products produced by the traditional flexible manufacturing. These communities of merchants established several modern manufacturing enterprises. The process began with two textile industries, followed by iron and steel, sugar, tea, cement, general engineering, chemical and processing of tobacco and indigo. It is also important to note that by the time India achieved independence, its manufacturing sector had already almost completed the stage of ***primary import substitution***.[55]

Chapter 6 of the book deals with the consequences of the British Raj on the Indian manufacturing sector. This chapter discusses the importance of endogenous factors as well as the role of the British Raj in influencing the nature, structure and extent of growth of manufacturing in the Indian sub-continent. The existence of small and fragmented markets for manufactured products due to the small size of the middle classes, differentiated along religion, regional and cast lines in terms of tastes and needs, created a narrow market for standardized products produced by modern large-scale manufacturing enterprises. These conditions provided opportunities for the continuity of traditional manufacturing, as well as for the birth of modern flexible manufacturing. Therefore, this chapter explains how the circumstances created by the colonial administration, factor endowment and market conditions created a complex form of manufacturing enterprises, which India inherited at the time of independence. This form of manufacturing sector posed many challenges for the governments that were elected in India after independence. How the political regimes formed by the governments of independent India addressed these challenges is the subject matter of another book, which I am writing separately.

Finally, in Chap. 7, I provide the summary of the main components of the story of Indian manufacturing and the lessons that we learnt from the this book.

Notes

1. These empirical investigations have been conducted by Clark (1940), Chenery (1960) and Kuznets (1966).
2. To know the process of transition of developed economies into three different phases see Fuchs (1968), Bell (1973), Castells (1996), Aoyama and Castells (2002) and Seth (2006).
3. See Aristotle (2001), pp. 936–940.
4. For a comprehensive understanding on jajmani system see Wilson (1970), Fukazawa (1972), Commander (1983), Fuller (1989), Mayer (1993) and Guha (2004).
5. The term protean stage of industrialization was used by noted economic historian Hobsbawn (1954).
6. See for this term Tilly and Tilly (1971).
7. This term was used by Mendels (1972).
8. The term nascent stage of capitalism has been used in Levine (1977).
9. Perlin (1983) used the term to describe a transition from subsistence manufacturing to commercial manufacturing.
10. Term proto-industry was used by Freudenberger and Redlick (1964).
11. See for details Freudenberger and Redlick (1964).
12. See Mendels (1972, 1982).
13. The anthropologists describe the cultures that include the knowledge of crops and animals as Neolithic cultures, see Daniel (1943).
14. These facts have been obtained from Singer et al. (1958) and Victor (1888).
15. For the understanding of the process of transformation of societies to different types of farming see Boserup (1965).
16. For the appropriate meaning of *kauf* and *verlag* system, see Kriedite et al. (1981).
17. For comprehensive understanding of the modern flexible manufacturing system see Kenny and Florida (1993), Knudsen (1996), William et al. (1987) Womack et al. (1990), and Womack and Jones (1996).
18. To know about the criticism of the term proto-industrialization being used as stage in the evolution of modern manufacturing see Freudenberger and Redlick (1964).
19. I have used this term earlier in my papers Seth (2002, 2003, 2008).
20. These ideas have been evolved by the scholars who have enriched the literature on Institutional Economics. To understand the arguments of institutional economist see North (1990, 2006).
21. To know the emergence of several technological innovations, which changed the character of British manufacturing, see Singer et al. (1958), Vol. III and Ashton (1948).
22. For the description of the idea of Industrial Revolution see Toynbee (1958), Ashton (1948), Hobsbawn (1968a, b) and Neff (1943).

23. For the understanding of Smithian formulation of division of labour see Chap. 4 of the book and Smith (1776 [2003]).
24. For this idea see, Wrigley (1988), pp. 4–7.
25. For the understanding of Smithian Process of Growth and Premethien Process of Growth see Wrigley (1988).
26. The concept of Schumpeterian Growth Process has been discussed in detail in Parker (1982).
27. For the fact given here regarding the East India Company see Gardner (1971). Factory at that time was referred to the office of the factor, i.e., agent of the company along with a warehouse, which was used to keep merchandise purchased from Indian sources to be shipped to England.
28. The incident has been narrated in greater detail in Malabari (1910).
29. These facts have been obtained from Powell (1892).
30. For the working of factories owned by the East India Company see Hussain (1958).
31. See for the ideas of Adam Smith, on East India Company, in Smith (1776 [2003]) Vol. I, p. 426 and Winch (1978).
32. The concept of de-industrialization has been used to describe the experience of Indian manufacturing. Contrary to British experience where industrial revolution occurred through the process of transformation of traditional flexible manufacturing into modern industry. In India the traditional flexible manufacturing experienced decline which led to the shifting of craftsmen to agricultural activities. This movement of work force from manufacturing to agriculture is described as de-Industrialization, see. Thorner and Thorner (1962) and Morris (1987).
33. The name of the mill established by Cowasjee Davar was Bombay Weaving and Spinning Company. Before the birth of cotton textile mills in Bombay, cotton manufacturing was experimented in Calcutta by Bereath Cotton Mills and also at Pondicherry, but both the cotton mills could not succeed see Gandhi (1930), p. 52 and Mehta (1954).
34. George Aukland was a Scot, who was an ex-officer of the British Navy and had experience of running coffee plantations in Ceylon (Sri Lanka), see Buchanan (1966).
35. Economists like Rostow evolved the concept of leading sector, leading sector is that part of the industrial economy of an economy which while growing also affect, the growth of other economic activates through technological linkages. See Rostow (1957) and Rostow (1963). Later on it was observed that the railways acted as a leading sector in the development of the American economy see Fogel (1964).
36. However there exists a long debate amongst the economic historian regarding the exact dating of the British industrial revolution. But some of the

historians believe that the industrial revolution in England began in 1760. For details regarding this debate see Chap. 3 of the present book.

37. American, Samuel Morse had made a perfect electric telegraph along with the codes in 1838. It was introduced along the stretch of railway out of London in 1839 see Cairncross (2001), p. 23.
38. In 1858 it took forty days for news of Indian Mutiny to reach London, by 1870 after the establishment of several telegraphic lines which connected India to London, the messages from India to England could be sent in hours, see Solymar (1999), p. 3.
39. The founding father of scientific management Fredrick Winslow Taylor took the of idea division labour to its logically stretchable level. He believed that complex manual operations could be divided into simple routine operations and furthers these divided tasks can be done scientifically by precise measurement of every movement involved and thus eliminating all inessential movement; see Taylor (1919, 1885), Nockolds (1959) and Seth (2000), pp. 70–71.
40. With the separation between ownership and control in large American enterprises, there emerged a class of professional managers who managed the enterprises. This phenomenon has been described as managerial revolution; see Marris (1964). It was observed by some of the scholars that the Visible Hand of the management has been replaced by the invisible hand of the market process, see Chandler (1977).
41. See Kondratieff (1979), pp. 519–562.
42. For the ideas of Schumpeter, see Schumpeter (1934, 1939).
43. For detailed analysis of the concept of long-waves and clustering of innovations and their different forums see Freeman (1995), Freeman and Louca (2001) and Louca (2007).
44. There are several examples in history to substantiate this point that sometimes new waves alter the course of polity as well as society. For instance the debates regarding Corn Laws in England during 1830–1840, and spread of Luddaism during industrial revolution in England to arrest the decline of old occupations (like the occupation of handloom weavers) from the emergences of machines.
45. See List (1841).
46. See Freeman (1981).
47. Several scholars have contributed to the literature on national system of innovation; see Freeman (1995), Dosi (1999) and Lundvall (2007).
48. For the concept "Social System of Innovation".
49. See Lundvall and Pyka (ed) (2007), pp. 872–881.
50. See Lundvall and Pyka (ed) (2007), pp. 882–895.
51. See Marx (1853a, b, c).

52. The justification for the policy of free trade was provided by Adam Smith in his theory of absolute advantage and by Ricardo in the theory of comparative advantage. Their theories are part of all important text books on Trade Theories.
53. These facts have been taken from New Cambridge Modern History of Vol. IX pp. 56–57.
54. For the understanding of the objectives of development of railways in India see Macpherson (1955), Thorner (1950) and Astha (1976).
55. The stage of primary import substitution refers to that particular stage in the process of growth of manufacturing of a country, when it starts processing raw materials produced by its primary sector, and develops capability when there is no need to import such products.

Bibliography

Aoyama, Y., & Castells, M. (2002). An empirical assessment of the information society: Employment and occupational structure of G-7 countries 1920–2000. *International Labour Review, 141*(1–2), 123–139.

Aristotle. (340BC/2001). *Basic works of Aristotle.* New York: Modern Library.

Ashton, T. S. (1948). *Industrial revolution 1760–1800.* Oxford: Oxford University Press.

Astha, M. C. (1976). *Railway rates and fares in India 1849–1922—A historical cum analytical study.* PhD thesis, Department of Economics, Delhi School of Economics, New Delhi.

Bell, D. (1973). *Coming of post-industrial society – A venture in social forecasting.* London: Arnold Heineman Publishers.

Boserup, F. (1965). *The condition of agricultural growth.* London: Allen and Union.

Buchanan, D. H. (1966). *The development of capitalist enterprises in India.* London: Frank Cass.

Cairncross, F. (2001). *The death of distance how the communications revolution will change our lives.* London: Texere.

Castells, M. (1996). *The rise of network society.* Cambridge, MA: Blackwell.

Chandler, A. (1977). *The visible hand; the managerial revolution in American business.* Cambridge, MA: Harvard University Press.

Chenery, H. B. (1960). Patterns of industrial growth. *American Economic Review, 53,* 220–335.

Clark, C. (1940). *The conditions of economic progress.* London: Macmillan.

Commander, S. (1983). Jajimani system in North India. *Modern Asian Studies, 17* (2), 283–311.

Daniel, G. (1943). *Three ages.* Cambridge: Cambridge University Press.

Dosi, G. (1999). Some notes on national system of innovations and production and their implication for economic analysis. In D. Archibugi, J. Howells, & J. Michie

(Eds.), *Innovation policy in global economy*. Cambridge, MA: Cambridge University Press.
Fogel, R. W. (1964). *Railroads and American growth: Essays in economic history*. Baltimonre: Johns Hopkins Press.
Freeman, C. (1981, August). *Technological infrastructure and international competitiveness*. Draft paper submitted to OECD, Paris (Mime).
Freeman, C. (1995). The national innovation system in historical perspective. *Cambridge Journal of Economics, 19*(1), 23–45.
Freeman, C., & Louca, F. (2001). *As time goes by: From industrial revolution to the information revolution*. Oxford: Oxford University Press.
Freudenberger, H., & Redlick, F. (1964). Industrial development of Europe, reality, symbols and images. *Kyklos, 17*, 372–402.
Fuchs, V. R. (1968). *The service economy*. New York/London: National Bureau of Economic Research.
Fukazawa, H. K. (1972). Rural servants in Maharashtrian village, demiurgic or jajmani system. *Hitotshubashi Journal of Economics, 13*, 108–124.
Fuller, C. J. (1989). Misconceiving the grain heap: A critique of the concept of jajmani system. In J. Parry & M. Bloch (Eds.), *Money and morality of exchange* (pp. 50–62). Cambridge: Cambridge University Press.
Gandhi, M. P. (1930). *The Indian cotton textile industry*. Bombay.
Gardner, B. (1971). *East India Company's history*. Report and Hart Davis, London.
Guha, S. (2004). Civilization markets and services-villages servants from seventeenth century and twentieth century. *Indian Economic and Society History Review, 41* (1), 79–101.
Hobsbawn, E. J. (1954). The general crises of European economy in the seventeenth century. *Past and Present, 5*, 33–53.
Hobsbawn, E. J. (1968a). *Labouring men*. London: Weidenfeld and Nicolson.
Hobsbawn, E. J. (1968b). *Industry and empire—An economic history of Britain since 1750*. London: Weiden Field and Nicolson.
Hussain, H. (1958). *The company weavers of Bengal: The East India Company and the organization of textile production in Bengal 1750–1813*. New Delhi: Oxford University Press.
Kenny, M., & Florida, R. (1993). *Beyond mass production*. Oxford: Oxford University Press.
Knudsen, D. C. (1996). *The transition to flexibility*. Honwell: Kulwer Academic Press.
Kondratief, N. D. (1979). Long waves in economic life. *Review, 2*, 519–562.
Kriedite, P., Medick, H., & Schlumbhar, J. (1981). *Industrialization before industrial revolution* (trans: Schen, B.). Cambridge: Cambridge University Press.
Kuznets, S. (1966). *Modern economic growth; rate, structure and spread*. New Haven: Yale University Press.
Levine, D. (1977). *Family formation in the age of nascent capitalism*. New York: Monthly Review Press.

List, F. (1841). *Das Nationale System der Politischen Okonomie (National systems of political economy)*. London: Longman.

Louca, F. (2007). Long waves, the pulsation of modern capitalism. In H. Hanusch & A. Pyka (Eds.), *Elgar companion of neo-schumipeterian economics* (pp. 766–774). Northampton: Edward Elgar.

Lundvall, B.-A. (2007). National innovation systems; From list to freeman. In H. Hanusch & A. Pyka (Eds.), *Elgar companion to neo-Schumpeterian economics*. Cheltenham/Northampton: Edward Elgar.

Macpherson, W. J. (1955). Investment in Indian railways c. 1845–1875. *Economic History Review, VIII*, 177–187.

Malabari, P. D. M. (1910). *Bombay in making*. London: T. Fisher and Uniwin.

Marx, K. (1853a, June 25). The Biritsh rule in India. *New York Daily Tribune*. In I. Hussain (Ed.) (2006). *Karl Marx on India* (pp. 11–17). New Delhi: Tulika Books.

Marx, K. (1853b, July). East India Company—Its history and results. *New York Daily Tribune*.

Marx, K. (1853c, June 24). East India Company its history and results. *New York Daily Tribune*.

Marris, R. (1964). *The economic theory of managerial capitalism*. London: Macmillan.

Meek, R. (1962). *Economics of physiocracy*. New York: George Allen and Unwin.

Mehta, M. (1991). *Indian merchants and entrepreneurship in historical perspective*. New Delhi: Academic Foundation.

Mendels, F. F. (1972). Proto-industrialization—First phase of industrialization process. *Journal of Economic History, 32*, 241–261.

Metcalf, B. (1993). Norms of comportment among imperial Mughal officers, in John. In F. Richards (Ed.), *Power administration and finance in Mughal India* (pp. 255–289). Eldershot: Variorum.

Morris, D. M. (1987). Indian industry and business in the age of laissez faire. In D. Tripathi (Ed.), *State and business in India* (pp. 123–156). New Delhi: Manohar Publication.

Nayyar, D. (Ed.). (1994). *Industrial growth and stagnation, the debate in India*. New Delhi: Oxford University Press.

Nockolds, H. (1959). *The magic of a name*. London: G.T. Foulis.

Noiel, D. (1949, 1958). *The history of sugar*. Chapman & Hall, London.

North, D. C. (1990). *Institutions, institutional change and economic performance*. Cambridge: Cambridge University Press.

Parathasarthi, P. (2001). *The transition to a colonial economy-weavers, merchants and kings in South India 1720–1800*. Cambridge: Cambridge University Press.

Pearson, M. N. (2007). Markets and merchants in the Indian Ocean. In F. Bethancourt & D. R. Curto (Eds.), *Portuguese oceanic expansion, 1400–1800* (pp. 88–108). New York: Cambridge University Press.

Powell, B. (1892). *Land-systems of British India*. Oxford: Clarendon Press.
Rosen, G. (1958). *Industrial change in India*. Glencoe: Jllionais.
Rostow, W. W. (1957). *Stages of economic growth—A non communist manifesto*. Cambridge, MA: Cambridge University Press.
Schumpeter, J. A. (1934). *The theory of economic development*. Cambridge, MA: Harvard University Press.
Seth, V. K. (1999). *Capacity utilization in industry theory and evidence*. New Delhi: Deep and Deep Publication.
Seth, V. K. (2000). Technology and labour in retrospect and prospect. *Managment and Change, 4*(1), 61–90.
Seth, V. K. (2002). Primitive flexible to modern flexible manufacturing and the process of skilling and deskilling of labor. *Indian Journal of Labor Economics, 45* (4), 1175–1194.
Seth, V. K. (2003). The beginning and the end of primitive flexible manufacturing system. *Journal of Management Research, 3*(2), 73–86.
Seth, V. K. (2006). *Economics of services: A story of transformation of Cinderella into the queen of hearts*. New Delhi: Ane Books.
Singer, C., et al. (Eds.). (1958). *History of technologies* (Vol. 2–4). Oxford: Clarendon Press.
Smith, A. (1776 [2003]). Introduction. In A. B. Krueger (Ed.), *The wealth of nations* (Bantam classic). New York: Bantam Books.
Taylor, F. W. (1885 [1919]). *Two papers on scientific management*. London: Routledge and Sons.
Thirsk, J. (1994). Industries in countryside. In J. A. Chractre (Ed.), *Pre-industrial Britain*. Oxford: Blackwell.
Thorner, D. (1951). Great Britain and development of Indian Railway. *Journal of Economic History, XI*, 326–350.
Thorner, D. (1980). *Shaping of modern India*. Bombay: Sameeksha Trust.
Tomilson, B. R. (1993). *The economy of modern India 1860–1970*. Cambridge: Cambridge University Press.
Victor, H. (1888). *Wanderings of plants and animals from their first home* (edited by James S. Stallybrass). London: Swan Sonnenshein.
Wild, A. (1999). *The East India Company: Trade and conquest from 1600*. London: Harper Collins.
Wilson, C. (1965). *England's apprenticeship 1603–1963*. London: Longman.
Wilson, B. R. (1970). *Religious sects: A sociological study*. London: Weidenfeld and Nicholson.
Wiser, W. H. (1969). *The Hindu jajmani system*. Lucknow: Lucknow Publishing House.
Womack, J. P., & Jones, D. J. (1996). *Lean thinking banish waste and create wealth in your corporation*. New York: Simon and Schuster.
Wright, R. (2000). *Stolen continents—A conquest and resistance in the America*. London: Phoenix Press.

CHAPTER 2

Traditional Flexible Manufacturing and the Mughal Empire

For a considerable period, manufacturing activities, in their embryonic form, were carried out by artisans in villages. Prior to the Industrial Revolution, which gave birth to machine-based modern manufacturing, many economies of the world had become manufacturing economies with the help of crafts-based manufacturing. Scholars believe that this phase of manufacturing, which depended on traditional technology organized in craft-based small workshops, contained the seeds for the birth of modern machine-based manufacturing. As mentioned, some economies, especially those of Europe and South and Southeast Asia, experienced this form of manufacturing during the fifteenth and, in some cases, the sixteenth century. This phase of manufacturing has been variously described by different scholars, as discussed in the introduction of the book, as the '*protean stage of industrialization*',[1] '*industry before industrialization*',[2] '*proto-factory*',[3] '*nascent capitalism*'[4] and '*proto-industry*'[5,6] (a concept quite popular amongst scholars.) This phase of manufacturing has been described in the present study as '*traditional flexible manufacturing*'.

The renaming of the term has been done, because, unlike other studies that consider proto-industry as a stage in the development of modern manufacturing, we are putting more stress on the way proto-industry was organized. Because of its unique organizational structure, it enjoyed several kinds of flexibilities. Since it used traditional technology and enjoyed several kinds of flexibilities, it has been described in the present study as traditional flexible manufacturing. It emerged as a result of loss of independence of craftsmen to merchants. The latter supplied the former with working capital

V.K. Seth, *The Story of Indian Manufacturing*,
DOI 10.1007/978-981-10-5574-4_2

and met their subsistence requirements to control the marketing of the final product. It was the merchant who supplied the final product to meet the needs of mass markets. The emerging bondage of craftsmen to merchants was a product of market expansion, arising as a consequence of centralized authority of state, and emergence of urban centres and towns, which were integrated with global networks of long-distance trade.

Looking at the preconditions that are necessary for the successful growth of traditional flexible manufacturing, one has to start the story of Indian manufacturing from c.1500. At this point in time, the polity of the Indian sub-continent was totally fragmented because its geographical spaces were ruled by different kingdoms. The fragmented polity had given rise to several regional economies. Each regional economy contained vast tracts of fertile land, which were irrigated by several perennial rivers. These fertile tracts of land provided preconditions for growing a variety of crops, fruits and spices. These regional economies also had a long tradition of production of handicrafts, which satisfied the needs of not only the common man but also the luxury needs of kings and their courtiers.

Most of these regional economies were connected, through maritime trading, by land and seas with other regions and with the rest of the world. The Indian sub-continent had several port cities bordering the Arabian Sea, the Indian Ocean and the Bay of Bengal. The southern part of the sub-continent was integrated through trading with the economies of the Gulf and Southeast Asia, including China. The Western and Eastern parts of the sub-continent also had long-standing trading relationships with the economies of Gulf and South-east Asia. The northern part of the sub-continent was connected through trading by land routes with the economies of Central Asia, China and Europe. European economies were also an integral part of the trading network of the Indian sub-continent by land routes. The land routes existed because of the existence of contiguous land mass, which facilitated the movement of caravans of traders from Venice, Italy to China.

In this kind of polity and economy that existed in the Indian sub-continent around c.1500, two important historical events occurred that altered the course of the story of Indian manufacturing. One was the entry of European merchants through sea route after the discovery of the route to India by Vasco da Gama in c.1498. The other was the beginning of the Mughal Empire in c.1526, with the entry of Mughals through land route. These two concurrent events set the stage for the emergence and spread of traditional flexible manufacturing in India. If historical importance of this period is viewed in the context of the global economy, the year c.1500 stands at the cusp in the history of Indian manufacturing. These two distinct forces of globalization of Indian economy were the Empire buildings by Mughals and the arrival of the trading companies from Europe. Here, the process of globalization has been viewed as a historical process of integration of fragmented spatial units, which facilitates political, economic, technological and cultural inter-connectivity amongst these integrated spatial units. The inter-connectivity achieved with the help of these diverse processes creates de-spatialization, which makes a unit of space totally transformed by compression of space. Once connectivity occurs between fragmented spaces, individual units of space did not remain the same; they became transformed and reinvented.

Empire building in this sense is also an important process of globalization, which creates connectivity between isolated political, economic and cultural regions through trade and commerce, transportation, language, knowledge and technology. Empires created connectivity across territories that may otherwise have remained isolated from each other.[7] Therefore, unlike the contemporary phase of globalization, globalization at this stage of human history could be understood as a consequence of some military or political event.

During the first century of the Christian Era, a major part of the global economy was ruled by four mighty empires—the Roman Empire (based around the Mediterranean region), the Parthian Empire (covering land of two rivers along Iran), the Kushan Empire (which at Pamirs spread along the western most outpost of China), and the empire established by the Han Dynasty of China. The peaceful existence between these four empires lasted more than a century, which created connectivity as well as inter-dependency between these contiguous empires. These stable and peaceful empires provided opportunities to risk-taking merchants to engage in trade beyond their borders. To meet the needs of Rome's rich nobility's ultimate desire for silk, a trading route emerged, starting from Rome to Chang'an, the capital of the Han Dynasty of China. For the first time in human history, a trade route was named after the commodity of ultimate conspicuous consumption—the *Silk Route*.[8] Subsequently, Genghis Khan's Mongol Empire restored peace and tranquility along the Silk Route, which facilitated European trade by defending and protecting it.

Merchants and markets existed much earlier than the emergence of empires, but frequent extortion of wealth of merchants by predatory warlords added to the uncertainties associated with risk-taking behaviour. Good empires created a legal administrative framework, which was necessary to protect private property and for enforcement of contracts.[9] Wherever these peculiar needs of merchants and entrepreneurs were met by the political regime, the development process became faster. Empires provided these essential pre-requisites of development of manufacturing. Most of the legal and administrative infrastructure, which is needed for supporting economic activities in market economies, emerged during the super empire created by Pope Gregory VII in c.1075. This empire emerged when supremacy of the Roman church was established above the European states, by declaring—'let the territorial kingdoms serve or be the slave of celestial state'.[8,10] After this declaration, for the first time, merchants and entrepreneurs did not depend on the patronage and mercy of predatory rulers or warlords. The impact of this transformation was experienced by the entire Christian Europe.

Most of the empires were created by the use of horses in wars. The use of horses facilitated the emergence of bigger empires.[11] However, with the use of gunboats, the British established an empire where the sun never set. This empire imposed the philosophy of free trade across South and Southeast Asia and Africa. Therefore, it will not be inappropriate to state that at this stage of human history that the process of globalization spread across different countries and continents as a consequence of 'barrel of maximum gun', gun powder, and the ferocity of nomadic horsemen.[12] It will not be incorrect to say that, in the modern context, due to the absence of empires, there has emerged a need to invent new international organizations to provide some of the public goods that empires provided in the past. Some examples of public goods are global peace, uniform laws and free flow of goods and services, on which the forces of globalization rest. The formation of the European Union and other regional trading blocs and economic unions, such as the U.N.O (United Nation Organization)., World Bank, International Monetary Fund (IMF) and World Trade Organization (WTO).

The second powerful force of globalization was initiated by Vasco da Gama, who discovered the sea route to India, when he reached Calicut coast in c.1498. His voyage was linked to the age of discovery, when attempts were being made to extend the commercial interaction of Europe across oceans of Africa, Asia, North and South America. The discovery of a sea route to India paved the way for establishing links with different European economies and facilitated its participation in an emerging rudimentary global economy, evolved by European trading companies.

These two inter-connected episodes of globalization performed a very vital role in conditioning the composition, pattern and growth of Indian manufacturing. The first episode created an integrated economy of continental size, and the second episode expanded the commercial horizon of the Indian economy to European economies also. The European trading companies, including the East India Company of England, until the Battle of Plassey c.1757, maintained a cordial relationship with the Mughals for conducting their trading activities in coastal areas because they were aware that the authority and military might of the Mughal Empire were too strong to be tested in the battlefield. Therefore, until c.1757, the Mughal Empire determined the socio-economic and political environment that influenced Indian manufacturing. Due to these obvious reasons, in the sections that follow, I attempt to explain the relationship between the economic environment created by the Mughal Empire and the traditional flexible manufacturing, which was emerging as a dominant form of organization of production in Indian manufacturing.

The rest of the chapter has been organized into four sections. In Sect. 2.1, I describe the socio-political and economic environment created by the Mughal Empire; Sect. 2.2 deals with the impact of environment created by the Mughal Empire on the composition of Indian manufacturing. Finally, in Sect. 2.3, I provide an economic analysis of the traditional flexible manufacturing as an organization of production.

2.1 Economic Environment of Manufacturing During the Mughal Empire

The Mughal Empire, like many other empires of Central, West and South Asia, grew as a fragment of Genghis Khan dispensation. The Mongol Empire created by Genghis Khan was geographically most expanded, and was chronologically the most stable empire that the world ever witnessed. The empire established an important contact between Asia and Eurasia.[13] Babur Mirza laid the foundation of the Mughal Empire in India after defeating Sultan Ibrahim Lodhi in the Battle of Panipat in c.1526. Babur Mirza was the son of Umar Shaikh Mirza, a great grandson of Timur, the ruler of the Ferghana Valley. Babur's mother was Qutlugh Nigar Khanum, who was the daughter of Yunus Khan. Yunus Khan was from the lineage of Chagatai Khan, the second son of Genghis Khan.[14] Mughals were perhaps the first to rule over an area that almost covered the geographical space later known as British India and afterwards became independent India.

The Mughal Empire provided an important public good—law and order—which is an important precondition for the growth of manufacturing. This important aspect of the Mughal Empire has been appropriately described by Abul Fazl Allami, a friend, philosopher and guide of the Mughal emperor Akbar. He writes—'A king is the origin of stability and possession. If royalty does not exist, the storm of strife would never subside nor do selfish ambitions disappear. Mankind, under the burden of lawlessness and lust, would sink into the pit of destruction; the world, the great market place, would lose its perspective and the whole earth would become barren waste'.[15]

Mughals believed in the divine theory of monarchy, which emphasized that kings are descendants of God, and they have been chosen by God to rule on lesser beings. To establish divine descent, the lineage of Mughal emperors begins with Adam followed by five biblical prophets before reaching Ikhnus (also known as Hermes), followed by Joseph and his son Turk, ruler of Turkistan. Thus, shifting from biblical to first Turko-Mongol figure. After discontinuity of several generations, the most important divine event occurred when Alanquwa (Alan–quwa), a Mughal princess married to Zubin Biyan, the king of Mughalistan, became a childless widow. But Alanquwa was a woman of high morality and spirituality. Once, while sleeping in her tent at night, a ray of light miraculously entered her body and impregnated her. The progeny from this conception were triplets, three brothers, who were collectively known as *Nairum* (i.e., light produced). From the eldest of the three brothers, who possessed the greater proportion of illumination of his mother, the hidden light passed through several generations until *shahenshah* of mankind, Akbar, was born.[16]

However, as long as the Mughal lived in their native land, the main source of revenue was essentially earned from '*plundering thy neighbor*'. When Mughal's established their empire in India, the king had the right to appropriate surplus. One of the Sanskrit words for the king is *shastpada* or the owner of the sixth share of produce. Manu, however, permits one-fourth of the produce as a share of the king.[17] Therefore, in India, when Mughal emperors created a hierarchy of *mansabdars*, they were given the right only to appropriate surplus. Hence, wars in India were primarily not to capture land but to appropriate surplus.[18] The Hindu view of social order was based on self-sufficient villages, or *autarkic villages*, organized into numerous endogamous hierarchically ranked occupational casts, and sub-casts, segmented into four broad *varnas* (castes). At the top of the social hierarchy were *brahmins* (priests), followed by *kshatriyas* (warriors), *vaishyas* (traders and merchant) and *sudras* (workers and ordinary peasants). The caste system played an important role in making available a constant labour supply for labour-intensive agriculture. However, the views held by Mughals were at variance with the Hindu worldview. Mughals derived their views on social order from *Naqshbandi* and *Chisti* order of *Sufism* and Islam. Abul Fazl Allami described this worldview in *Ain-i-Akbari*. He divides social order into four broad categories; he does not rank them in any hierarchical order. According to his interpretation, all societies for their existence need people belonging to these four categories. However, the order in which he describes them may provide implicit ranking, in terms of their relative ranking in the social order.

Fazl starts his description with '*warriors*'. According to him, *warriors* in political body have the nature of '*fire*'. Their flames, directed by understanding, consume the straw and rubbish (i.e., rebellion and strife), but kindle the lamp of the rest in a world of disturbances. By first giving the description of warriors, perhaps he is signifying the importance of this class in expanding and maintaining a vast empire. This might also be due to the fact that the Mughal's shared their lineage with warriors like Genghis Khan and Timur. Description of warrior class is followed by the class of *artificers* and *merchants*. In Fazl's worldview, a person belonging to this category holds the place in the society like '*air*', because from their labour and travels, God's gifts become available and the breeze of contentment nourishes the rose tree of life. Fazl then talks about the class of the '*learned*' who, according to him, resemble '*water*'. From their pen and their wisdom, a river rises into draught of the world and garden of creation receives from their irrigation power of peculiar freshness. Finally, he mentions '*husbandmen*', whom he compares with '*earth*'. It is because by their exertions, the staple of life is brought to perfection and strength and happiness flow from their work.[19]

Before the Mughal Empire, in the social order in India, crafts were still an integral part of agriculture and were not distinct from the rural economy. The social division of labour based on caste lines was extended to include artisans also. Certain crafts became caste specific and these were prohibited form cultivation. Instead of creating a separation between craftsmen and cultivators, and creating a market-mediated process of exchange between craftsmen and cultivators, a new kind of symbiotic relationship was created between artisans and cultivators by introducing a distinct institution known as the *Jajmani system*. This institution was evolved by creating unique *gift-exchange* relationships amongst different occupational groups to provide the availability of different kinds of labour needed in the village economy. The gift-exchange relationship eliminated the market-mediated exchange and thus postponed the emergence of markets in the rural economy. The system was evolved primarily as a customary system of distribution of village produce amongst different sections of the village community and not as a system of production for exchange.[20]

The *Jajmani system* evolved as a social organization from symbiotic relations among patron households and different occupational groups, like barbers, laundrymen, carpenters, blacksmiths, masons, watchmen, folk artists, domestic servants and priests.

In this system, different occupational groups received share from the village produce based on customary payment, in lieu of market-determined rewards. The customary payments were made on occasions like harvest, birth, marriage, death and on important festivals. This system took the form of the *baluta system* in western India (especially Maharashtra). In the *baluta system*, hierarchy of functionaries in villages supplied services to all other members of the village community in exchange for equally specific payments in kind, but sometimes accompanied by a certain amount of cash.[21] The *Jajmani system* in Punjab was known as *vand*.[22]

Hegel described the *Jajmani system* as 'the most degrading spiritual serfdom' because it organized village society in such a way that it suffered from numerous rigidities based on religious beliefs. He observed that this system has created a social order, where 'the arrangement is fixed and immutable, which is subject to no one's will. All political revolutions, therefore, are matter of indifference to common Hindu, for his lot is unchanged'.[23] This formulation of Hegel provided basis for the concept of '*Asiatic Mode of Production*' evolved by Karl Marx.[24] This conceptualization was used by Marx to explain why Asian economies will remain stagnant unless and until they experience a strong exogenous force, which takes them out of this stagnation. He did not realize that the premises on which he was building his concept had been significantly altered by strong exogenous forces created by the Mughal Empire.

The village autonomy created by the *Jajmani system* and its various forms in different parts of India registered a decline during the Mughal Empire. The decline occurred due to monetization of the village economy, created by collection of land revenue in cash. The system also lost grip on social order because of migration of several artisans to emerging urban areas to pursue their vocations. It has been observed by historians that during the Mughal reign, the entire production of manufactured items meant for long-distance and medium-distance trade was carried out by artisans who were fully weaned away from the *Jajmani system*.[25] This suggests that the weakening of the *Jajmani system* increased the mobility of artisans, which increased the availability of artisans to participate in traditional flexible manufacturing.

The Mughal Empire also helped in increasing the size of the market, which is an important precondition for the existence of traditional flexible manufacturing. The size of the market increased due to integration of several fragmented markets into a single political unit and by eliminating trade barriers amongst them. Mughal rule also included an interregnum of seven years (c.1538–c.1545) when Sher Shah Suri ruled India while Mughal emperor Humayun was in exile in Iran. He developed a network of roads, including the major road that is now popularly known as Grand Trunk Road (National Highway No. 1). He constructed roads that connected Punjab to the town of Sonargaon, near the Bay of Bengal (*daraya-i-shor*), Agra to Burhanpur, Agra to Jodhpur and Chittor, and Agra to Lahore. Subsequently, under the Mughal rule, these roads were further extended to Cambay, Surat, Ahmedabad and Multan.[26] These roads intensified the connectivity between different markets and facilitated the movement of goods across different parts of India.

The administration of the sprawling Mughal Empire was based on extraction of surplus from agriculture to maintain the vast army of the empire. To appropriate agricultural surplus, a military administrative hierarchy of functionaries was evolved based on a Mongolian system of organization of army and administration. These functionaries were called *mansabdars*. The rank of each official was described numerically, which determined pay as well as rank in administration. For *mansabdars* who were performing military functions, their ranks were also specified numerically into trooper (*suwars*) ranks. The hierarchy followed in ranking *mansabdars* reflected the prevailing social hierarchy. Similar to the ranks of officials of the administration in the Mongol Empire, and in the Mughal Empire also, at the top were generally persons belonging to the royal family called *Tuman Begi* (i.e., ten thousands), followed by close relatives and nobles, *Ming Begi* (i.e., thousands) and captains, *Yug Begi* (hundreds). The *mansabdars* with *suwar* ranks were required to retain fully equipped cavalrymen equivalent to their numerical rank, which were made available to the emperor at the time of war.[27] Along with *mansabdars*, the Mughal Empire also created a skilled lower middle-level bureaucracy to look after imperial administrative machinery, consisting of trained persons in bookkeeping, auditing, correspondence, procurement and supply, record keeping and maintenance of stores. The formation of this politico-military administration created aristocracy and nobility, which generated demand for manufactured goods.

To meet the financial needs of the vast Mughal royalty, aristocracy, nobility and army, exorbitant land revenue rates were levied on peasantry by the Mughal Empire. These high rates of land revenue were paid by farmers in cash. The payments of land revenue in cash needed the introduction of official currency. Mughals created a centrally controlled network of imperial mints. The coins were made of gold (*muhar*), silver (*rupia*) and copper (*dam*). Most of these metals were obtained through trade of Indian products in foreign markets and also from local mines. When massive new supplies of silver and gold emerged from America and Japan, which were used to buy Indian goods, India became the 'ultimate sink of these metals'.[28,29,30] The trimetallic currency system of gold, silver and copper was originally based on copper *dam*. However, the Mughal monetary system maintained silver, *rupia*, as its principal coin. The Mughal silver rupee coin was elven grams in weight and was stamped from meticulously engraved imperial dyes.

According to historical records of early c.1600, Akbar's treasure hoard, calculated on the basis of information obtained from two different sources—the *Ain-i-Akbari* and *Tabaqat-i-Akbari*—consisted of slightly less than seven million gold *muhar* (weight 76 metric tons), seventy-two hundred million silver rupees (minimum 802 metric tons). Moreover, the Mughal emperor retained a surplus income over expenses between 3.9 million silver coins equivalent in cash each year, which was divided into three important heads; (i) salaries of mansabdars; (ii) military expenses and (iii) expenses on imperial household. These items of expenses also generated additional demand for manufactured goods. Monetization of the economy and expanding size of the market led to extensive trading, and created historical premises for the formation of a class of merchants, traders and moneylenders. Each region gave birth to certain trading communities, which subsequently developed a network of artisan's workshops to evolve traditional flexible manufacturing and later on became entrepreneurs when the process of modern manufacturing began in India. Communities like *marwaris*, *jains* and *banias* became important in Rajasthan; *khatris*, *aroras* and *banias* in Punjab; Tamil-speaking *chettiars* and Telugu-speaking *komatis* in South India; and *parsees*, *gujaratis*, *bohras*, *khojas* and *bhatias* in Western India dominated trading activities.[31]

These trading communities were involved in commodity trading, usually financing the working capital needs of artisans through the putting-out system (i.e., traditional flexible manufacturing), and providing supply to the army. Some of the big traders and moneylenders formed big banking firms that were engaged in transfer of large volumes of money from one place to another, and through a wide variety of financial instruments and transactions. The house of Jagat Seth was founded by *marwari* merchants Manikchand and Fatehchand belonging to the Oswal community of *marwaris.* One of the important financial instruments that was extensively used during the Mughal Empire was *hundi.* This instrument was used by the communities of merchants as a letter of credit and promissory note. This instrument helped merchants in transactions meant for long-distance trade and travel, when travelling, carrying and transferring of cash was risky. The *hundis* were issued by reputed merchant banking firms and their acceptability amongst merchants depended on the reputation of the issuing banking firm.[32] The use of *hundi* as a financial instrument can be traced back to the remote past. Legend tells us that it was the *hundi* issued by the banking firm of the *nagarseth* of Ahmedabad to Vastupal Tejpal that helped in financing the construction of the Dilwara Temple, located at Mount Abu.[33]

The widespread use of *hundi* as a financial instrument emerged during the Mughal rule, when it played an important role in facilitating the transfer of cash across different *subas* of the empire, and as cash used for long distances. Traded *hundies* were generally of two types: (i) *Darshani hundi*, which was treated like a bearer cheque, where payments were made immediately at the time it was presented and (ii) *Muddati hundi* or *Miadi hundi*, which involved deferred payment, after a stipulated time period, mentioned in the *hundi.*[34] It is paradoxical that the financial instrument that was associated with reputation and honour of the person who issued the *hundi* during the Mughal rule is currently associated with transactions made by the underworld and an illegal economy. It is now popularly known as *hawala.*[35]

These big banking firms had branches in different parts of the country. The Mughal state depended on them for procuring commodities, as well as cash and credit on a regular basis, because the main source of revenue during the Mughal rule was land revenue. This source was seasonal and was also affected by the vagaries of nature. The monetization of the economy and system of credit in Mughal India made its essential to evolve alliance with large trading banking firms. The house of Jagat Seth was transferring the annual collection of the Nawab of Murshidabad (Bengal *suba*) in advance to the Mughal emperor at Delhi. The annual values of remittance were more than ten million rupees.[36] Mughal emperor Aurangzeb had personally honoured the head of the firm, Mr Manekchand, for his timely loan of a large amount of money to the emperor with title of *Jagatseth*. Similarly, the richest *sheroff* of Gujarat, Shantidas, received the title of *Nagarseth* from Jahangir. The Mughal treasury accepted all his *hundis*.[37] Similar influence on local authorities was exercised by merchants like Krishna Arjunj Nathji and Jagannath Laldas Parekh in Surat, Gopaldas Manohardas in Benaras, and Lala Kashmiri Mal in Awadh.[38]

Increased reliance on a network of banking firms to transfer revenue from provinces to the capital city of the Mughal Empire adulterated the fiscal mechanism. The reliance on bankers was accompanied by direct involvement of the Mughal emperor, princes, his relatives and nobility in foreign trade. These conditions suggest that at that time, a sort of state mercantilism was emulated by the Mughal state in India. This has emerged as a consequence of combining *imarat* and *tijarat*. By combining these characteristics, the Mughal state developed state mercantilism and was committed to increasing the production of manufacturing by protecting employment of artisans.[39] The increasing appropriation of agricultural surplus facilitated the growth of urban agglomerations, which could be compared in size with any big city of the world. These cities were either administrative cities like Delhi, Agra and Lahore, or centres of production of luxury handicrafts, following the traditional flexible manufacturing system of organization of production like Patna, Banaras, Murshidabad, Dacca, Rajmahal and Surat. Some of the cities of the Mughal Empire were located on trade routes evolved by imperial highways or navigable rivers. The textile industries of Patna and Banaras had reaped immense advantage from possibilities of transporting their products through the Ganga river to the European trading centre at Hugli. These networks of cities and urban centres provided appropriate economic conditions for the growth of inland and overseas trade in manufactured goods, and agricultural commodities. The Mughal economy was organized around a network of market towns. In this network, commodities flowed from rural areas to urban centres, accompanied by a return flow of cash and currency to villages, to facilitate the payments of land revenue in cash by farmers.[40]

Urbanization increased during the Mughal Empire due to migration of artisans from villages to cities and urban centres of handicrafts. Urbanization also increased due to migration of landless workers to urban areas, who took advantage of employment opportunities in low end service occupations. According to available evidence, India at the time of Akbar had 120 large and small cities, 3200 *qasbas* (small towns) and 40,000 villages. During this time, 15 per cent of the total estimated population between 107 and 115 million lived in urban areas.[41]

The ruling class of the Mughal nobility had an extravagant lifestyle; their need for luxurious goods was satisfied by artisans manufacturing luxury handicrafts like high-quality cotton (Dacca muslin), jewellery (gold as well as silver), cutlery, decorative swords and weapons. The nobles maintained households with large numbers of wives, concubines, slaves and domestic servants. The Mughal emperor's own *harem* consisted of 5000 females, divided into two categories: (i) *mahin-bano* (i.e., high-ranking women), totalling 300 and (ii) *parastaran-i-huzur* (under the protection of the king or concubines). The total value of cloth consumed by the household of the emperor was valued at 1.6 million dams, and total value of gold pots, silverwares, dishes, cutlery, figurines, chandeliers, bedsteads, copper wares and porcelain crockery was 1.49 million dams according to *Ain-i-Akbari*.[42] If we add to this amount the consumption of thousands of nobles and *mansabdars* of the Mughal state, the total domestic demand for luxury handicrafts will be quite substantial.[43]

The expenditure pattern of Mughal nobility created urban culture and the pattern of their demand manifested into an occupational pattern, where a large proportion of the working population was engaged in the manufacturing of handicrafts and providing services to the elite. It has been estimated that the village economy gave employment to 72 per cent of the total working population and contributed 45 per cent of the national income. The non-village economy provided employment to 18 per cent of the population, while contributing 52 per cent of the national income. The non-village economy of the Mughal empire consisted of the Mughal emperor, court, *mansabdars*, *jagirdars*, native princes and hereditary *zamindars*, merchants, bankers, shopkeepers, traditional professionals, soldiers, urban artisans, construction workers and domestic helpers.[44]

To maintain regular supply of commodities and manufactured goods among different towns and cities during the Mughal Empire, the entire supply chain was managed by the community of *banjaras*. They were a migratory community and owned large stock of bullocks and bullock carts, wagons and packs of horses, and had intimate knowledge about the routes to different places in the country. It is speculated that the word *banjara* is derived from the Sanskrit work *vanij*, which also lies in the root of Sanskrit word *vanijaya* (i.e., *business*).[45] Apart from *banjaras*, *komati* and *khoja* communities were also performing the same function in Southern and Western India, respectively. It has been recorded that during the seventeenth century *banjaras* alone transported, on average, 821 million ton miles a year. Conversely, in 1882, the entire Indian railway network transported only 2500 ton miles.[46] There are also records regarding caravans consisting of Oxen numbering 40,000, where each oxen carried between 100 and 150 kilogram of merchandise.[47] There existed caravan *sarai* at a days' distance from each other on the trade routes, which could accommodate thousands of traders and their animals.[48]

Mughal India emerged as a nodal centre of a dynamic world trading system with various levels of commercial interactions and interdependencies between urbanized and commercial locations spread along the Indian Ocean (China, Central Asia), Middle East (Iran, Iraq, Afghanistan) and Europe (England, France, Spain, Portugal). Apart from sea routes, trade also occurred through land routes. Since overland routes passing through India—Persia up to the Ottoman Empire—were stable and relatively safe for trade, they provided ample opportunities for creating wealth by traders from India, Baghdad, Mosul and Damascus. A great deal of trade occurred between India and Central Asia through pastoral nomads, who extensively travelled territories between Indus and Oxus rivers. These trading nomads were known as *powindas*; the word is derived from the word *pawwal* from Pastho language (spoken in Afghanistan), or the word *puydian* of Persian language (spoken in Iran). These words mean 'to graze and roam'. The *powinda* traders carried with them merchandise loaded on thousands of camels annually. Similar routes were followed by the nomadic communities of *qazags* and *kalmuk* of South Russian steppes and the *bhotiya* tribe of the Western Himalayas.[49]

There were several port cites located along the vast coastline of the Indian economy. These port cities connected it with different economies through trade and commerce and created additional demand for Indian manufactured goods. Trade was dominated by traders from Central Asia and Arabia, but subsequent to the discovery of the sea route by Vasco da Gama, trading links were established by Portuguese, Dutch, French and British trading companies. Due to their trading and military rivalries, the Indian ports were monopolized by different European nations from time to time. Due to these reasons, India had become a theatre of battles for market shares amongst rival European trading nations. This has been described by scholars as 'globalization with gun boats'.[50]

The expansion in the domestic market and integration of the Indian economy in the global trading system provided positive preconditions for the growth of manufacturing. Looking at the size of the economy of Mughal India, it is believed that the 'Mughal empire was a wealthy empire, which dwarfed European states. In c.1700 population of India was twenty times that of United Kingdom. India's share in total world output at that time has been estimated at 24 per cent, while Britain's share was just 3 per cent'.[51]

The analysis given previously clearly establishes that during the Mughal rule, the prevalence of law and order and absence of regional trade barriers led to the integration of markets. This period also witnessed the process of growth of urbanization leading to growth in towns and cities. These towns and cities were inhabited by the elites and nobles, who were used to higher standards of living and conspicuous consumption. These developments had significant impact on Indian manufacturing. The increases in the extent of monetization of the economy, on the one hand, reduced the magnitude of customary mode of exchange based on the *Jajmani system* and intensified the market-mediated exchange relationship. On the other hand, it helped in the creation of a new class of traders and merchants. The emergence of this class was an important condition for the birth of traditional flexible manufacturing. In the section that follows, I focus on the consequences of these developments in determining the composition of Indian manufacturing.

2.2 Composition of Manufacturing During the Mughal Empire

During Mughal rule, India emerged as an important manufacturer of some agro and metal-based products. The most important industry, which experienced high growth rates, was the textile industry. Indian textiles were first reported in the *Perilus Maris of Erytherien Sea*, written by an unknown Greek writer. It is believed that wild cotton (Arabic quoton) was harnessed by the Indians living in the Indus river valley around 2000 BC. The cultivation of cotton, along with skills sharpened over centuries by Indian artisans in weaving, dying and printing of textiles, provided Indian cloth a distinct competitive advantage. Indian cotton and silk textiles were traded along most of the port cities located in the Indian Ocean, Arabian Sea and Red Sea, and through caravan routes to Central Asia. During Mughal rule, manufacturing of textiles received a big push and its production spread to different parts of India. Each region produced a distinct and specialized variety of textiles. These regional centres of textile manufacturing were interconnected through trade routes to cover domestic as well as foreign markets. In western India, Gujarat, Sindh and Rajasthan were centres of textile production. In south India, these centres were concentrated around the Malabar Coast, whereas in eastern India, centres of textile manufacturing were located at Banaras, Bengal, Orissa and Assam. North Indian towns like Agra, Lucknow and Amritsar were also important centres of textile manufacturing.

Muslin of Dacca, known as *malmal-khas*, was the finest muslin in the world. An expert spinner of muslin in Dacca could spin an even thread upward of four miles in length from *sicca* (rupee) weight of cotton. Watson had found that in comparison to muslin produced in England and France, Dacca muslin was of exceptional quality. Muslin in Dacca was available in different varieties; the most celebrated variety was *jamdani*, which derived its name from the Persian language—*jam* (flower) and *dani* (pot) (i.e., flower pots). Apart from *jamdani* the other varieties that were produced in Dacca were *tanjib, abroan, albalee, sarbate junglee, eshac* and *dorea*. Weavers in south India were also producing very fine muslin, which was comparable in quality with Dacca muslin, known as *Arni muslin*.[52] Muslin was also produced in Seerampore and Bankibazar areas of Bengal. These areas were also famous for the manufacture of cotton-stuff or *Calicoe*, Lucknow was famous for *chintzs*, and Ahmedabad was known for *dhotis* and *dupattas*. The towns of Nagpur, Umaer and Pooni were specializing in silk-bordered cotton textiles, which were dyed with popular dyestuff called *chay*.[53] Textile producers of the Coromandel Coast produced sailcloth, bedcovers, pillowcases, handkerchiefs and mattresses. Sindh in particular was renowned for embroidered quilts.

Manufacturing of silk textiles also spread to different regions, and each region specialized in the production of a specialized variety of silk. Surat was known for *Patola silk*, whereas *choppahas, bandena* and *corah* silk were produced at Murshidabad, Malda, Seerampore and Bankibazar areas of Bengal. These areas were also important centres of manufacturing velvet, satin, taffetas and tasar. *Mugga* silk was produced in Assam. Cities like Banaras and Ahmedabad specialized in silk brocade work. Kashmir, Amritsar and Ludhiana were important centres for the production of *shawls* and woollens.[50] Indian weavers were using a variety of looms to weave different kinds of textiles. *Vertical looms* were used to manufacture carpets. Fancy pattern cloth was manufactured with the help of ***drawlooms***. Weavers who were mobile used ***pit looms***, which were light in weight and could be assembled, disassembled and transported easily.[54]

An important and distinct characteristic of Indian textile manufacturing during Mughal rule was that some of the processes that are an integral part of textile production emerged as separate specialized industries, like printing, painting, bleaching, dying of textiles, production of dyestuffs and indigo. India had mastered the art of making vegetable dyes. Each region had evolved its own dyestuff, which was derived from local plants and flowers. Indigo manufacturing was localized at Bayana, Sarkhej, Jaipur, Bulandshahr and Khandesh. However, indigo produced at Bayana was considered the best.

Production of textiles was totally embedded in the social structure of the period, because division of labour practiced in the production of textiles was organized on caste lines. Here, processes involving the production of textiles were allocated to different castes, depending on their place in the caste hierarchy. The characteristics embodied in textiles in terms of '***purity***' and '***pollution***', and '***sacred***' and '***profane***', determined not only the status of the persons who were involved in its production, but also the social status of the person who consumed it. Therefore, ritualistic status of weaver, spinner, dyer, printer, tailor and washerman depended on the ritualistic quality of cloth that he produced.[55]

In market economies, the process of commoditization of products results in their alienation from social existence, which makes them an impersonal object of consumption. In India, production as well as consumption of textiles was embedded in the social framework of the existing social order, which was based on caste lines. It was not only the texture of the textile, or the type of fibre being used in its production that determined its consumption by different castes; even colours were caste specific. In this kind of social order, persons copied the pattern of consumption dictated by their caste affiliations or by their ancestors. This process was quite at variance to the consumption pattern observed in market economies, which is guided by individual choices, imitation of strangers, and current trends in fashion.

Mughal rule played a significant role in diminishing caste-based consumption of textiles. It was the duty of the Mughal emperor to consume the textiles and other commodities produced by their subjects. This role of Mughal king led to the diversity in the consumption pattern of textiles. The same role was played by the local administration of the Mughal Empire (***mansabdars*** and ***jagirdars***). The consumption pattern of textiles by emperor and nobility transformed the textiles from being social products to impersonal commodities.[56]

The other major change, which emerged during the Mughal Empire, in the consumption of textiles, was the changes in the attitude of orthodox Hindus regarding the use of stitched cloth. Since stitched cloth can be made only after cutting the textiles, it was considered ritualistically impure or profane by orthodox Hindus. Therefore, use of stitched cloth was not permitted in Hindu rituals. On the contrary, Islamic values based on the *Koran* required that a woman's entire body be fully covered by cloth. To have such a garment, cutting of cloth and tailoring was inevitable. Hence, in orthodox Islam, unstitched clothes are viewed as primitive and barbaric. Due to these divergent attitudes towards stitched clothes, Hindu women believed the *sari* to be the ultimate form of female garment and Muslims considered 'sari as an unstitched garment and relic of primitive time'.[57] However, the attitude of Hindus towards stitched cloth also changed because of social necessity, and even Hindu women began to cover their entire bodies.

The process of de-castification of textiles by the Mughal Empire, and the use of stitched cloth, led to the evolution of fashion. This became possible because under the changed circumstances, an individual need not copy the dress code dictated by caste affiliation or by his ancestors; he could make individual choices regarding texture, type of fabric, colour of textile and style of stitching. These changes helped in integrating segmented markets of textiles based on caste lines into a mass market. During this period, in Deccani Muslim states, new variety of textiles were created to satisfy the desire of Indian Muslims who wanted to use silk textiles. Since Islam forbids the use of silk, a new variety of textile evolved, which used cotton as well as silk fibres. These varieties were known as *mashroo* and *himroo*. These textiles were in great demand amongst Mughal nobility as well as in European markets.[58] Amongst the craftsmen involved in the production of textiles, as new specialized skills emerged with expanding market and variations in consumer preferences, new castes and sub-castes were created. Weavers known as *caingaloom* and *mambaloom* are such examples. Caingaloom weavers were those who were making textiles that were traded in the Malay Archipelago. The word *kain* or *cain* means cloth in Malay. Similarly, *mambalooms* were specialized in making painted chintz. *Mooree* weavers produced cloth that formed the base for producing chintz.[59]

During the Mughal period, demand for textiles grew significantly in foreign markets also. India contributed 25 per cent of the world's total output of textiles. When English merchants began to buy Indian cloth to sell it in England, the result was nothing less than a 'national makeover'. While describing the extent of this makeover of the British lifestyle, Daniel Defoe wrote in 1708 that 'Indian textiles crept into our houses, our closets, own bedchambers, curtains, cushions, chairs and at beds themselves, were nothing but calicoes or Indian stuff'.[60] English traders purchased huge quantities of Indian textiles that were largely used in England, and a portion of this was re-exported to different European countries. At that time, English traders had nothing to offer in exchange for Indian textiles, apart from bullion earned from their trade with rest of the world or brought from America.

English traders were earning bullion by entering into multiple markets simultaneously by arbitrage—the differences in the exchange rates between gold and silver—by placing themselves as middlemen in different trading locations.[61] The importance of bullion brought from America to buy Indian textiles has been very appropriately described by Andre Gunder Frank, when he mentions that 'Europe used American money to buy itself a ticket on Asian train'.[62] Apart from the British, the Dutch were also exporting a special indigo-dyed blue colour cloth from India to Africa to buy and clothe slaves. This cloth has been very aptly described as the 'cloth of sorrow'.[63]

The art of smelting iron was known to Indians since ages. It is said that India was the original producer of carbon steel (wootz iron).[64] The most important physical manifestation of the skills of the iron makers of India is the iron pillar located near Qutub Minar in Delhi.[65] There are historical records to substantiate the fact that the best steel used in Greek and Hellenistic monarchies was imported from India. Idris-ul-Mushtaq mentions that Indian iron makers were using compounds while making iron.[66] Marco Polo mentions the *ondanique* (Indian iron), which was imported in Kirman to make several metallic products like swords, bows and quivers.[67]

During the Mughal rule, demand for iron for making weapons, armours and other household products increased substantially. Therefore, this industry also experienced higher rates of growth. Mughal India was known for making steel that enjoyed worldwide repute. Iron produced in Kutch was used for making the famous *Korij* swords. Iron smelting was widespread in Monghyr and Birbhum in Bihar, Satghud and Rajahmundry areas bordering Andhra Pradesh, and Gwalior in Madhya Pradesh. The Chota Nagpur Plateau was popularly known as *Loharmahal* (i.e., palace of iron). Here, low-caste Hindus and the tribal population were involved in mineral production. These communities were *Pharias*, *Santhals*, *Agerias* and *Kola*.[68] Indelwai was an important centre for making swords, daggers and lances. These products were made from the iron mined in Kalaghat Hills.

Abul Fazl Allami mentions that in the royal *karkhanas* (workshops) of the Mughal king, several kinds of weapons were made for the royal army. He mentions that guns were made in these karkhanas of such sizes that a ball used in these guns weighed 12 *mans* (one *man* is equal to approximately 40 kilograms). Such guns were transported to the battlefield with the help of several cattle and elephants. These *karkhanas* were also making guns that could be taken in separate pieces to the battlefield, and were easily reassembled for use. He also mentions guns that could be carried by a single elephant, known as *Gajnals*, and guns that could be carried by a single soldier, known as *narnals*. These *karkhanas* were also making matchlocks of several types, like the one designed by Ustad Kabir and Husayn known as *damanak*. Fazl in his *Ain-i-Akbari* mentions names of several kinds of swords with different sizes and shapes (like *Khasa*, *kotal*, *jamadhar*, *yakbandi* and *khapwas*). He also describes different kinds of daggers, spears and lances that were made in the royal *karkhanas*. In these *karkhanas*, several kinds of armours were also made to protect soldiers, horses and elephants during war. In *Ain-i-Akbari*, it has been recorded that 69 such items were manufactured in these *karkhanas*. Looking at the size of the private armies maintained by several *mansabdars* of different ranks during this period, it appears that there might have been enormous demand for this defence equipment. If we add to this the demand for these defence products in foreign markets, one can imagine the size of this manufacturing. The Mughal emperors who have evolved hierarchy of *mansabdars* also provided scope for the existence of hierarchy of *karkhanas* owned by high-ranking *mansabdars*. These *karkhanas*, to some extent, reduced the role of market-mediated exchange in the products produced by these *karkhanas*. This happened because their products were largely consumed by families of nobles.[69]

Apart from weapons, several other products, household utensils, equestrian equipment (horseshoes, stirrups, curb bits, etc.) and farm equipment were also made from iron. Manufacturers produced several other iron products that were used in construction, shipbuilding, shoe making, clamps, door knockers, rings and needles, and so on. Apart from iron, products made from other metals like brass, copper and bell metals were also manufactured. Units manufacturing different metal products were spread in places like Poona, Nasik, Banaras, Hyderabad, Tanjore, Moradabad and Aligarh.

India is considered to be the house of sugarcane (saacharun). Indian stories are full of myths, which also gives us the idea of the sugarcane origin in our country. Due to abundant availability of cane, sugar manufacturing also has a long history in India. It becomes obvious when we notice that the words that are used to describe sugar in different languages are quite close to the word '*shakkar*' used in India for sugar in common language. Sugar is called *schkar* in Persian language, *sukkar* in Arabic, *seachrum* in *Latin*, and *sucre* in French.[70] During the Mughal rule, manufacturing of sugar experienced significant expansion due to its increased domestic as well as foreign demand. During this period, sugar was traded through the famous caravan routes of Khaiber and the Bolān Pass, and through sea routes to Europe and other countries located on the coasts of the Indian and Arabian Oceans. Manufacturing of sugar was widely spread, and followed the geographical spread of cane cultivation. It was highly concentrated in Gangetic India, Burdwan, Murshidabad and Gorakhpur, and in certain parts of southern India. Sugar during this period was sold in different forms—*jaggrey* and *gur*.[71]

During the reign of the Mughals, demand for vegetable oils increased substantially because these oils had become important items of consumption amongst the elite and urban consumers. The increased demand led to the spread of the oil milling industry across different regions in India, and led to the mushrooming of regions that specialized in the production of certain kinds of vegetable oils. The regional specialization was based on the availability of a particular oilseed crop for processing. The major varieties of vegetable oils produced in India were mustard oil (produced in north India), groundnut oil (western India), coconut oil (southern India) and linseed oil (southern India). Along with these oil mills, there were also manufacturers of aromatic oils, which were used in making traditional medicines, perfumes and *itra*.

Saltpetre was another important mineral product manufactured in India during Mughal rule. It was processed from nitrogenous organic matter and potassium nitrate deposits found on the surface of the soil. Those who were engaged in the manufacture of saltpetre scrapped the upper crust of the salt earth. Saltpetre was used in the production of gunpowder; also, the urban rich used it for cooling drinking water during long summers. Saltpetre was manufactured around Agra (Shergough, Kagarol, and Sadbad), Ahmedabad, Orissa, Bihar and Bengal. The saltpetre produced around Agra was used for preparing gunpowder for the royal army, whereas saltpetre produced in other parts of the country was meeting export demand of French, Dutch and English trading companies.[72]

Apart from these major industries, there were some other industries that existed during Mughal rule and provided income and employment to thousands of households. For instance, opium was grown and processed in the towns of Malwa and Bihar. This assumed importance during the rule of the East India Company when it became an important item of exchange for Chinese tea. Tobacco was also processed for oral consumption. Alcoholic drinks were also prepared from local produce like palm and mahua fruits, and from molasses in sugar-growing areas. The manufacture of shoes and footwear was also widespread due to the availability of leather in most areas in India because of the existence of large population of domesticated animals. Leather was also used in making saddles. Good quality shoes were made at Lahore, Sindh, Agra, Multan, Kolhapur and Jaipur.[73]

2.3 Traditional Flexible Manufacturing During the Mughal Empire

The Mughal Empire had inherited a long tradition of craft-based systems of petty production, where manufacturing was organized in artisanal household workshops. This system of production was not suited to the needs of consumers during the Mughal period. The integration of markets had created a substantial increase in the demand for products of manufacturers. The growing size of the market was accompanied by a growing demand of Indian manufactured goods in foreign markets. To meet the needs of such a growing market, the scale of production had to be increased. Since artisans, during this period, did not have resources to increase the scale of operation, the growing market needs could not be solely met by them. The objective conditions prevailing at this time, therefore, necessitated structural transformation in the organization of production in manufacturing. This transformation was initiated by traders who had a substantial amount of financial resources in their hand, whereas scattered, small craft producers did not have enough capital to expand the scale of operation to meet growing demand for manufactured goods. To meet the growing demand, merchants evolved a network of artisanal household workshops through the putting-out system. This system facilitated in meeting the capital requirements for production as well as the selling of products in the market.

In this kind of system of production, technology and scale of production did not change at all, but isolated units of small producers were made part of networks evolved by merchants who provided to small household producers, working capital or raw materials, as well as means of their subsistence, in lieu of commitment by the artisans that they will supply the final products to the merchants. In this way, large-scale marketing networks, created by merchants, roped in the small producers to supply products for mass markets. Since traders purchased raw materials in bulk to supply to artisans who were a part of their networks, they saved costs due to the advantages of '*economies* of *bulk purchase*'. They also sold final products in bulk quantities in mass markets, making substantial profits.

In this form of organization of production, small artisanal workshops became dependent on traders. This was because merchants had achieved the advantage of large-scale marketing networks over scattered petty-production units. This stage of production of manufactured products has been defined as '*proto-industry*' by economic historians. However, we are using here the term traditional flexible manufacturing to describe it as a method of organization of production. During the Mughal period, this system of organization of production was known as the *dadni* system (system of advance). This system created a particular kind of relationship between merchants and artisans. It was based on an agreement between the two, which specified terms regarding quantity, quality, price and date of delivery of output. It also specified the amount of advance given by the merchant. The agreement was compatible with the law of *sillim* (sales), which derived its jurisdiction from Quran. Here it means delivery on stipulated date.[74] Perhaps in the absence of law of contracts, the law of *sillim* provided religious legitimacy to contracts between artisans and merchants. The traders, who were participating in the *dadni* system, were assisted by *paikars* (travelling representatives of the traders for giving advances), and *mukeems* (merchandisers), who helped them by inspecting the quality of the final product supplied by the artisans.

In the dadni system, when a merchant entered in a relationship with artisan household workshops through the putting-out system and appropriated final output to sell in the market, it created preconditions for the emergence of modern industry. In this relationship, the producer (i.e., artisan) did not sell the final output directly in the market but produced it for the merchants, which confirmed to the specifications specified by the merchants regarding quality and quantity. This suggests that the merchant purchased the work effort, skill and services of equipment owned by the artisans in lieu of working capital or materials and their subsistence needs. Thus, in a way, merchants provided employment to the artisans. In the modern factory system, the owners of factories provide equipment, materials and wages to the workers and control the selling and marketing of the product. Therefore, the difference between organization of production of the putting-out system and the modern factory form of production of manufacturing is quite different. In the modern factory, the ownership of equipment, process of production and marketing are the responsibilities of the owner of the factory; whereas in the putting-out system, the small producer still controlled the production process, and instead of working in the factory, he worked at his household production unit with equipment owned by him. Tchitchirov, in his book, has recorded several instances of such relationships between artisans and merchants across different industries and regions, for example, in the production of chintz in Patna, pig iron in Bihar, amongst the weavers of the Coromandel Coast, Bengal and Gujarat, and in the oil pressing industry at Shahabad.[75]

In the new form of organization of production, merchants knew about market fluctuations, consumer preferences, and controlled the decision regarding expansion or reduction in the capacity of the network to produce. Merchants, through this process, made use of supply of cheap labour from the countryside. People were shifting from villages to city because of higher rate of land revenue imposed by the Mughal Empire and monetization of the rural economy had reduced returns on smallholdings. Therefore, small peasants with smallholdings had to make a choice between *land-intensive* farming and *labour-intensive* craft production. Hence, traditional flexible manufacturing can be viewed as a first historical experience of using an unlimited supply of labour for the growth of manufacturing.[76]

The economic implications of this system were increased division of labour and accumulation of capital. The significant advantage of the traditional flexible method of organization over other kinds of methods of production of handicrafts had been division of labour and specialization. The extent of division of labour depended on the expansion of the market. Perhaps after observing these traditional flexible manufacturing workshops of Kirkcaldy in Scotland, Adam Smith developed the axiom that the division of labour is limited by the size of the market.[77] In this form of organization of production, there occurred division of control on two important strategic managerial decisions. The decisions regarding '*what to produce*' were made by merchants due to their better understanding about market and customer preferences. Therefore, they significantly influenced the process of *product development* in terms of quality, characteristics and design of the product. Decisions regarding '*how to produce*' remained with the artisans. Since merchants had control over final product and its marketing, accumulation occurred outside the unit of production, in the hands of merchants. The consequence of this division of decision making was that artisans remained poor and individual production units remained small; whereas merchants became more resourceful.[78]

In the pre-machine age of manufacturing, productivity and efficiency depended on the extent of division of labour. In Indian traditional flexible manufacturing, production was organized in such a manner that involved several intermediate stages in the production of a product. The decomposition of product into several intermediate stages was also possible due to the existence of division of labour based on, not only technological factors, but also on social factors like caste.

The separation of functions between craftsmen and merchants provided opportunities to the merchants to take advantage of several kinds of flexibilities that were inherent in the system. While merchants were divorced from risks and problems associated with the organization of production, they enjoyed the flexibility of expanding and contracting the level of output in response to market fluctuations. They could adjust their network according to market fluctuation by including or excluding craftsmen in their network, without incurring costs of lay-off and retrenchment, if these artisans had been working in their factories. This system also provided to the merchants flexibility to supply a variety of products (i.e., *economies of scope*) customized according to the needs and requirements of different set of customers. This could be achieved just by adding to the network additional artisans to produce products with varied specifications. The system also provided opportunities to the merchants to take advantage of *economies of scale*. The scale of operation of each merchant depended on the amount of capital employed in stocks of raw materials, finished products and number of artisan households that were part of his network. The sum of capital employed by some of the merchants in their network was much larger than employed in fixed assets by early modern factories.[79]

One trader, Kasi Viranna of South India, had created a network of around 2200 of the best weavers around Jahanabad. He controlled the entire network of weavers, and settlements, which were part of his network, called viranna villages. He, along with his family members, controlled a substantial share of purchases made by the East India Company.[80] It has been recorded that in Bayana in the 1620s, two local traders, Mirza Sadiq and Ghazi Fazil, had established a monopoly in the indigo trade. Another wealthy trader, Khemchand, had established his monopoly in the textile trade in Sohroha near Balasore in Orissa. Similarly, Gurivi Setti and Kunigiri Setti had emerged as dominant traders of textiles in Vijaynagar.[81] There were perhaps many other big traders who operated their networks of artisans in different parts of India, but their names have not been recorded in historical documents.

Separation of functions also proved to be advantageous to producer-artisan households because they did not have to bear the risk of market fluctuations. They were making the products according to the orders placed and materials supplied by the traders. The system also provided *pace flexibility* to the artisans because they worked in their own household workshops, which provided them freedom from discipline and freedom to change pace and rhythm of work. Artisans could follow work stoppages for rest as and when they liked. Craftsmen generally followed a leisurely pace during weekdays and worked harder during weekends to make deliveries of their final output, so that they could collect working capital necessary for the next cycle of production.[82]

The traditional flexible manufacturing system of production created a household workshop economy based on vertically disintegrated inter-industry and intra-industry, specialized and dispersed production units, localized in clusters in specific areas. These production units produced specific products with close networking amongst these units. Such clusters had emerged in different parts of India during Mughal rule, specializing in the manufacture of certain products at diverse locations. Such clusters existed at Dacca, Patna, Murshidabad, Kasimbazar, Sonargaon, Banaras, Balasore and Assam in Eastern India; Surat, Baroda, Ahmedabad and Jaipur in Western India; Agra, Aligarh, Gwalior, Moradabad in Central India; Amritsar and Ludhiana in North India; and the Coromandel and Malabar coasts in Southern India. Such organization of production has been conceptualized as *flexible specialization*.[83] These clusters of units organized into the traditional flexible manufacturing system of production are historical roots of the contemporary just-in-time production complexes that have emerged as a consequence of *modern flexible* manufacturing. Locations like Route 128 in Massachusetts and the Silicon Valley and Los Angeles basin of California in the United States; the M5 between London and Oxford in England; Baden Wurthernberg in Germany; Toulosse in France; and Emilia-Romagna in Italy are examples of cluster of production units, organized into modern flexible production. Both sets of cluster of units, based on either the traditional flexible manufacturing system of organization production or the modern flexible system of production, have a commonality that they achieve competitive advantage by reducing overhead and inventory costs and receive '*external economies*' from the cluster, sufficient enough to compensate for '*internal economies*', which are advantageous for vertically integrated large production units.

One of the most important characteristics of the Indian traditional flexible manufacturing was that it neither registered any change in the size of production units nor did it experience technological change. One important reason for the absence of changing scale of operation and technological change was that its organization of production was such that the craftsmen never had sufficient resources that they could use to increase the scale of operation or improve technology. Accumulation in this form of manufacturing occurred outside the production unit, in the hands of merchants. Even merchants were not interested in bringing about technological change because they had ability to expand output by increasing the network of artisans, without investing resources in technological change. Despite the absence of technological change, Indian craftsmen made wonderful, manufactured items with simple tools and equipment. For instance, Marx provides the example of Dacca muslin, which, according to him, was woven 'without capital, machinery, division of labour' and states sarcastically—'it is only the special skills accumulated from generation to generation and transmitted from father to *son that gives to the Hindoo, as it does to spider this proficiency*'.[84] By equating skills of Indian craftsmen with that of spiders, he has downgraded the dexterity of Indian craftsmen, which was based on improvement of skills and knowledge of crafts. He equates this acquired knowledge to the genetically transmitted skill of a spider that helps it weave its web, or weaver bird to weave its nest. This view is held by several scholars who have similarly described the craftsmenship of the Indian artisans in their writings. Since knowledge of skills remained within the community of artisans, the knowledge evolved in complete isolation. Therefore, it was believed that 'India affords evidence for the theory of genetic transmissibility of acquired characteristics amongst humans'.[85]

Some other scholars believe that it was the absence of accumulation of capital that discouraged the process of innovation and technological change in the process of production during the Mughal period. This view is advanced by those scholars who do not understand the role that capital plays in the process of production. The importance of capital emerges when expansion in production is constrained by the supply of factors of production (i.e., land, labour). Supply-side constraints imposed by any factor that is limited in supply in the production process, compels the producers to improve efficiency of that limiting factor through technological changes. Use of capital facilitates in improving the efficiency of the limiting factor. 'During Mughal Empire, land and labour were abundant, therefore there was no need for capital'.[86] Some historians attribute the absence of technological change in traditional flexible manufacturing, during Mughal rule, to the phenomenon called 'high-quality trap'. According to this view, the high-quality trap emerged in Indian manufacturing because specialization had gone to such an extent that 'it would have required great incentive to induce these craftsmen to change their methods of production and habits drastically'.[87]

Despite the fact that the Indian economy did not have basic institutional infrastructure to provide conditions for the efficient working of the market, like property rights, business laws and modern banking institutions, it evolved into a vibrant commercial and manufacturing nation during the Mughal rule. The success story of Indian traditional flexible manufacturing depended on the pattern of organization of two important communities—the community of merchants and community of craftsmen. The merchants and traditional banking family firms not only possessed but also protected their capital as well as trade-related intellectual capital accumulated over generations. Communities of merchants also developed rules regarding its smooth transfer within the boundaries of the sub-continent, as well as to the next generation. The membership of the community provided these merchants a code of conduct and institutional support for resolving interpersonal conflicts through community *panchayats* (councils). Therefore, unlike the risk-taking entrepreneurs of the Western societies, the entrepreneurs engaged in the production of traditional flexible manufacturing were members of the 'risk reducing community'.[88] The membership to the community provided the Indian merchants much needed protection of their capital and intellectual capital, availability of cheap credit, common community-specific rules and regulations, as well as protection against entry of merchants who did not belong to the community.

Similarly, craftsmen and artisans belonging to different crafts were members of caste-based communities, which helped in accumulation of skill-specific knowledge and practices. These communities also facilitated in the transfer of craft-specific knowledge and skill to the next generation by giving craft training to the new members of the community, and helped in protecting skill-specific intellectual capital by denying its spread to the persons belonging to other castes. Therefore, in the Indian subcontinent, division of labour was not dependent on the size of the market as envisaged by Adam Smith, but on the organization of crafts into specialized and dedicated caste-based communities.

The system of community-based specialization of merchants and craftsmen, and their community-specific rules and regulations, acted as substitutes for the role of state to legislate property rights and commercial and business laws. This happened because the Mughal state accepted the supremacy of communities forming their own laws and governing conduct of their members. The Mughal state intervened only in instances when there emerged inter-community conflicts. Due to this reason, the Mughal Empire did not evolve elaborate legal structure.

Despite the fact that communities have played a significant role in different societies, mainstream economics associates them with several kinds of distortions in the efficient working of markets. Its dislike for communities emanates from the fact that existence of communities creates different kinds of market imperfections, which affect the optimum or desirable outcome of market-mediated decisions made by different economic agents. This happens because membership of a community distorts preference of economic agents; it results in the principal-agent problem and dichotomy between insider and outsider. Communities also affect social mobility, occupational mobility and regional mobility of their members. Due to these reasons, it is believed that the existence of communities creates several types of market imperfections.[89]

Contrary to their dislike for communities, economists prefer institutions. This is based on the premise that institutions define the context in which the market operates; they also determine the rules of the game, which have to be observed by the economic agents while participating in the market. Moreover, good institutions protect individuals from the interference of communities and make markets efficient. This suggests that institutions help in creating order as well as stability in the market.

In the literature on '*social capital*', communities are valued because the bonds between individuals that emerge as a consequence of being members of a community, based on trust and reciprocity, create positive impact on development.[90] The role of communities has been rehabilitated in the social sciences by concepts like '***communities of practice***'.[91] This is defined as 'a group of people bound together by their interest in common working practice'.[92] The literature on communities of practice has identified several such communities; task-based communities are distinguished on the bases of particular art facts, tools and work environment. These communities are also studied in the context of awareness acquired through mastery of certain tasks in shared division of labour.[93] The caste-based task communities have another important characteristic in addition to non-cast communities that they make membership totally exclusive with strict entry barriers.

Historical facts suggest that the organization of production based on traditional flexible manufacturing led to the birth of the Industrial Revolution in Britain by creating in the society two distinct classes that are necessary to achieve successful transformation of an economy into a modern industrial nation. One, it provided historical conditions for the accumulation of capital in the hands of merchants, who assumed the role of entrepreneurs or capitalists. Two, it helped in skill formation amongst large masses in the absence of any organized institution dedicated to the formation of skills. These skilled workers consisted of that segment of the society that was totally divorced from agriculture and fully committed to manufacturing. These fully committed craftsmen could not protect their household workshop against the capital-intensive, energy-driven modern factories. These craftsmen committed to traditional manufacturing, supplied working classes for modern factories after losing their household workshops. This is how, in Britain, craftsmen transformed from being *price workers* to *wage workers.*

It is speculated that traditional flexible manufacturing, which flowered during the Mughal rule and developed capacity of the system to supply Indian manufactured products throughout the world, as an irony of history, experienced systematic decline with the slow and systematic decline of the Mughal Empire. While this was happening in India, traditional flexible manufacturing in Britain was transforming into modern manufacturing. Indian traditional manufacturing, despite having qualitatively better potentialities than the British traditional manufacturing, experienced decline.

The Mughal state had created preconditions for the birth of traditional flexible manufacturing; it is suggested that these conditions slowly disappeared with gradual erosion in the authority of the Mughal state. After the death of Aurangzeb, gradually, one region after another declared autonomy from the Mughal Empire. The fragmentation of the Mughal state into several new political entities intensified military conflict, plunder, banditry, and decline in trade and commerce and financial infrastructure. Markets fragmented due to the emergence of barriers on inter-regional trade and multiple currencies issued by new political dispensations. The need to finance frequent wars increased conflict between rulers and big financial firms with the new ruling elite. Raids on merchants' wealth by local rulers had become quite frequent. During Mughal rule, merchants depended on local rulers and nobles for the protection of their property and businesses.

Travellers who have left written records of their travels during the Mughal Empire state that there were tensions between merchants and local rulers. It is mentioned that since merchants' 'wealth consisted of cash and jewels....they preserved it as close private as they can, lest Mughal exchequer should be made their treasury'. It has been mentioned in historical records of the Mughal period that when Sultan Parviz was appointed the governor of Patna, there was so much panic amongst merchants of Patna that no one wanted to appear to have wealth. This caused massive disappearance of trading activities. Historical records also tell us that when Akbar died, his death caused so much panic and insecurity amongst merchants that it became almost difficult to distinguish between high and low and rich and poor.[94] The biography of Banarsidas (*Ardhakathanak*) mentions how the local governor of Jaunupur, Nawab Qilich, arrested all the rich merchants of the area to take away their wealth. He also mentions another incident that occurred during the rule of Jahangir, when one of the *Umrao*, Agha Noor, imprisoned several merchants.[95] Under the new circumstances, when the Mughal Empire had been fragmented into several different political entities, new rulers and nobles wanted financial support from the merchants not only for maintaining their territories but also, if possible, to expand it. Since a strong centralized state was now missing, conflicts between new political dispensations were inevitable.[96] There are recorded instances when local rulers confiscated properties of big merchants because they wanted financial support without honouring their contracts, by refusing to pay back loans and payments of interest.[97] The property of Banarsidas, one of the wealthy merchants of the time, was confiscated by local rulers.[98] The example of Jagat Seth is also quite relevant to understand the condition of merchants in the new political scenario. Jagat Seth had played a key role in the Battle of Plassey, and he eventually lost his life.[99]

It is a historical fact that when the state is incapable of providing protection to private property and lacks legal strength to enforce contracts, it creates circumstances that discourage merchants from using their resources and entrepreneurial skills to create more wealth. The insecurity regarding private property rights created by deteriorating law and order, and rise in banditry, put the wealth of merchants into a bottomless pit. In such circumstances, the merchant class withdrew from productive activities like financing of networks of craftsmen. Therefore, some scholars suggest the emergence of disconnect between merchants and craftsmen led to disconnect between craftsmen and mass markets. This facilitated the slow and systematic decline of traditional flexible manufacturing. Therefore, these scholars evolved a stylized historical fact that the decline of the Mughal Empire led to the decline of traditional flexible manufacturing.

The decline in traditional flexible manufacturing is considered the beginning of the process of de-industrialization that was experienced by Indian manufacturing. This also meant that the large proportion of population that was engaged in the production of handicrafts lost its occupation and went back to work in the villages to follow occupations that are associated with primary activities.[100] However, it should be noted that traditional flexible manufacturing is just one of the several modes of organization of production of handicrafts to cater to mass markets. History gives us several possibilities in which production of handicrafts could be embedded. Scholars who have been writing about the process of de-industrialization have been laying stress on the process of decline of handicrafts.

Several studies conducted by scholars on different regions that assumed autonomy from the Mughal Empire, amply suggest that the aforementioned stylized fact is not tenable. This is because, in certain regions, the preconditions that are necessary for the continuity of traditional flexible manufacturing continued to exist. In some regions, dependence of local rulers on merchants increased substantially. In fact, local rulers began to share revenue and earned from new territories they acquired through war with the merchants who financed their war effort. They even promoted trades and crafts to increase revenue.

Therefore, continuity of preconditions, which are necessary for the existence of traditional flexible manufacturing, arrested its decline. These regional studies establish that the decline in the authority of the Mughal state led to a variety of regional experiences with varying repercussions for the survival of traditional flexible manufacturing. The emergence of successor states of the Mughal Empire in Awadh, Hyderabad, Maharashtra, Punjab, Rajasthan, Mysore and Kashmir, provided necessary conditions for continued existence of traditional flexible manufacturing.[101] Hence, scholars who associate the process of de-industrialization with the decline of the Mughal Empire are operating on slippery ground. This is because they ignore that the decline of the Mughal Empire created political space for the ascendency of the rule of the East India Company. Scholars have not given due recognition to the economic forces that prevailed between the period subsequent to the death of Aurangzeb in c.1707 and the end of the

Mughal Empire in c.1857. This is an important historical period, when the authority of the Mughal state slowly eroded and led to conditions that facilitated the process of ascendency of the rule of the East India Company. Therefore, there is need to assess the role of the East India Company in influencing traditional flexible manufacturing. This is the subject matter of the chapter that follows.

Notes

1. This term has been used by Hobsbawm (1954).
2. To understand this term see Tilly and Tilly (1971).
3. Mendels (1972) has used this term in his studies.
4. For the concept of nascent capitalism see Levine (1977).
5. Freudenberger and Redlick (1964) have used the term proto-industrialization, initially in their study.
6. For the concept of primitive flexible manufacturing, see Seth (2002, 2003, 2008).
7. To understand the economic consequences of empire building, see Anthony (2001), Furguson (2002, 2004) and Lal (2004).
8. For the historical importance of the Silk Route, see Frank and Brownstone (1986) and Liu (1996).
9. For this idea, see Hicks (1969).
10. For the knowledge regarding the creation of legal and administrative framework evolved by Pope Gregory VII, see Berman (1983) and Lal (2006).
11. With the use of horses in the war, the size of empire grew from 0.6 megametres (1 megametre is equal to one lack square metres) under Saragon of Akkad, to less than 3 megametres under Maurya's in India, 4 megametres under the Roman Empire, 6 megametres under the Han Dynasty of China, and 25 megametres was the size of the Mongol empire; see for details Christian (2004), pp. 316–317.
12. To understand the process of globalization initiated by empire building, see Mc Neill (1989) and Findley and O'Rourke (2008).
13. For detailed understanding of Mongol Empire, see Brent (1976), Kahn (1984), Juvaini (1997) and Weatherford (2004). To know the role of Mongol Empire in the process of globalization, see Findley and Laundahl (2006).
14. For the biographical details of Babur Mirza, see Zahiruddin Muhammad Babur's *Baburnama* p. 8.
15. See of this Abul Fazle Allami (1927), p. 2.

16. For the genealogy of Mughal emperors, see Allami (1956) and Richards (1976), pp. 262–263.
17. For understanding the role of king in Hindu society, see Walker (1968) and Bhatia (1974).
18. For details regarding this idea, see Shelvankar (1940), pp. 96–102, 139–143.
19. For the knowledge regarding this worldview, see Allami (1927), p. 4.
20. For details regarding Hindu *Jajmani system*, see Wiser (1969).
21. Hindu *Jajmani system* and its various variants, practiced in different parts of India, have been studied in detail in the writings of Fukazawa (1972), Commander (1983), Fuller (1989), Mayer (1993) and Guha (2004).
22. For the practice of *Vand* in Punja, see Punjab, District Gazetteer (1904).
23. See Hegel (1959), p. 154.
24. For Marx's ideas on India's social system, see Hussain (ed.) (2006), p. XXXI.
25. See the work of Raychaudhuri (1982).
26. For the details regarding construction of network of roads during Mughal Empire, see Sarwani (1974) and Alam and Subrahmanyam (eds.) (1998).
27. For details regarding Mughal administration and organization of army, see Moorland (1923, 1962) and Metcalf (1993).
28. To know the details regarding Mughal monetary system, see Richard (1983), p. 26.
29. For understanding the trimetallic currency system followed by Mughal rulers, see Habib (1987).
30. For details regarding these facts, see Moosavi (1987).
31. To know in detail the activities of different merchant communities, which were important in indifferent parts of India, see Darling (1926), Pillai (1930), Pavlov (1964), Mishra (1965) and Levkovsky (1965).
32. For historical facts regarding the house of Jagat Seth, see Little (1920), Hunter (1888,1974) and Karen (1998).
33. See Jain (1929) and Habib (1963).
34. See Jain (1929), Subramaniam (1987), Siddiqi (1995) and Markovits (2000a, b).
35. See Martin (2009).
36. For these facts, see Pavlov (1964).
37. For details regarding these business houses, see Tripathi (1991), Das Gupta (1994), and Subramaniam (1996, 2012).
38. To obtain understanding regarding trade and commerce in India during Mughal Empire, see Habib (1963), and Hambly (1982).
39. For this aspect of Mughal Empire, see Subhramanyan (1995) and Alam and Subramanya (1998).
40. These estimates have been taken from the work of Habib (1982).

41. For these facts, see Abul Fazl Allami's *AIn-i-Akbari* pp. 45 and 59–100.
42. For greater details, see Moosavi (1987).
43. For details regarding the occupational pattern of the working population and composition of the national income of India, during Mughal Empire, see Maddison (1971) and Maddison (2003), p. 110.
44. For details regarding the community of *Banjaras*, see Naqvi (1986).
45. To know about the role of *Komati and Khoja* communities in facilitating the domestic trade in India, see Tchitchirov (1965).
46. These facts are given in Habib (1990).
47. For the importance of caravan trade in the Indian economy, see Burton (1993) and Brenning (1990).
48. These facts are given in Burton (1993).
49. For details regarding the extent of trade through caravan routes to different countries, see Rizvi (2000).
50. For idea of globalization with gunboats, see Furguson (2004).
51. These facts are given in Maddison (2003) and Furguson (2004).
52. For details regarding Dacca muslin, see Watson (1879) and for Arni silk, see Parthasarthi (2001), p. 11.
53. For the use of dyestuff known as *chay*.
54. For learning about regional specialization in the manufacturing of textiles, see Gadgil (1924), Pavlov (1976) and Raychaudhri (1982).
55. For knowledge regarding the social context in which textiles were produced in India, see Bayley (1986).
56. For these ideas, see Bayley (1986), p. 298.
57. This statement is attributed to a writer from Lucknow, Mr. Abdul Halim Sherer, who is quoted in Bayley (1986), p. 296.
58. Statement of Daniel Defoe. This appeared in the Weekly Review of Jan 31, 1708 and reported in Furguson (2004), pp. 16–17.
59. See Ramaswamy (2008), p. 277.
60. See Ramaswamy (1980, 2008), p. 283.
61. For understanding the business practices used by Europeans to earn bullion, see Frank (1998), p. 177.
62. See Frank (1998), who has used these statements while negating the notion 'that Europe built a World around itself', preface, pages XXIV and XXV.
63. For this quotation, see Chaudhuri (1990).
64. For details regarding practices of making iron and steel in India, see Heath (1839).
65. See Bell (1881).
66. For these facts, see Idrish-a-shariff (1960).
67. For this, see Marco Polo (1998).
68. For historical facts regarding iron mining in the Chota Nagpur and Birbhum area, see Sanyal (1968) and Gupta (1980).

69. These facts have been reported from Allami's Ain-i-Akbari, p. 120.
70. This information has been mentioned in Noiel (1949, 1958) and Mintz (1986).
71. For the history of sugar manufacturing in India, see Khan (1923) and Gandhi (1934).
72. For the manufacturing of saltpetre, see Allami (1927), Gadgil (1924), Habib (1982) and Trivedi (1998).
73. See Gadgil (1924), Pavlov (1964) and Tchitchirov (1965).
74. For details regarding *dadni* system, see Chaudhury (1972 1979).
75. These facts are reported in Tchitchirov (1965).
76. Lewis (1954) evolved a model of development that suggested how developing economies can use unlimited supply of surplus labour for accumulation.
77. See Smith (1776, 2003).
78. See Freudenberger and Redlick (1964).
79. For comparing these circumstances with European economies, see Pollard (1968).
80. See Tchitchirov (1965) for these facts.
81. For these facts see, Tchitchirov (1965).
82. Similar facts have been observed by Landes (1998) regarding British experience.
83. The term flexible specialization has been evolved by Piore and Sabel (1984).
84. Marx (1867, 1978), vol. I p. 340.
85. See Dietman (1988), pp. 3–11, Derett (1976) and Roy (2010).
86. See Chaudhri (1972, 1978), p. 274.
87. See Roy (2010).
88. See Roy (2010), p. 1.
89. See Storper (2008) for understanding the attitude of economists regarding communities.
90. For the literature on social capital based conceptualization of communities, see Olson (1965) and Putnam (2000). Social scientists now distinguish between Olson-defined communities and Putnam-defined communities.
91. For the concept of community of practice, see Wenger (1998) and Amin and Roberts (2008).
92. See Dugrid (2008), p. 1.
93. See Amin and Roberts (2008), p. 17.
94. See for details Banarsidas's *Ardhakathanaka*, its translated version from Braj Bhasa by Banarasidas (2009), *Chaupai* No. 110–114, pp. 49–50.
95. See for this Banarasidas (2009) Chaupai No. 467–469, pp. 195–1975.
96. For elaborate description of the working of this system, see Weber (1929), Tamilson (1993), Stein and Subramanyan (1996) and Lal (2007).

97. Ovington (1689, 1929) mentions the conditions of merchants as prevalent in c.1689.
98. This incident has been recorded by Thevnot, a traveller, who was present in India during 1665–1667; see Thevenot (1949).
99. For details regarding the house of Jagat Seth, see Little (1920, 1921), Hunter (1974), Devera (1987) and Karem (1998).
100. For details regarding the phenomenon called de-industrialization, see Thorner and Thorner (1962), Morris (1975) and the Chap. 3 of the book.
101. For regional studies on Awadh, see Alam (1986), Cole (1988) and Mourad (2012); for Hydrabad, see Richard (1975); for Bengal, see Om Prakash (1998) and Chaudhry (1995); and for Punjab, see Singh (1991); Banga (1978), and Grewal (1995).

Bibliography

Alam, M. (1986). *The crisis of empire in Mughal North India-Awadh and Punjab 1707–1748*. New Delhi: Oxford University Press.

Alam, M., & Subhramanya, S. (Eds.). (1998). *The Mughal state 1526–1770*. New Delhi: Oxford University Press.

Allami, A. F. (1927). *The Ain-i-Akbari* (trans: Bluchmann, H. B. & Ed. Leiut Colonel D. C. Phillot from Persian, Vols. 1 & II). New Delhi: Oriental Book Preprint Corporation.

Allami, A. F. (1956). *Akbar Nama* (trans: Bluchmann, H. B., Vol. 1). Calcutta: Bibliotheca.

Amin, A., & Roberts, J. (2008). *Community economic creativity and organization*. New York: Oxford University Press.

Anthony, P. (2001). *People and empire—A short history of European migration, exploration and conquest: From Greece to present*. New York: Modern Library.

Ball, V. (1881). *Manual of geology of India (economic geology)* (Vol. III). Calcutta: Government of India Press.

Banarsidas. (2009). *Ardhakathanak* (trans: Chawdhury, R.). London: Penguin Books.

Banga, I. (1978). *Agrarian system of Sikhs*. New Delhi: Oxford University Press.

Bayley, C. A. (1986). The origin of swadeshi (home industry): Cloth and Indian society 1700–1930. In A. Appadurai (Ed.), *The social life of things—Commodities in cultural perspective*. Cambridge, UK: Cambridge University Press.

Berman, H. (1983). *Law and revolution*. Cambridge, MA: Harvard University Press.

Bhatia, B. M. (1974). *History and social development* (Vol. 1). New Delhi: Vikas.

Brent, P. (1976). *The Mongol Empire, Genghis Khan; his triumphs and his legacy*. London: Weildenfield and Nicolson.

Brenning, J. A. (1990). Textile producers and production in late seventeenth century Coromandel. In S. Subhramanyan (Ed.), *Merchants, markets and the state in early modern India* (pp. 66–89). New Delhi: Oxford University Press.

Burton, A. (1993). *Bukhara trade 1558–1718.* Paper on Inner Asia No. 23, Indian University Institute of Inner Asian Studies, Bloomington.

Chadhury, S. (1972). Merchants, companies, and rulers; Bengal in eighteenth century. *Journal of Economic History of the Orient, 31*, 71–85.

Chaudhuri, K. N. (1978). *The trading world of Asia and the East India Company 1660–1760.* Cambridge: Cambridge University Press.

Chaudhuri, K. N. (1979). Markets and traders in India during seventeenth and eighteenth centuries. In K. N. Chaudhry & C. J. Dewey (Eds.), *Economy and society-essays in Indian economic and social history.* New Delhi: Oxford University Press.

Chaudhuri, K. N. (1990). *Asia before Europe, economy and civilization of the Indian Ocean from the rise of Islam to 1750.* Cambridge: Cambridge University Press.

Chaudhury, S. (1995). *Form prosperity to decline; eighteenth century Bengal.* New Delhi: Manohar.

Christian, D. (2004). *Maps of time, an introduction to big history.* London: University of California Press.

Commander, S. (1983). Jajimani system in North India. *Modern Asian Studies, 17* (2), 283–311.

Darling, M. (1926). *Punjab peasantry in prosperity and debt.* London: Oxford University Press.

de Thevenot, J. (1687 [1949]). In S. N. Sen (Ed.), *Indian travels of Thevenot and Careri*, New Delhi: National Archives of India.

Derrett, J. D. M. (1976). Rajdharma. *Journal of Asian Studies, 35*(4), 597–609.

Detmar, R. (1988). Problem of India's arrested development under British rule. In C. Dewey (Ed.), *Arrested development in India-historical dimensions* (pp. 3–11). New Delhi: Manohar.

Devra, G. S. L. (1987). The Mughal state and Indian business. In D. Tripathi (Ed.), *State and business in India—A historical perspective* (pp. 1–24). New Delhi: Manohar.

Duguid, P. (2008). Community of practice then and now. In A. Amin & J. Roberts (Eds.), *Community, creativity and organization* (pp. 1–10). New York: Oxford University Press.

Findley, R., & Kevin, H. O. R. (2008). *Power plenty: Trade war and the world economy in second millennium.* Princeton: Princeton University Press.

Findley, R., & Mats, L. (2006). The first globalization episode: The creation of Mongol empire or the economics of Genghis Khan. In G. Thorborn & H. H. Khonodker (Eds.), *Asia and Europe in globalization: Continents, regions and nations* (pp. 13–54). Leiden: Brill.

Frank, I. M., & Brown Stone, D. M. (1986). *The silk road a history, facts.* Oxford/New York: File Publications.

Frank, A. G. (1998). *Reorient – Global economy in the Asian age.* New Delhi: Vistar Publication.

Freudenberger, H., & Redlick, F. (1964). Industrial development of Europe, reality, symbols and images. *Kyklos, 17*, 372–402.

Fukazawa, H. K. (1972). Rural servants in Maharashtrian village, demiurgic or jajmani system. *Hitotshubashi Journal of Economics, 13*, 108–124.

Fuller, C. J. (1989). Misconceiving the grain heap: A critique of the concept of jajmani system. In J. Parry & M. Bloch (Eds.), *Money and morality of exchange* (pp. 50–62). Cambridge: Cambridge University Press.

Furguson, N. (2002). *Empire: The rise and demise of the British world order and lessons for global power.* New York: Basic Books.

Furguson, N. (2004). *Empire: How Britain made the modern world.* London: Penguin.

Gadgil, D. R. (1924). *The industrial evolution of India in recent times: 1860–1939.* Oxford University Press.

Gandhi, M. P. (1934). *Indian sugar: Its past, present and future.* Calcutta: Government of India Press.

Grewal, J. S. (1995). *The Sikhs of Punjab.* New Delhi: Manohar Publishers.

Guha, S. (2004). Civilization markets and services-villages servants from seventeenth century and twentieth century. *Indian Economic and Society History Review, 41* (1), 79–101.

Gupta, R. K. (1980). Iron manufacturing industry in Birbhum. A study of growth and evolution. *Journal of Indian History, 38*, 94–106.

Gupta, A. D. (1994). *Indian merchants and the decline of Surat: 1700–1750.* New Delhi: Manohar Publishers.

Habib, I. (1963). *The agrarian system of Mughal India: 1506–1707.* Bombay: Asia Publishing House

Habib, I. (1982). *An atlas of the Mughal empire.* New Delhi: Oxford University Press.

Habib, I. (1987). A system of trimetalism in the age of price revolution, effects of silver influx on the Mughal monetary system. In J. F. Richard (Ed.), *The imperial monetary system in Mughal India.* New Delhi: Oxford University Press.

Habib, I. (1990). The merchant communities in pre-colonial India. In J. D. Tracy (Ed.), *The rise of the merchant empire: Long distance trade in the early modern world 1350–1750.* Cambridge: Cambridge University Press.

Hambly, G. R. G. (1982). Towns and cities—Mughal India. In T. R. Chaudhari & I. Habib (Eds.), *The Cambridge economic history of India* (Vol. 1, pp. 434–457). New Delhi: Orient and Longman.

Haynes, D. (1996). The logic of artisan firms in capitalist economy: Handloom weavers in Western India c.1880–1947. In B. Stein & S. Subramanyan (Eds.),

Institution and economic change in South Asia. New Delhi: Oxford University Press.
Heath, J. M. (1839). Indian iron and steel industry. *Journal of Royal Asiatic Society, 5*(3), 390–399.
Hegel, G., & Friedrich, W. (1959). *The philosophy of history* (trans: Sibree, J.). New York.
Hicks, J. R. (1969). *A theory of economic history*. Oxford: Clarendan Press.
Hobsbawn, E. J. (1954). The general crises of European economy in the seventeenth century. *Past and Present, 5*, 33–53.
Hunter, W.W. (1888, 1974). Statistical accounts of Bengal, Vol. IX, Concept Publishing, New Delhi.
Hussain, I. (Ed.). (2006). *Karl Marx on India from New York daily tribune*. New Delhi: Tulika Book.
Idrish, A.-S. (1960). *Kitab Nuzahat-ul-mushtaq* (trans: Ahmad, S. M.). Leiden: Macmillan.
Jain, L. C. (1929). *Indigenous banking in India*. London: Macmillan.
Juvani, A.-M. (1997). *Genghis Khan: The history of the world conqueror* (trans: Boyle, J. A.). Manchester: Manchester University Press.
Kahn, P. (1984). *The secret history of Mongols: The origin of Genghis Khan*. San Francisco: North Point Press.
Khan, S. A. (1923). *The East India trade in 18th century: In its political and economic aspect*. Oxford: Oxford University press.
Karen, L. (1998). The great firm theory of the decline of Mughal empire. In M. Alam & S. Subhramanyam (Eds.), *Mughal state 1526–1750* (pp. 398–148). New Delhi: Oxford University Press.
Lal, D. (2007). *Hindu equilibrium*. Oxford: Oxford University Press.
Lall, D. (2004). *In praise of empire—Globalization and order*. New York: Palgrave Macmillan.
Landes, D. (1998). *The wealth and poverty of nations*. New York: Abacus.
Levine, D. (1977). *Family formation in the age of nascent capitalism*. New York: Monthly Review Press.
Levkovsky, A. I. (1965). *Capitalism in India*. Bombay: Popular Prakashan.
Lewis, W. A. (1954). Economic development with unlimited supply of labor. *Journal of Manchester School of Economics and Social Studies, 22*, 139–192.
Little, J. H. (1920). The house of Jagat Seth. *Bengal Past and Present, 20*, 1–119.
Liu, X. (1996). *Silk and religion*. New Delhi: Oxford University Press.
Maddison, A. (1971). *Class structure and economic growth: India and Pakistan*. London: Allen and Unwin.
Maddison, A. (2003). *The world economy—A new millennial perspective*. New Delhi: Overseas Press.
Marco, P. (1998). *The book of Ser Marco Polo* (Ed. & trans: Yule, H.). New Delhi: Oriental Reprints (Vols. I and II).

Markovits, C. (2000a). Colonialism and traditional crafts. *Economic and Political Weekly, 35*, 3939–3940.

Markovits, C. (2000b). *Global world of Indian merchants 1750–1947; traders of Sind from Bukhara to Panama*. Cambridge: Cambridge University Press.

Martin, M. (2009). Hundi/Hawala: The problem of definition. *Modern Asian Studies, 43*(4), 909–937.

Marx, K. (1867, 1978). *Capital* (Vol. I). Progress Publishers: Moscow.

Mc Graw, T. K. (1988). *Essentials of Alfred Chandler—Essays towards a historical theory of big business.* Boston: Harvard Business School Press.

Mehta, M. (1991). *Indian merchants and entrepreneurship in historical perspective.* New Delhi: Academic Foundation.

Mendels, F. F. (1982). *Proto-industrialization: Theory and reality.* General Report, International Economic History Congress, Budapest.

Metcalf, B. (1993). Norms of comportment among imperial Mughal officers, in John. In F. Richards (Ed.), *Power administration and finance in Mughal India* (pp. 255–289). Eldershot: Variorum.

Meyer, P. (1993). Inventing village trading, the late 19th century origin of North Indian jajmani system. *Modern Asian Studies, 27*(2), 378–386.

Mintz, W. (1986). *Sweetness and power; The place of sugar in modern history.* London: Penguin.

Mishra, B. B. (1965). *The Indian middle class.* London: Oxford University Press.

Moorland, W. H. (1923). *From Akbar to Aurangzeb.* London: Macmillan.

Moorland, W. H. (1998). Ranks (Mansab) in Mughal state service. In M. Alam & S. Subramanayam (Eds.), *Mughal state 1526–1750.* New Delhi: Oxford University Press.

Morris, D. M., & Dudley, C. B. (1975). Selected railway statistics for Indian subcontinent (India, Pakistan, Bangladesh) 1853–1946. *Artha Vijnana, 17*(3), 185–298.

Mun, T. (1621 [1930]). *A discourse on trade from England unto East India.* New York: Facsimile Text Society.

Noiel, D. (1949, 1958). *The history of sugar.* Chapman & Hall, London.

Oglivie, S., & German, M. (Eds.). (1996). *European proto-industrialization.* Cambridge: Cambridge University Press.

Olson, M. (1982). *Rise and decline of nations; economic growth, stagflation and social regulation.* New Haven: Yale University Press.

Om Prakash (1998). Trade and politics in eighteen country Bengal. In L. Blusse & F. Gaastra (Eds.), *On the eighteen century as a category of Asian history.* Aldershot: Ven Lein in Retrospect Ashgate.

Ovington, J. (1929). In H. G. Rawlison (Ed.), *A Voyage to Surat in the year 1689.* London: Hakluyt Society.

Parthasarathi, P. (2011). *Why Europe grew rich and Asia did not—Global divergence 1600–1850.* Cambridge: Cambridge University Press.

Pavlov, V. I. (1976). *Historical premises for India's transition to capitalism*. Moscow: Nauka Publishing House.
Perlin, F. (1983). Proto-industrialization in precolonial South Asia. *Past and Present, 98*, 30–95.
Pillai, A. S. (1930). *Monograph on Nottuketti Chettiers banking system*. Madrass Province Banking Enquiry Commission Report, Vol. III.
Piramal, G., & Herdeck, M. (1986). *India's industrialist*. Bombay: India Book House Distributors.
Pugh, P. (2000). *The magic of a name—The Rolls-Royce story—The first fourty years*. Cambridge/New York: Icon Books/Totem Books.
Rajan, R., & Zingales, L. (2003). *Saving capitalism from capitalists*. Princeton: Princeton University Press.
Ramaswamy, V. (1980). Notes on textile technology in Medieval India with special reference to South. *Indian Economic and Social History Review, 17*, 227–241.
Ray, T. (2006). *Economy history of India 1857–1947*. New Delhi: Oxford University Prsss.
Ray, T. (2010). *Company of kinsmen—Enterprise and community in South Asian history 1700–1940*. New Delhi: Oxford University Press.
Rich, E. E. (1952). *Cambridge economic history of Europe* (Vol. II). Cambridge: Cambridge University Press.
Richard, J. F. (Ed.). (1976). *Kingship and authority in South Asia, Center for South Asian Studies*. Madison: University of Wisconsin.
Rizvi, J. (2000). *Trans-Himalayan caravans: Merchant princes and peasant traders in Ladakh*. New Delhi: Oxford University Press.
Sampson, A. (1996). *Company man: The rise and fall of corporate life*. London: Harper Collins.
Sarkar, S. (1983). *Modern India 1885–1947*. New Delhi: Macmillan.
Seth, V. K. (2000). Technology and labour in retrospect and prospect. *Managment and Change, 4*(1), 61–90.
Seth, V. K. (2002). Primitive flexible to modern flexible manufacturing and the process of skilling and deskilling of labor. *Indian Journal of Labor Economics, 45* (4), 1175–1194.
Seth, V. K. (2006). *Economics of services: A story of transformation of Cinderella into the queen of hearts*. New Delhi: Ane Books.
Shavell, S. (1979). Risk sharing and incentives in principal-agent relationship. *Bell Journal of Economics, 10*, 53–73.
Siddiqi, A. (Ed.). (1995). *Trade and finance in colonial India* (pp. 250–264). New Delhi: Oxford University Press.
Singh, B. (1966). *European agency houses in Bengal*. Calcutta: Firma Mukhopadhyay.
Smith, B. M. D. (1967). The Galtons of Birmingham Quacker gun merchants and bankers. *Business History, 9*(2), 135–148.
Smith, A. (1776, 2003). Introduction. In A. B. Krueger (Ed.), *The wealth of nations* (Bantam classic). New York: Bantam Books.

Stern, P. (2004). *One body corporate and politic—The growth of East India Company state in the late seventeenth century*. Unpublished PhD thesis, Columbia University, Columbia.

Stroper, M., & Christopherson, S. (1987). Flexible specialization and regional agglomeration. *Annal of the Association of American Geographers, 77*(1), 25–48.

Subhramanyam, S. (1995). Of 'Imarat and Tijarat': Asian merchants and state power in Western Indian Oceans—1400–1730. *Comparative Studies in Society and History, 37*(4), 225–247.

Subramaniam, L. (1987). Banias and business: The role of indigenous credit in the process of imperial expansion in Western India in the second half of eighteen century. *Modern Asian Studies, 21*(3), 455–520.

Subramaniam, L. (2012). *Three merchants of Bombay—Trawadi Arjunji, Nathji, Jamsetjee Jeejeebhoy and Premchand Roychand*. New Delhi: Penguin Books.

Sutherland, L. (1952). *The East India Company in the eighteenth century politics*. Oxford: Clarendon Press.

Taylor, F. W. (1947). *Scientific management*. New York: Harper and Row.

Thorner, D. (1951). Great Britain and development of Indian Railway. *Journal of Economic History, XI*, 326–350.

Thorner, D. (1980). *Shaping of modern India*. Bombay: Sameeksha Trust.

Tchitchirov, A. L. (1965). *India-changing economic structure in sixteenth to eighteenth centuries outline history of craft and trade*. New Delhi: Manohar.

Tripathi, D. (1987). Business and princely states—The case of Baroda. In D. Tripathi (Ed.), *State and business in India—A historical perspective* (pp. 274–301). New Delhi: Manohar.

Tripathi, D., & Mehta, M. (1990). *Business houses of Western India—A study of Entreprenerurial response, 1850–1956*. New Delhi: Manohar.

von Schmoller, G. (1902). *The mercantile system and its historical significance: Illustrated chiefly from Prussian history*. New York: Macmillan.

Washbrook, D. (1990). South Asia in the world system and world capitalism. *Journal of Asian Studies, 49*(3), 479–508.

Watson, J. F. (1887). *The textile manufactures and costumes of the people of India*. London: Indian Office Library.

Weber, A. (1929). *Theory of location of industry* (trans: Freidrich). Chicago: Chicago University Press.

Weiss, A. (1990). *Efficiency wages—A model of unemployment lay-offs and wage dispersion*. Oxford: Clarendon Press.

Wink, A. (1986). *Land and sovereignty in India: Agrarian society and politics under the eighteen century Maratha Swarajya*. Cambridge: Cambridge University Press.

CHAPTER 3

The Process of Transformation of Traditional Flexible Manufacturing into Industrial Revolution in Britain

It is an important historical fact that while traditional flexible manufacturing was experiencing a crisis, after the gradual erosion in the authority of the Mughal Empire in India, this form of manufacturing was not only thriving, but also being systematically transformed into modern manufacturing in Britain. This process of transformation of traditional flexible manufacturing into modern manufacturing has been described as the Industrial Revolution. The understanding of this process of transformation is quite relevant to have greater insight into the story of Indian manufacturing as it will provide answers to several pertinent questions that are being raised by inquisitive minds. To satisfy the needs of such inquisitive minds, we have to visit Britain to understand how the process of transformation of traditional flexible manufacturing into modern manufacturing occurred during the same period.

We are conscious that it will be quite a substantial digression for the readers who want to remain focused on the story of Indian manufacturing. However, this digression is quite relevant because there exists a strong inter-relationship and inter-dependence between the story of Indian manufacturing and the Industrial Revolution of Britain. The relationship between the two is so intimate that it was the British Industrial Revolution that largely conditioned the story of Indian manufacturing. We are quite certain that the reader will find this digression worthwhile.

As has been described earlier, some of the economies of the world, especially those of Europe and South and Southeast Asia, experienced the birth of a particular kind of manufacturing that has been variously

V.K. Seth, *The Story of Indian Manufacturing*,
DOI 10.1007/978-981-10-5574-4_3

conceptualized by different scholars.[1] This form of manufacturing has been described in the present study as traditional flexible manufacturing. Despite the fact that there existed this form of organization of production simultaneously in several economies, only in Britain did its transformation to modern manufacturing occur for the first time in human history. Since industrial revolution of Britain was first of its kind, it retained some of the characteristics of traditional flexible manufacturing. These characteristics remained embedded in British manufacturing even after the Industrial Revolution. This has happened because there existed continuity in the process of transformation from traditional flexible manufacturing to modern manufacturing. The British experience of industrial revolution thus provides a unique example for historical analysis. For the purposes of analysis, this chapter has been divided into two broad sections. Section 3.1 describes traditional flexible manufacturing of Britain, and Sect. 3.2 deals with the process of transformation of traditional flexible manufacturing of Britain into modern manufacturing.

3.1 Traditional Flexible Manufacturing of Britain

The adoption of traditional craft skills by a growing number of rural households for producing products, either for self-consumption or for the market, using traditional organizational patterns, emerged in several forms in England. These several forms existed much before the advent of industrial revolution in Britain. Several forms in which crafts-based methods were organized were (i) peasant manufacturing for household consumption; (ii) village and local specialized manufacturing; (iii) urban crafts-based manufacturing, organized into craft guilds; (iv) material-oriented manufacturing, located near sources of materials; and (v) merchant-organized manufacturing.[2] However, due to several socio-economic and technological changes occurring after the fifteenth and sixteenth centuries, there emerged institutions based on private property rights, and expansion in the size of the market due to emergence of towns and large urban agglomerations. The consolidation of centralized authority of state vested in the hand of the prime minister, rather than the Crown, after the glorious revolution of c.1688.[3] These changes affected diverse forms of manufacturing differently.

Britain, at that time, was enjoying the benefits of demographic dividend, which resulted from rapid population growth. This happened as a result of lowering of the age of marriage and an increase in the rate of marriages, in both rural and urban areas.[4] The demographic dividend provided to the British economy a classical precondition that is necessary for the emergence of traditional flexible manufacturing. The existence of widespread open and disguised unemployment made available workers at zero opportunity cost. In an economy where there are shortages of labour, attracting labour to work in manufacturing necessitates the payment of higher wages, which also provide compensation for their opportunity cost. The existence of surplus labour provided objective conditions for the growth of traditional flexible manufacturing.[5] By the year c.1707, Britain had become an important member of the common market consisting of Ireland, Britain, and British colonies in America. The existence of this common market removed demand-side constraints also, by providing to the traditional flexible manufacturing of Britain access to markets, without having to pay for entry barriers, like tariffs.[6]

At the same time, inspired by the ideas of Martin Luther and John Calvin, which were transforming the way of thinking amongst Europeans in general and the British in particular, a new attitude was emerging towards earning profit. The new wisdom suggested that the time spent in the pursuit of profession is equivalent to time spent in the prayer of God. 'Work' therefore became a form of worship. Hence, profit and accumulation of wealth that emerged as a consequence of work was accepted as an answer to worship. These ideas were articulated in the famous poem written by Bernard Mandeville, which appeared in c.1714, entitled, *The Fable of Bees*.[7]

In the new circumstances, village manufacturing registered a decline whereas urban workshops, mines and smelting works, prospered and strengthened. The erstwhile independent craft shops lost their independence to merchants who supplied them with raw materials, working capital and their subsistence. In lieu of all these services given to the craft workshops, they took away their final products to sell them to distant markets. The emerging bondage between independent craftsmen and merchants was a product of expansion of the market. Independent artisans earlier sold their product to local customers whom they knew personally.

As markets expanded, craftsmen did not have investable funds to expand the size of their operations; in these circumstances, merchants formed networks of craftsmen, from whom they collected products to sell in distant markets. This process created disconnect between producer (the artisan) and consumer, because as markets expanded, customers became more distant and stranger to the craftsmen. This led to dependence of craftsmen on merchants, who were in direct contact with the customers, for obtaining useful knowledge about customers' tastes and preferences. Moreover, as markets expanded, they became more volatile and less predictable, which further increased their dependence on merchants. Once craftsmen entered into a relationship of dependence on merchants, they were continuously under debt, for not only working capital needs but also for subsistence. Due to their increased indebtedness, their output was already mortgaged to the merchants. Therefore, the craftsman 'was in fact, if not in principle, a proletarian selling not commodities but labour'.[8] The increasing dependence of craftsmen on merchants for meeting their various needs led to the birth of traditional flexible manufacturing.

At the stage of traditional flexible manufacturing, there was dominance of craft-based methods of production, and importance of circulating capital instead of fixed capital due to simplicity of tools and lower costs of equipment used by artisans. There also existed sub-division of tasks amongst craftsmen along different lines of specialized jobs. These characteristics of the traditional flexible manufacturing provided it with several kinds of flexibilities. This has already has been explained in Chap. 2 of the present study.

In the new circumstances, only the merchant knew about market fluctuations and changes that were occurring in consumers' tastes and preferences. He also had the capacity to adjust the supply of product in response to market fluctuations by altering the size of the network of craftsmen that he had organized in the putting-out system. By using the organization of production based on traditional flexible manufacturing, merchants were able to use the supply of cheap labour from the countryside, which were free from guilds. This process was facilitated by demographic growth amongst rural households. The increasing specialization practiced by rural households created a new class of underemployed workers, consisting of marginal and landless workers. Declining returns on small holdings left small peasants and landless workers with only one choice—to shift from *land-intensive farming* to *labour-intensive craft-based production*. The transition of small peasant households to rural manufacturing occurred at a critical phase in which they could not earn even their subsistence needs from agricultural operations.

The process of integration of small peasants and underemployed workers of rural areas into the networks developed by merchants, to produce products on a mass scale to cater to the needs of distant markets, was spread over a very long historical period. However, the transition occurred faster after the emergence of the age of discovery and establishment of global markets by trading companies like the East India Company. The integration of rural artisanal households through the putting-out system in the network of merchants created an organizational structure of production that developed capability to produce products on a large scale to cater to mass markets.[9]

This process of production can also be viewed as the first historical experience of using an unlimited supply of underemployed workers for growth and accumulation.[10] The traditional flexible manufacturing system can be described as that form of organization of production in manufacturing in which manufacturing was based on close cooperation between household production and circulating capital of merchants. The integration between these two established linkages between crafts-based household production with market through organized trade and the putting-out system helped in evolving this new way of organizing production. The economic consequences of this form of organization production were increased division of labour and accumulation of capital.

The most important peculiarity of this system of organization of production in manufacturing was that the '*seller*' was divorced from production, and the '*producer*' from selling and marketing of the product. However, due to their better understanding of markets and customers, merchants (sellers) exercised significant control on quality, characteristics and design of the products. The increasing influence of merchants on products shifted the control on product development from craftsmen to merchants. However, in this form of organization of production, artisans retained their control on production process and know-how. Therefore, this system of production facilitated the emergence of division of sphere of control on two important strategic managerial decisions, which are essential for the proper running of the business of manufacturing. The division in the sphere of control occurred between merchants and craftsmen. In this division of the decision-making process, decisions regarding '*what to produce*' were appropriated by the merchants, while decisions regarding '*how to produce*' remained with artisans. The emerging separation between spheres of decision making on questions like 'how to produce' and 'what to produce' resulted in alienation of producers from their ultimate consumers. Since, in this form of organization of production, it was the merchant who controlled the final product as well as markets, accumulation occurred outside the unit of production, in the hands of merchant out-putter. Hence, this form of organization of production did not affect the unit of production, but provided more resources to merchants to expand level of output by expanding their network of craftsmen households, without causing any change in the size of unit of production or level of technology being used.[11]

The separation of functions between craftsmen and merchants provided ample scope for different kinds of operational flexibilities to the craftsmen as well as to the merchants. In this form of organization of manufacturing, producer did not bear the risk of market fluctuations, because he was making products according to the orders placed and the quantity of raw materials or working capital provided to him by the merchants. On the other hand, merchant did not experience the problems of management associated with the organization of productions. The putting-out merchant enjoyed the flexibility in expanding or contracting the level of output by including, or excluding, craftsmen's households from his network. This form of organization of production provided flexibility to the merchant to adjust the level of output in response to market conditions. It also provided possibilities of '*economies of scope*', because it provided scope of producing a variety of products, customized according to the preferences of different sets of customers. This opportunity was possible by including in the network additional artisans to produce products with varied specifications.

Organization of production of manufacturing in the traditional flexible mode also provided several kinds of flexibilities to the producers (i.e., to the craftsmen). It provided '*pace-flexibility*' as well as '*time-flexibility*' to the craftsman's household production unit by giving it freedom to vary pace and rhythm of work. These flexibilities were available to craftsmen because they worked in their household workshops, where they had freedom to follow their own work stoppages for rest, as and when they needed.[12] As described in Chap. 2, the traditional flexible manufacturing form of organization of production also provided several other opportunities to save costs. For instance, merchants were taking the advantage of *economies of bulk purchase*, because they were buying raw materials in bulk quantities, which they supplied to several artisans. This helped them in negotiating lower prices with material suppliers. This form of organization production provided possibilities for *economies of scale*. The scale of operation of each merchant consisted of the amount of capital employed in stocks of raw materials, finished goods and number of craftsmen employed in his network. The sum of capital employed by British merchants in traditional flexible manufacturing was larger by any standards. It was in fact much larger than employed in fixed assets in early centralized factories that emerged during the Industrial Revolution.[13] In hosiery, worsted, woollen, flannel, silk, cotton and woollen blanket manufacturing, an average British merchant employed around 300 weavers. One of the merchants, James Key Burry, employed between 2500 and 3000 weavers. In Warrington, England a company of merchants dealing in shoes had around 400 artisans employed in its network.[14]

Traditional flexible manufacturing helped in creating a network of vertically disintegrated household workshops on the bases of inter-industry and intra-industry specializations. These workshops produced specific products or components of a product, based on division of tasks, with very close networking amongst units located in specific locations. These vertically disintegrated networks of units of production, integrated through the putting-out merchant, helped in reducing '*overhead costs*', '*inventory costs*' and generated '*external economies*' through '*economies of agglomeration*'.[15] All these economies in costs were sufficient enough to compensate for '*internal economies*', which are enjoyed by the vertically integrated large production units. This suggests that at that time, in Britain, traditional flexible manufacturing was providing advantage to smaller units of production with several kinds of economies. It also suggests that the organization of production in the manufacturing sector was very close to the organization of production described by the term '*flexible specialization*'.[16]

During the period when traditional flexible manufacturing was the dominant form of organization of production in Britain, its expansion was constrained by the availability of circulating capital. The nature of this form of capital is that its gestation period is the same as that of the production cycle. Scholars have identified that the basic differentiating factor that separated modern manufacturing from traditional flexible manufacturing was the distinction between the natures of capital—in the latter, it was the circulating capital, whereas in the former, it is the use of *fixed capital*.[17] This suggests that during the traditional flexible manufacturing phase, growth occurred in each period of the production cycle from the *wage fund*, which was available with the merchants from the previous production cycle's net output. Therefore, it is worth emphasizing that the wage fund theory of wages, which was evolved by classical economists, was more relevant during the period when manufacturing was organized into traditional flexible form.

On the contrary, in the case of modern manufacturing, which needs substantial investments in fixed assets for buying plant, machinery and equipment, life of capital goods is spread over several production periods. The availability of capital with production units at a point in time results from past decisions on investments, and for subsequent periods, it depends on expected rate of return on capital, over its future lifespan. Therefore, differences between the nature of capital being used in traditional flexible manufacturing and modern manufacturing can be explained by the differences in the nature of capital implicit in the Smith–Ricardo and Harrod–Domar models of growth. In the Smith–Ricardo model, equilibrium always occurs between supply and demand, at a point in time, depending upon productivity, wages and wastage of funds into unproductive uses. Whereas, in the Harrod–Domar model, there exists only one possible rate of growth, which equates demand with supply, where stock of capital is fully utilized and expectations of investors are also fully realized.[18]

A significant advantage created by the traditional flexible method of production of manufacturing over earlier forms of handicraft-based systems of manufacturing had been the division of labour and specialization, which resulted in increasing its efficiency and the possibilities of innovations in the methods of production. The division of labour increased with the expansion of the market. Perhaps, after observing the workshops organized in the traditional flexible manufacturing format, Adam Smith concluded that the 'division of labour is limited by the size of the market'. In fact, Smith believed that division of labour was not the product of any great wisdom; it emerged as a consequence of human propensity to truck, barter and exchange. Exchange relationships give occasion to the division of labour. It is through exchange relationships that human beings get what they want from others, in exchange for what they can offer to others. The absence of an exchange relationship in other species also results in absence of division of labour. Smith states, 'nobody ever saw a dog make a fair and deliberate exchange of bone with another dog'. 'A *mastiff* is quite different from *greyhound*, a *greyhound* from *spaniel* and *spaniel* from *shepherd* dogs. The strength of *mastiff* is not supported by the sagacity of *spaniel* and docility of *shepherd* dog. The effect of these differences, in geniuses and talents, for want of power of disposition to barter and exchange, cannot be brought in common stock'.[19]

Smith's knowledge regarding the advantages of division of labour came from his observations of pin manufacturing workshops in Kirkcaldy, Scotland. These workshops were using division of labour to manufacture pins with the help of simple tools.[19] It will not be inappropriate to describe that his ideas on division of labour pertain to pin-making workshops, when its manufacturing was organized in the traditional flexible mode of manufacturing. This becomes obvious from the fact that his pin-making workshops represent an organizational form where they are using better organization of production, without any fundamental changes in the level of technology.[20] Despite the fact that the master craftsman could perform all the functions and jobs that were necessary to manufacture a product, no master craftsman would perform all the jobs, especially those that required simple operations and lower level of skills, because that would result in waste of his skills, experience and time. Not only this, unless and until the job required the strength of an adult male worker, women and children were used to perform those jobs. This kind of division of labour was guaranteed by the household production, because traditional flexible manufacturing units were household workshops.

It is also important to mention that the division of labour and specialization, which was embedded in traditional flexible manufacturing, was not only necessary for achieving efficiency in the process of production, but was also used as an instrument for providing instructions and training to apprentices. Through division of labour, the master craftsman also divided skills and knowledge required to produce a product into several stages. The young apprentices acquired skills by using these stages. In this scheme, each successive stage of division of task improved his knowledge and skills. New entrants were initiated in a trade by being asked to perform simple functions and rough jobs. As they reached different stages of operations, they graduated to perform more complex and difficult tasks. Unlike the craft guild system, where new entrants to the craft training were drawn essentially from the family, the traditional flexible manufacturing necessitated the recruitment of apprentices from outside the family. Hence, traditional flexible manufacturing helped in skill formation in the absence of any organized institution for skill formation.[21]

The meaning and advantages of division of labour changed with the introduction of machines during the phase of modern manufacturing. Machines are tools that are run with the help of inanimate sources of power. The division of labour embedded in machine based-manufacturing makes workers simply an appendage of machines, and their use in the process of production eliminates the influence of division of labour on dexterity and learning.[22] It is because of this reason that the concept of division of labour, as evolved by Adam Smith, ceases to be compatible with modern manufacturing.

The economic implications of division of labour that are compatible with machine-based process of manufacturing have been analysed in greater detail by Babbage.[23] Babbage's main contribution to the body of knowledge was to identify principles that determine the use of machines. He observed that division of labour in the process of production was facilitated by the emergence of large-scale enterprises. This led him to believe that extent of division of labour in an enterprise was related to scale of its operations. This idea is quite at variance with the idea of Smith who believed that the extent of division of labour is determined by the size of the market. Babbage has explained the logic underlying his observations regarding the relationship between scale of operation of an enterprise and the division of labour in one of the chapters of his book titled *Causes and Consequences of Large Factories*.[24] He established a principle that states that in order to minimize costs, it will be inevitable to increase the size of the firm, by some multiple, where value of the multiple will be determined by the division of labour. He suggested that 'when number of processes, in to which, it is most advantageous to divide it, and number of individuals to be employed by it are ascertained, then all factories, which do not employ a direct multiple of this latter number will produce the article at a greater cost'.[25]

The idea of division of labour, as evolved by Babbage, also differs from its conceptualization by Smith, in terms of its consequences. Smith explained that the division of labour facilitates in dividing tasks into specialized jobs, which increases the learning and dexterity of workers. This learning helps in the process of innovation. Babbage, on the other hand, explains that the master manufacturer, by dividing the work to be executed into different processes, each requiring different degree of skills or force, can purchase exactly the precise quantity of both necessary for each process. Whereas, if the whole work were executed by one workman, that person must possess sufficient skills to perform most difficult, and sufficient strength to execute most labourious of operations, into which the work is divided'.[26]

However, when machines are employed in the process of production, the worker is deprived of making even part of a pin or a head of a nail, because machinery makes it possible to transform a coil of wire into millions of pins. Under these circumstances, one can realize that Smith's concept of division of labour becomes totally irrelevant.[27] It is also interesting to note that it was the steam engine, invented by James Watt, which helped in converting tools into machines. The steam engine was invented in the same year in which *Smith's Wealth of Nations* was published. Ironically, it was Watt's engine that made the concept of division of labour evolved by Smith totally irrelevant. The other interesting historical fact that needs to be mentioned is that both James Watt and Adam Smith were fellow Glaswegians.[28]

The principle established by Babbage, clearly shows that the use of machines makes it inevitable to invest in significantly higher fixed capital that can be recovered only through large volume of output. This formulation of the manufacturing process is quite at variance with the concept of traditional flexible manufacturing. In a production unit, organized in the traditional flexible manufacturing mode, the production process was based on specialization, where each task was broken down into a series of discontinuous steps, and each step was assigned to a separate worker. Therefore, the Smithian concept of division of labour pertains to this form of manufacturing process. In this form of manufacturing, output could be expanded only by increasing workers, their skills and capabilities. Hence, to increase output, it was necessary to increase human resources. However, in the formulation evolved by Babbage, machines have the capability to replace not only workmen, but also jobs. Hence, technology is capable of providing indefinite possibilities. Therefore, in machine-based manufacturing, expansion in output can be achieved with, or without, additional units of labour.[29] This suggests that in modern machine-based manufacturing, machines can replace some tasks performed by workers. This differentiation makes it quite obvious that Smith was describing division of labour, which was embedded in traditional flexible manufacturing. He is not mentioning, anywhere, whether he is referring to workshops, wherein workers use simple tools, or large manufacturing units that use machines driven by inanimate sources of power.[30]

Similarly, it seems that the theory of profit, developed by classical political economists, also pertains to historical reality, which existed at the time of traditional flexible manufacturing. The basic premise on which the classical economists' profit theory is based, is that there exists an inverse relationship between profit and wages. This inverse relationship is valid only under the putting-out system. The prime function of a merchant in traditional flexible manufacturing was to bridge the gap between the start of the production process and its end, by giving advances of materials and subsistence of artisans for the production cycle. Under these conditions, profits emerge as a return paid for the use of stocks. Therefore, in traditional flexible manufacturing, where merchants only collected final output to sell in the market against advances, they did not direct the process of production from the beginning to its successful completion. Due to these reasons, classical economists could not derive the theory of profit, either from productivity of capital (because at that stage, production was financed by circulating capital) or from element of risk borne by entrepreneur.[31] This formulation of profit, which established an inverse relationship between wages and profit, opened the possibilities for the emergence of exploitation theory of profit, described by Schumpter as '*depredation theory*'.[32]

There also exists another important theoretical construct in the writing of Adam Smith that lends support to the fact that he evolved his understanding regarding this construct also from the organization of production in manufacturing that was prevalent in Britain during the traditional flexible manufacturing era. This construct was the shiftability of business endeavours from industry to industry in response to changes in profitability across industries. This shiftability of investments from one industry to another is possible only when capital is in the form of circulating capital. As the proportion of fixed capital increases in the process of production, it becomes difficult to shift business endeavours from one line of production to another, because fixed capital is invested in specific plants and equipment dedicated to the production of a product. This also suggests that Smith's theory regarding shiftability of capital pertains to the historical period when traditional flexible manufacturing was prevalent in Britain. Economic theory emerged with the writings of Ricardo, which he wrote when full-fledged modern factories were in existence in Britain. Even Ricardo added a chapter on machinery (chapter 32) in the third edition of his book, which appeared in c.1821.[33]

It is also an important historical fact that the traditional flexible manufacturing system of organization of production derived its sustenance from the nature of pre–Industrial Revolution household and family. Some of the scholars have tried to explain the symbiotic relationship that existed between traditional flexible manufacturing and nature of pre-industrial household. To explain how each one helped in sustaining the other, they have relied on some of the important characteristics of observed behaviour of pre-industrial household production units. These characteristics have been systematically analysed by Chenov.[34] Some other scholars have extended his analysis regarding pre-industrial peasant household to explain the nature of household production associated with traditional flexible manufacturing.[35]

According to this strand of thought, the central feature of a craftsman's household, during the period of traditional flexible manufacturing, was that its economic behaviour was not guided by objectives of profit maximization or accumulation, but to secure subsistence for the family. Therefore, each family will not supply work effort beyond the point where possible increase in output outweighed the disutility of work. This suggests that each family tried to achieve balance between its subsistence needs and subjective dislike for manual work (i.e., disutility of work). This behavioural pattern has been described as '*labour consumption balance*' (i.e., balance between family subsistence needs and expenditure of family labour).[36] This kind of behavioural pattern existed in the family economy due to lack of alternatives, which also made the opportunity cost of labour zero as the family's subsistence needs were not guaranteed. Therefore, opportunity cost of additional unit of labour increased only after subsistence needs of the family were achieved. This suggests that when return on family economy increased due to favourable conditions, the family had no compulsion to increase work efforts. Hence, during better and favourable times, family, instead of increasing work effort to maximize return on family labour, increased consumption and leisure. Similarly, in an event when return on family economy declined, the artisan's family increased its expenditure of work effort, even beyond the amount that is expected in an economic system, which depends on the wage-labour relationship. This behaviour pattern of household family workshops during the period of traditional flexible manufacturing gave birth to the famous '***backward sloping***' supply curve of labour.

The behavioural characteristics of the pre-industrial household became embedded in traditional flexible manufacturing. The self-regulating mechanism of the family economy provided institutional support to the family of craftsmen during bad periods to fall back on self-exploitation. The self-exploitation was inevitable when it was necessary to guarantee subsistence of the family. Willingness of artisans to supply work effort, even at below subsistence level, provided built-in flexibility to traditional flexible manufacturing to realize higher rates of profits during bad periods. This kind of flexibility could not be achieved in market-mediated wage-labour relationships. The emergence of symbiosis between pre-industrial household workshop and traditional flexible manufacturing created objective conditions where 'Lilliputian cottage industry choked-off large industry, preventing its emergence as a dominant form of production'.[37]

3.2 Process of Transformation of Traditional Flexible Manufacturing into Modern Manufacturing

Despite the fact that the traditional flexible system of manufacturing enjoyed continued existence for a considerable historical time, it contained within itself the seeds of its own transformation into modern manufacturing. To describe alternatively, the traditional flexible manufacturing contained the preconditions that were necessary for the birth of the Industrial Revolution. Traditional flexible manufacturing facilitated the birth of modern manufacturing by creating in the society two distinct classes, which are required for achieving successful transformation of the manufacturing. These classes were a class of workers and a class of owners of capital. The expansion of market brought more and more rural households in the networks created by putting-out merchants. On the one hand, it helped in the formation of skills amongst large masses in the absence of any organized institution engaged in the process of skill formation. On the other hand, it also created circumstances that slowly divorced these skilled craftsmen from the agricultural sector. Hence, it facilitated the creation of a class of artisans that was fully committed to manufacturing. It was this segment of the working population that provided the supply of workers for factories during industrial revolution in Britain.

The traditional flexible system of manufacturing simultaneously helped in the accumulation of capital in the hands of merchants. As mentioned earlier, the capital employed by merchants in developing their networks of artisans was quite large. This capital accumulated by merchants was waiting for necessary conditions to be transformed from circulating capital into fixed capital.[38] Merchants had not only accumulated capital during the prevalence of traditional flexible manufacturing, but had also utilized opportunities to acquire sufficient knowledge about production processes as a consequence of their interaction with craftsmen. Merchants had also developed the knowledge about markets, and intricacies of inter-regional and international trade. These advantages collectively equipped them to become entrepreneurs to provide leadership during industrial revolution in Britain.[39]

There were also certain other factors that contributed towards hastening the process of transformation of traditional flexible manufacturing into modern manufacturing. As described earlier, the traditional flexible production system of manufacturing worked through the network of household workshops organized by merchants. However, this form of organization did not provide them power to supervise the working of the individual unit of production. Despite the fact that the merchants had employed large volume of their resources as circulating capital to maintain their networks, they had no control on the quality of the product, efficiency of the production unit and work schedules followed by each individual household workshop. These conditions of helplessness of merchants were bound to change the system. Merchants were seeking to evolve a new method that would use their managerial capabilities and provide them control, along with ownership of assets, so that they could protect their interests and minimize risk.

To explain how the convergence between ownership and control occurred, historical experience provides two different paths. To understand these alternative paths, one needs to distinguish between '*capital extensive*' and '*capital intensive*' manufacturing industries.[40] Let us first understand extensive manufacturing industries. As traditional flexible manufacturing system of production expanded, it led to geographical spread of networks created by merchants. The spatial spread of production units made it difficult for merchants to keep time schedules of deliveries of products, and also gave them less control over wastage of raw materials. This happened because the owners of household workshops had conflicting interests with the merchant out-putter; the former had custody of raw materials that were supplied by the latter. They converted these materials into final products at their own pace and in their own place of work.

This created a kind of principal-agent problem between craftsmen and merchants. Due to the agency problem, bonds between merchants and craftsmen were quite loose and uncertain. Craftsmen always tried to cheat merchants vis a vis raw materials. Such embezzlement of materials was usually at the cost of quality of products.[41] To control this embezzlement, the merchants tried to put in place legislation through their influence on British Crown and Parliament. They succeeded in enacting several legislations, some of which covered all the manufacturing industries (e.g., the legislations of c.1749, c.1774 and c.1777), whereas some others were specific to a particular manufacturing industry (e.g., for worsted in c.1777, and for silk manufacturing in c.1774). Merchants were also slowly realizing that if they intended to minimize embezzlement of materials and regulate the pace of work, they'd have to shift artisans to a building where they could work under merchants' close supervision.[42]

To shift craftsmen from their household workshops to work in centralized workshops was not so simple, because merchants had to motivate them to come out of their cottages to work in factories. To an artisan, who was enjoying several kinds of flexibilities that gave him freedom, working in a mill amounted to the loss of freedom. To artisans, these mills symbolized prisons. These negative consequences, on craftsmen if they worked in factories, could be overcome only by giving them higher wages. These higher costs of skilled labour made factories non-competitive in comparison to the workshops of traditional flexible manufacturing.

To increase the competitiveness of their centralized workshops, merchants were expected to invest in better tools and equipment. In heat-using technology, like brewing, glass making, iron making and metal smelting, there were enormous savings of fuel costs due to the emergence of centralized factories. These savings more than compensated the higher costs of capital equipment and higher wages. In other manufacturing industries, emergence of centralized factory had to wait until the availability of power-driven machines, which could make it possible to undersell the products produced by traditional flexible manufacturing. Wherever the process of adoption of new techniques and use of power-driven machines was slow, factories used cheap labour of children and women to obtain cost advantages. Hence, to make modern factories more competitive than household workshops, investment in capital goods became inevitable. The necessity to invest in fixed capital facilitated the process of transformation of circulating capital into fixed capital. As the proportion of fixed capital in the production process increased, the control of owners of capital on the production process increased. This is how '*convergence*' occurred between '*ownership*' and '*control*'. Through this process, the owner of the capital appropriated control over the production process. Therefore, the system of production, which was based on centrality of capital and was managed by the owners of capital, was defined by political economists as the '*capitalist system*'.

In the case of capital-intensive manufacturing, convergence between ownership and control occurred much faster. As these manufacturing units needed large investments in equipment (viz. in metal processing, paper, glass and printing), investments needed to buy land and buildings, and power-supplying equipment to run the machines, household workshops became totally uncompetitive. Innovations like the blast furnace for smelting ore and use of porters in the brewing industry substantially increased the magnitude of investment in fixed capital. The proportion in which the need for fixed capital increased in these manufacturing industries, was the same proportion in which, simultaneously, traditional flexible manufacturing lost its competitive advantage. Hence, in capital-intensive manufacturing industries, centralized modern factories emerged much earlier than other manufacturing industries, because in these industries the circulating capital of merchants used in the putting-out system was transformed into fixed capital much earlier.[43]

In Britain, most mining districts were located in remote mountainous or hilly regions, where households generally practiced pastoral farming along with mining. In these areas, it was difficult for the household to rely on agriculture alone for their subsistence. Therefore, the extractive industries were carried on in conjunction with farming. Historians have observed that mining and farming were ancient bedfellows, and in most places, it is difficult to state whether the beginning of settled farming preceded or followed mining of tin, lead, copper, iron and stone-chalk.[44] The symbiosis between mining and farming was so intimate that it described in *Snow White and the Seven Dwarfs* as the dwarfs' occupation of raising '*their own food and operating their own independent shallow mine in the middle of the forest*'.[45]

However, from seventeenth century onwards, widely dispersed small-scale shallow surface deposits were approaching exhaustion. This necessitated digging to greater depths for the survival of the mining industry. This could not be achieved with the help of simple tools, which were being used by small mining units. Therefore, most people associated with the mining industry realized the need for investment in better equipment. This necessity resulted in the mechanization of the mining industry, through the use of water power pumping technology, gunpowder blasting and use of atmospheric pumping engines. These technologies transformed mining into a large-scale production.

These two aforementioned intertwined processes led to convergence between ownership and control in the hands of merchants, both in the capital-extensive and capital-intensive manufacturing industries. These symbiotic processes transformed the traditional flexible network of workshops into centralized factories. In the new circumstances, merchants assumed control over production processes as well. Due to this process of transformation, master craftsmen who looked after their own production unit, lost control over the production process and joined as foremen in centralized factories. The artisans who worked in his workshops became workers. What this means is that towards the end of the era of traditional flexible manufacturing, the merchant out-putter who had lost control over fragmented units of household workshops that were part of his network assumed total control over production units by integrating fragmented units of production in the modern factory form of organization.

Alternatively, the process of transformation of traditional flexible manufacturing into modern manufacturing can be explained with the help of a scheme evolved by Chayanov. His scheme is based on the understanding of the behaviour of households that were engaged in traditional flexible manufacturing. As described earlier, the family-based production units of craftsmen were not working with the motive of accumulating wealth but to achieve labour–consumption–balance; their behaviour was not compatible with the emerging market economy. This incompatibility was reflected when return on family production increased with expansion of market and favourable economic conditions. In such a situation, when market forces necessitated the increase of supply of work effort by the craftsmen, the family economy motivated them to increase leisure and consumption. Hence, precisely when profit-making opportunities were greater, merchants observed with great frustration the unreasonable inversion of sensible economic behaviour—the supply of work effort declined when returns increased.[46] The household economy associated with traditional flexible manufacturing made the organization of production downward flexible (i.e., supply of work effort increased during adverse economic conditions) and upward rigid (i.e., supply of work effort declined during better economic conditions). This behaviour of household workshops made it difficult to expand output during a boom. The only possible alternative available to the merchants was to extend the network of artisans. This process expanded the traditional flexible production system to all possible geographical areas 'like liquid seeking its level in this case the lowest possible wage rate'.[47] However, while expanding the network to all possible locations, merchants realized the shortcomings and limitations inherent in the organization of manufacturing into traditional flexible manufacturing.

To overcome the built-in weaknesses of the traditional flexible manufacturing process, merchants had to evolve a new method of organization of production that had the capability to out-compete the low-wage household economy. The textile manufacturing industry of Britain evolved this kind of advantage much earlier than any other industry. This advantage occurred when there sprang up, within a short period of time, a series of inventions regarding processing of cotton. These inventions, along with the invention of the steam engine, provided opportunity to create a competitive advantage for modern manufacturing. This happened when inanimate power, which was harnessed by James Watt, was used to run the '*spinning Frame*' designed by Ark Wright. This was the first instance when integration was achieved between tools and inanimate sources of power, and it gave birth to machines. This integration changed the meaning of manufacturing from 'made of hand' to 'made of machines'. This transformation that the 'shuttle dropped from the fingers of the weaver and fell into iron fingers'[48] can be accepted as the beginning of modern manufacturing, which is also described as the beginning of the Industrial Revolution.

In c.1712, Thomas Newcomen's steam engine was used for the first time in the process of production. Far-reaching changes were also occurring in the manufacturing of textiles. The process of technological change in manufacturing of cotton textiles began with the invention of the '*flying shuttle*', invented by John Kay in c.1733. This invention made it possible to weave cloth larger than arm's length. This machine also increased the speed of weaving quite substantially. The increase in the speed of weaving created structural imbalance in the manufacturing of cotton textiles. This mismatch between the speed of weaving and spinning of yarn put pressure on innovators to develop better spinning machines with better speed. Richard Arkwright invented the '*spinning frame*' in c.1764, which patented in c.1769. It was also known as the '*water frame*' because it used the waterwheel. However, after the invention of the steam engine by James Watt and Hargreev's '*spinning jenny*' in c.1769, cotton textile manufacturing became more productive in comparison to handicraft-based manufacturing of cotton textiles. The invention of the power loom in c.1784 by Edmund Cartwright completely transformed British cotton textile manufacturing.

The first modern cotton textile factory was established by Richard Arkwright for making cotton hosiery products at Cramford in Derbyshire in c.1771. In c.1793, Arkwright, along with two other partners, floated the first cotton textile mill to produce calicoes, which were imitations of calicoes produced in India. He also established other cotton textile mills in other locations in Britain, for instance, Chorley in Lancashire, which employed around 500 workers, and another mill at Manchester, which employed around 600 workers.

To transform the traditional flexible manufacturing of Britain into modern manufacturing, technological changes also occurred in other complementary manufacturing industries such as iron and steel and mining of coal. Coal was emerging as a prime mover of modern manufacturing and steel was emerging as a metal for developing modern machines and equipment. Later on, expansion of railways depended on the availability of steel. Abraham Darby evolved a process to convert coal into coke, which reduced most of its impurities. Coke also helped the iron master to attain much higher levels of temperature for smelting iron. He used this technique in his iron work at Coalbrookdale. In c.1760, the steam engine was used, at first, in the coal mine for pumping out ground water, which facilitated the digging of coal from deeper underground. Manufacturing of steel also experienced technological changes during the same period. Benjamin Huntsman invented cast iron in c.1740. Attempts were also made by innovators to reduce high silicon from coal-pig iron. For this purpose, brothers Charles and John Wood developed the 'stamping and polling' process in c.1761, which was rapidly adopted by the iron and steel industry in most parts of Britain by c.1770. The most important technological change, which completely transformed the manufacturing of steel, was the invention of the Bessemer converter by Henry Bessemer.[49] Before the invention of the Bessemer converter, steel was produced by mixing wrought iron with carbon. This process required continuous processing for weeks, using vast amounts of fuel. Instead, Bessemer used pig iron, which was developed by smelting iron ore, which already had carbon content. The steel produced with the help of the Bessemer converter could be used in construction for making machines and bridges.

These revolutionary technological changes that occurred in the British manufacturing industries are collectively described as the Industrial Revolution. This revolution helped in shifting the location of cotton textile manufacturing from the Indian sub-continent to Britain. This shift establishes links between the story of Indian manufacturing and the Industrial Revolution of Britain, as it created a process by which the key manufacturing industry of the Indian sub-continent experienced decline, while it experienced rapid growth in Britain. The process of relocation of cotton textile manufacturing gave birth to a 'trajectory of divergence' between the Indian sub-continent and Britain. It was this divergence that led to riches in Britain and underdevelopment in India.[50]

The process of divergence began as a consequence of adoption of cotton textile by consumers in most European economies. At that point in time, European consumers were shifting their preference from textiles made of other materials like wool, flex and hemp, in favour of cotton textiles. Availability of cotton textiles in different qualities, varieties and in different prices, in comparison to local woollen textiles, facilitated the process of adoption of consumption of cotton textiles by different strata of British society. Indian cotton textiles had certain qualities that were not present in the woollen textiles produced in Britain, such as fast colours and light and washable fabrics. These characteristics were compatible with the notion of cleanliness.[51] Moreover, demand for cotton textiles was global in nature. Due to these obvious reasons, manufacturers of Britain undertook a demand-led process of import substitution. The increased demand for cotton textiles created objective conditions for the mechanization of different processes, like spinning, weaving and finishing of textiles.

The manufacturing of cotton textiles in Britain began as import substitution with the production of *fustians*. Fustians was developed as a form of cotton textile, which was evolved from the mixing of linen and cotton.[52] Its production spread quite fast in the East Anglia and Lancashire regions of Britain. This kind of import substitution was based on understanding of market rather than improvement in technology as Britain was still at the stage of traditional flexible manufacturing (proto-industrialization).[53] The success of fustians signalled to the British manufacturers that any improvement in technology would further increase the demand for British cotton textiles. This realization motivated British manufacturing to evolve perfect substitutes for Indian cotton textiles.[54]

The process of import substitution of cotton textiles in Britain was also facilitated by the availability of cheap raw cotton from cotton plantations, which British expatriate entrepreneurs had established in the Caribbean, Virginia, Georgia and the Carolinas. In these plantations, cotton was cultivated with the help of African slaves. These plantations provided supply-side support to the efforts of British textile manufacturers to initiate the process of import- substitution.[55] Taking advantage of these supply- and demand-side factors, British cotton textile manufacturers transformed the oriental luxury product (the Indian cotton textiles) into a product of mass consumption, which could be consumed by consumers belonging to different strata of British society. This process led to the transformation of 'luxury product' into 'fashionable product' and became the new incentive for technological change.[56] Luxury products are *exclusive* by nature; the process of transformation made them more *inclusive*. Thus, what was a source of '*pleasure for few*' became '*pleasure for many*', through technological change.[57]

The process of import substitution of Indian cotton textiles by British cotton textiles was supported by the British government, as well as by prevailing circumstances in Europe. The ideological premises for protecting the emerging infant modern cotton textile manufacturing industry from foreign competition was articulated in the work of John Pollexfen, *England and India: Inconsistent in their Manufacturers*, which was published in c.1697. In this text, it was argued how imports of cotton textiles from the Indian sub-continent were acting as impediments to the growth of the British cotton textile industry. In c.1700, the British government began to discourage the imports of foreign-produced cotton textiles in Britain. During the rule of William IV, British Parliament enacted laws that prohibited the wearing of silk and cotton textiles manufactured in other countries by British citizens. A penalty of £200 was imposed on persons having or selling imported textiles. These laws were made more stringent during the rule of George I, II and III.

British manufacturers of cotton textiles also received protection from foreign competition when Napoleon imposed the continental system to limit European trade with Britain in c.1806. This ban continued until Napoleonic war came to an end at the Battle of Waterloo in c.1815. Despite the fact that the Napoleonic war was protecting British cotton textile manufacturing, British government continued to provide protection to it. In c.1813, the government imposed duty on calicoes at the rate of £3, 6 s, 8d per cent upon importation, and if the merchandise was meant for domestic consumption, there was further duty of £6,6 s, 8d per cent. On muslin, the duty on imports was £10 per cent and if it was meant for domestic selling, it was £27, 6 s, 8d per cent.[58] The emergence of cotton textile manufacturing created objective conditions for the growth of other complementary manufacturing industries because modern cotton textile manufacturing assumed the role of leading sector.[59] This led to a series of technological changes in the other related manufacturing industries.[60]

There are differences amongst scholars regarding the exact dating of the Industrial Revolution as they have approached it differently. Some scholars view industrial revolution as revolutionary changes in the manufacturing of products, whereas others interpret it as a process of economic development. Scholars who lay emphasis on the former accept it as a period of discontinuity. They explain the phenomenon by listing technological innovations that transformed the methods of manufacturing that manifested in quantitative changes in the rate of growth of output in the manufacturing sector. On the contrary, scholars who view it as a process of transformation of the aggregate economy explain it with the help of qualitative changes that occurred in the British economy.[61] Since we are more concerned with the emergence of modern manufacturing sector in Britain rather than with the transformation of the economy, we can accept c.1760 as the turning point in the manufacturing sector of Britain.[62] The new technological possibilities provided new opportunities to the merchants to become industrialists. However, to take advantage of new possibilities, merchants had to convert their circulating capital into fixed capital. This is because in new circumstances, they had to invest substantial resources on buying of new machines and equipment.

For the first time in human history, there emerged specialized manufacturers of machines and equipment who were designing them for use by others. This trend was revolutionary in nature because prior to this time, craftsmen used to make their own tools. The emergence of patent laws, which provided return on intellectual capital of craftsmen who designed new machines and equipment, encouraged some of the master craftsmen to devote their time and energy to develop these machines. Thus, new manufacturers of capital goods who developed machines that could be used across several manufacturing industries sprang up in the British manufacturing sector. In new circumstances, merchants could buy these new machines and equipment to establish new enterprises. Hence, ownership of capital became quite crucial. Since new entrepreneurs required substantial ownership of capital to float new factories, political economists began to call them '*capitalist*' However, the transformation of merchant into manufacturer made him 'prisoner of his investment'.[63] Some of the new industrialists also came from the ranks of equipment manufacturers.

In traditional flexible manufacturing, most of the investment made by the merchants was in the form of circulating capital (i.e., variable costs). The nature of his capital provided the merchant several kinds of advantages. He had freedom to enter and exit from the market. He also had flexibility to alter the level of output in response to market conditions. Moreover, he could also shift his investment from one line of operation to another. Despite these advantages, merchants transformed themselves into manufacturers to become prisoners of their investment. They made this choice because new techniques of production, accompanied by a new form of organization of production in the form of factory or mill, gave him immense power due to convergence of '*ownership* and *control*'. This new format gave merchants the opportunity to undersell and outcompete small household workshops.

The convergence between ownership and control also altered the social reality for the craftsman. The new circumstances were a great departure from the past. The introduction of machines took away his autonomy on the production process. For the first time in history, artisans had to come out from their household workshops to work in the factories. In altered circumstances, the factory became a '*place of work*' and household became a '*place of consumption*'. In the new dispensation, instead of the craftsman employing their own tools and equipment, now tools and equipment owned by entrepreneurs began to give him employment. Therefore, in the organization of production, craftsmen lost their autonomy on work schedules and work stoppages. Instead, machines now determined the place of work and work stoppages. In the factories, new craftsmen were working in teams, following similar time schedules, determined by machines. For each craftsman, work began and stopped simultaneously. Moreover, craftsmen, for the first time, had to work under the watchful eye of the supervisor's foreman, or overseer. It has been very appropriately described by one of the leading economic historians of the Industrial Revolution, that in the new situation for workers, 'factory was a new prison, the clock new kind of a jailor'.[64] The process described clearly shows how the centralized factory in the *'embryonic stage'* was growing in the womb of traditional flexible manufacturing.

The process of mechanization varied across different manufacturing industries and across different processes in each manufacturing industry. Therefore, mixed-firms or mixed-industry or what is described as '*technological dualism'* continued to persist for a much longer period. The uneven spread of machines provided unique characteristic of flexibility to British manufacturing. For instance, in some industries like clothing, cabinet making, cutlery, gun making and shoe making, small producers remained dominant for a considerable period. Historical evidence suggests that use of machines in the process of production in British manufacturing enterprises was a long, drawn-out process rather than a distinct event. This process was neither linear nor continuous. In reality, it was constrained by complex historical forces including the anti-machinery '*Luddite Movement*'.[65] Mechanization in some manufacturing industries was the result of growing demand, which necessitated the use of machines to achieve mass production. In some other manufacturing industries, it was emerging as a consequence of contradiction between owners of factories and working classes. In such instances, mechanization empowered owners against working-class resistance.[66]

The extent of mechanization in British manufacturing enterprises was largely guided by the interest of mill owners. In some mills, production was carried out without sourcing of certain jobs to household workshops. For example, sugar was produced in big factories like Messers Tate Lyle Silvertown, while sweets were being manufactured in backstreet kitchens. Timber was processed in power-driven sawmills, where machines were used, but up to c.1860, the wood was shaped at the carpenter's bench or at cabinetmakers' trestles. Even iron-making giant furnaces of Black County, co-existed with backyard workshops of blacksmiths. Cutlery manufacturing of Sheffield had a similar kind of organization of production, where big rolling mills supplied materials to 60 small crafts-based workshops.[67]

In the textile manufacturing industry of Britain, which felt the impact of mechanization much earlier than other industries, the extent of use of machines was quite uneven 'though production of textiles was mechanized, clothing trade remained largely dependent on "needlewomens" fingers'. Mechanization was more prevalent in worsted, less in woollen, more widespread in cotton, and less in the production of linen and silk textiles. Technological dualism continued to exist in textile mills for a considerable period. Even in large textile mills, hand mules as self-actors coexisted in the spinning department, while handlooms and power looms coexisted in the weaving section of mills, manufacturing silk and woollen textiles.[68] In metallurgy industry, steam power and machines were used in primary processes like puddling and rolling, while moulding remained primarily a craft-based skill. In the engineering industry also, despite the emergence of mechanization of most processes, fitting remained in the hands of craftsmen.[69]

The system of production based on standardization of products and use of interchangeable parts, which was being used in the manufacturing of sewing machines in the United States, did not spread to Britain. For instance, Enfield Rifles were manufactured with the help of machines, using interchangeable parts, while ammunition was made by hand. In Birmingham gun quarters, small-arms manufacturing was widespread amongst small producers, whereas lock, stock and barrel remained separate handicraft trades. Gun manufacturing in Britain remained primarily a craft-based production, produced in decentralized small workshops. This was quite at variance with the practice of manufacturing guns in the United States where guns were being manufactured since c.1820, using interchangeable parts at the Eli Whitney Gun Manufacturing Unit. British gun manufacturing has been described as a 'slice of eighteenth century industrial organization, which has survived entirely intact into the world of Great Exhibition'.[70]

As we know, the great exhibition of industrial products, to showcase the strength of the British manufacturing sector, was organized in c.1851 in Hyde Park, London. The exhibition continued from May 1, c.1851 to October 15, c.1851. The building that was the venue for the exhibition was made of 2,50,000 glass panes and was appropriately called the Crystal Palace. In this exhibition, 14,000 exhibits were displayed. All these exhibits were the luxury products of yesteryear, which modern manufacturing had transformed into products of mass consumption. The continued existence of traditional flexible manufacturing along with modern manufacturing was a unique characteristic of the British manufacturing sector. The reason for this 'especially in the case of gun manufacturing in Birmingham' was the prevalence of irregular demand for guns in Britain. Unlike the United States, where possession of firearms was a necessity amongst frontier farmers, in Britain, it was part of conspicuous consumption of upper class elites of the society. Due to uncertainty of demand, accompanied by need for customization of guns to suit the needs of elite customers, traditional flexible manufacturing continued to persist. These factors also inhibited the manufacturers of guns to invest in expensive capital goods to increase the extent of mechanization.[71]

Similarly in Sheffield, even up to c.1860, steel manufacturing was split into two distinct groups, one consisting of large steel makers and the other consisting of older crafts-based workshops producing cutlery. Historical evidence amply demonstrates that small-scale workshops continued to exist, using vertically disintegrated division of functions along with multi-plant operations. Another inherent characteristic of the British manufacturing was that each manufacturing industry existed in clusters at one important location. Existence of cluster of firms in locations like Manchester, Lancashire, Birmingham and Sheffield, producing similar products, also shows how the concentration of specialized, industry-specific skilled craftsmen facilitated the growth of a localized network of different and related manufacturing enterprises, with strong technological linkage in cluster form.[72]

In Britain, skilled workers not only supplied their own skills in the process of production, they also provided training to apprentices and junior workers. This was a unique process of skill formation at that time, when firms depended on skilled workers to organize production at the shop floor and also generate workforce at a lower cost. The massive process of skill formation did not depend on state support. Since reproduction of skill was associated with particular manufacturing skills, industry-specific skills were available at specific locations. Once an entrepreneur made a choice regarding what to produce, his choice of location of facility of production depended on from where he would be able to hire requisite skills.[73] This phenomenon gave birth to regionally concentrated firms, vertically disintegrated along narrow specializations, in a range of manufacturing activities. This led to inter-firm interdependence amongst manufacturing enterprises due to dependence on other firms to supply their necessary inputs and for the sale of their output. These regional clusters of manufacturing firms kept growing not because of the growth of existing firms, but because of the entry of new firms.

These vertically specialized firms became horizontally fragmented or experienced what is described as 'plant disintegration but local integration'.[74] This unique characteristic of British manufacturing has received expression in several concepts evolved by different scholars. Marshall coined the term '*industrial district*' in his book *Industry and Trade*, which he published in c.1919.[75] Another scholar, P. Sargent Florence, evolved the term '*industrial swarming*' in 1948.[76] It is the existence of firms that are interdependent in cluster form that created '*external economies*' for small producers, which made them capable of facing competition from large vertically integrated enterprises based on maximizing '*internal economies*'. This phenomenon has also been described as '*economies of agglomeration*'. Some scholars have also described this form of organization of production as '*flexible specialization*'.[77]

This form of organization of production provided opportunities for entrepreneurs to establish family firms. This was possible because entry level or threshold level of capital required was very small. Due to this reason, families with small and medium levels of wealth could also establish these firms. These conditions made entry quite easy. Perhaps looking at this peculiar characteristic of British enterprises, Alfred Marshall evolved the concept of '*perfect competition*'.[78] According to Marshall, perfect competition exists when '*entry*' and '*exit*' of firms in the market is free, and each producer is supplying an insignificant proportion of the total supply of the industry. This characteristic of British manufacturing has been described as '*small producer capitalism*' or '*personal capitalism*'.[79]

The integration between ownership and control in small family businesses imposed managerial constraints on the growth of the size of the firm. Due to limited availability of managerial resources, depending upon the size of the family, these firms remained small, single-plant, single-location firms. They derived their strength from external economies, which emerged because these firms were part of cluster of firms. Managerial constraint emerges, in firms, when after attaining a certain size, the firm's further growth can continue only if the family is prepared to delegate decision-making power, responsibility or control to hired managers. During the eighteenth and nineteenth centuries, amongst British business families, there prevailed a general belief that one should avoid hiring of professional managers because they cannot be trusted. Being salaried persons, where they managed the businesses owned by others, it was expected that they would not manage businesses with the same degree of commitment, vigilance and care with which owners of businesses will themselves manage the enterprises. This belief led to a general acceptance of a managerial principle that enterprises that hire professional managers are committing suicide. This was accepted as a principle because it was expected that these businesses were bound to fail due to 'negligence and profusion'.[80]

This suggests that the prevailing thought amongst the owners of family businesses against hired managers was guided by what is now known to contemporary economists as the *'agency problem'*.[81] This problem arises in professionally managed enterprises where a divorce has emerged between the owners of capital (principals) and the owners of control (agents)—when the agents (professional managers) follow objectives to guide their decisions, which are not necessarily compatible with the objectives of the owners (principals). The objective of the owners of the firm is to maximize profits, whereas behaviour of professional managers is guided by status, power or influence-enhancing objectives, which compromises the objectives of owners of firms.

The acceptance of this stylized fact about professional managers by the British owners of family businesses, kept the size of their businesses small. The economic consequence of this belief was that British manufacturing evolved its structure through a market-mediated '*invisible hand*' rather than through a managerially coordinated '*visible hand*'.[82] However, this form of organization of firms provided advantages to entrepreneurial business families to retain ownership as well as control of their enterprises. This peculiar characteristic of the British manufacturing firms has been conceptualized by business historians as *proprietary capitialism*'.[83] The distinguishing feature of proprietary capitalism is integration of ownership and control.

Continuity of flexibility in the production process in British manufacturing enterprises, achieved with the help of integrating traditional flexible manufacturing with modern manufacturing techniques, provided a unique advantage to British manufacturers; they now had the opportunity to produce customized products according to the preferences of the consumers, and this remained an important determinant of competitive advantage. In ship building, production of consumer durables, and engineering products, customization was significantly prevalent. Ramsone's of Ipswich, which was known for manufacturing of farm machines and equipment, maintained 700 models in their catalogues. The use of customization rather than standardization of products remained an essential characteristic of British manufacturing until the twentieth century. It remained embedded in British manufacturing, despite the distinct advantage that the Fordist method of production provided to the manufacturing firms of the United

States, which depended on vertical integration of production processes to produce standardized products. This process was quite different from British manufacturing, which practiced vertical disintegration and customization. The Fordist method also produced products on assembly line, using interchangeable parts for mass markets, in large enterprises, employing hierarchy of managers to maximize internal economics of scale. In contrast, in British manufacturing, family businesses were engaged in the production of products in small batches, without the use of assembly line-based production process. Instead they used inter-firm division of functions to maximize external economies. It is interesting to mention that while business historians symbolize Fordism as a paradigm of the American method of organization of production, they do not realize that the British Rolls-Royce, can similarly be described as the epitome of the British manufacturing paradigm, which kept on producing cars manufactured according to the preferences given by their elite customers up until recently.[84,85]

There is great relevance in studying the process of transformation of traditional flexible manufacturing of Britain into modern manufacturing for the story of Indian manufacturing. In the entire process of transformation, two dates are quite important—the years c.1760 and c.1851. It was the year c.1760 when the process of growth of modern manufacturing began in Britain. The year coincided with the beginning of the geographical expansion of the rule of the East India Company in c.1757, after the Battle of Plassey. This coincidence tells us that when a company of merchants of Britain was consolidating its political power in the Indian sub-continent, at that point in time the British manufacturers were beginning to influence British economy and polity.

The second important date is c.1851, which symbolized the emerging supremacy of the British manufacturers, when they showcased their success story at the Great Exhibition. The consequence of this important historical event on the story of Indian manufacturing was that it coincided with the end of the rule of the East India Company and beginning of the British Raj in c.1857. The end of the rule of the East India Company was a manifestation of the end of mercantilism and the beginning of the influence of British manufacturers on the process of policy making towards the economy of India. Subsequent chapters deal with these two important historical events that altered the story of Indian manufacturing.

Notes

1. This form of organization of manufacturing has been described variously by scholars; viz. 'protean stage of industrialization' by Hobsbawn (1954); 'industry before industrial revolution' by Tilly and Tilly (1971); 'Proto-industrialization' by Mendels (1972) and 'Proto-factory' by Freudenberger and Redlick (1964).
2. For understanding regarding these diverse forms of manufacturing, which were prevalent in Britain before the Industrial Revolution, see Cippola (1974), Burke (1979), Postan and Rich (1952), Thirsk (1994) and Weber (1927).
3. To understand the impact of socio-economic and technological changes that were experienced by different forms of manufacturing enterprises, see Dobb (1963); Hill (1967) and Cippola (1974).
4. See for this, Wrigley and Schofield (1981).
5. See Wrigley (1988) and Hudson (1996).
6. See Oglivie and Carmen (1996).
7. See Mandeville (1714, 1989) and Hoppit and Wrigley (1994).
8. For this, see Landes (1969), p. 44.
9. See Mendels (1972).
10. In 1954, W. Arthur Lewis, one of the leading development economists, evolved a model that explained how developing economies can use the unlimited supply of labour to achieve economic growth; see Lewis (1954).
11. For details regarding these ideas, see Freudenberger and Redlick (1964) and Landes (1998).
12. For this kind of analysis, see Landes (1998).
13. Pollard (1968)
14. Pollard (1968), p. 46.
15. The concept of *economies of agglomeration* was introduced in economic analysis by Alfred Weber to show how firms, which are located at close proximity to each other in cluster form and have strong technological linkages, get benefits from external economies. Therefore, all kinds of saving in costs, which firms enjoy due to their location at the same place, he described as economies of agglomeration; see Weber (1929).
16. For the complete understanding of the concept of flexible specialization, see Piore and Sabel (1984).
17. To understand the differences between the nature of capital being used in traditional flexible manufacturing and modern manufacturing, see Hicks (1969) and Pollard (1968).
18. For understanding the difference between the Smith–Ricardo and Harrod–Hicks models regarding concept of capital see Hicks (1965) and Fei and Gustav (1989).

19. See for this Smith (1776, 2003).
20. See West (1976).
21. See Pollard (1968), p. 19 and Caton (1985).
22. For understanding the process as to how traditional flexible manufacturing helped in skill formation in Britain, see Seth (2002).
23. To understand how the use of machines influences the learning of workers, see Berg (1980), Braverman (1974, 1989) and Seth (2000, 2002).
24. See Babbage (1835/1971).
25. See Babbage (1835/1971), Chapter 22.
26. See Babbage (1835/1971), p. 212.
27. See Babbage (1835/1971), p. 175.
28. See Rosenberg (2000) and Seth (2002).
29. For the understanding of Babbage's formulation, see Rosenberg (1976).
30. For detailed understanding about Smith's concept of division of labour, see Kindleberger (1976), Caton (1985) and Landes (1986).
31. For understanding Smith's theory of profit, see Smith (1776, 2003), Ch. IX and X.
32. For detail about the 'depredation theory', see Schumpeter (1954), p. 333.
33. For understanding regarding Ricardo's views of machinery and its implications for the process of production, see Ricardo (1863/1882), Chapter 32. See Chayanov (1925/1968).
34. See Chayanov (1925/1968).
35. For knowledge regarding the relationship between the behaviour of pre-industrial household and traditional flexible manufacturing, see Kriedite et al. (1981).
36. For understanding the concept of *labour-consumption balance.*
37. To know how the behaviour of the household of the pre-industrial family economy provided advantage to traditional flexible manufacturing, see Medick (1994), p. 416.
38. For understanding regarding the process of transformation of circulating capital of merchants into fixed capital of modern factories, see Defoe (1928).
39. For the knowledge regarding these different advantages, see Mendels (1982) and Wilson (1994).
40. To have complete understanding regarding the process of transformation from traditional flexible manufacturing into modern manufacturing, as experienced by capital-extensive and capital-intensive manufacturing industries, see Freudenberger and Redlick (1964), p. 379.
41. See Ashton (1926).

42. To know the legislations that were enacted by the British government to control embezzlement of raw materials by the artisans, see Pollard (1968), pp. 46–48.
43. For these facts, see Landes (1994), p. 123.
44. To know about the process of transformation of traditional flexible manufacturing into modern factory form of organization in British manufacturing industries, see Mathias (1959).
45. To know the relationship between mining and farming, in historical perspective, see Thirsk (1994), pp. 72–73.
46. See. Burt (1998), p. 104.
47. For understanding of 'inversion of behaviour' amongst the members of household production units of traditional flexible manufacturing, see Chayanov (1925/1968), Miller (1969) and Kriedite et al. (1981).
48. See, Landes (1994), p. 124.
49. For the understanding regarding the process of technological change that occurred during this period, see Singer et al. (1958), vol. I and II, Rider (2007), vol. I and II and Horn (2007).
50. See Prasannan Parthasarthy (2011), Ch. IV, in which he has described this process of divergence in detail.
51. For a detailed analysis of this argument, see Lemire (1991, 2004).
52. See Coleman (1973) and Jenkins (1994).
53. See Walton (1989).
54. See Chapman (1965).
55. For further details regarding this strand of thought, see Parthasarthy (2011), Riello and Parthasarthy (2011) and Riello and Roy (eds.) (2009).
56. See Berg (2005).
57. See Berg (2005), pp. 21–46, 85–111.
58. See Seth (1987).
59. Rostow (1957) has described the textile manufacturing industry of Britain as the leading sector of the economy because it provided the initial impetus to several other manufacturing industries. Also, see Rostow (1963).
60. For details regarding the series of inventions that transformed the traditional flexible manufacturing into the modern factory form of organization of production, see Singer et al. (1958) and Ashton (1960).
61. The scholars who believe that there was continuity in the process of transformation for traditional flexible to modern manufacturing, rather than any revolutionary change, see Hobsbawn (1968) and Wilson (1965).
62. The scholars who have accepted c.1760 as the turning point in the growth of output in the manufacturing sector of Britain are Neff (1943) and Ashton (1948).
63. See Landes (1998), p. 43.

64. See Landes (1998), p. 43–44.
65. These ideas are available in Frank (1998).
66. Anti-machine rhetoric is an on-going theme in the working-class movement. Machine breaking amongst Luddites has provided enough material for the historians of the British labour movement, see Berg (1980), pp. 15–17.
67. See Marx (1853), Dutt (1956), p. 103 and Seth (1987), Ch. II.
68. For details regarding these facts, see Raphael (1994), Hoppit and Wrigley (eds.) (1994), vol. II, Part V pp. 196–263.
69. See Berg (1993).
70. See Treblicock (1969) and Behagg (1194).
71. See, Berg (1980), p. 4.
72. See Behagg (1998), p. 6.
73. See Habhakuk (1962), Smith (1967), Treblicock (1969) and Behagg (1998).
74. See Habhakuk (1962), Zeitlin (1995), Lazonick (1990) and Berg (1993).
75. See Lazonick (1991), p. 26.
76. See Marshall (1919).
77. See for details Florence (1948).
78. See Weber (1929).
79. See Marshall (1929).
80. See Piore and Sabel (1984).
81. See for this Smith (1776, 2003).
82. The problems associated with separation between ownership and control has led to the origin of behavioural as well as managerial theories of firms. These theories are discussed in detail in the text books of Micro-Economics and Managerial Economics.
83. To know the distinction between markets coordinated (working of invisible hand of market) and managerially coordinated (visible hand of managerial hierarchies), see Chandler (1962, 1977) and Mc Graw (1988).
84. Term used by Chandler (1962, 1977) for the British form of capitalism, which was based on organization of production, where firms enjoyed integration between ownership and control.
85. To know in detail the history and characteristics of the Rolls-Royce Motor Car company of Britain, see Nocklods (1959), Bird and Hallow (1984) and Pugh (2000).

Bibliography

Ashton, T. S. (1926). The domestic system in early Lancashire tool trade. *Economic History Review, 1*(1), 120–143.

Ashton, T. S. (1948). *Industrial revolution 1760–1800*. Oxford: Oxford University Press.

Ashton, T. S. (1960). *Industrial revolution, 1760–1800.* Oxford: Oxford University Press.

Babbage, C. (1835/1971). *Economics of machinery and manufactures.* Fairfield: A. M. Kelley.

Behagg, C. (1998). Mass production without factory; craft producers, guns and small firm. *Business History, 40*(3), 1–15.

Berg, M. (1980). *Machinery question and making of political economy.* Cambridge, UK: Cambridge University Press.

Berg, M. (1993). Small producers capitalism in eighteenth century England. *Business History, 35*(1), 17–39.

Berg, M. (2005). *Luxury and pleasure in eighteenth-century Britain.* Oxford: Oxford University Press.

Bired, A., & Hallow, I. (1984). *Rolls-Royce motor car.* London: B.T. Batsford.

Braverman, H. (1974). *Labor and monopoly capital – The degradation work in twentieth century.* New York: Monthly Review Press.

Braverman, H. (1989). The degradation work in twentieth century. *Monthly Review Press, 41*(5), 35–47.

Burke, P. (1979). *The new Cambridge modern history* (Vol. XIII). Cambridge: Cambridge University Press. (companion volume).

Burt, R. (1998). Proto-industrialization and stages of growth in mining industry. *Journal of European Economic History, 27*(1), 85–104.

Caton, H. (1985). Pre-industrial economics of Adam Smith. *Journal of Economic History, 45*(4), 833–853.

Chandler, A. (1962). *Strategy and structure: Chapters in the history of industrial enterprise.* Cambridge, MA: MIT Press.

Chandler, A. (1977). *The visible hand; the managerial revolution in American business.* Cambridge, MA: Harvard University Press.

Chapman, S. D. (1965). The transition to the factory system in midlands cotton spinning industry. *Economic History Review, 18*, 526–543.

Chayanov. (1925/1968). *The theory of peasant economy* (D. Thorner, B. Kerbalay and R. E. F. Smith. Eds.). Homewood: American Economic Association.

Cippola, C. M. (Ed.). (1974). *Fontena economic history of Europe – 16th & 17th century.* Glasgow: Collins Fontana Books.

Coleman, D. C. (1973). Textile growth. In N. B. Harte & K. G. Panting (Eds.), *Textile history and economic history.* Manchester: Manchester University Press.

Dobb, M. H. (1963). *Studies in the development of capitalism.* London: Rentledge and Kegen Paul.

Dutt, R. C. (1956). *The economic history of India under early British rule* (p. 103). London: Rentledge and Kegan Paul.

Fei, J., & Gustav, R. (1989). Economic development in historical perspective. *American Economic Review (Papers and proceedings), 44*, 356–400.

Florence, P. S. (1948). *Investment location and size of plant: A realistic inquiry in to the structure of British and American industries.* Cambridge: Cambridge University Press.

Freudenberger, H., & Redlick, F. (1964). Industrial development of Europe, reality, symbols and images. *Kyklos, 17,* 372–402.

Habhakuk, H. J. (1962). *American and British technology in the nineteenth century.* Cambridge, MA: Cambridge University Press.

Hill, C. (Ed.). (1967). *Reformation to industrial revolution* (Vol. 2). Harmondsworth: Penguin Books.

Hicks, J. R. (1969). *A theory of economic history.* Oxford: Clarendan Press.

Hobsbawn, E. J. (1954). The general crises of European economy in the seventeenth century. *Past and Present, 5,* 33–53.

Hobsbawn, E. J. (1968). *Industry and empire—An economic history of Britain since 1750.* London: Weiden Field and Nicolson.

Hoppit, J., & Wrigley, E. A. (1994). Introduction. In J. Hoppit & E. A. Wrigley (Eds.), *The industrial revolution in Britain, the economic history society* (pp. X–XXXI). Oxford: Black Well.

Horn, J. (2007). *The industrial revolution—Milestones in business history.* London: Greenwood Press.

Hudson, O. (1996). Proto-industrialization in England. In S. C. Ogilivie & M. Cerman (Eds.), *European proto industrialization* (pp. 49–66). Cambridge: Cambridge University Press.

Jenkins, D. T. (Ed.). (1994). *The textile industries.* Oxford: Blackwell Publishers.

Kindleberger, C. P. (1976). Historical background: Adam Smith and industrial revolution. In T. Wilson & A. Skinner (Eds.), *Market and the state – Essays in honour of Adam Smith.* Oxford: Oxford University Press.

Kriedite, P., Medick, H., & Schlumbhar, J. (1981). *Industrialization before industrial revolution* (trans: Schen, B.). Cambridge: Cambridge University Press.

Landes, D. S. (1969). *Unbound prometheus.* Cambridge: Cambridge University Press.

Landes, D. S. (1986). *Revolution in time: Clocks and making of modern world.* Cambridge: Cambridge University Press.

Landes, D. (1994). *The wealth and poverty of nations.* New York: Abacus.

Landes, D. S. (1998). *Wealth and poverty of nations.* New York: Abacus.

Lazonick, W. (1990). *Competitive advantage of the shop floor.* Cambridge, MA: Harvard University Press.

Lazonick, W. (1991). *Business organization and myth of market economy.* New York: Cambridge University Press.

Lemire, B. (1991). *Fashion favorite: Cotton trade and consumer in Britain, 1650–1800.* Oxford: Oxford University Press.

Lemire, B. (2004). Fashion and tradition: Wearing of wool in England during consumer revolution. In G. L. Fantana & G. Gayot (Eds.), *Wool products and markets 13th to 20th centuries* (pp. 577–594). Patna: Clerandon University Press.

Lewis, W. A. (1954). Economic development with unlimited supply of labor. *Journal of Manchester School of Economics and Social Studies, 22*, 139–192.

Mandeville, B. (c.1714, 1989). *The fable of the bees.* Penguin Classics, Penguin Books, London.

Marshall, A. (1919). *Industry and trade.* Cambridge: Cambridge University Press.

Marshall, A. (1929). *Principles of economic analysis.* London: English Language Book Society.

Marx, K. (1853, June 24). East India Company its history and results. *New York Daily Tribune.*

Mathias, P. (1959). *Brewing industry in England, 1700–1830.* Cambridge: Cambridge University Press.

Mc Fetridge, D. (1995). Knowledge, market failure and multinational enterprise—A comment. *Journal of International Business Studies, 26*(2), 409–406.

Medick, H. (1994). Proto industrial family economy – The structural function of household and family during transition from rural society to industrial capitalism. In R. A. Church & E. A. Wrigley (Eds.), *Industrial revolution* (pp. 72–98). Oxford: Blackwell.

Mehta, M. (1991). *Indian merchants and entrepreneurship in historical perspective.* New Delhi: Academic Foundation.

Mendels, F. F. (1972). Proto-industrialization—First phase of industrialization process. *Journal of Economic History, 32*, 241–261.

Miller, J. R. (1969). A reformulation of a. V. Chayanov's theory of peasant economy. *Economic Development and Cultural Change, 28*(2), 209–220.

Nayyar, D. (Ed.). (1994). *Industrial growth and stagnation, the debate in India.* New Delhi: Oxford University Press.

Nehru, J. (1947). *The discovery of India.* London: Meridian Books.

North, D. C. (2006). *Understanding the process of economic change.* New Delhi: Academic Foundation.

Parthasarathi, P. (2001). *The transition to colonial economy weavers, merchants and kings in South India 1720–1800.* Cambridge: Cambridge University Press.

Pillai, A. S. (1930). *Monograph on Nottuketti Chettiers banking system.* Madrass Province Banking Enquiry Commission Report, Vol. III.

Piramal, G., & Herdeck, M. (1986). *India's industrialist.* Bombay: India Book House Distributors.

Prescott, W. H. (1847). *History of the conquest of Mexico and Peru.* New York: Random House.

Ramaswamy, S. (2011). *The goddess of the nation—Mapping mother India.* New Delhi: Zuban.

Raychaudhri, T. (1982). Non-agricultural production in Mughal India. In T. Raychaudhri & I. Habib (Eds.), *Cambridge economic history of India, Vol. 1 (c.1200–c.1750)* (pp. 261–307). New Delhi: Orient Longman.

Ricardo, D. (1863/1882). *Principles and political economy and taxation*. Homewood: Irwin.

Richards, J. F., & Rao, V. N. (1998). Banditry in Mughal India: Historical and folk perception. In M. Alam & S. Subhrahmanyam (Eds.), *The Mughal state: 1526–1757* (pp. 491–519). New Delhi: Oxford University Press.

Rider, C. (Ed.). (2007). *Encylopaedia of the age of the industrial revolution, 1700–1920* (Vols. I and II). London: Greenwood Press.

Rosenberg, N. (1976). *Perspective on technology*. Cambridge, MA: Cambridge University Press.

Rosenberg, N. (2000). Charles Babbage in complex world. In D. Colander (Ed.), *Complexity in history of economic thought* (pp. 37–57). London/New York: Routledge.

Rostow, W. W. (1957). *Stages of economic growth—A non communist manifesto*. Cambridge, MA: Cambridge University Press.

Schumpeter, J. A. (1954). *History of economic analysis*. London: Oxford University Press.

Sen, S. (1998). *Empire of free trade: East India Company and the making of colonial market place*. Philadelphia: University of Philadelphia Press.

Seth, V. K. (1999). *Capacity utilization in industry theory and evidence*. New Delhi: Deep and Deep Publication.

Seth, V. K. (2000). Technology and labour in retrospect and prospect. *Managment and Change, 4*(1), 61–90.

Singer, C., et al. (Eds.). (1958). *History of technologies* (Vol. 2–4). Oxford: Clarendon Press.

Sinha, N. K. (1968). *The economic history of Bengal*. Calcutta: Firma K. Mukhopadhyaya.

Smith, A. (1776, 2003). Introduction. In A. B. Krueger (Ed.), *The wealth of nations* (Bantam classic). New York: Bantam Books.

Smith, B. M. D. (1967). The Galtons of Birmingham Quacker gun merchants and bankers. *Business History, 9*(2), 135–148.

Tchitchirov, A. L. (1965). *India-changing economic structure in sixteenth to eighteenth centuries outline history of craft and trade*. New Delhi: Manohar.

Thirsk, J. (1994). Industries in countryside. In J. A. Chractre (Ed.), *Pre-industrial Britain*. Oxford: Blackwell.

Thorner, D. (1980). *Shaping of modern India*. Bombay: Sameeksha Trust.

Traverse, R. (2007). *Ideology and empire in eighteen century India—The British in Bengal*. Cambridge: Cambridge University Press.

Wallace, D. R. (1928). *Romance of jute—A short history of the Calcutta jute mill industry 1855–1927*. London: W. Thacker.

Weber, M. (1927). *General economic history* (trans: Knight, F.). New York: Greenberg.

West, E. J. (1976). *Adam Smith: The man and his work*. Indianapolis: Liberty Press.

Willam, K., et al. (1987). The end of mass production. *Economy and Society, 16*, 405–439.

Wilson, C. (1994). Entrepreneurship in industrial revolution. In J. Huppit & E. A. Wriglay (Eds.), *The industrial revolution in Britain* (pp. 305–321). Oxford: Blackwell.

Wright, R. (2000). *Stolen continents—A conquest and resistance in the America*. London: Phoenix Press.

Wrigley, E. A. (1988). *Continuity chance and change, the character of industrial revolution in England*. Cambridge: Cambridge University Press.

Zeitlin, J. (1995). Flexibility and mass production at war-aircraft manufacturing in Britain, United States and Germany, 1939–1945. *Technology and Culture, 36*(1), 46–79.

CHAPTER 4

Traditional Flexible Manufacturing and the East India Company

As described in the Introduction, the Indian economy experienced the impact of two different forces of globalization, in tandem, around c.1500. One was the entry of Mughals through land route, and the other was the arrival of European trading companies through sea route. Also discussed was the fact that from c.1526 to c.1707 (i.e., until the death of Auranzeb), it was the centralized authority of the Mughal emperor that determined the economic environment experienced by the Indian sub-continent. The kind of economy that evolved during the Mughal Empire provided necessary and sufficient conditions for the emergence of traditional flexible manufacturing. In due course, a symbiotic relationship emerged between the Mughal Empire and traditional flexible manufacturing. The implications of this symbiotic relationship have been discussed in detail in Chap. 2 of the present study.

After the death of Aurangzeb, centralized authority of Mughal state was fragmented into several regional centres of power, competing with each other to fill the political space vacated by the erosion of authority of the Mughal state. After the Battle of Plassey (c.1757), the East India Company began to fill the vacuum created by the decline of the Mughal Empire in the political space of the Indian sub-continent. The Battle of Plassey was followed by the Battle of Buxar (c.1764), which led to the Treaty of Allahabad between the Mughal Emperor Shah Allam II and the East India Company. Through this treaty, the emperor granted '*diwani right*'(i.e., right to collect land revenue) to the East India Company on Bengal Subah, which, at that time, consisted of territories of present-day Bihar, Bengal and

V.K. Seth, *The Story of Indian Manufacturing*,
DOI 10.1007/978-981-10-5574-4_4

Orissa. This was the beginning of a process of transformation of a trading company into a territorial empire in the Indian sub-continent.

Subsequent to this historical event, the dictates of the East India Company were enforceable in a vast geographical area, far beyond its earlier trading centres located at Bombay (Mumbai), Calcutta (Kolkata) and Madras (Chennai). The company had its own army and administrative setup, which had developed capability to influence not only the destiny of millions of people living in the Indian sub-continent, but also to influence its economy. The organizational transformation of the East India Company into an agency of the empire became so paramount for the British Crown that it continued to control territories in India, on behalf of the Crown, up to c.1857. Therefore, the company continued to perform the '*Business of empire*[1]' from c.1757 to c.1857; despite the fact that the East India Company ceased to be a trading company in c.1833 (when the British parliament stripped it of its commercial functions).

The transformation of the East India Company into an agency of the British Empire has been variously described by scholars. Some describe it as '*imperium in imperio*',[2] the '*East India Company state*',[3] and some others describe it as '*sovereign merchant*'[4] or '*corporate state*'.[5] The government of this trading company was most powerful in Asia; it was the government whose revenues were larger than that of the country from where it had obtained chartered status. This was a peculiar kind of a government, which was owned by a trading company, whose shares were traded daily at the London Stock Exchange.

The East India Company's rule, as long as it lasted, represented the interest of the merchants of England. Until the beginning of the Industrial Revolution, trade was the foundation of the British economy. Therefore, for a considerable period, interests of the company were the same as those of dominant elements of the British society. With the growth of modern manufacturing in England, a new social structure evolved, which facilitated the emergence of a new class of *industrialists*, whose interests were at variance with the interests of landed gentry and traders. The new social structure replaced the ideological dominance of mercantilism with classical political economy. Hence, the period during which the Indian sub-continent experienced the territorial expansion of the East India Company coincided with the period of transformation of British traditional flexible manufacturing into modern manufacturing. This suggests that during this period, the interest of Indian traditional flexible manufacturing became entangled in the basic conflict between merchants and industrialists of Britain.

In the present chapter, an attempt has been made to understand how the transformation of the East India Company into an agency of the British Empire impacted the economic environment of business in India in general and functioning of traditional flexible manufacturing in particular. This chapter, in fact, deals with how the second force of globalization (i.e., the East India Company) influenced the story of Indian manufacturing. For purposes of our narrative, this chapter is divided into four sections. In Sect. 4.1, I describe the conditions and circumstances that led to the transformation of the East India Company into an agency of the British Empire. In Sect. 4.2, I analyse the impact of the rule of the East India Company on traditional flexible manufacturing in India. In Sect. 4.3, I present the debates regarding the process of de-industrialization that occurred as a consequence of the rule of the company. In Sect. 4.4, I explore the emerging contradictions between the interests of merchants and the new emerging class of industrialists in Britain, which became manifest in the birth of classical political economy, and how new ideology impacted the rule of the East India Company in India.

4.1 The Process of Transformation of the East India Company

We have described, in Chap. 2 of the present study, that around c.1500, European economies depended on the imports of spices and luxury manufactured goods produced by the traditional flexible manufacturing of India to satisfy the needs of European nobles and rich gentry. The imports from Asia were financed by exports of bullion to balance the trade. This pattern of trade continued to exist for centuries. The trade in spices and luxury handicrafts was dominated by the local Asian trader, while European merchants enjoyed a subordinate position by performing their operations at the lowest level of the supply chain. The occupation of Constantinople by the Ottoman Empire in c.1453 denied the access to trade in spices and luxury handicrafts to European traders. The use of pepper was necessary for Europeans to make preserved meat edible, making it an essential part of their lives. This necessity compelled the Europeans to find alternative sea routes to get access to spices like cinnamon, nutmegs, pepper, clove and mace.[6] Also the need to search for gold and silver, which were required to buy spices, provided preconditions for the beginning of the '*age of discovery*'. These twin needs converted European rulers into '*venture capitalists*' because they began to finance very risky voyages for the discovery of sea routes by European navigators. The competition amongst European nations to reach the source of spices was so intense that it would not be misplaced to view it like the contemporary competition amongst the nations to reach the moon. Encouraged by the king of Spain, Christopher Columbus began his voyage for discovery of India by going west across the Atlantic and instead discovering America, which provided ample supply of gold and silver. The Portuguese sailor Vasco da Gama began his journey southward along the coast of Africa and the Cape of Good Hope and succeeded in reaching Malindi. The popular perception is that here Vasco da Gama met one Indian merchant and ship pilot, Kanji, who guided him to reach Calicut, the 'City of Spices'. Despite the fact that the American and Indian sub-continents were geographically distant from each other, they sustained each other for a considerable period. The gold and silver obtained from America helped Europeans to buy spices and luxury manufactured products from India. It is also not surprising that Columbus referred to the natives of America as Indians.

The Portuguese were the first Europeans to establish direct trade links with the Indian sub-continent, after Vasco da Gama's ship reached Calicut at the Malabar Coast in c.1498. The Portuguese wanted to make Calicut a dedicated port for their trading operations. For this purpose, they waged war with the ruler of Calicut, Zamborin, in c.1502. However, they did not succeed. They were able to negotiate with the ruler of Cochin to construct a port in his kingdom. Subsequently, Portuguese were able to occupy the island of Goa from the Sultan of Bizapur in c.1510 under the command of Alfonso de Albuquerque. Since then, Goa remained the Portuguese maritime capital in the Indian sub-continent until it was made and integral part of India. After establishing their foothold in Goa, Portuguese also developed several fortified trading settlements in Hubli and Chittagaon in Bengal. The total maritime empire of Portuguese in the Indian sub-continent was known as '*Estado da India*'. The arrival of Portuguese in India transformed the nature of trade at the trading centres where Portuguese started conducting business. It led to the demise of free trade because with the help of their superior naval power, the Portuguese were able to monopolize trade.[7] This was challenged by the Dutch company known as Vereenigde Oost Indische Companie. This company was formed in c.1602. The Dutch company established its trading posts in Muslipatnam, Tegepatnam, Pulicut, Nagapatnam and at Surat.[8]

The first serious attempt made by a French trading company to enter the Indian market is dated c.1664, when Colbert formed the Companies des Indes Orientales. The French company started its trading operations from its trading base at Machilipatnam in c.1667. Soon it had to abandon Machilipatnam, where its forces were defeated by the forces of the Dutch company. However, one of the agents of the French company, Bellanger de Lespinay, was able to get a grant of a small village from the Sultan of Bijapur in c.1673. This village, Pondicherry, now called Puducherry, later became the main centre of trading of the French trading company. Subsequently, the French company also started its operations at Chandernagore in Bengal.[9]

India's poignant contact with Britain began when Queen Elizabeth I signed the Royal Charter that granted monopoly of trade with India to the East India Company on New Year's Eve (December 31) of c.1600. The company began its operations from Surat in c.1608. In the year c.1618, the East India Company was able to negotiate its first trade treaty with Mughal emperor Jahangir.[10] After establishing its roots in Surat, it opened another factory (Office of the Factor) at Madras (now known as Chennai) in c.1640 and constructed here its first fort, St. George.[11,12] The East India Company also acquired territorial rights over the territory that grew into what came to be known as Bombay (Mumbai) on lease from the British King Charles II in c.1668. Charles II had received this territory as dowry from the Portuguese, when he married Portuguese Princess Kathrine of Aragon in c.1662. This area consisted of seven islands: Wadala, Colaba, Mazgaon, Old Women's Island, Mahim, Parel and Mutunga-sion. At that point in time, this area was inhabited by tribes like Koli, Bhandari, Agri, Prabhu and Pachkalashis, who used to speak dialects of Maratha or Konkani languages. Bombay had an excellent harbour and natural isolation, which protected it from land attacks. The East India Company also received permission from the Mughal emperor Aurangzeb to open a new factory in villages, Kolikata, Govindpore and Sutanati, located on the banks of river Hugli in c.1690. It is here that a new settlement of the company grew with the name Calcutta (now known as Kolkata). The East India Company constructed here a fortified trading centre in Fort William.[13] This place subsequently became the capital of the empire created by the East India Company.

There existed serious conflict of interest amongst different European trading companies because each one of them wanted to corner the largest share of market in the trade with India. So long as the Mughal Empire was strong, their conflicts remained confined to coastal regions. These companies were operating in India through patronage and goodwill of the Mughal emperors. Therefore, these companies remained primarily trading companies until the death of Aurangzeb in c.1707. After his death, the Mughal Empire began to disintegrate into several autonomous regions as a consequence of factionalism, rebellion and threat from across the borders. The emerging political uncertainty provided preconditions to European trading companies to mobilize military and naval power at unprecedented levels. The defense mobilization by these companies was also emerging as a consequence of globalization of conflict amongst European countries, especially between France and Britain.[14] The increased militarization of European countries and globalization of their conflicts were laying the foundation of an epochal transition to a new world order, which was termed by historians and political analysts as '*imperialism*'.[15]

The weakening of the centralized authority of the Mughal state after the death of Aurangzeb created opportunities for some of the provincial rulers to carve out independent regional bases of power. Murshid Kuli Khan, the governor of Bengal Subah, established his own autonomous rule in Bengal, and shifted its capital from Dacca to a town named after his own name, Murshidabad. Similarly, Asaf Jan, Nizam-ul-Mulk, evolved a power base in southern India with its capital at Hyderabad. Another provincial ruler, Sadat Khan (Burhan-ul-Mulk), carved out an autonomous regional power in Awadh with its capital at Lucknow.

Some of the non-Muslim warrior groups also established their regional power bases in different parts of the Indian sub-continent. Marathas formed an empire in western India, Sikhs in Punjab and Jammu and Kashmir, Rajputs in Rajasthan, Jats in territories around Delhi, and Hyder Ali and Tipu Sultan in Mysore. Amongst all these political entities, Marathas were emerging as strong contenders to replace the centralized authority of the Mughal state. However, after their defeat in the third Battle of Panipat, the Maratha Empire was segmented into several smaller political entities. Bhonsales were ruling from Poona, Holkars were rulers at Indore, Scindias at Gwalior, and Gaikwards in Baroda. However, after the death of the Bhonsale King Shahu in 1749, his prime minister, Balaji Bajirao Peshwa, took control of his territories and established the Peshwa Dynasty.

For a considerable period of time, historians believed that the fragmentation of polity, economy and legal environment, after the death of Aurangzeb, led to systematic decline in traditional flexible manufacturing in India.[16] However, several studies conducted by scholars on specific regions amply suggest that in certain regions, the preconditions that are necessary for the continuity of traditional flexible manufacturing continued to exist, arresting its decline. These regional studies show that the decline of Mughal authority and emergence of autonomous successor states led to a variety of regional experiences with varying consequences for the continuity of traditional flexible manufacturing.[17]

The emergence of autonomous regional powers in Awadh, Hyderabad, Maharashtra, Punjab, Rajasthan and Mysore created certain objective conditions that facilitated the continuity of the traditional flexible manufacturing, chief being the consolidation of power of merchants in these new autonomous states. These autonomous centres of power created new elites through the process of '*gentrification*' of society.[18] The revenue needs of these successor states also increased the importance of revenue farmers. These different varieties of elites described as '*portfolio-capitalists*'[19] sustained themselves by performing several important economic functions in the society. Their activities had given birth to what is now known as '*tributary commercialism*'.[20] These elites provided supply-side as well as demand-side support to arts and crafts. It was this support that provided sustenance to traditional flexible manufacturing.

Despite the fact that the fragmentation of the polity did not completely wipe out the existence of traditional flexible manufacturing, it did increase the intensity of conflict amongst regional rulers. These conflicts were not just confined to war for territorial occupation, but also for settling successions amongst rival claimants of power. In all these conflicts, demand for the might of the forces maintained by different European trading companies increased significantly. Rival political entities started *outsourcing* the military power of rival European trading companies to settle their conflicts. The demand for military power of the companies provided them not only revenue to maintain larger armies at the cost of local rulers, but also opportunity to increase their sphere of influence. The possibilities to maintain large armies with little costs provided added advantage to the European trading companies to expand their military power, which had become a necessity as a consequence of emerging globalization of conflict amongst European states.

The process of outsourcing of the armies of different European companies began with the death of the ruler of Hyderabad, Nizam-ul-Mulk, in c.1748, which led to a war of succession amongst his sons. A similar war of succession occurred after the death of Nawab of Arcot. In both the wars of succession, French and British trading companies' forces supported opposing claimants. When successions were finally settled after the wars, it so happened that the French-company-supported prince became the ruler of Hyderabad and the British-company-backed Prince Muhammad Ali Wallajah became the ruler of Arcot.[21] These two incidents increased the participation of the forces of European trading companies in the conflicts amongst local rulers.[22]

In the case of Bengal, there emerged a direct confrontation between the Nawab of Bengal and the British East India Company. In c.1717, the East India Company received a *firman* (permission) from the Mughal emperor, Shaharyar, to carry on free trade in Bengal Subah. The East India Company, ignoring the true intent of the *firman*, started issuing *dastaks* (entry passes) with financial consideration to the officials of the East India Company for their private trade, as well as to other European and local traders. This practice substantially reduced the revenue of the Nawab of Bengal.[23] The agitated Nawab Siraj-ud-Daulah threw out the contingent of the East India Company from Bengal. However, after receiving enforcement from Madras under the command of Robert Clive, the army of the Nawab of Bengal fought a battle at a place called Plassey in c.1757. Clive won the Battle of Plassey by entering into a secret deal with the commander of the army of the Nawab, Mir Zaffar, who deserted the army of the Nawab in the battlefield. The Battle of Plassey was the most successful 'business deal' of the East India Company because it won the war by entering into a deal with the commander of the forces of the Nawab. It resulted in a massive transfer of wealth gotten from the systematic plunder of Bengal.[24]

Nawab of Bengal, along with the forces of Mughal Emperor Shah Allam, tried to recapture the lost territories of Bengal from the East India Company. The forces of the East India Company and the combined forces of the Nawab of Bengal and Mughal emperor faced each other in the Battle of Buxar c.1764, which ended with the Treaty of Allahabad between the East India Company and the Mughal Emperor Shah Allam II. This treaty granted *diwani rights* (rights to collect taxes) to the East India Company. This transformed the company into the ruler of Bengal.[25] This transformation converted the '*John Company*' into a Frankenstein known as '*Company Bahadur*'.[26,27] After integrating monopoly of trade with monopoly on political power in Bengal, the East India Company obtained unrestricted access to the best products produced by the traditional flexible manufacturing of Bengal. After controlling the centres of traditional flexible manufacturing in Bengal, the East India Company followed the same strategy to consolidate its political control over other centres of traditional flexible manufacturing located in other parts of the Indian sub-continent. The process began with Bihar (c.1765), followed by Eastern Uttar Pradesh (c.1775), Malabar Coast and Mysore (c.1792–1801), and Surat and Ahmedabad (c.1800–1803). In new circumstances, the nature of interaction between the East India Company and the Indian economy underwent qualitative change.

In order to understand the consequences of the symbiotic relationship that evolved between the East India Company and traditional flexible manufacturing, after the company became the ruler of the Indian economy, it is essential to know its organizational structure; to understand its objectives (mission), its work culture and nature of its operations.

4.2 Consequences of the Rule of the East India Company on Traditional Flexible Manufacturing

The transformation of the East India Company into an agency of the British Empire resulted in restructuring of the relationship between the company and the economy of the Indian sub-continent. The new relationship restructured several macro- and micro-level processes that had significant impact on the traditional flexible manufacturing of India. The most important macro-level change was that the East India Company began to use substantial parts of income that it earned from land revenue on financing its purchases of Indian merchandise for exports. It began to buy Indian silk, muslin, cotton textiles, spices and other products manufactured by the Indian traditional flexible manufacturing, without making payments in bullion.

The new arrangement reversed the historical process in which Western countries, including Britain, had nothing to offer apart from bullion to buy manufactured items. Once the tax revenue of the company increased substantially, its directors decided that the revenue would be used to finance the purchases of merchandise in India for export, and whatever surplus remained would be used to purchase tea at Canton in China. The new arrangement helped the company to finance its purchases in India for exports without any corresponding flow of bullion or merchandise. This form of trading arrangement has been described as '*unrequited trade*'.[28] The new trading arrangement brought an end to the advantage that the Indian sub-continent had enjoyed since the days of the Roman Empire—to receive bullion in exchange for its products. The economic consequences of this arrangement were discussed by Burke who was the author of the Ninth Report of the Select Committee, which appeared in June 25, 1783. He called the new arrangement a '*drain of wealth*'[29] from India to Britain. According to Burke, the drain of wealth from India took place (i) through the use of land revenue to buy Indian manufactured items for exports, which was described by the East India Company as '*investments*' and (ii) £ 4,00,000 was paid annually to the British Crown, which was described by Burke as '*tribute*'. While describing the drain of wealth, Burke wrote 'Numerous fleets of large ships, loaded with most valuable commodities of the east annually arriving in England gave rise to an opinion of happy condition and growing opulence of a country whose surplus production occupied so vast a space in commercial world. This export, from India, seemed to imply reciprocal supply, by which the trading capital employed in the production was continuously strengthened and enlarged. But a payment of tribute, and not a beneficial commerce to that country, wore this specious and divulsive appearance'.[30,31,32]

The understandings of Burke regarding the magnitude of drain of wealth from India were based on the calculations made by Philip Francis. Similar opinions were also expressed by William Bolts, Alexander Dow and Harry Verelst.[33] The idea of '*drain of wealth*' was also highlighted by nationalist historians like R.C. Dutt and Dadabhai Naroji who used it to explain the deteriorating state of the Indian economy.[34] The drain of wealth also became an important argument in the writings of Marxist historians in India to explain the economic consequences of British imperialism. A noted Marxian historian, Irfan Habib, also calculated the extent of drain of wealth from India to Britain. According to his estimates, the amount of drain was more than two million pounds during c.1789–1790 and more than £ 4.7million during the years c.1790–1801. This amount, according to Habib, was 9 per cent of the GNP of India at that time.[35]

It is quite revealing that two decades after the Battle of Plassey (c.1757), in c.1778, a new painting was installed on the ceiling of the revenue committee room, where meetings regarding the changing fortunes of the East India Company were held. The painting was painted by the artist Spiridione Roma and was entitled *The East Offering Her Riches to Britannia*. The painting shows three women, each of them representing a different country. The fair women is Britania, who is looking down upon the other two women. A kneeling black woman represents India, who is offering rubies and pearls to Britannia. The painting clearly illustrates how the wealth of the East was being drained into Britain.[36]

In c.1818, five years after the trading monopoly of the East India Company had come to an end (c.1813), a marble relief was made in the memory of an expatriate British merchant of Calcutta, named Alexander Colvin, by a sculptor Richard Westmacott. In the sculpture, India is depicted by a seated woman, her body shown without any ornaments and her hands resting on an empty pot. The relief clearly show that by this time the material wealth of the country had been taken away to Britain.[37] This also shows how the transformation of the Indian economy had captured the imagination of not only scholars but also artists.

Historically, European economies paid bullion in exchange for Indian manufactured products. Since the Indian economy always suffered from shortages of precious metals, the bullion earned from trade with Europeans performed an important function in the economy of the Indian sub-continent. It provided metal for minting coins, which were necessary for the smooth functioning of the economy of the sub-continent. The East India Company created shortages of precious metals by taking over the treasuries of local rulers and by financing trade from India by land revenue collected from the Indian sub-continent. These activities of the company resulted in shortages of currency, which caused virtual stagnation of business activities in India.[34] The disappearance of bullion in exchange for merchandise reduced the gains from trade, as expressed by trade theorists in terms of income and employment. A group of merchants of Calcutta submitted a petition to the Calcutta council, complaining how the lack of currency had brought virtually all businesses to a halt.[38]

Before the Battle of Plassey, company officials were performing trading activities through their dependence on local merchants for purchasing product produced by traditional flexible manufacturing. For this purpose, local merchants had organized networks of artisanal household workshops, based on the *Dadni* system, to procure products for selling in the markets. These merchants facilitated the company officials in procuring products produced by traditional flexible manufacturing for exports. The trading houses of Indian merchants, like the House of Jagat Seth and Umichand, were far bigger in size of operations. The dependence on local merchants was far greater amongst the British employees of the East India Company, who were involved in private trade in India. These company employees were always in need of credit. The dependence on local merchants for credit needs had increased the incidence of indebtedness amongst the officials of the company to the local merchants.[36] Moreover, the East India Company also had to share trading space with other European trading companies. The phenomenon is described as '*mutual mercantilism*'.

The transformation of the East India Company into Company Bahadur entirely altered the circumstances. Through its newly acquired political power, it attempted to marginalize the role of Indian merchants. To make Indian merchants ineffective in the markets, the company used market as well as non-market strategies to evolve its own monopoly over trade in the products produced by traditional flexible manufacturing.

The organizational structure, which had provided competitive advantage to Indian manufacturing to become one of the largest suppliers of manufactured goods, had emerged as a consequence of nexus between merchants and craftsmen, and has been described here as traditional flexible manufacturing. So long as this nexus between merchants and craftsmen continued to exist, Indian manufacturing had the capability to produce abundant quantities of quality handicrafts to satisfy the needs of the market. The East India Company deliberately created conditions to break this nexus. The moment this nexus was broken, it led to the beginning of de-industrialization. To begin with, the company declared its direct monopoly of trade in certain products. In these products, company became the sole buyer or monopsonist. The process of monopolization of trade, using its newly earned political power, began in c.1750 when Mir Zafar gave away monopoly of trade in saltpetre to the East India Company. This was followed by monopolizing of trade in salt and opium in c.1793. The appropriation of monopoly rights in trading of these products eliminated the role of Indian merchants in the trading of these commodities. In new circumstances, when the company enjoyed twin monopolies (i.e., monopoly of trade and monopoly of political power in the market), it also developed capacity to manipulate prices in these products. Once the East India Company gained political dominance in the Indian sub-continent, the native merchants had to pay duties at various internal checkpoints, whereas British merchants and their agents were exempted from such duties.

The East India Company also minimized the role of Indian merchants by reducing their role in procurement of merchandise. For this purpose, the company developed independent channels to procure goods produced by Indian craftsmen directly through *gomashtas* employed by the company. To facilitate the process of procurement of Indian manufactured items like textiles, the company divided markets into *aurangs*, consisting of centres of traditional flexible manufacturing of certain specific products. Each *aurang* was working under the supervision of chief *gomashtas*, who in turn employed *gomashtas* and *paikars* for the purposes of procurement of products. These g*omashtas* replaced the role of Indian merchants in each market.

After the occupation of Awadh, to exploit its political power for commercial gains, the company employed 500 *gomashtas* in Awadh alone. European traders who were close to top army officers or administrators of the company like '*residents*' took maximum advantage of the new circumstances. Under the patronage of Resident of Awadh Mr. Nathaniel Middleton, Job Scott, a free merchant, was able to establish a commercial empire of cotton piece goods in *aurang* at Tanda. Here he exercised absolute power, which he used to appropriate output of weavers of the area at lower prices without facing any competition. Another trader who gained from the patronage of the Resident of Awadh was John Hyde.[39] Such nexuses between local officials of the East India Company and British traders existed in all the markets in India, where the East India Company assumed political power. Such nexus between local officials of the army and administration and British traders emerged in the most important centres of production of traditional flexible manufacturing.

Finally, to deal a final blow to the role of Indian merchants in the production and distribution of products produced by Indian traditional flexible manufacturing, the East India Company banned the *Dadni* system in c.1770, which was the basis of the emergence of traditional flexible manufacturing as a form of organization of production. It is by using all these aforementioned strategies that the emergence of the East India Company as a monopolist of the political power eliminated the symbiotic relationship with local merchants, which was based on mutual mercantilism.[40]

During the same period, the British manufacturers developed competitive advantage over the products produced by the traditional flexible manufacturing of the Indian sub-continent. The increasing importance of British manufacturers, and their emerging influence on the political economy of Britain, forced the East India Company to share trading space with British manufacturers. It allowed the entry of free merchants who acted as agents of the British manufacturing companies to sell their products in the Indian markets. These agents formed agency houses, which were generally partnership companies, working under the patronage of the officials of the East India Company. Some of these companies were formed by the relatives of the officials of the company. Slowly, these companies began to function as banks and became important channels to facilitate remittance sent by British expatriate workers to their relatives in Britain.[41] Most of these agency houses operated from the presidency towns, because these agency houses had evolved a symbiotic relationship with the British citizens working in India.

The first agency house, which was founded in Bombay Presidency, was Forbes Company in c.1767, followed by Alexandra & Company in Calcutta Presidency in c.1767 and Parry and Company in Madras Presidency in c.1789. Apart from Alexandra & Company, which also established the Bank of Hindustan in Calcutta in c.1770, Calcutta Presidency had the largest number of agency houses, including Ferguson & Company, Colive & Company, and Pickard & Company. The largest agency house of Calcutta was Palmer & Company, which was established in c.1810. Similarly, Bombay Presidency also had several other agency houses like Bruce Fawcett & Company established in c.1790, Mackintosh & Company in c.1805, Richi Stewart & Company in c.1810 and Leaking & Company in c.1810.[42,43]

The East India Company not only marginalized the role of local merchants in the process of organization of production and distribution of handicrafts, it also imposed several kinds of restrictions on Indian craftsmen. These restrictions were imposed to reduce their bargaining power, to procure their products on a preferential basis and at substantially lower prices. The company prevented weavers from weaving cloth for other merchants, either local or foreign, until they had produced enough to meet the requirements of the company. The East India Company introduced indenture (*khatbandi*) regulation in (c.1770–1786), which turned weavers into bonded labour, because this regulation helped in binding craftsmen exclusively to the agent of the company. Through this regulation, the company denied artisans the right to sell their products in the open market; it also reduced their bargaining power against the agents of the company. What began as a coercion or bullying of Indian craftsmen by the company agents, after the Battle of Plassey, became official with the *khatbandi* regulations. These regulations affected the life of *tanti* (weavers), *nunias* (salpetre makers) *mulungi* (salt workers), and *nacauds* (silk reelers).[44]

Using its political power, the East India Company also forced craftsmen to work in the production centres organized by the company. In these centres, their wages were quite low and they were forced to work for longer hours. The working conditions in these centres were also much lower than the Mughal *karkhanas*, which were owned by the Mughal emperors, royal family members and *subedars* of different provinces.[45] The company organized centres of production of Indian manufacturing goods in each of the important centres to meet where traditional flexible manufacturing was clustered. For example, in Madras Presidency, for its requirement of muslin, it set up a company factory near Madras, and for calicoes, it evolved centres of production in Cuddalore-Inegram and Madapollam.[46] The company also imposed *motarfa* tax on weavers. Although this tax was subsequently abolished in Bengal Presidency in c.1793 and in Bombay Presidency in c.1844, it continued to be in operation in Madras Presidency. From this tax, the company earned. 1 million rupees in c.1850. According to the estimates of Dodwell, due to the imposition of motarfa tax during the period c.1800–1875, employment in the manufacturing sector declined by almost 50 per cent.[47]

Bolts, who has provided a detailed account of the company excesses on weavers, remarks that the company used all possible methods of exploitation to exploit weavers 'such as fines, imprisonment, flogging and forcing bonds on them'.[48] The impact of the tyranny of the East India Company was so devastating for the weavers and silk winders of Dacca that they began the practice of self-mutilation by cutting their thumbs to avoid exploitation by the officials of the company. This practice slowly eclipsed the whole industry, which was known to produce the best quality muslin or *malmal khas*.[49]

Manufacturing of silk in the Indian sub-continent was organized into three separate activities: (i) cocoon rearing and harvesting (i.e., sericulture), (ii) spinning or reeling of silk yarn (i.e., making of raw silks), and (iii) weaving of silk. Sericulture was a highly localized activity due to its dependence on mulberry trees for rearing of cocoons. Since cocoons also need processing within a period of four to five days after harvest, the spinning and reeling of silk was also localized near the places where cocoons were harvested. Due to this reason, weaving of silk was widespread in all the corners of the Indian sub-continent, from Bengal in the East to Surat in the west and Coromandel Coast in the South.

The interests of the East India Company were related to raw silk only, which it was exporting to provide inputs to British manufacturers. However, the Bengal raw silk was of poor quality because it was produced by *nacauds* using traditional technology of reeling. In order to protect the interest of thousands of workers 'working under British throwsters, who totally depended on Bengal raw-silk',[50] the first thing that the East India Company did, as soon as it assumed political power in Bengal, was to improve the quality of raw silk.

The East India Company introduced 'filature silk' in Bengal around c.1790; it had borrowed this technique from the Italian silk industry. This method used the mechanism of double crossing and twisting silk thread, which gave the silk yarn roundness and even quality of thickness. By c.1823, the volume of silk yarn wound on filature machines had replaced the silk yarn wound by traditional *nacauds*. The filature silk became quite popular in areas like Rampur, Jangipur, Kumarhatti, Malda, Ganutia and Purlia.[46] The filature silk machine required investment in fixed assets. The volume of fixed investment required for the adoption of new technology was beyond the means of craftsmen. These machines were generally owned by the East India Company or by its officials who were engaged in private trade. In these factories, *nacauds* were employed as wage workers. The introduction of the filature method of silk reeling led to the decline of the traditional silk reeling industry.[51] However, it should be noted here that while the East India Company introduced new technology for reeling of silk, it did not introduce new techniques, which were emerging in Europe, for the weaving of silk.

With the increasing monopsnistic behaviour of the East India Company, the contracts between weavers and the company agents were redesigned. Traditional flexible manufacturing largely depended on the contract between weavers and merchants. The nature of contract determined the nature of the organization of production. During Mughal rule, weavers enjoyed freedom to cancel their contract with the merchants by selling their cloth to any other merchant and returning the advance taken from the merchants. Weavers enjoyed this freedom because there existed competition amongst merchants (i.e., buyers). As the company was able to monopolize trade in textiles, it emerged as the single and largest buyer of textiles.

The new circumstances provided immense bargaining power to the officials of the company, who also introduced new a format of contract. In the new format, the company took away the power to break the contract from the craftsman and gave itself the right to cancel the contract. It also abolished the practice that forced the buyers to buy the entire cloth from the weavers for which the contract was signed. The company also assumed the right to reject the cloth that did not meet the quality standards. These revisions in the contract marginalized the bargaining power of weavers. The company agents also started misusing the provisions laid down in the new contracts. One of the most misused provisions was the rejection of large proportions of cloth that the weavers were weaving for the company against quality standard. These rejected pieces of cloth were again procured by the company agent at throwaway prices.[52]

As a result of these strategies, the East India Company was able to break the nexus between local merchants and craftsmen's household workshops, which was the precondition for the existence of traditional flexible manufacturing. The nexus survived the decline of the Mughal Empire in some of the successor states of the Mughal Empire. However, its existence was made difficult by the emerging political power of the East India Company. It is interesting to mention that the *Dadni* system ended in c.1770, when it was discontinued by the orders of the East India Company. This was prompted by the conflict between Indian merchants and the agents of the company regarding advances that were to be paid to Indian merchants. The conflict became so intense that the two big merchants of Bengal, Shobha Ram Basak and Ramkrishan, refused to take business on behalf of the East India Company.[53]

The emerging disassociation of merchants with the organization of production of handicrafts led to a disconnect between craftsmen and mass markets. This facilitated the slow and systematic decline of traditional flexible manufacturing. Therefore, it would not be inappropriate to state that traditional flexible manufacturing experienced a decline with the territorial occupation of the Indian sub-continent by the East India Company. Decline in traditional flexible manufacturing does not *pari passu* mean decline of handicrafts. Since primitive flexible manufacturing is just one of the methods by which handicrafts can be to supplied to products in mass markets, crafts-based methods of production can be embedded into several kinds of organizational structures. This suggests that after the process of territorial expansion of the East India Company, craftsmen lost the support of merchants who had been organizing them into networks to sell their products in the mass markets. The disengagement of merchants with crafts-based manufacturing exposed craftsmen to several kind of risks and uncertainties.

After the emerging political power of the East India Company, traditional flexible manufacturing was reorganized in a new relationship between craftsmen and the agents of the company. In the new relationship, craftsmen largely became wage workers. This major shift in the relationship between company agents and artisans has been described as a shift 'from market determined to coercion based transaction'.[54] Some historians justify the use of coercion by the agents of the East India Company as a necessary evil, to enforce contracts with craftsmen in the absence of third-party institutions, which would have helped in enforcing contracts. The process of decline of traditional flexible manufacturing and emerging political power of the company reduced the bargaining power of the craftsmen. Similar to what exists between '*parasite and host*' where parasite uses the host for its own sustenance'.[55,56]

It is worth mentioning here that while the East India Company broke the nexus between craftsmen and merchants in the manufacturing sector, it developed a strong symbiotic relationship with Indian merchants who were engaged in the trading of commodities. This was necessary, because by this time, the Indian economy had been transformed from a manufacturing economy to a supplier of raw materials. India had slowly become a net exporter of primary products like cotton, after the Civil War in America, and exports of jute after the Crimean War. India has also emerged as an important exporter of opium. In order to export these commodities, the East India Company and agency houses developed strong ties with Indian merchants. These merchants helped them to maintain a supply chain of these commodities from farmer onward up to shipping of these commodities to Britain.

The decline in traditional flexible manufacturing, along with the fall of royalty and nobility with the territorial expansion of the East India Company, resulted in large-scale migration of Indian merchants to Bombay, Calcutta and Madras. Bombay became an important destination for *Gujaratis*, *Parsees* and *Banias* because it provided them protection from Maratha invasions. It provided them the same public good, which was provided by the Mughal Empire during its best period—the protection of private property and rules of law. This shift also transformed them from producers of handicrafts to traders in commodities. These merchants earned a substantial amount of wealth by selling cotton and opium. They subsequently became entrepreneurs and established modern cotton textile mills in Bombay.

The merchants from different southern locations converged at Madras, while *Marwaries* migrated to Calcutta.[52] This process of migration of merchants from old port cities to presidency towns led to a decline of trade in old ports like Surat, Calicut and Coromandel Coast.[57] These old port cities experienced isolation from their sources of supply because political instability disturbed the entire supply chain and weakened the existing networks evolved by merchants.

The period during which the Indian sub-continent was experiencing territorial expansion of the East India Company coincided, in Britain, with the process of transformation of its traditional flexible manufacturing to modern manufacturing (which Marxists describe as a change in the mode of production discussed in Chap. 3). The emergence of new techniques accompanied by organization of production into modern factories created conditions where machine-made goods could compete with Indian handicrafts. In fact, it gave rise to unequal competition between '*machine-facture*' of Britain and '*manufacture*' of India. The use of machines made it impossible for cheap Indian workers to compete with Lancashire's mules and throstles. Consequently, so many centres of traditional flexible manufacturing just disappeared. The advantage achieved by the British manufacturing industries facilitated a reverse flow of textiles and other manufactured items to Indian markets, which provided preconditions for the emergence of de-industrialization or re-ruralization. The new transformation reduced the significance of the monopoly of trade with India given to the East India Company, which the British government ended in c.1813, and in c.1833, all trading activities of the East India Company were terminated.

From this analysis, it is quite apparent that the end of traditional manufacturing came along with the territorial expansion of the East India Company. This happened because the company used its political power to marginalize the role of Indian merchants, which broke the nexus between merchants and craftsmen. However, it seems that the speed of the process of de-industrialization increased after c.1813 with the end of monopoly of trade given to the East India Company. For the sake of simplicity, we can take c.1813 as a crucial year when British manufacturers began to influence government policies.

4.3 Debate on De-industrialization

As described in Chap. 2, until c.1750, the Indian sub-continent was producing about 25 per cent of the manufacturing output of the world. This proportion declined significantly by c.1900s when its share was just 5 per cent. This significant decline in the output of manufacturing sector of India is described as de-industrialization. The fall in the output of the manufacturing sector also resulted in decline in the contribution of the manufacturing sector to the national income as well as employment. The notion that the Indian sub-continent experienced de-industrialization is an important event in the story of Indian manufacturing.

The episode also symbolizes an important period in the story of Indian manufacturing when Indian skilled craftsmen experienced mass de-skilling and were forced to work on land. The skills, which they had mastered over centuries to make the best manufactured products of the world, suddenly became irrelevant because they went back to work in the primary sector. This process was contrary to the experience of British manufacturing where traditional manufacturing was replaced by modern manufacturing. In the case of India, traditional manufacturing declined without being replaced by modern manufacturing. Therefore, in the context of the story of Indian manufacturing, de-industrialization also means a movement of the economy one stage backward rather than one stage forward. The decline in traditional flexible manufacturing was not just a loss in the income and employment in the manufacturing sector; it was a loss of potential source of transformation of traditional manufacturing into modern manufacturing. If the process of de-industrialization had occurred due to replacement of traditional flexible manufacturing by modern manufacturing, as had happened in the case of Britain, it would not have been cause for concern. This happened in India because the decline in manufacturing was not accompanied by the birth of modern manufacturing. On the contrary, it was believed that the birth of modern manufacturing was deliberately postponed or arrested.

Due to the aforementioned reasons, the episode of de-industrialization assumes greater importance in the biography of Indian manufacturing. This event has occupied an important place in the writings of nationalist as well as Marxian economists and historians. Nationalist scholars evoked it as the negative consequence of foreign rule. Therefore, achieving independence from foreign rulers was seen as an important solution to the problem. Marxian scholars, on the other hand, used this episode to paint a darker picture of capitalism, because de-industrialization in India emerged as a consequence of imperialism. Marxians believed that imperialism is a higher stage of capitalism.

The gravity of the process of de-industrialization experienced by the manufacturing sector of the Indian sub-continent was revealed for the first time by William Bentinck who was governor general of India during the period c.1833–1835. He mentioned, in great detail, the consequences of the imported mill cloth of British textile mills on the decline of Indian cotton textile manufacturing. His views were quoted by Karl Marx in his book when he mentions 'the bones of cotton-weavers are bleaching the plains of India'.[58] He mentions again the process of de-industrialization experienced by the manufacturing sector of the Indian sub-continent in his article dated June 25, 1853 entitled 'British Rule in India', which appeared in the *New York Daily Tribune*. In his words, 'It was the British intruder who broke up the Indian handloom and destroyed the spinning wheel. England began with driving the Indian cotton from the European market, it than introduced twist into Hindustan and in the end inundated the very mother country of cotton with cotton'. He further adds, 'from c.1818 to c.1836, the exports of twist form Great Britain to India rose in proportion to 1: 5200' and this resulted in 'British steam and science uprooted, over the whole surface of the Hindoostan, the union between agriculture and manufacturing'.[59]

Marx, though, mentions the process of de-industrialization, but accepts this event as a necessary jolt that the Indian society and economy needed. This is because he viewed de-industrialization as an important push towards a social revolution. Marx believed that the Indian economy was operating under the conditions he mentioned while describing the Asiatic mode of production; these conditions had made the economy self-sustained but stagnant, a form of low-level equilibrium trap. He writes that in India 'family communities were based on domestic industry in peculiar combination of hand weaving, handspinning and hand tilling agriculture, which gave them self-supporting power ... English interference dissolved these semi-barbarian, semi-civilized communities, by blowing their economical basis, and thus produced the greatest and, to speak the truth, the only social revolution ever heard of in Asia'. Thus he glorified the historical role of de-industrialization—'If not, whatever may have been the crimes of England, she was the unconscious tool of history in bringing about a revolution, and concludes the article, by saying "whatever bitterness the spectacle of the crumbling of an ancient world may have for our personal feelings, we have a right, in point of history to exclaim'.[60]

Unlike Marx, who viewed de-industrialization as a big push to a stagnant society, it became quite an important issue in the narrative of nationalist scholars, who used it as an important example of the failure of the British rule. According to them, British manufacturing interest forced the British government to expose the traditional flexible manufacturing of India, without providing any protection against the products of modern manufacturing of Britain. It is the lack of protection to native manufacturing that had caused de-industrialization. The nationalist scholars did not accept the argument that it was the improvement in technology, which British manufacturing has achieved, that provided it competitive advantage over handicraft-based methods of production used by Indian traditional flexible manufacturing.

The nationalist scholars argued that de-industrialization emerged as a consequence of lack of protection given to India manufacturing. The argument implicitly assumes that, if at that point in time, India was ruled by a national government that had provided protection to Indian manufacturing, de-industrialization would not have occurred. This line of thinking was necessary to add fuel to the desire of achieving political independence. It is for this reason that the episode of de-industrialization appears again and again in the writings of nationalist scholars like Dadabhai Naroji in c.1901,[61] and R.C. Dutt in 1906.[62] In the writing of Gandhi, it is almost a running theme. Finally, it appeared in the book written by Nehru.[63] In fact, one of the casualties of the process of de-industrialization, *charkha* (the spinning wheel), became an important symbol for the Gandhi-led independence movement in India.

Despite the fact that the process of de-industrialization had assumed significance in the imagination of the scholars, representing ideologically all colours of the rainbow, it did not result in the settling of some of the important issues related to the debate. The first contentious issue is when the process of de-industrialization began. The knowledge of the timing of this event is of great importance, because the timing of this historical event will tell us who culprits of this crime were. Regarding the timing of the process of de-industrialization that was spread between c.1750 and c.1860, historians have identified two different sub-periods. According to some scholars, it was the decline of the Mughal Empire that caused de-industrialization. The other set of historians believe that de-industrialization occurred because of penetration of cheaper products produced by British mills.

The second important issue on which there is no consensus amongst scholars is regarding the factors that caused de-industrialization—whether the process was caused by factors that were endogenous to the economic structure of the economy or exogenous to the economy. Scholars who lay stress on endogenous factors believe that de-industrialization was inevitable because the factors that caused de-industrialization were embedded in the structure of the economy. On the other hand, scholars who emphasize the role of exogenous factors argue that it emerged as a consequence of economic interference of the East India Company and vested interests of the British Raj.

The contemporary debate on de-industrialization began after the availability of data reported in the population census, national income estimates and occupational structure of the working population after c.1870s. The earliest empirical evidence regarding the occurrence of de-industrialization was reported by Colin Clark. He used data reported in the census reports of c.1881–1911. His analysis established that the proportion of population engaged in manufacturing, mining and construction declined from around 28 per cent to 12 per cent.[64] Daniel Thorner challenged his estimates and used revised data on census from c.1881 to c.1931. His estimates show that sectoral composition of workforce engaged in manufacturing registered insignificant change between c.1881 and c.1901 and it remained stationary between the period c.1901–1931. According to his estimates, there is no evidence that de-industrialization in India occurred after c.1881. Therefore, he suggests that if at all de-industrialization occurred in India, it must have happened prior to c.1881.[65]

Harnetty's observations regarding handloom weaving show that the handloom weaving, which supported the entire domestic demand, along with exports, peaked in value terms in c.1800 and in terms of volume in c.1802. Afterwards, the exports of textiles to Britain declined due to emerging competition from British mills.[66] Simmons used data reported by Paul Bairoch regarding the process of de-industrialization experienced by economies that were on the periphery of the European economies. According to him, the Indian sub-continent in c.1750 accounted for 25 per cent of the global output in manufacturing. By c.1800, its share declined to 20 per cent, in c.1860 to around 10 per cent and by c.1880 it was less than 3 per cent. However, critics of these estimates argue that decline in the global share of Indian manufacturing might have occurred not necessarily due to decline in the rates of growth of manufacturing in India, but perhaps due to faster rates of growth in manufacturing experienced by other countries. In order to overcome this criticism, Bairoch also provides estimates on per capita level of industrialization in India. The index of per capita level of industrialization declined from 7 in c.1750 to 6 in c.1830 and to 2 in c.1880. This suggests that there was no dramatic de-industrialization between c.1750 and c.1830. The de-industrialization is more pronounced after c.1830, when mill-made products from Britain began to enter the Indian market.[67]

Krishnamurty, who studied the problem of de-industrialization with the help of shifts in the occupational pattern of the population for the period c.1881–1951, found that the occupational pattern was almost stationary. He did not observe significant changes in the occupational pattern of the population, with the share of primary sector remaining at about 70 per cent, that of manufacturing around 10 per cent and services accounting for about

15–20 per cent. His exercise also suggests that if there had been any event of de-industrialization in India, it must have occurred prior in c.1881.[68]

The empirical exercise conducted by Clingingsmith and Williamson, covering the period between c.1750 and c.1913, shows that the Indian sub-continent underwent the process of de-industrialization during the two sub-periods between the eighteenth and nineteenth centuries. The first phase of de-industrialization occurred between c.1750 and c.1810, resulting from the decline of the Mughal Empire. The second epoch covers the period between c.1810 and c.1860, resulting from the competition from the factory system of production in Britain. They suggest that India had a century-long, two-part transition from being net exporter to net importer of textiles. According to them, the process of de-industrialization was complete by c.1860. It is perhaps due to this reason that almost all the studies that cover the period after c.1860 do not find any evidence of de-industrialization.[69]

According to Roy, if there existed reliable data on the occupational pattern of Indian population for the period c.1750 or c.1800, the proportion of workforce engaged in manufacturing would have been around 25–30 per cent, which declined to about 8–9 per cent by c.1900. This drastic fall in proportion of persons engaged in manufacturing shows de-industrialization or peasantization of the economy. However, he suggests that acceptance of this fall as de-industrialization will amount to 'exaggeration of reality'.[70] This is because during the early colonial period, manufacturing used abundant supply of slack labour time.[71]

It is evident from the studies conducted on timing of the process of de-industrialization, that the studies that cover the period after c.1860 did not find any evidence of de-industrialization. Therefore, many scholars have suggested that de-industrialization might have occurred much earlier.[72] It is precisely for this reason that we have included the debate on de-industrialization during the period when the East India Company ruled the Indian sub-continent (i.e., between c.1757 and c.1857).

Scholars who believe that de-industrialization occurred prior to c.1860 attribute it to the collapse of the Mughal Empire. The empire provided political stability, integrated market, single currency, single economic environment, and consumption of traditional product produced by Indian traditional flexible manufacturing. Therefore, decline of the Mughal Empire led to the process of de-industrialization. (This process has been described in greater detail in Chap. 2.) I also concurred with this conclusion, as it is clear from my paper on 'Primitive Flexible Manufacturing and Mughal

Empire',[73] 'it would not be inappropriate to state, that the dawn and the dusk of primitive flexible manufacturing is intimately linked with the rise and fall of the Mughal empire'.[74] However, studies conducted by scholars on specific regions show that the preconditions that are necessary for the continuity of traditional flexible manufacturing continued to exist in some of the successor states of the Mughal Empire, like Awadh, Hyderabad, Maharashtra, Punjab, Rajasthan and Mysore.

Our understanding about the timing of the process of de-industrialization is based on the organizational structure of the traditional flexible manufacturing. Most scholars who have participated in the debate regarding timing of the process of de-industrialization have ignored that traditional flexible manufacturing, which catered to the needs of mass markets, including exports, survived due to the nexus between merchants and craftsmen, which was working through the *dadni* or putting-out system. In this form of organization of production of manufactured items, merchants played a strategic role. This nexus between merchants and artisans did not break after the decline of the Mughal Empire. Instead, it continued to exist in successor states. However, little understanding of history clearly shows that the nexus between merchants and craftsmen broke down with the emergence of the East India Company as a ruler of the sub-continent. Therefore, according to our understanding, de-industrialization occurred during c.1757–1857.

It was the emergence of the East India Company as a political entity that encouraged it to monopolize trade in the manufactured products. In the process of monopolization of trade, it systematically followed policies and strategies to marginalize the role of the Indian merchant in the organization of production of Indian handicrafts. This process has already been described in greater detail in an earlier section. Hence, we will associate the beginning of de-industrialization with the territorial occupation of the Indian sub-continent by the East India Company. Due to this reason, spatial variation in the timing of de-industrialization also varies, because different territories of the Indian sub-continent were politically occupied by the company at different points in time.

The process of de-industrialization began with the occupation of Bengal by the East India Company after the Battle of Plassey in c.1757. Therefore, c.1757–1800 could be the period when Bengal experienced de-industrialization. It is also a historical fact that during this period the most prosperous province of the Mughal Empire became the poorest region of the Indian sub-continent and experienced a famine in c.1770. Subsequently, the spatial spread of de-industrialization became related to the time when different spatial units of the Indian sub-continent were politically occupied by the company. However, certain regions remained under native rulers like Hyderabad, Mysore, Punjab, Rajasthan and Kashmir. In these regions, traditional flexible manufacturing continued to exist, especially in certain manufactured products that were catering to *niche* domestic markets. Craftsmen who lost the support of merchants for marketing their products in mass markets, and also in terms of receiving working capital needs, either started working as wage labour in the factories organized by the company, or worked at reasonably disadvantaged contracts for the *gomastas* employed by the company, or had to give up their craft.

Massive decline was also experienced by the traditional flexible manufacturing that was engaged in the production of luxury handicrafts. The emergence of the East India Company as a political power eliminated the hierarchy of *Mansabdars* (nobles and aristocrats) of the Mughal period and replaced them by salaried administrators of the company. This shift in the structure of elite led to a decline in the demand for luxury handicrafts. It is reported that 'perhaps three quarter of the domestic demand of luxury handicrafts, with magnitude of about 5 per cent of the Mughal national income, disappeared'.[75]

Along with the decline of domestic and export demand of the products of traditional flexible manufacturing, Indian handicrafts faced competition from machine-made goods. Some anthropologists have raised a very pertinent question regarding the increasing demand of machine made goods in India—'How, then, do culturally determined patterns of consumption change to accommodate alien manufacturers in a society in which the market itself is not powerful enough to create taste'.[76,77]

It is suggested that the period after c.1750, when English machine-made goods began to penetrate the Indian markets, coincided with the period of fragmentation of the Mughal Empire and emergence of new regional elites. These new regional political entities were trying to assert their identities. They intended to differentiate themselves from the Mughal aristocracy; to these elites, the British manufactured products provided excellent substitutes. This was reflected in the differentiated consumption pattern of regional elites, who gained power in the successor states of the Mughal Empire.

The consumption pattern of the new power elite also was manifested when they constructed their new palaces. They decorated their new residences with European clocks, curtains, mirrors, art wares and other novelty items. Lucknow became an important place for the demand of British glass wares. Huge quantities of sheet glass from Britain was imported, which was used in the construction of *chota* and *Bara Imambara*.[78] The use of British products by new power elites can be seen in the items displayed in the Salar Jung Museum of Hyderabad, and other museums showing the collection of rulers of Rajasthan, Punjab and Awadh. One can say with certain amount of brevity that the consumption pattern of new elites was transformed from being socially embedded to *commodity fetishism,* which is based on the premise that the value of a product lies in itself.

The other powerful influence, which reduced the demand for the products of traditional flexible manufacturing, was that of European freebooters, who were appointed by the local rulers to train and modernize their armies. These army advisors and trainers began to dress their armies in scarlet English broadcloth. Under their influence, local rulers adopted British textiles to prepare uniforms for their armies. Thus, the demand for British textiles received initial support from these local rulers, According to Faiz Baksh, who recorded historical facts of the period, 'when eunuch administrator of Lucknow Jawahir Ali Khan's personal soldiers (*sabit-khanis*) were seen in green livery, irregulars in black, and Mewati soldiers in white, it made other rich people envious'.[74]

The most important weakness of the debate on de-industrialization is that it has remained focused on associating decline in textile manufacturing in the Indian sub-continent with de-industrialization. It may be because textile manufacturing was the most important manufacturing industry of that period. However, in the process, they have ignored other manufacturing industries. Perhaps these manufacturing industries might have undergone different experiences. It seems that the flexible manufacturing that was engaged in the production of domestic utensils, metal wares, agricultural implements, and silver and gold jewellery might have experienced an upward trend because these products were not experiencing any competition from British factory-made products. Similarly, manufacturing industries like hide and skin, leather and leather products might have also experienced positive growth rates because demand for leather products by the army of the East India Company and local rulers for saddles, breaches and shoes must have increased substantially. The demand for shoes and footwear might have also increased due to westernization of dress code of emerging new professions.

Until the beginning of the Industrial Revolution, trade was the foundation on which the entire structure of the British economy depended. Therefore, for a considerable period, there existed symbiosis between the interest of the East India Company and the dominant sections of the British society. The emergence of modern manufacturing in Britain led to the creation of a new social structure, where the interests of manufacturers became more important. These were at variance with the interests of the landed gentry and merchants. The new social order replaced the supremacy of mercantilism, the ideology that supported the interests of traders and the company, with classical political economy, which articulated the interests of manufacturers. The shift in the ideological premise of the British social order reduced the historical relevance of the East India Company. The consequences of this ideological shift was that the East India Company ceased to be a trading company in c.1833 and as an agency of British imperialism in c.1857. In the section that follows, I narrate the emerging contradiction between the East India Company and new social order that was emerging in Britain.

4.4 Contradiction Between the East India Company and New Social Order

The shift in the ideological supremacy of classical political economy to mercantilism emerged as a consequence of emerging contradiction between the interests of merchants and industrialists. In the new circumstances, the antagonists of the East India Company included a British landed gentry, which was losing economic and social space to the wealthy ex-employees of the East India Company. These employees, on returning from India, were buying their estates and were also competing for the women belonging to the same strata of society for marriage. These ex-employees were also entering the British Parliament, changing the social and economic composition of the members of the parliament. This, is in turn, was altering the nature of debates and issues, which were being taken up in the Parliament.

The more articulate and formidable antagonist of the East India Company was emerging as a new class with a new ideology. This new class was emerging as a consequence of transformations, which British society and economy was experiencing after the invention of the steam engine. The emergence of steam power as an inanimate source of power was transforming the technology being used in British manufacturing and means of transportation. These new inventions in technology altered the British traditional manufacturing into modern manufacturing, which has been described by historians as the Industrial Revolution. The process of transformation of British traditional flexible manufacturing into modern industry has been analysed in greater details in Chap. 3 of the present study.

Before the emergence of the Industrial Revolution, foreign trade was the foundation on which the structure of the British economy depended. Therefore, interests of the East India Company were the same as those of the dominant sections of the British society. However, the transformation of the British flexible manufacturing into modern manufacturing resulted in a new social order, which was manifested in the emerging dominance of interest of manufacturers. Their interests were at variance with the interests of the landed gentry and merchants. The emergence of a new class of manufacturers gave birth to new ideology and economic laws, which were articulated by Adam Smith in his famous book, *Wealth of Nations*. The new ideology evolved into a new school of thought known as classical political economy. The new thought began to challenge the premises on which mercantilist ideas were based. Mercantilism provided historical justification for the existence of companies like the East India Company, in particular, and social dominance of merchants in general. The earlier schools of thought like physiocracy considered manufacturing activities as 'sterile' economic activity. This school of thought was evolved to protect the interest of capitalist farmers (i.e., peasants).[75] While mercantilist school protected and propagated the interests of merchants, for the first time, manufacturing was accepted as a productive economic activity by Adam Smith and other classical economists.

It was during the unfolding of the age of discovery that the commercial horizons of the economies were vastly extended and economic intercourse of Europe, which was restricted to Europe and part of the Mediterranean, expanded thousands of miles across the oceans of Asia, North and South America, Australia and Africa. The connectivity provided access to new markets and interaction between various centres of trade and crafts through long-distance trade. It was trade in Asian commodities like spices, cotton and silk textiles, tea and coffee that expanded the volume of trade. European demand for these products was immense and Europeans had to pay for these commodities in bullion. The bullion that financed Europeans' desire for these products came essentially from America. This process led to the emergence of a trade triangle, in which gold and silver from America reached Asia through European intermediaries, in exchange for Asian products.

These historical conditions provided preconditions for the existence of trading companies like the East India Company. At the same time, the increasing contribution of trade to the prosperity of the European economies provided historical conditions for the birth of mercantilism. Since expansion of trade at that time was constrained by the availability of bullion, the followers of this school believed that it is the stock of gold in the possession of an economy that is the real wealth. The views of this school of thought are spread in the writings of pamphleteers like Thomas Mun (c.1571–1596) and Josiah Child (c.1630–1699) in England, and Antoine de Montchrétien (c.1576–1621) and Jean Bodin (c.1530–1536) in France.

Following the spread of the influence of this school of thought, gold became the obsession of Europeans. It is a well-known historical fact that their discoveries of new lands were primarily motivated by the stores of large volumes of gold and silver in the possession of people inhabiting these lands. Christopher Columbus, who opened the route to America for Europeans, wrote 'Gold is wonderful thing. Whoever owns it is the lord of all wants. With gold, it is even possible to open the soul of the way to paradise'.[79] There also exist historical accounts regarding the life of Atawallpa, the ruler of Peru, who was held hostage by the Spanish conquistador, Castilian, Francises Pizarrow, and was killed despite paying the ransom of a roomful of gold and two roomfuls of silver.[80] The lust for gold amongst Europeans, at that time, was quite surprising to the people of other civilizations, where gold and silver had only ritualistic or iconic value. For instance, 'for Peruvians gold was sweat of sun and silver the moon's tears.'[81]

The glitter of gold was so overbearing for the Europeans during this period that they undermined the importance of all other economic activities that did not help in increasing the stock of gold in an economy. Therefore, there emerged an understanding that economic activities like trade and shipping are most important because they help in earning gold. They also accepted the role of nation state, which helped in increasing the stock of gold by manipulating terms of trade by imposing protective tariffs. These historical facts became the essential premises of the mercantilist philosophy.[82]

The term mercantilism was used for the first time by one of the disciples of Qusney Mirabeau.[83] There are several versions of mercantilism that can be traced to different pamphlets published by different authors who enriched the contents of the school of thought. It is interesting to mention that some of the important British thinkers who contributed to the development of the idea of mercantilism, like Misselden, Mun and Child, were associated with the East India Company; the former two occupied the position of directors of the company. In common parlance, mercantilism stands for policies that promote favourable balance of payment by encouraging exports and discouraging imports, with the intention of earning bullion for the country. In the writing of some of the scholars of mercantilism, it is a system of thought that was focusing on nation building by transforming the polity and economy of middle ages into a system of nation state.[84] However, ideas on mercantilism presented by Thomas Mun in his two important pamphlets entitled *England's Treasure by Foreign Trade*,[85] and *Discourse of Trade from England into East Indies*, presented theoretically more profound versions of mercantilism. His presentation on mercantilism was accepted by Adam Smith as an important description of mercantilism. Subsequently, most of the arguments of Smith against mercantilism were based on the ideas articulated by Mun. The economic logic that Thomas Mun evolved to justify the preeminence of foreign trade as an instrument to enrich an economy was based on the premise that a nation cannot become rich just by getting gifts or by purchase taken from some other nation. The only solution to increase wealth of a nation is foreign trade 'where in we must ever observe this rule, to sell more to strangers yearly than we consume of their value'.[86] He further adds that as an individual enriches himself by not spending as much as he gains, the same maxim should be used by a country. He also suggested that natural resources should be developed to reduce dependency on imports of resources, and also advocated for the protection of local manufacturing enterprises. He also advocated the use of tariffs to increase trade surplus (i.e., the stock of bullion).

If one followed the norms mentioned by Mun for creating national wealth, then logically it was difficult to justify the massive exports of bullion by the East India Company to India to buy merchandise. Many of the contemporary scholars of the time considered this act of the company as a drain of wealth from Britain. However, Mun justified the export of bullion by the East India Company on the ground that though the company sends bullion to India, it re-exports the merchandise to other countries on profit, which increases the possession of bullion of Britain. Therefore he explained how the export of bullion was advantageous to the country. He also suggested that enlightened citizens of the country should view the entire process of transactions rather than limit their vision to initial transactions between Britain and India. He justified the export of bullion to India by telling that when people view export of bullion to India in isolation 'which misinforms their judgement and leads them to error, For, if we only behold the action of a husbandman in the seed time, when he casteths away good corn into ground, we will rather accompt him as a madman rather than husbandman; but when we consider his labour in *harvest time*, which in end of his endeavour, we find the worth of his actions'.[87]

It was following the mercantilist policies that the British government provided chartered status to several companies, including the East India Company, to enjoy monopoly of trade with the East, enacted English Navigation Act in c.1651 to protect British shipping from Dutch competition, and Corn Laws in c.1689 to regulate the trade of food grains. The objective of the Corn Law was to encourage British farmers to continue to grow corn by guaranteeing them protection from cheaper imports and allowing them to export when there was glut in the market. For the first time, the ideological premises of mercantilism, on which the foundation of the East India Company rested, were challenged by Smith in his book. His book was a manifestation of the socio-economic transformation of the British economy from the dominance of trade to dominance of manufacturing in the process of creating wealth.[88]

Smith explained that the belief that wealth of a nation consists of gold and silver is based on the premise that they play dual role in the functioning of an economy (i) as a facilitator of transactions, and (ii) as a measure of value. If we view gold and silver as facilitators of transactions, then having more of them will help in buying more of other commodities. While viewing them as a store of value, it becomes quite obvious that having more of gold and silver in possession provides more claim on other commodities. It is because of its dual role that the Mercantilist school justified the possession of gold. Mun justified the exportation of bullion to buy foreign goods on the ground that if consumption of foreign goods does not increase in the country, these goods are re-exported on higher profits, which helps in increasing the treasure, than if it was spent on purchase. Smith writes that it appears a solid and sophisticated argument that export of bullion for trade may be advantageous for the country, but the argument was more to draw the attention of the government to preserve quantity of metals, and ignores 'how preserving and augmenting the quantity of any other useful commodity, which freedom of trade without any such attention never fails to supply in proper quantity'.[89] He emphasizes that these arguments were given by the traders to the Parliament that trade enriched the country. Merchants know that this process enriches them, and in what way it enriches the country they do not explain, because it was not part of their business.

Smith provides an entirely a new role to foreign trade in comparison to the role provided to it by the mercantilists. We are presenting here his view in his own language to put to the readers not only his ideas, but also the words that he uses to present his argument with force.

> Between whatever places foreign trade is carried on, the all of them derive two distinct benefits from it. It carries out that surplus part of the produce of their land for which there is no demand among them, and buys back in return for it something else for which there is demand. By means of it the narrowness of the home market does not hinder the division of labour in any particular branch of art or manufactures from being carried to the highest perfection. It encourages them to improve its productive power and augment its annual produce to the utmost, and there by increases ***real revenue and wealth of the society*** (emphasis added).[90]

To put new perspective on the role of foreign trade, he gives the example of trade between Britain and Asia—'the discovery of a passage to the East Indies opened perhaps a still more extensive range of foreign commerce than that of America, notwithstanding the greater distance the rest were mere savages. But the empires of China, Indostan, Japan as well as several other in the East Indies, without having rich mines of gold and silver, were in every other respect much richer, better cultivated and more advanced in arts and manufactures rich and civilized nations can always exchange to a much greater value with one another'.[91] However, he stressed that due to the influence of mercantilism, Britain and other European economies have not been able to maximize the benefits from foreign trade with the East because of the existence of monopoly of trade given to companies like the East India Company. He concluded that 'no great nation in Europe had ever yet had benefited from free commerce'.[92]

Smith systematically demolished the premises on which the old social order dominated by merchants was based. Smith's criticism of merchants is well known. He writes in his book, 'the capital of merchants, on the contrary, seems to have no fixed or necessary residence anywhere, but may wander about from place to place according as it can either buy cheap or sell dear'. He makes an important comment—"A merchant, it has been said, very properly, is not necessarily citizen of any particular country. It is in great measure indifferent to him from what place he carries on his trade, a very trifling disgust will make him remove his capital. No part of it can be said to belong to any particular country till it has been spread, as it were, over the face of that country". He makes this point more emphatically, when he states—"capital……that is acquired by commerce…… is very precarious and uncertain possession until some part of it has been secured and realized in the civilization and improvement of land."[93]

The aforementioned ideas of Smith formed the premises on which his arguments against the East India Company in particular, and against merchants in general, were based. He did not approve of the founding of a great empire by the East India Company based on mercantilist principles. He remarked that this kind of empire was a project 'fit for nations whose government is influenced by shopkeepers'.[94] The consequences of the new ideology on the East India Company, articulated by the classical political economy, under the leadership of Smith, were that the British parliament passed Pitt's Act of c.1784, which increased parliamentary control over The East India Company. Subsequently, the Charter Act of c.1813 ended the trading monopoly of the East India Company. Finally, in c.1833, the East India Company was stripped of its commercial functions, but was allowed to remain the governing agency of the British empire in India. Hence, the period during which India experienced territorial expansion of the East India Company coincided with the rise and subsequent dominance of manufacturers in British society. This resulted in structural change in the British social order. In this strange way, the Indian economy became entangled in the basic conflict between merchants and manufacturers that was taking place in Britain. This conflict came to an end with the end of the rule of the East India Company in c.1857, which also signified the end of mercantilism in Britain. For the people living in the Indian sub-continent it was the beginning of the British Raj.

In this chapter an attempt has been made to study the consequences of the second force of globalization, the European trading companies, in the story of Indian manufacturing. We have analysed the process of transformation of the East India Company from a trading company to an agency of the British Empire. The process has been explained with the help of different historical events that helped the East India Company to fill the political space that was left empty by the decline of the Mughal Empire.

The emerging political dominance of the company, after the Battle of Plassey, for the first time provided opportunity to the East India Company to consolidate its monopoly of trade. The symbiotic use of these twin monopolies facilitated the process whereby the East India Company was able to break the nexus that existed between merchants and craftsmen, the nexus that had facilitated the birth of traditional flexible manufacturing. Through the use of its political power, it was able to marginalize the role of Indian merchants in the process of traditional flexible manufacturing. It simultaneously also evolved policies that adversely affected the bargaining power of craftsmen. These new circumstances broke the nexus between the merchants and craftsmen, which was the basis of organization of production in traditional flexible manufacturing in India. This nexus worked through the *dadni* or putting-out system. Once divorced from the support of Indian merchants, Indian artisans either had the option to become wage workers in the centres of production organized by the East India Company or work on highly disadvantaged contracts with the *gomasthas* of the company. If craftsmen did not want to exercise these alternative options, the third option was to give up their craft and work in the agricultural sector.

It is due to the strategies used by the company that created separation between Indian merchants and craftsmen that I have suggested that the beginning of the process of de-industrialization should be dated with the political occupation of India by the East India Company. It is the territorial expansion of the company that created historical conditions for the decline of traditional flexible manufacturing, which had provided the organizational structure where Indian craftsmen could get access to mass markets including exports. Although crafts could exist in other organizational structure, also, craftsmen could not supply their products in distant markets. It is also explained in this chapter that the new power elites in different regions in India after the decline of the Mughal Empire also facilitated the process of de-industrialization. These new regional elites adopted Western goods in their consumption pattern, to differentiate themselves from the Mughal elites. By adopting Western products, they reduced the demand for luxury handicrafts. The final blow to Indian handicrafts was dealt by the potent force of machine-made goods produced by British modern manufacturing, which replaced the products produced by traditional handicrafts. Despite these forces at work, Indian handicrafts continued to exist in regions where local rulers continued to rule. These crafts also remained operational in the products that satisfied *niche* markets. Moreover, the emergence of integrated markets with single currency and similar regulations provided opportunities for the growth of those manufactured products that did not experience competition from British manufactured products like metallic domestic utensils, agricultural implements, shoes and footwear, silver and gold jewellery, and so on.

The chapter also provided an analysis of changes that were occurring in Britain, which were facilitating the emergence of new ideology based on classical political economy. The new ideology was challenging the existence of companies like the East India Company and was making them totally irrelevant in the new social order. In the chapter that follows, this process is described in greater detail. The chapter deals with how organizational structure of the East India Company provided conditions for the opportunistic behaviour on the part of management and employees of the company and resulted in several kinds of compromises with the quality of corporate governance of the company. The acts of omission and commission performed by the management also created hostility between the company and the civil society of Britain. It was the changing attitude of the civil society towards the East India Company that forced the British Parliament to intervene in the functioning of the company. This aspect of the rule of the company will be discussed in detail in the next chapter.

Notes

1. The process that transformed the East Indian Company from conducting business to conducting business of empire has been explained by Bowen (2006).
2. This term has been used by Travers (2007).
3. This particular representation of the rule of the company is given in Stern (2004).
4. The terms 'merchant sovereign' or 'sovereign merchant' have also been used by Travers (2007).
5. The term 'corporate state' has been used by Robins (2006).
6. For the understanding regarding different spices and plants that influenced the course of history, see Hobhouse (1999).
7. To know the nature, extent and implication of trade with Portuguese in India, see Whiteway (1893), Hall (1996) and Fruber (2004).
8. To know the story of the Dutch East India Company, see Om Prakash (1985) and Pearson (2007).
9. For a detailed knowledge about the operation of the French company, see Malleson (1901).
10. For the entry of the East India Company in Surat.
11. Balpours (1885).
12. For the occupation of Bombay by the East India Company and the subsequent evolution of Bombay, see Malabari (1910).
13. To know in details the history of the East India Company in Bengal, see Travers (2007).
14. For the understanding of the process of globalization of conflict amongst different European economies, see Travers (2007).
15. For the understanding regarding the transition of the world order towards *imperialism*, see Bayly (1988).

16. The scholars, who believed that there emerged 'dark ages' after the decline of the authority of the Mughal Empire, explained the process from different perspectives. Sarkar (2014) explains the political social and religious decline of the Mughal Empire. Some other scholars explained the decline of Mughal Empire due to fiscal constraint or crises in the Jagirdari system; see Chandra (1998), Ali (1975) and Karen (1979).
17. For economic and social consequences of the decline of Mughal Empire on different regions, see the study of Wink (1986); for Maharashtra, Stein (1985); for Mysore, Grewal (1995); and for Punjab, Banga (1978) and Marshall (1975a).
18. The term *gentrification* of the society for the Indian economy has been used by Bayly (1983) and for the process of gentrification of Bengal, see Om Prakash (1998).
19. The concept of 'Portfoliocapitalists' has been used by Bayly and Subrahmanyan (1988).
20. For details regarding the origin of 'tributary commercialism' in Mysore during the period of Hyder Ali and Tipu Sultan, see Ludder (1996).
21. For details regarding the war of successions, see Bayly (1988).
22. The reasons that led to the participation of European trading companies in the local conflicts have been discussed in Stein (1998).
23. For the economic implication of the *firman* issued by Mughal Emperor Farrukhsiyar and the use of *dastak* by the East India Company, see Robins (2006), pp. 64–65.
24. The Battle of Plassey was quite an important business deal for the East India Company. Just by giving Mir Jafar, the commander of the forces of Nawab of Bengal Shiraj-ud-Daula, promise to appoint Nawab of Bengal. It is recorded that 'the company loaded the treasury of gold and silver on a fleet consisting of around hundred boats was sent to Calcutta. In this business deal Clive has earned £ 2.5 million for the company and £ 2,34,000 *for him*self'; see Robins (2006), p. 3.
25. The Imperial decree, which provided Diwani right to the East India Company, by Mughal Emprie Shah Allam II, provided monopoly of political power as well as trading monopoly to the East India Company. This status of the East Indian Company violated its own relationship with respect to British Crown and Parliament; see Dirk (2006), p. XIII.
26. The East India Company was popularly called the John Company; see, Furber (1970).
27. Details regarding the occupation of different centres of traditional flexible manufacturing have been obtained from Powell (1892).
28. See Robins (2006), p. 76.
29. See Burke (1783 [1981]), vol. V.
30. See Burke (1783 [1981]), vol. V, p. 224.

31. See Robin (2006), pp. 2–5 and Ramaswamy (2011), pp. 76–78.
32. See Ramaswamy (2011), Ch. IV and Grioseclose (1993), pp. 50–53.
33. For the ideas on economic drain of persons described in the text, see Bolt (1772 [1998]), Dow (1792) and Verlest (1776).
34. See Dutt (1906), pp. 29–31.
35. For the estimates on economic drain made by Habib, see Habib (1990).
36. Hunder (1868) and Hunter (1874).
37. See Marshall (1975b), p. 466.
38. See Om Prakash (2007), p. 151.
39. See Robinson (1912), p. 70 and Sinha (1968), vol. III.
40. See Robins (2006).
41. See Singh (1966) and Hodgson (1938).
42. See Tripathi (2004).
43. See Bhattacharya (1982), p. 273.
44. See Bhattacharya (1982), p. 288.
45. For details regarding the organization of factories by the East India Company and conditions of craftsmen working in these factories, see Hussain (1958).
46. For these facts, see Parthasarathi (2001), p. 84.
47. Dodwell (1922) p. 2 and also quoted in the Cambridge Economics History of India, p. 369.
48. See Bolt (1772 [1998]), p. 74.
49. See Robins (2006), p. 60.
50. Bhattacharya (1982).
51. Bhattacharya (1966).
52. See Arasaratnam (1980) and Parthasarathi (2001).
53. Chaudhary (1972), pp. 81–83.
54. See Om Prakash (2007), pp. 133–168; Roy (2010), pp. 135–139 and Roy (2005), p. 345.
55. See Dewey (1988), p. 5.
56. See Markovits (2008) and Subramanian (2012), p. 5.
57. See Sinha (1968), vol. I p. 360, Hunter (1874), vol. IX, pp. 252–268 and Seth (1987), pp. 25–26.
58. See Marx (1868), vol. I, p. 275.
59. See Marx (1813), in Iqbal Hussain (ed) (2006), p. 14.
60. See Marx (1853a, b, c, 2006; pp. 16–17).
61. See for the ideas of Dadabhai Naoroji, Naoroji (1901/1962).
62. See Dutt (1906).
63. See Nehru (1947).
64. See Clark (1950).
65. See Thormer (1962).
66. Harnetty (1991).

67. See Simmons (1985) and Bairoch (1982).
68. Krishnamurthy (1983), p. 548.
69. See Clingingsmith and Williamson (2004), pp. 20–21.
70. See Roy (2005), p. 118.
71. See Roy (2005), p. 118.
72. See Roy (2005), Markovits (2000), Sarkar (2014) and Clingingsmith and Williamson (2004).
73. See Seth (2008).
74. See Seth (2008), p. 92.
75. See Maddison (1971), pp.34–35, Maddison (2003), Deavy (1975) and Gadgil (1924).
76. See Appadurai (ed) (1986), p. 303.
77. See Faiz Baksh (1889), ii p. 97, quoted in Appadurai (1986), p. 304.
78. Physiocratic tradition took roots in France in the seventeenth and eighteenth centuries in the writings of Francois Quesney (1694–1774), Mercier de la Riviere (1770–1793) and Du Pont de Nemours (1739–1817). In order to establish economic importance of land-owners, Physiocrats accepted agriculture as the only productive activity because it had the capacity to produce *economic surplus* over and above the cost of capital. Therefore, non-agricultural economic activities were described as *sterile activities*; see Meek (1962), Robins (1979) and Baumol (2002).
79. The statement of Columbus is mentioned in Wright (2000), p. 11.
80. See for these facts Prescott (1847) and Wright (2000).
81. See Wright (2000), p. 81.
82. See Robbins (1998), p. 48.
83. See Heckshler (1935).
84. See Schomoller (1902), pp. 51–52.
85. See Mun (1664 [1928]).
86. See Mun (1621 [1930]), p. 17.
87. See Mun (1664 [1928]), p. 19.
88. Adam Smith's ideas against mercantilism are available in his book Wealth of Nations, Book IV chapter I, See Smith (1776, 2003).
89. See Smith (1776, 2003), pp. 539–540.
90. See Smith (1776, 2003), p. 546.
91. See Smith (1776, 2003), p. 554.
92. See Smith (1776, 2003), p. 500.
93. See Smith (1776, 2003), p. 563.
94. See Winch (1978), p.32.

Bibliography

Appadurai, A. (1986). *The social life of things commodities in cultural perspective*. Cambridge: Cambridge University Press.

Arasaratnam, S. (1980). Weavrs, merchants and company—The handloom industry in South India 1750–1790. *Indian Economic and Social History Review, 17*, 257–281.

Athar Ali, M. (1975). The passing of empire—The Mughal case. *Modern Asian Studies, 9*(3), 385–396.

Bairoch, P. (1982). International industrialization levels from 1750 to 1980. *Journal of European Economic History, 11*, 10269–10333.

Bal Pours, E. (Surgeon General) (1885). *Encyclopedia of India*, Austria.

Banga, I. (1978). *Agrarian system of Sikhs*. New Delhi: Oxford University Press.

Bayly, C. A. (1983). *Townsman and bazars: North Indian society in the age of British expansion*. Cambridge, UK: Cambridge University Press.

Bayly, C. A. (1988). *The British empire and the world 1780–1830*. Oxford: Oxford 89 University Press.

Bayly, C. A., & Subrahmanyan, S. (1988). Portfolio capitalists and political economy of early modern India. *Indian Eco and Social History Review, 25*(4), 401–424.

Bhatacharya, S. (1982). Regional economy 1757–1857, the Eastern India. In D. Kumar & M. Desai (Eds.), *Cambridge economic history of India* (Vol. II, pp. 270–331). Hyderabad: Orient Longman.

Bhattacharya, S. (1966). Cultural and social constraint on technological innovations—Some case studies. *Indian Economic and Social History Review, 3*, 242–256.

Bolt, W. (1772 [1998]). Considerations on Indian affairs. In P. Tuck (Ed.), *The East India Company 1600–1858*. London: Routledge.

Boumol, W. J. (2002). Services and leaders and leaders of services. In J. Gadrey & F. Gallonz (Eds.), *Productivity, innovation and knowledge in services* (pp. 147–163). Cheltenham/Northampton: Edward Elgar.

Bowen, H. V. (2006). *The business of empire—East India Company and imperial Britain 1756–1833*. Cambridge, UK: Cambridge University Press.

Burke, E. (1783 [1981], June). Ninth report of select committee. In P. J. Marshall (Ed.), *The writings and speeches of Edmund Buske* (Vol. V). Oxford: Clarendon Press.

Chadhury, S. (1972). Merchants, companies, and rulers; Bengal in eighteenth century. *Journal of Economic History of the Orient, 31*, 71–85.

Chandra, S. (1998). Review of the crises of the Jagirdari system. In M. Alam & S. Subrahmanyan (Eds.), *Mughal state* (pp. 347–360). New Delhi: Oxford University Press.

Clark, C. (1950). *Conditions of economic progress*. London: Macmillan.

Clingingsmith, D., & William, J. G. (2004). *India's de-industrialization under British rule, new ideas and new evidence*. Working Paper No. 10586 National Bureau of Economic Research, Cambridge, MA.

Daw, A. (1792). *The history of Hindoostan from the death of Akbar to complete settlement of the empire under Aurangzeb*. Dublin: Luke White.

Deavy, B. (1975). *The economic development of India: A Marxist analysis*. Nottingham: Spokesman Books.

Dewey, C. (Ed.). (1988). *Arrested development in India—The historical dimension*. New Delhi: Manohar Publication.

Dirk, N., & B. (2006). *The scandal of empire—India and the creation of Imperial Britain*. Ranikhet: Permanent Black.

Dodwell, H. (1922). *Madras weavers under the company*. Proceedings of the Indian Historical Research Committee.

Dutt, R. (1906). *The economic history of India in the Victorian age*. London: Kegan Pal.

Furber, H. (1970). *John company at work: A study of European expansion in the late eighteenth century*. New York: Octagon.

Furber, H. (2004). Rival empires of trade 1600–1800. In *Maritime India*. New Delhi: Oxford University Press.

Gadgil, D. R. (1924). *The industrial evolution of India in recent times: 1860–1939*. Oxford University Press.

Grewal, J. S. (1995). *The Sikhs of Punjab*. New Delhi: Manohar Publishers.

Grioseclose, B. (1993). *British sculptures and Company Raj*. Newark, New Jersey: University of Delaware Press.

Habib, I. (1990). The merchant communities in pre-colonial India. In J. D. Tracy (Ed.), *The rise of the merchant empire: Long distance trade in the early modern world 1350–1750*. Cambridge: Cambridge University Press.

Hall, R. (1996). *Empire of monsoon*. New York: Harper Collins.

Harnetty, P. (1991). De-industrialization revised: The handloom weavers of the central provinces of India. *Modern Asian Studies, 25*(3), 455–510.

Heckscher, E. R. (1935). *Mercantilism* (trans: Shapiro, M.). London: George Allen and Unwin.

Hobhouse, H. (1999). *Seeds of change—Six plants that transformed the mankind*. London: Papermac.

Hodgson, G. H. (1938). *Thomas parry: Free merchant*. Madras: Higginbrothers.

Hunder, W. W. (1868). *Annal of rural Bengal*. New York: Leypoldt and Holt.

Hunter, W. W. (1874). *Famine aspects of Bengal districts*. London: Turner.

Hussain, H. (1958). *The company weavers of Bengal: The East India Company and the organization of textile production in Bengal 1750–1813*. New Delhi: Oxford University Press.

Hussain, I. (Ed.). (2006). *Karl Marx on India*. New Delhi: Tulika Books.

Karen, L. (1979). The great theory of the decline of the Mughal empire. *Comparative Studies in Social History, 21*(2), 161–167.

Krishnamurty, J. (1983). Occupational structure. In D. Kumar (Ed.), *The Cambridge history of India c. 1757–c. 1970* (Vol. II). New Delhi/Cambridge: Orient Longman/Cambridge University Press.

Ludden, D. (1996). Caste society and unit of production in early modern South India. In B. Stein & S. Subrahmanyan (Eds.), *Institutions and economic change in South Asia* (pp. 105–133). New Delhi: Oxford University Press.

Maddison, A. (1971). *Class structure and economic growth: India and Pakistan*. London: Allen and Unwin.

Maddison, A. (2003). *The world economy – A new millennial perspective*. New Delhi: Overseas Press.

Maikovits, C. (2008). *Merchants, traders, entrepreneurs–Indian business in colonial era*. Renikhet: Permanent Block.

Malabari, P. D. M. (1910). *Bombay in making*. London: T. Fisher and Uniwin.

Malleson, G. B. (1901). *History of French in India c.1674 to c.1961*. Edinburg: John Grant.

Markovits, C. (2000). *Global world of Indian merchants 1750–1947; traders of Sind from Bukhara to Panama*. Cambridge: Cambridge University Press.

Marshall, P. J. (1975a). Economic and political expansion: The case of Awadh. *Modern Asian Studies, 9*(4), 465–482.

Marshall, P. J. (1975b). Economic expansion—The case of Awadh. *Modern Asian Studies, 9*(4), 465–482.

Marx, K. (1853a, June 25). The Biritsh rule in India. *New York Daily Tribune*. In I. Hussain (Ed.) (2006). *Karl Marx on India* (pp. 11–17). New Delhi: Tulika Books.

Marx, K. (1853b, July). East India Company—Its history and results. *New York Daily Tribune*.

Marx, K. (1853c, June 24). East India Company its history and results. *New York Daily Tribune*.

McNiell, W. H. (1989). *The age of gunpowder empire: 1450–1800*. Washington, DC: American Historical Association.

Mukund, K. (1992). Indian textile industry in the seventeenth and eighteenth centuries: Structure organization and response. *Economic and Political Weekly, 19*, 2057–2065.

Mun, T. (1664 [1928]). *England's treasure by foreign trade*. Oxford: Basil Blackwell.

Mun, T. (1621 [1930]). *A discourse on trade from England unto East India*. New York: Facsimile Text Society.

Naoroji, D. (1901/1962). *Poverty and UnBritish rule of India*, (Reprint), New Delhi: Government of India Press.

Naqvi, H. K. (1986). *Agriculture, industrial and urban dynamics under sultans of Delhi*. New Delhi: Oxford University Press.

Neff, J. U. (1943). Industrial revolution reconsidered. *Journal of Economic History, 3*, 2–29.

Nick (2006) – Author's name is Robbins, Nick – In the biblography, the name is with 'Robbin'

Om Prakash (1985). *Dutch East India Company and economy of Bengal 1630–1720*. Princeton: Princeton University Press.

Om Prakash (1998). Trade and politics in eighteen country Bengal. In L. Blusse & F. Gaastra (Eds.), *On the eighteen century as a category of Asian history*. Aldershot: Ven Lein in Retrospect Ashgate.

Om Prakash (2007). From negotiation to Coersion—Textile manufacturing in the eighteen centuries. *Modern Asian Studies, 41*(6), 1331–1368.

Parthasarathi, P. (2001). *The transition to colonial economy weavers, merchants and kings in South India 1720–1800*. Cambridge: Cambridge University Press.

Pavlov, V. I. (1976). *Historical premises for India's transition to capitalism*. Moscow: Nauka Publishing House.

Powell, B. (1892). *Land-systems of British India*. Oxford: Clarendon Press.

Prescott, W. H. (1847). *History of the conquest of Mexico and Peru*. New York: Random House.

Ramaswamy, V. (2008). Traditional crafts, technology and society in pre-colonial Peninsular India. In R. Dutta (Ed.), *Rethinking a millennium, perspectives in Indian history from eight to eighteenth century* (p. 27). New Delhi: Akar.

Roy, T. (2005). *Rethinking change in India-labour and livelihood*. London/ New York: Rontledge.

Roy, T. (2006). *Economy history of India 1857–1947*. New Delhi: Oxford University Prsss.

Riello, G., & Parthasarathi, P. (Eds.). (2009). *The spinning world—A global history of textiles 1200–1850*. Oxford: Oxford University Press.

Robins, N. (2006). *The corporation that changed the world—How the East India Company shaped modern multinationals*. Hyderabad: Orient Longman.

Robbins, L. (1998). In S. G. Medema & W. J. Samuels (Eds.), *A history of economic thought—The LSE lectures*. Oxford: Oxford University Press.

Royden, H., & Zeitlin, J. (Eds.). (1985). *Division of labour, skilled workers and technological change in nineteenth century Britain*. Sussex: Harvest.

Sarkar, S. (2014). *Modern times: India 1880–1950, environment, economy, culture*. Ranikhet: Permanent Black.

Sen, S. (1998). *Empire of free trade: East India Company and the making of colonial market place*. Philadelphia: University of Philadelphia Press.

Seth, V. K. (2006). *Economics of services: A story of transformation of Cinderella into the queen of hearts*. New Delhi: Ane Books.

Siddiqi, A. (Ed.). (1995). *Trade and finance in colonial India* (pp. 250–264). New Delhi: Oxford University Press.

Singer, Chargles et al. (Eds.). (1958). *History of technology* (Vols. II–IV). Oxford: Clarendon Press.

Singh, R. (1998). *A despotism of law: Crime and justice in early colonial India*. New Delhi: Oxford University Press.

Smith, A. (1776, 2003). Introduction. In A. B. Krueger (Ed.), *The wealth of nations* (Bantam classic). New York: Bantam Books.

Smith, B. M. D. (1967). The Galtons of Birmingham Quacker gun merchants and bankers. *Business History, 9*(2), 135–148.

Solymar, L. (1999). *Getting the message: A history of communication*. Oxford: Oxford University Press.

Stein, B. (1985). State formation and economy reconsidered. *Modern Asia Studies, 19*(3), 387–341.

Stein, B. (1998). *A history of India*. New Delhi: Oxford University Press.

Subramaniam, L. (1987). Banias and business: The role of indigenous credit in the process of imperial expansion in Western India in the second half of eighteen century. *Modern Asian Studies, 21*(3), 455–520.

Thorner, D. (1951). Great Britain and development of Indian Railway. *Journal of Economic History, XI*, 326–350.

Toynbee, A. (1958). *Industrial revolution*. Boston: Becon Press.

Tripathi, D. (2004). *Oxford history of Indian business*. New Delhi: Oxford University Press.

Verdier, D. (2002). *Moving money, banking and finance in industrialised world*. New York: Cambridge University Press.

Vijayabhaskar, M. (2001). *Industrial formation under conditions of flexible accumulation—The case of global knitwear node, in Southern India*. Unpublished PhD thesis, Centre of Development Studies, Thiruvananthpuram (Mimio).

Wenger, E. C. (1998). *Community of practice: Learning, meaning and identity*. Cambridge: Cambridge University Press.

Wilson, C. (1965). *England's apprenticeship 1603–1963*. London: Longman.

Winch, D. (1978). *Adam Smith's politics—An essay in historiographic revision*. Cambridge: Cambridge University Press.

Womack, J. P., et al. (1990). *Machines that changed the world—The story of lean production*. Temecula: Rawson Associates.

CHAPTER 5

Structure of Organization and Nature of Corporate Governance of the East India Company

In order to understand the consequences that the economy of the Indian sub-continent, in general, and traditional flexible manufacturing, in particular, experienced during the rule of the East India Company, it is necessary to know its organizational structure. The consequences that the economy of the Indian sub-continent experienced during the rule of the company is described in greater detail in Chap. 4. It is due to its organizational structure, that the company evolved a unique symbiotic relationship with the economy of the sub-continent. Most consequences the economy of the sub-continent experienced can be traced back to the structure of the company's organization. This is because its organizational structure determined its (i) objectives (vision and the mission), (ii) its work culture, and (iii) the nature of its corporate governance. Hence, it is important to understand the pattern of its organization and the nature of its corporate governance to assess the impact of different kinds of interactions that the East India Company had with the economy of the Indian sub-continent. In the present chapter, an attempt has been made to study the relationship between the organizational structure and nature of corporate governance of the East India Company.

The East India Company was the first joint-stock multinational corporation of the world. It had continuous existence as a commercial corporation from c.1600 to c.1833; and as purely an agency of the British Empire from c.1833 to c.1857. Looking at this corporation in historical perspective reveals that it had the capacity to put to shame some of the largest contemporary corporations that are listed in the Fortune 500; in terms of persons

V.K. Seth, *The Story of Indian Manufacturing*,
DOI 10.1007/978-981-10-5574-4_5

employed, volume of merchandise handled and extent of market power enjoyed by them. Due to its organizational structure, it was perhaps the first business organization that experienced the consequences of separation between ownership (stockholders) and control (managers); this provided scope for opportunistic behaviour on the part of its staff and management. Historical details of the same show that there existed striking similarities between the behaviour of management of the East India Company and management of some of the contemporary corporations. It is also important to mention that some of the acts of omission and commission of the management of the East India Company also provoked similar responses from its contemporary civil society, policymakers and intellectuals, as have been received by contemporary corporations.

Despite the fact that the first joint-stock company was registered in Britain, this form of business organization became more prevalent in the United States. Due to this reason, most of the theoretical works to understand the consequences of separation between ownership and control originated in the United States, beginning with the pioneering study done by Berle and Means.[1] The wider issues that are associated with the implications of separation between ownership and control have become the subject matter of a new specialization known as 'corporate governance'. This specialization covers the relationship between managers and stockholders of joint-stock companies. The main focus of this discipline is to study the nature of power-sharing between stockholders and management of the company and its implication on corporate performance.

It has been observed by business historians that the nature of corporate governance varies in time as well as in space, because corporate property relationship in each time and in each space is embedded in the institutional arrangements that have emerged due to interaction between society, polity and business.[2] The studies on corporate governance have established that the most important determinant of the nature of corporate governance is corporate property relationship. This relationship varies across nations as a consequence of (i) financial arrangement viewed in terms of bank versus market driven,[3] (ii) objectives followed by the corporate sector, in terms of stockholders versus stakeholders value maximization,[4] and (iii) policy regime of the time, expressed in terms of regulated or market driven.[5]

This suggests that corporate governance is related to the formulation of a framework that can minimize the consequences of separation between ownership and control. Therefore, it is a study of the institutions, regulations and socially and legally accepted business practices, which give context to the relationship between owners and stakeholders of a corporation. Over time, the literature on corporate governance has evolved around two different streams of thought. One strand of thought is trying to study the nature of corporate governance based on the nature of the relationship between owners of the company and those who control the process of decision making. Different theories of corporate governance have evolved on the basis of the nature of the relationship between two group of stakeholders. The second strand of thought has developed the theories regarding corporate governance, based on the extent of concentration or dispersal of ownership of stocks of the company.

Agency theory of corporate governance is based on the premise that corporations are formed through nexus or contracts amongst different interests.[6] Since contracts are not able to anticipate all the consequences that might arise in the future, the agents who control the decision-making process start exercising discretion on the use of resources of the company. This provides them opportunity to maximize their own personal objectives, which may be different from the objectives of the owners. The implicit premise of the agency theory of corporate governance is that there exists conflict of interest between owners and the persons who control the company. Managers are guided by self-interest and they do not miss any opportunity to maximize their personal gains.[7]

Some scholars have evolved an alternative view of corporate governance, which accepts that business firms are a nexus of a well-established trust relationship rather than as a nexus of people with conflicting interests. This theory is known as the stewardship theory of corporate governance.[8] The theory is based on the assumption that managers are not just agents of owners of the company but work as stewards of the company. Because their objective of maximization of personal wealth is constrained by several other higher level human motives. These other motives emerge in managers due to the existence of a market of managers. The market of managers rewards personal achievements, recognition and reputation. Hence, managers' greed to maximize personal wealth is constrained by other motives, which command a premium in the market of managers. Contrary to these premises, there has emerged another premise in the study of corporate governance, which has evolved from the understanding that organizations emerge as a consequence of a 'social contract' between different stakeholders. Therefore, this strand of thought believes that the purpose of the business firm is to create value for its stakeholders.[9,10,11]

The literature on corporate governance has also identified another strand of thought based on the pattern of ownership of stocks of companies (i.e., whether stocks are concentrated in the hands of few individuals or are widely dispersed amongst them). It has been observed that the economies in which stocks are concentrated in the hands of few individuals in the form of large blocks generally have weak and inefficient stock markets. There is also lack of transparency in corporate disclosure, and the stock market fails to change the ownership of the companies. On the contrary, economies that have a dispersed pattern of stockholding generally have a well-developed and efficient stock market, higher standards of corporate disclosure and transparency, and a market for corporate control. Different patterns of stockholding may emerge as a consequence of the 'polity' or 'legal system' prevalent in different economies.

The conduct of the East India Company has been the subject matter of hundreds of books and research papers. However, most of these writings have remained preoccupied with the study of its conduct as an agency of the British Empire, to put it mildly, or as an instrument of British imperialism, to put it strongly. Therefore, most historical writings on the East India Company have remained focused on implication of its conduct on the economy, polity and society of the Indian sub-continent. There exists a paucity of studies that have analysed the East India Company purely as a corporate entity to understand its behaviour and identify how far the behaviour of this first joint-stock company contrasted with the behaviour of contemporary corporations. This exercise is essential to know the answers to the questions regarding the nature of corporate governance of the East India Company. The present case study of the East India Company is a modest attempt in this direction. It is relevant to mention here that while exploring historical writings to evaluate the working of the East India Company as a corporate entity, its history was, in fact, unfolding the contemporary debates on corporate governance. This is how one realizes that the past has the capacity to explain the present.

For the purposes of the analysis the present chapter has been divided into three sections, where each section deals with a specific aspect of the problem. Section 5.1 deals with the history of the East India Company as a corporation and its organizational structure. It also deals with the internal contradiction that existed in its organizational structure. Section 5.2 deals with the relationship between its organizational structure and the nature of corporate governance. Section 5.3 deals with the responses of its contemporary civil society, policymakers and intellectuals that influenced the nature of its corporate governance.

5.1 The East India Company

The East India Company was formed on New Year's Eve c.1600 by a royal charter granted by Queen Elizabeth I of Britain. It was formed with the help of cooperation between three different strata of British society. Persons belonging to these three strata of the society hardly interacted with each other in the normal course, but were bound together to satisfy their private interests. The East India Company was financially provided by the rich merchants and bankers of London. It received logistical support to transport its merchandise across different oceans by sailors and navigators. Its merchandise was protected by the soldiers, in the seas from different pirates, and at different locations across the world, from local rulers and administrators. Therefore, its organizational structure had to accommodate and harmonize the interest of these three important constituents.

The status of 'Royal Chartered Company' was given by the British Crown as a specific privilege. The chartered company status was given to a company that was able to establish harmony between private profit motive and public interest, or the interest of the Crown.[12] The status of chartered company also endowed it with quasi-sovereign rights, like the right to mint its own coins, exercise law and order, and provide justice in its own overseas settlements.[13] This status did not provide any additional advantage to its managers who were operating in Britain. Naturally, this status provided more scope for opportunistic behaviour by its overseas managers. This is because the chartered status, which gave them quasi-sovereign rights, provided them with opportunities to bribe or enter into a treaty with local rulers. At times, they had to take decisions to fight wars with them as well. Therefore, overseas operations of the company necessitated the empowerment of overseas managers of the company. These additional powers given to overseas branches of the company helped them evolve their own distinct identity.

Since the beginning of its operations, the East India Company maintained its own army and naval force. Because of this scholars believe that the East India Company represented the idea of 'commerce with sword in your hand'. The main objective of the British Crown to grant royal charter to the East India Company was to give the company monopoly of trade with the East so that it could corner a substantial market share in the lucrative Asian trade for the British Crown. Around the same period, several companies were formed by merchant adventurers, which were also granted royal charter, to exploit the trading opportunities created by the discovery of new sea routes. Such companies were Muscovy, Levant, Royal African Company and Hudson Bay Company. Most of these companies created by the royal charter have become part of history, apart from Hudson Bay Company, which is still operating in Canada as the largest chain of stores, and the British Broadcasting Corporation (BBC).[14] It also appeared in the newspaper that Sajiv Mehta, a British citizen of Indian origin, re-launched the East India Company on Indian Independence Day by opening a store in London in 2010.[15]

Granting of the royal charter to some selected companies led to a debate amongst contemporary scholars. The scholars who were antagonist to the practice of granting royal charter to companies believed that the chartered companies were organized primarily to maximize and appropriate monopoly rent. In this formulation, rent emerges as a consequence of creating entry barriers by royal charter, which helps the company to earn monopoly profits.[16] However, protagonists of the practice viewed that the granting of royal charter to a company amounted to giving the company a right over its intellectual capital (like granting a patent right, which also gives monopoly right to the company). This is because royal charters were granted to only those companies that created information and knowledge about new overseas routes, markets or source of supply of new products. The granting of royal charter provided to the company a monopoly, which reduced the risk of venturing into uncertain business.[17] However, such property rights had the potential to encourage rent-seeking behaviour amongst them The protagonists of the chartered status argued that the status did not automatically yield rents, unless and until the company transacted goods and services. During the course of transaction, rent could be earned either by minimizing costs or maximizing surplus earned from the transaction.[18]

Despite the existence of alternative views on granting of chartered status to the company, contemporary scholars of the East India Company, by and large, accepted the appropriateness of granting it the said status. These scholars accepted the validity of practice, on the grounds that the opening of trade in new markets and products is as important as innovation, and therefore such companies are entitled to enjoy monopoly right for a certain period of time. Hence, these scholars believed that conferring of chartered status by the Crown is equivalent to granting intellectual property right or patent right on new knowledge.[19] Even Adam Smith, who was one of the most influential critics of the East India Company, did not dispute the granting of the status of chartered company to the East India Company. He writes that 'when a company of merchants undertakes at their own risk and expense to establish a new trade with some remote and barbarious nation ... It is the easiest and most natural way in which the state can recompense them... A temporary monopoly of this kind may be vindicated upon the same principles on which like monopoly of a new machine is granted to its inventor, and that of a new book to its author'.[20]

Unlike the trading companies of other European countries, the East India Company was a joint-stock company. This was a unique innovation in its organizational structure, unlike, say, a Portuguese company that was owned by the Portuguese government, and the Dutch company that was also owned by the state but accepted private investment from individual investors without giving them voting rights. The concept of joint-stock has its roots in an Italian invention, *Compagnia*, a name based on the Latin phrase *cum panis*, which means a 'sharing of bread'.[21]

Initially, the East India Company raised separate joint stocks for each voyage, beginning with the first voyage in c.1603. Each stockholder subscribed to stocks voyage by voyage and profits made from each voyage were shared by the stockholders after the completion of each voyage, based on the value of their stocks. In the first voyage, 218 investors came together and raised total capital of £68,373, to finance the fleet of four small ships. From c.1613 onwards, joint stocks were raised on an annual basis. The first annual stock valued £420,436, to finance fleets for four years c.1613–1616. The second annual stock raised a total investment of £1.6 million to finance annual voyages between c.1617 and c.1622.[22] Several annual joint stocks were raised subsequently, however, after c.1653, joint stocks became permanent.[23] The stocks of the company, in the beginning, were traded at its head office located at Leadenhall Street in London. Later, trading in its shares shifted to London's Royal Exchange. However, after c.1773, its shares were traded at the London Stock Exchange.[24]

Although modern joint-stock companies enjoy unlimited lifetimes, lasting beyond the lifetime of original promoters, the East India Company needed periodic revalidation of its charter from the British Crown. Hence, its continued existence depended on its goodwill and relationship with the British Crown and Parliament. To guarantee its existence, the East India Company followed the policy of bribing people in authority to obtain their patronage. This policy was followed strategically when Sir Josiah Child became a member of the board of directors of the company in c.1674 and remained with the company till c.1699. He strictly believed in the tenants of *mercantilism* and wrote series of articles and pamphlets to project his ideas, including the famous pamphlet, *A New Discourse on Trade.* He was instrumental in consolidating the position of the company with the British Crown.

Child gave 10,000 guineas to Charles II, the king of Britain, to obtain continuity of the charter of the company in c.1681. This eventually became an annual payment, until the rule of Charles II ended. One of the important benefits of the relationship between the British Crown and the company was that the company received the occupation of Bombay (now Mumbai), which Charles II had received from the Portuguese as dowry when he married the princess of Portugal. When James became the king of Britain, the practice of bribing the Crown was revived when charter of the company was renewed in c.1693. Following the receipt of the royal charter by the company in c.1693, British Parliament appointed a committee made up of British parliamentarians to investigate the charges of corruption. The committee discovered multiple incidents of bribing by the company to secure continuity of the charter. Each act of bribery was traced back to the governor of the company, Sir Thomas Cook, son-in-law of Josiah Child. This amply demonstrates that the company followed the policy of bribing to obtain the patronage of the British Crown.[25]

The stockholders of the company were big and small investors from diverse socio-economic backgrounds. These thousands of stockholders were spread over each and every part of British geography. As recorded by historians of the East India Company, there were 3084 individuals who owned the East India Company stock in 2826 separate holdings. These shares were owned either jointly, individually, or in the name of partnerships or corporations.[26] The interests of these thousands of shareholders were looked after by a very small group of persons—twenty-four directors of the company and their supporting staff at the London office, and the governors of three presidencies in three different corners of the Indian sub-continent—Bombay (Mumbai), Calcutta (Kolkata) and Madras (Chennai). These governors, along with their supporting staff of writers, factors, and so on were actually running the business of the company.

In c.1785, one of the secretaries of the East India Company, Thomas Morton, compiled a detailed description of its administrative structure. This is the only surviving detailed account of the organizational structure of the company. It has been reported in the recent work of Bowen.[27] According to this description, the company was governed by a court of directors who functioned with the help of various committees, headed by one of the directors. The committees formed by the court were divided into three different classes. The first class of the committee was described as 'correspondence', which consisted of committees of correspondence, lawsuits, military funds and treasury. The second class of committees included the warehouse, accounts, buying, and house committees. Finally, the third class included shipping, troops, stores, private trade and prevention of private trade committees.

In c.1776, a separate office of 'examiner' was created in addition to the office of secretary of the company. This office assumed significant importance over time because it emerged as a vital bureaucratic link between the company office in London and different presidencies in India. This office was occupied by some of the leading intellectuals and experts on India. For instance, Nattaniel Brassey, the author of *Code of Gentoo Laws* published in c.1776, Halhead, the author of *Grammar of Bengal Languages*, published in c.1778, James Mill, the author of *British India*, and poet Thomas Love Peacock occupied the office of examiner at different points in time.[28] The employees of the East India Company were strictly promoted on the basis of seniority. The lower bureaucracy of the company consisted of examiners of different ranks and writers. In addition to its London office, the East India Company's offices in the presidencies in India followed a hierarchical structure, for example, a new employee in the company could enter as a writer. After serving as a writer for at least five years, he was promoted to factor. After serving for another three years, he was designated as junior merchant, and then a few years later he was promoted to senior merchant. Some employees of the company were also elevated to work in presidency councils and were even appointed as governors of the presidencies.

Functioning of the East India Company was similar to modern corporations because it was also managed by a hierarchy of paid managers. These managers coordinated the buying and selling of merchandise, and processing of information at the London office as well as over long distances between the company's operational offices in different presidencies and its various trading centres. The administrative structure of the company was evolved to make effective control of the head office over managers and other employees working in overseas locations. This was necessary to reduce the costs of transactions made by the company because it relied on the 'managerially coordinated transactions rather than market coordinated transactions'.[29] The dependence on managerially coordinated transactions necessitated the emergence of a managerial hierarchy and networks of coordination to establish appropriate control mechanisms.

In the contemporary economic analysis, a firm or a company is conceptualized as a collection of contracts that determine the nature of the relationship between the organization and different stakeholders. The East India Company had also entered into multiple contracts to work efficiently. For example, there existed a contract between the British Crown and Parliament and the board of directors of the company, through Royal Charter, to run the company according to the objectives of the charter. Similarly, there existed a contract between the stockholders of the company and its management. Finally, there was a contract between management located at the London office and management and staff working at overseas locations. To reach the appropriate and desired outcome from each contract, there was need for proper design and enforcement of contracts. Institutional economics have given due importance to the process of evolution of third-party institutions to enforce contracts.[30]

The organization of the East India Company into a joint-stock company created a separation between ownership and control. This form of organization provided, on the one hand, several kinds of advantages to the company, like the possibility of raising capital from a wider pool of investors and the ability to hire professional managers instead of relying on hereditary owners. On the other hand, this form of organization had a built-in tendency to breed double disadvantages to the company. The limited liability of the owners (stockholders) liberates them from bearing full consequences of their investment decisions, because consequence of each of their investment decisions is limited to the extent of their investment in stocks. The limited liability of investors makes them indifferent to the functioning of the company on a daily basis. The separation between ownership and control also creates opportunities for the management to resort to opportunistic behaviour and pursue their own objectives, which may or may not be compatible with the objectives of the owners.

The opportunistic behaviour on the part of management has been described as the principal (owner)-agent (management) problem.[31] The analysis of the principal-agent problem in economics is as old as the subject is. The first systematic writing in economics by Adam Smith highlighted the problem by stating that the organizational structure of a joint-stock company has inherent tendency towards mismanagement because of 'negligence, profusion and malversation of their own servants'.[32] However, the analysis of this problem has now been developed in greater detail in principal-agent models. During the time of the East India Company, the joint-stock form of organization was an exception not the rule. The separation between ownership and control became all-pervading in America during the 1920s and 1930s.[33] The existence of principal-agent problems at a much wider scale led to the origin of theoretical literature on this problem. The main objective of the theoretical writings on principal-agent models is to establish harmonious relationships between the interests of different stakeholders. In a typical principal-agent model a principal (owner) hires an agent (manager) to act on his behalf to achieve objectives laid down by the owner. But the agent, after being hired, often pursues objectives other than those proposed by the owner. This happens because incentives of the principal are at variance with the incentives of the agents, leading to 'incentive conflict'. This poses two important challenges to the principal (i) selection of an appropriate agent; otherwise it leads to 'adverse selections' and (ii) once the agent has been hired, the principal has to evolve an appropriate incentive structure to motivate the agent to perform and achieve objectives of the principal; otherwise it will lead to 'moral hazards'.[34] In principal-agent problems, the principal can reduce the agency costs (the costs of opportunistic behaviour of the agents) by monitoring and collecting information to minimize adverse selection as well as moral hazards.

5.2 Problems of Corporate Governance in the East India Company

Principal-agent problems can arise in any corporation between any contracting parties. Since in the East India Company there existed multiple contracts, they created multiple kinds of principal-agent problems. In case of corporations that are confined to national boundaries, such problems are less likely in comparison to companies with multinational operations. The most serious principal-agent problem, which the East India Company experienced, was between the management and staff located at the head office of the company in Britain and managers and employees working in the Indian sub-continent. The problem was much more serious due to the additional disadvantage of information asymmetry, which existed due to temporal and spatial distances that separated the principals from the agents. The temporal and spatial distances were insurmountable due to the existing means of transportation, which were not only slow but also hazardous, and means of communication that could not provide information rapidly.[35] Moreover, as has been previously described, the chartered status of the company provided to the overseas managers certain quasi-sovereign rights, which allowed them to follow their own objectives. This special status provided ample opportunities to its overseas employees, especially its armed wing, to occupy territories in foreign lands to maximize their personal wealth. The East India Company therefore contained, in its organizational structure, the integration of interest of merchants and conquistadors. This is because it was built into its DNA to make the East India Company an agency of British imperialism. Due to the aforementioned reasons, we are focusing on the principal-agent relationship that existed between the head office of the company at London and its overseas officials working in the Indian sub-continent. This has been done because other forms of relationship between principal and agents were more or less the same as experienced by other joint-stock companies.

To minimize agency costs, companies have to minimize moral hazard and the impact of adverse selection. To minimize moral hazard, a company can pay to its staff 'efficiency wages'. An interesting example of efficiency wages was the introduction of a five-dollar-a-day minimum wage by Henry Ford in his Ford Motor Company in 1914. Ford stated while introducing this raise in wages, 'There was no charity in raising minimum wages ... we are building for future ... A low wage business is always insecure'.[36] This means that the company has to pay to its employees wages higher than the market-clearing level. This practice is at variance with mainstream neo-classical understanding of the functioning of labour markets, which envisages that wages are flexible and have a tendency to settle at the market-clearing level. The efficiency wage criterion is based on the premise that higher wages, to some extent, improve the efficiency of workers. This happens because paying efficiency wages increases loyalty of workers; it motivates them and help in reducing work shirking and attracting better workers.[37] However, paying higher wages than the market-clearing level increases costs to the company.

The alternative policy that can help a company to minimize the problems associated with moral hazard is to impose a bond or entry-level fees on employees.[38] This strategy increases cost to the employee rather than cost to the company. However, indirectly, it may also increase cost to the company if workers reduce their work efforts to the level mentioned in the bond. Some economists view efficiency wages and bonding as alternative strategies, but some others view them as complementary strategies.[39] The extent to which bonds and efficiency wages can reduce the opportunistic behaviour on the part of employees depends on the risk of detection and conviction, and quantum of punishment (i.e., the value of the bond). Therefore, in addition to paying efficiency wages and imposing bond, a company has to invest in resources for foolproof supervision of its employees, failing which, despite receiving higher wages, employees can indulge in opportunistic behaviour by cheating as well as shirking work.

To minimize the impact of adverse selection, bonding as a method can be used when a company hires employees at different ranks. However, it is not an optional strategy because execution of a bond does not guarantee that the company hires the best workers. A better strategy would be to hire employees at entry-level positions and design an appropriate promotion scheme to allow the employees to grow in ranks within the company according to their skills and capabilities. By following this strategy, a company can acquire more information about the employees, which can help in minimizing the consequences of adverse selection caused by information asymmetry. It is a historically validated strategy because most of East India Company's employees who made significant contributions to the growth of the company came from this group. There does not exist similar evidence regarding the appropriateness of the strategy of bonding. The East India Company had also designed an appropriate employment contract, which it entered into with its overseas employees, to minimize the possibilities of opportunistic behaviour. This necessitates a careful analysis of the contract that was executed between the East India Company and its overseas employees.[40]

Economic historians who have studied the nature of contract and its enforcement between the East India Company and its overseas employees, observed that the contract consisted of two important parts. The first part consisted of 'covenant of indenture' and the other part consisted of 'bond' duly signed by two persons standing as guarantee to the bond. Employees generally entered in the service of the company at the age of seventeen as writers. The covenant of indenture was given in the form of a lengthy document that was filled by the overseas employees. The contract also had a clause for 'dismissal' of the employees in case they breached the contract. The contract lasted for five years, which was generally automatically renewed. The contract also incorporated the scheme of rewards and incentives that were given to its employees. To minimize agency costs of contract, the company was giving to its employees a fixed annual salary, which was quite low, promotion based on seniority and permission to trade on their own account to any port or place in India northwards of the equator. Since the company permitted its employees to carry on trade on their own account also, it provided them facility to remit their earnings from private trade to Britain. The remittance was sent through a bill of exchange. The bills functioned like a deposit wherein employees deposited rupees in the treasury of the company in India. The company paid sterling equivalent amount to the beneficiaries of the payment in Britain.[41]

The analysis of the contract between the company and its overseas employees amply demonstrates that the company made serious attempt to minimize the consequences of moral hazards and adverse selection while designing its contract. The analysis also suggests that the company hired its employees at the entry level and gave them opportunity to grow within the company according to seniority and capabilities. This policy provided, as has been mentioned, better information to the company about the employees and helped in reducing the chances of adverse selection. Although the company paid lower wages to its overseas staff, it provided them incentive to earn more through private trade. Through this strategy, the company reduced its cost yet provided opportunities to employees to earn a decent income. The company did not pay efficiency wages to its employees because, in the absence of effective supervision, probability of detection of inefficiency was very small. Effective supervision was quite difficult due to constraints imposed by geographical distance and quality of means of transport and communication. To complement the advantages of these two provisions in the contract, the company threatened dismissal as well as asked its employees to provide cash guarantees by two persons. These provisions in the contract worked as constraints on the opportunistic behaviour of its overseas staff.

The appropriateness of the provisions of executing a bond and payment of low-wage strategy of the company has already been discussed in respect to their consequences on the principal-agent problem. What has not yet been discussed are the repercussions of permission given to its employees to conduct trade on their own account, in the Indian sub-continent, on the principal-agent problem. Initially, private trade was conducted by the overseas employees of the company because the company did not have operational and administrative capability to prohibit it. This suggests that the private trade to its employees was allowed unofficially, as a result of managerial incapacity of the company to regulate the opportunistic behaviour of its overseas staff in carrying out private trade. Therefore, it is quite obvious that private trade by its overseas employees must have existed since the inception of the operations of the company in the Indian sub-continent. However, permission to trade on their own account was formalized in the contract signed by them since c.1675. Even the company's employees working on company ships were accepting low wages because their employment provided opportunity to make profits from private trade. This form of trade was described as 'privileged trade', which was a 'proxy' for smuggling. The foreign free-cargo allowance ranged from six tons for a commander of the ship to a few cubic feet of space for a skilled tradesman.

Accepting that the private trade by its employees could not be regulated, the company used it as a clever strategy to reduce its cost by paying low wages and permitting its employees to engage in private trade to compensate the loss of income due to low wages. This practice also helped in maintaining the company's monopoly of trade in Britain because its employees could trade only on the ports of Asia. For middle- and senior-level officials of the company, this practice of permitting private trade gave opportunity for making fortunes before retirement from the company, and enjoying standard of living in London comparable to nobles and gentry of Britain. The servants of the company were driven by unregulated greed to maximize their private wealth because the company did not pay them decent wages. The private greed of the employees 'made roguery necessary' for them, otherwise it was difficult to survive on low wages in India. '*Nawab*' was the term used by the British for persons who returned from India with a huge fortune. This fortune allowed them to live life like '*nawabs*' (the term used for the provincial governors during Mughal rule). In c.1730, Edward Stephen became the first Bengal-returned '*nawab*' to enter the House of Commons. These employees, who returned to Britain with substantial wealth earned from India, became the object of envy of the British gentry.[42]

The opportunity to pursue simultaneously 'trade for the company' and 'trade on their own account' led to the practice of 'multitasking' by its employees.[43] This had serious implications on the principal-agent problem. The practice of multitasking is akin to the job of a bell boy in restaurants, who receives a fixed income from the employer and in addition receives 'tips' from customers. In order to maximize their tips, bell boys can compromise the interests of their employer. Several attempts have been made to study the repercussions of multitasking on the principal-agent problem. The studies have observed that if the wages paid to the employees are insensitive to the company's level of profit (i.e. absence of profit-sharing incentives), the employee will spend more effort on private trade. It has also been observed that when principal permits agents to pursue 'company trade' as well as 'private trade', private trade may draw more effort from the employees irrespective of whether their wages are high or low. This happens because the work for the company gives him a fixed salary, but private trade results in additional reward, which he does not have to share with the company. Therefore, an employee may spend more time and energy on private trade, and just enough on company trade, which rules out the possibility of dismissal.[44]

In the case of the East India Company, its employees had the added advantage of conducting private trade because there existed synergy between private and company trade. The synergy existed because employees could leverage the benefits of the trade-related infrastructure created by the company, as well as benefits from the intangible assets of the company, in terms of trading privileges given to the company by the local rulers in the Indian sub-continent, and protection provided by the private army and navy of the company. Despite the fact that the organizational structure of the East India Company created a principal-agent problem between the head office of the company in London and the operational offices in India, the company was able to achieve its objective of maximizing trade with India. The credit goes to the nature of the contract it had evolved between the company and its overseas employees. This is because the contract provided, on the one hand, a very lucrative incentive in the form of private trade as a 'carrot' and threat of dismissal as a formidable 'stick'. This threat, to a great extent, was able to put constraint on the opportunistic behaviour of the overseas employees. This suggests that the contract, as designed by the company, kept the private greed of company officials within acceptable limits.

The extent of opportunist behaviour of the employees of the East India Company and their chances for rent-seeking behaviour considerably increased after c.1717 when the company received a *farman* (permission) to carry free trade in the Bengal Subah from Mughal Emperor Shaharyar. The officials of the company, ignoring the true intent of the *farman*, started selling *dastaks* (entry passes) to the officials of the company for private trade, and to local and foreign traders as well. This practice increased the private earnings of its officials at the cost of revenue of the Nawab of Bengal,[45] which subsequently led to the Battle of Plassey. The opportunistic behaviour of the employees of the East India Company became quite obvious when its forces were able to install a *nawab* of their own choice, Prince Mohammad Ali Wallajah, in Arcot in c.1752.[46] In supporting the *nawab*, Robert Clive made a substantial fortune along with other company officials like Paul Befield. Private wealth was also created by the company officials who were advancing personal loans to the Nawab of Arcot and to the Rajah of Tanjore at very high rates of interest. This whole process had assumed the attributes of a scandal, after which, the company quite rapidly changed the governors of the Madras Presidency. However, all of them continued the practice and returned to London with substantial wealth.

The principal-agent problem again became alarming when the company appointed another puppet, Mir Zaffar, as the Nawab of Bengal, after the Battle of Plassey in c.1757. To take advantage of the new circumstances, company officials plundered the treasury of Bengal. The presents received by the chief architect of the battle, Robert Clive, alone amounted to £ 200,000, excluding his share in military spoils and income from *Jahgire* (land grant), which he received from the *nawab*. The annual revenue of the land grant was £ 27,000.

The opportunistic behaviour of the management of the East India Company in the Indian sub-continent became more glaring when the company entered the business of empire along with trade. This happened when the company was granted *diwani* right (right to collect land revenue) by the Mughal Emperor Shah Alam II after the Battle of Buxar in c.1764. *Diwani* right made the company richer because of the earnings from the land revenue from one of the richest provinces of the Mughal period, Bengal. Clive also exploited this opportunity by indulging in insider trading in company stocks. He did this before informing the company office in London about the grant of *diwani* right and its financial gains to the company. Clive purchased a substantial number of shares of the company through his attorney John Welsh. When the news of *diwani* reached in April, the company shares were traded at £165, by the middle of June, the price of each share increased to £187, and by Christmas, the share price was £223. In the longest meeting of the board on March 18, 1767, shareholders demanded an increase in dividends from 10 to 12.5 per cent. But the British government also wanted a share of the enhanced earnings. The company was forced to pay £400,000 to the government to cover war debt. Simultaneously, the Parliament passed the Dividend Act in June 1767, which fixed the dividend at 10 per cent. Clive also sold his stock in 1767, which doubled his investments in stocks.[47] However, when Clive came back to London in c. 1770, he was invited by the Parliament to defend his case on corruption. He was charged with appropriating £ 234,000 of the company exchequer. Although, instead of punishing him, the British Parliament considered him as the hero of Plassey. However, being humiliated by the charges of corruption, he committed suicide in c.1774.[48]

It is important to mention here that all the governors, or governor-generals who succeeded Clive, were the victims of 'objective conflict' with the board of directors of the company. The objective of the board of directors was to maximize the earnings from trade to have more and more resources at their discretion, and also to please stockholders. They had no interest in financing expensive battles against the native rulers. On the contrary, the officials of the company, who were located in the sub-continent, had the incentives to wage wars. Each war provided them access to the treasury of the local rulers, which provided them opportunity to accumulate personal wealth. Moreover, in new circumstances, directors of the company did not matter much because these actions, even though morally incorrect, were politically correct. They knew that the British Crown, as well as Parliament, would consider them as heroes who were laying the foundation of the British Empire in India.

The consequences of intertwining of these two monopolies—trade and political power—were experienced by the people of Bengal. The monsoon failed in Bengal in c.1769 and c.1770. The drought transformed into a famine. The human tragedy caused by this famine was much larger in proportion than the Bhopal gas tragedy caused by another corporation, Union Carbide in India in 1984. It had become clear that the Bengal famine was caused by corruption in the trading of rice by the East India Company servants, who were manipulating the market to make a fortune.[49] These greedy company officials transformed a food scarcity into a famine. Adam Smith, while describing the causes of the occurrence of famine in Bengal, writes in his book, *The Wealth of Nations*, which appeared six years after the famine, that 'The drought in Bengal a few years ago might probably have occasioned a great dearth. Some improper regulations, some injudicious restraints, imposed by the servants of the East India Company, upon rice trade, contributed, perhaps to turn that "derth" into famine'.[50]

So long as the East India Company remained primarily a commercial corporation, its activities interested only stockholders of the company or the British Crown. As soon as it also became the agency of the British Empire, its activities began to draw the attention of common British citizens. Due to this, lapses committed by the company in corporate governance attracted the attention of the British civil society and intellectuals, as well as the British state.

5.3 Responses of State and British Civil Society on the Issues Related to the Corporate Governance of the East India Company

After the Battle of Plassey (c.1757) and Battle of Buxar (c.1764), the East India Company was transformed into an agency of the British Empire. The transformation led to the entry of the company into the business of empire also.[50] This new role of the East India Company has been variously described by scholars. Some scholars describe it as 'imperium in imperio',[51] the 'East India Company state',[52] 'sovereign merchant', a 'company state',[53] or 'corporate state'.[54] The East India Company now ceased to be only a corporation, and began to draw attention of the British civil society and intellectuals as well as policymakers. It is because the British citizens, as well as the Crown and Parliament, became sensitive to a basic issue—whether a private trading company can be permitted to wage wars, follow independent foreign policy, and enter into treaties with sovereign powers in the Indian sub-continent. It was generally accepted by the people in authority that the East India Company could not be given complete autonomy to pursue foreign policy that is independent of Crown as well as Parliament. Authorities were also concerned about the opportunistic behaviour of the officials of the company. These issues led to a prolonged struggle between the state and the East India Company to arrive at a new equilibrium in their relationship.

The state intended to regulate the company's governance at the London office and at the offices located in different presidency towns in the Indian sub-continent. Attempts were being made to impose an environment on the workings of the East India Company, which may induce the practice of better governance. The protagonists of the company argued that the chartered status granted by the British Crown to the company provided it immunity from parliamentary interventions. Not only this, many enlightened citizens of the country were also of the view that giving right to the state to intervene in the affairs of the company might result in compromising the private property rights of the company.

Initially, attempts were made by the British Parliament to reform the corporate governance of the company at the London office through legislations. Reforms were initiated to alter the electoral process for the election of the company's court of directors. In c.1765, an act was passed (7 Geo III c. 48) that changed the voting qualifications of the stockholders of joint-stock companies. It was legislated that a stockholder should have, in his possession, stocks of the company worth £500 for at least last six months to qualify as a voter to elect the board of directors. In c.1767, another legislation was introduced (7 Geo III c. 49) that prohibited stock splitting. These two legislations were not able to reform the process of election for the court of directors. This was because well-informed electoral cartels were able to split holdings six months before the annual elections, which were held in the month of April every year. Therefore, massive activities occurred in the stock market during the months of September and October to change the ownership of the stocks.

The failure to reform the election process to the board of directors of the company by these two legislations, and occurrence of the 'Bengal Bubble' in the stock prices of the East India Company caused by the Bengal famine in c.1770, prompted the British Parliament to act. In the Bengal Bubble, some influential persons of British civil society had lost their lifetime savings, including Edmund Burke and William Burke. Under these circumstances, the Regulation Act was passed in c.1773. According to this act, only a quarter of the directors retired from their positions every year. Therefore, annual elections were held only to fill a quarter of the 24 director positions. The qualifications of the electoral club, which could participate in the election, were also changed to avoid stock splitting. According to the act, only those stockholders could participate in the election who possessed stocks of worth £1000, that too for a period of twelve months. The legislation also introduced the practice of proportionate voting, in proportion to the value of stocks possessed by the stockholders. Those who possessed the minimum qualifying amount of stocks (i.e., stocks worth £1000) were entitled to cast only one vote. Stockholders who owned stocks valued up to £3000 were able to cast two votes, and up to £6000 three votes. The maximum entitlement of four votes was given to stockholders who possessed stocks worth more than £10,000. This new legislation provided stability to the court of directors because on the one hand, elections were being held only for a quarter of positions, and on the other hand, the importance of small stockholders in the process of election was eliminated. The small stockholders constituted 55.4 per cent of the total stockholders of the company in c.1773. Through these reforms, this act provided more powers to big shareholders to influence important company decisions. The act reduced the incentive of stock splitting practiced earlier by voting cartels. The Regulation Act also put a ceiling on the amount of dividends that could be passed onto the stockholders of the joint-stock companies at 10 per cent.

To effect changes in corporate governance at company offices located in the Indian sub-continent, the Regulation Act of c.1773 curtailed the autonomy of the company by creating a post of governor-general of India based in Calcutta Presidency whose position was senior to the governors of the Madras and Bombay presidencies. To run the affairs of the company in India, a council consisting of five members was constituted. Of the five members, two members were nominated by the company and three members were to be appointed by the Parliament. These provisions in the law provided majority to the government-appointed members. However, this provision also led to hostility between company, appointed members and the government-appointed members of the council. The Regulation Act also tried to curtail the opportunistic behaviour of company officials because it brought trade in salt and opium under the monopoly of the company, which reduced the opportunity of private trade by company officials in these two important commodities.

Following the changes brought by the Regulation Act, Warren Hastings was appointed the first governor-general of the Calcutta Presidency and three councillors, Philip Francis, General John Clavering, and George Monson were appointed by the Parliament. These three councillors were totally ignored by Warren Hastings and other company officials. However, out of these three members, Philip Francis took a strong stand against company officials, especially Hastings.

William Bolts was perhaps the first person of civil society who brought to the notice of the civil society malpractices of the officials of the East India Company. He published his book, *Consideration on Indian Affairs*, in c.1772[55] in which he highlighted the conduct of Harry Verelst, who was the governor of Bengal after Robert Clive. Following the publication of his book, he was harassed by the governor and forced to leave India. He travelled back to London to seek justice from the British judicial system. The British courts provided him justice in c.1774, when the court accepted that Verelst was guilty. This was perhaps the first recorded example of fixing extraterritorial liability of a multinational corporation. Bolts' case established in the eighteenth century that a CEO of a company can be tried thousands of miles away from the place of crime. Why can similar action not be taken against the CEO of Union Carbide in the United States using a similar provision of law?

The information that Edmund Burke received from Philip Francis, and his relative William Burke, about the operations of the officials of the East India Company in India, shaped his views about the company. William

Burke was working with the Rajah of Tanjore, who provided information to Edmund Burke about the company officials who were involved in the scandal of providing personal loans to the Nawab of Arcot and Rajah of Tanjore. The two Burkes collaborated to bring notice to Britain's civil society the scandal amounting to £ 3 million in c.1779. Edmund Burke also received inputs from Philip Francis, who had returned from India, when he published the Select Committees 9th report in c.1783. He mentioned in this report that, after the territorial occupation of India by the East India Company, 'trade ceased to be exchange, and has been transformed into plunder'. He further wrote that company factors have turned in one stroke, the whole trade of the company in their own hands, on their own capital, at their own risque (risks) and the company has become factor for them.[56]

Edmund Burke was an influential thinker of the time. His book on French revolution had provoked ideological debate amongst conservative and liberal thinkers. He did not appreciate state intervention in the activities of the company so long as it did not violate the mandate given by the royal charter. Due to this reason he initially objected to the efforts of Prime Minister Lord North to regulate the company. However, once firsthand accounts of the activities of the officials of the East India Company, from Philip Francis and William Burke, became available to him, his thinking about the company changed significantly. During this time, the coalition government of Lord North and Charles James Fox was ruling the country. Edmund Burke helped the coalition government in drafting the East India Bill, which is popularly known as the Fox India Bill. The bill was presented in the Parliament in c.1783. This bill was intended to make the East India Company accountable to the British Parliament. For this purpose, the bill intended to replace the company's court of directors with seven commissioners to be appointed by the parliament. In December c.1783, when this bill was presented in the House of Commons of the British Parliament, it received massive approval.

Fearing that the Fox India Act may also receive positive response from the House of Lords, the East India Company provoked the British Crown by explaining how this bill would provide more powers to the prime minister and the Parliament on the affairs of the East India Company in comparison to the Crown. The explanation motivated King George III to issue a kind of whip to the members of the House of Lords by telling them that any member of the House who supports the bill would be considered as against the Crown. The results of the voting were quite obvious. The bill was defeated in the House of Lords. This led to the fall of the coalition government led by North and Fox. In the following year c.1784, elections were held to the British Parliament, in which the East India Company financed the election of the opposition party. After election, William Pitt the Younger formed the government.

Despite the fact that the new government headed by Pitt favoured the company, it was compelled, by the mounting pressure of the civil society and the parliamentarians, to regulate the East India Company. He introduced the Government of India Act, drafted by Henry Dundas, to Parliament in July 1784. The act accepted the appropriateness of the chartered status of the company, but proposed a five-member board of control appointed by the king, as well as gave the king powers to recall any company executive in India. This act provided more powers to the Crown in the affairs of the East India Company at the cost of the British Parliament.[57]

All these incidents left Burke a depressed and dejected person. His frustration is reflected in his statement that 'all the tyranny, robbery and destruction of mankind practiced by the company in the East are popularized and pleasing to the country'. However, when Hastings came back to England after completing his term as governor-general of Bengal in c.1788, Edmund Burke initiated an impeachment trial against him because he knew that the British legal system did not provide the legal provisions to try the misdeeds of the company officials; neither had he believed in getting any positive gain from the impeachment. He knew that whatever may be the merit in the evidence against Hastings, the outcome of the impeachment would eventually rest on the voting in the British Parliament. What happens at the time of voting, he had already experienced. The only possible gain in the entire impeachment process was that the debate in the Parliament did influence public opinion on the case. As it was expected, in c.1795, Warren Hastings was acquitted at the impeachment trial.

The impeachment trial and the accompanying debate that took place in the Parliament reveals that contemporary British citizens and intelligentsia were agitated about the impunity that the East India Company enjoyed from consequences of its misdeeds. Its officials were not only indulging in opportunistic behaviour, but were also resorting to all sorts of unethical and inhuman practices in full knowledge of the public and government. However, at that point in time, there was an absence of a legal framework to regulate the activities of corporations and their managers. The courts were not at all equipped to bring them to books and have them pay for their crimes. This is also the problem with contemporary legal systems because the process of globalization of businesses has not been accompanied by commensurate efforts to globalize institutions of corporate governance and justice redressal systems. However, what the incompetent legal system could not achieve, was made possible by socio-economic and political transformation experienced by the British society, polity and economy.

British Parliament made several attempts to improve the quality of corporate governance of the East India Company. These attempts were in the forms of legislations that were presented by the Parliament, such as the Regulation Act, Fox India Act, and Pitt's India Act. The Fox India Act became a casualty of the power struggle between the Crown and the Parliament. The acts that succeeded in getting approval from Parliament gave more power to bigger stockholders and excluded small stockholders from influencing the governance of the company. These reforms reflected the nature of British democracy of the period, when 'the electorate was narrow and consisted of scores of rotten borough'.[58]

The most articulate and formidable antagonist of the East India Company was emerging as a new class, with a new ideology, as a consequence of a transformation that the British economy and society were experiencing after the invention of the steam engine by James Watt, and several other inventions that occurred in the processing of cotton. This new ideological premise formed the basis of the book written by Adam Smith, *Wealth of Nations*, in c.1776. This book was in fact a manifestation of the socio-economic transformation of the British society from the dominance of mercantilism to the birth of classical political economy. The new school of classical political economy challenged the very foundation that provided logical premise for the existence of companies like the East India Company.[59]

The new ideology and new social order, which epitomized virtues of free trade, freedom of enterprise and free market, compelled the British state to end the trading monopoly of the East India Company through the Charter Act of c.1813 and finally, c.1833, the East India Company was stripped of its commercial functions. However, it remained as an agency of British Empire until c.1857.

Notes

1. In this study, the authors explain the phenomenon of separation between ownership and control, which was emerging as an important paradigm of organization of businesses in the United States; see Berle and Means (1932).
2. See, for this strand of thought, the writings of Roe (2002, 2003) and La Porta (1999, 2003).

3. The authors who have established the relationship between the financial sector and the nature of corporate governance are Boots (2000) and Verdier (2002).
4. For the difference between stockholders' interest and stakeholders' interest, see Blair (1995) and Harsmann (1996).
5. To understand the relationship between policy regime and the nature of corporate governance, see Hall and Soskice (2001) and Rajan and Zingales (2003).
6. For understanding the approach to corporate governance based on the principle-agent approach, see Arrow (1985), Fama (1980) and Alchian and Demsetz (1972).
7. See Jensen and William (1976) and Fama and Jensen (1983).
8. For the understanding of the stewardship theory of corporate governance, see Donaldson and Davis (1991) and Davis et al. (1997).
9. For details regarding the stakeholders approach to corporate governance, see Jones and Wicks (1999). Freeman (1984) and Hill and Jones (1992).
10. The scholar who established the relationship between polity and nature of corporate governance is Roe (2002, 2003).
11. La Porta (1999, 2003) has explained the impact of a country's legal system on the nature of corporate governance of the country's corporation.
12. For understanding Royal Charter Companies, see Gower (1969), pp. 23–25.
13. For details regarding quasi-sovereign rights of a charted company, see Robins (2006), p. 25.
14. See for these facts, Carlos and Nicholas (1988), Bowen (2006) and Robins (2006).
15. See, Times of India (2010), August, 16.
16. For the documents that challenged the granting of chartered status to the companies, see Jones and Ville (1996).
17. See, Jones and Ville (1996).
18. For this idea, see McFetridge (1995) and Jones and Ville (1996).
19. The scholars who believe that granting of Royal Charter amounted to granting of intellectual property right to the corporation are Scott (1909–1912) and Carlos and Nicholas (1996).
20. For this, see Smith (1776, 2003), Book IV, Chapter I, p. 957.
21. For this fact, see Brendel (1982), p. 436.
22. For these facts, see Mukerjee (1974), p. 393 and Keay (1993).
23. For this, see Lawsan (1993), p. 21.
24. These facts have been taken from Robins (2006), p. 11.
25. For these facts, see Robins (2006) and Bowen (2006).
26. To know the pattern of shareholding of the East India Company, see Bowen (1987, 1989).

27. To know the description of the administrative structure of the East India Company, see Bowen (2006), Ch. VII, pp. 185–186.
28. These details are given in Bowen (2006).
29. To understand the difference between managerially coordinated transactions and market-coordinated transactions, see Chandler (1977).
30. For these ideas, see Alchian Demsetz (1972); Grief (1993) and Hajeebu (2005).
31. To understand the nature of opportunistic behaviour of the managers, see Arrow (1985), Fama (1980) and Alchian and Demsetz (1972).
32. This quotation has been taken from Smith (1776, 2003), Ch 1, p. 957).
33. See, Bearle and Means (1932).
34. The use of incentives to influence the behaviour of the agent has been described in greater detail in the writings of Grassmann and Hart (1983), Holmstrom (1979, 1982) and Shavell (1979).
35. For information asymmetry between employees, and the London office and employees working in the Indian sub-continent of the East India Company, see Carlos (1991, 1992a, b), Carlos and Nicholas (1990) and Anderson et al. (1982).
36. For the ideas of Henry Ford on giving higher wages to the employees of the Ford Motor Company, see Raff and Summers (1987). The quotation used here is taken from Raff and Summer (1987), p. 559.
37. For why employers pay efficiency wages to their employees, see Salop (1979), Stiglitz (1984, 1988), Weiss (1990), Akerlof (1982, 1984) and Seth and Aggarwal (2004).
38. The use of bond or entry-level fees on employees to minimize the problem associated with moral hazards has been discussed in the works of Carlos (1992a, b).
39. See, Carlos (1992b).
40. For detailed analysis of the contract that was executed between the East India Company and its overseas employees, see Hajeebu (2005) and Grief (1993).
41. For the process of remittance of earnings of overseas employees of the company to their families and relatives in Britain, see Hajeebu (2005).
42. To know how employees of the company working in India became wealthy individuals after completing their services in India, see Holzman (1926) and Marshall (1976), p. 229.
43. For the practice of multi-tasking by the overseas employees of the East India Company, see Holemstorm and Paul (1991).
44. To understand the consequences of the practice of own account trade by the employees of the East India Company, see Hajeebu (2005).

45. The use of ***dustaks*** and its consequences on the relationship between the Nawab of Bengal Subah and the East India Company, see Robins (2006), pp. 64–65.
46. For installing the Prince Mohammad Ali Wallajah at Arcot by the East India Company, see Bayly and Subrahmanyan (1988).
47. For insider buying of stocks by Robert Clive, see Robins (2006), pp. 85–86.
48. For these facts, see Stein (1999), p. 208.
49. For this, see Roy (1998), pp. 512–515.
50. See, Bowen (2006).
51. For this, see Travers (2007).
52. For this, see Stern (2004).
53. See, Travers (2007).
54. See, Robins (2006).
55. To know how Bolts described the practices of the officials of the East India Company, see Bolt (1772 [1998]).
56. This quotation has been taken from Robins (2006), p. 128.
57. See, Marx (1853a, b, c).
58. See, Desai (2009), p. 48.
59. See, Smith (1776, 2003), Book IV, Ch. I.

Bibliography

Akerlof, G. A. (1982). Labour contracts as partial gift exchange. *Quarterly Journal of Economics, 47*(4), 43–69.

Akerlof, G. A. (1984). Gift exchange and efficiency wage theory four views. *American Economic Review, 74*(2), 79–83.

Alchian, A., & Demsetz, H. (1972). Production information costs and economics organization. *American Economic Review, 62*(5), 777–795.

Anderson, G. M., McCormick, R. E., & Tollison, R. D. (1982). The economic organization of the English East India Company. *Journal of Economic Behavior and Organization, 4*(2–3), 221–238.

Arrow, K. J. (1985). The economics of agency. In J. Pratt & R. Zeckhauser (Eds.), *Principals and agents: The structure of business* (pp. 37–51). Boston: Harvard Business School Press.

Berle, A. A., & Means, G. C. (1932). *Modern corporation and private property.* New York: Macmillan.

Blair, M. (1995). *Ownership and control—Rethinking corporate governance for twenty first century.* Washington, DC: Brooking Institute.

Bayly, C. A., & Subrahmanyan, S. (1988). Portfolio capitalists and political economy of early modern India. *Indian Eco and Social History Review, 25*(4), 401–424.

Bolt, W. (1772 [1998]). Considerations on Indian affairs. In P. Tuck (Ed.), *The East India Company 1600–1858.* London: Routledge.

Boot Arnold, W. A. (2000). Relationship banking: What do we know. *Journal of Financial Intermediation, 9*, 7–25.

Bowen, H. V. (1987). Lord Clive and speculation in the East India Company stocks 1766. *The Historical Journal, 30*, 890–927.

Bowen, H. V. (1989). Investment and empire in the later eighteenth century: East India stock holding 1756–1791. *Economic History Review, 42*, 186–206.

Bowen, H. V. (2006). *The business of empire—East India Company and imperial Britain 1756–1833*. Cambridge, UK: Cambridge University Press.

Brendel, F. (1982). *Civilization and capitalism; wheels of commerce* (Vol. II). London: Collins.

Carlos, A. M. (1991). Agent opportunism and the role of Royal African companies compared. *Business and Economic History, 20*, 142–151.

Carlos, A. M. (1992a). *Bonding versus internal promotion and trust: Evidence from the Royal African Company 1672–1691*. Colorado: Mimeo Department of Economics University of Colorado.

Carlos, A. M. (1992b). Principal-agent problem in early trading companies—A tale of two firm. *American Economic Review, 82*(2), 140–145.

Carlos, A. M., & Nicholas, S. (1988). Giants of early capitalism—The chartered trading companies or modern multinational corporation. *Business History Review, 62*(3), 398–419.

Carlos, A. M., & Nicholas, S. (1990). Agency problem in early chartered companies—The case of Hudson Bay Company. *Journal of Economic History, 50*, 853–875.

Carlos, A. M., & Nicholas, S. (1996). Theory and history: Seventeenth century, joint-stock trading companies. *Journal of Economic History, 56*, 140–153.

Chandler, A. (1977). *The visible hand; the managerial revolution in American business*. Cambridge, MA: Harvard University Press.

Davis, J. H., Schoorman, F. D., & Donaldson, L. (1997). Towards stewardship theory of management. *Academy of Management Review, 15*(3), 369–381.

Desai, M. (2009). *The rediscovery of India*. New Delhi: Allen Lane and Penguin Books.

Donaldson, L., & Davis, J. H. (1991). Stewardship theory of agency theory-ECO, governance and shareholders returns. *Australian Journal of Management, 16*(1), 49–64.

Fama, E. (1980). Agency problem and the theory of firm. *Journal of Political Economy, 88*(2), 288–307.

Fama, E. F., & Jensen, M. C. (1983). Agency problems and residual claims. *Journal of Law and Economics, 26*(2, Corporations and Private Property: A Conference Sponsored by the Hoover Institution), 327–349.

Freeman, R. E. (1984). *Strategic management: A stakeholders approach*. Boston: Pitman.

Gower, L. C. B. (1969). *The principles of modern company law*. London: Stevens and Son.

Grief, A. (1993). Contract enforcementability and economic institutions in early trade: The Maghribi traders coalition. *American Economic Review, 83*(3), 525–548.

Grossman, S., & Hart, O. (1983). An analysis of the principal agent problem. *Econometaica, 51*, 7–45.

Hajeebu, S. (2005). Contract enforcement in English East India Company. *Journal of Economic History, 65*(2), 496–523.

Hall, P., & Soskice, D. (2001). *Varieties of capitalism—The institutional foundation of corporate advantage*. New York: Oxford University Press.

Hill, C. W., & Jones, T. M. (1992). Stakeholders agency theory. *Journal of Management Studies, 29*(1), 131–154.

Holmstrom, B. (1979). Moral hazard and observability. *Bell Journal of Economics, 10*, 74–91.

Holmstrom, B. (1982). Moral hazard in teams. *Bell Journal of Economics, 13*, 324–340.

Holmstrom, B., & Paul, M. (1991). Multi—Task, principal—Agent analyses: Incentives and contracts, asset ownership and job design. *Journal of Law, Economics and Organisation, 7*(1), 24–52.

Holzman, J. M. (1926). *The nabobs in England*. New York: Columbia University Press.

Housmann, H. (1996). *Ownership of enterprise*. Cambridge/New York: Harvard University Press.

Jensen, M. C., & William, H. M. (1976). Theory of the firm: Managerial behavior, agency costs and ownership structure. *Journal of Financial Economics, 3*(4), 305–360.

Jones, S. R. H., & Ville, S. P. (1996). Efficient transactions or rent seeking monopolists? Rationale for early chartered trading companies. *Journal of Economic History, 56*(4), 898–915.

Jones, T. M., & Wicks, A. C. (1999). Convergent stakeholders theory. *Academy of Management Review, 24*(1), 200–221.

Keay, J. (1993). *The honorable company*. London: Harper Collins.

Lawson, P. (1993). *The East India Company—A history*. London: Longman.

La Porta, R. (1999). Corporate ownership around the world. *Journal of Finance, 54* (2), 471–517.

La Porta, R. (2003). Agency problem and dividend policies around the world. *Journal of Finance, 55*(1), 1–33.

Marx, K. (1853a, June 25). The Biritsh rule in India. *New York Daily Tribune*. In I. Hussain (Ed.) (2006). *Karl Marx on India* (pp. 11–17). New Delhi: Tulika Books.

Marx, K. (1853b, June 24). East India Company—Its history and results. *New York Daily Tribune*.

Marx, K. (1853c, June 24). East India Company its history and results. *New York Daily Tribune*.

Marshall, P. J. (1976). *East India fortunes—The British in Bengal in the eighteenth century*. Oxford: Clarendon Press.

Mazumdar, S. K. (2012). *India's late, late industrial revolution—Democratization of entrepreneurship*. New Delhi: Cambridge University Press.

Mourad, K. (2012). *In the city of gold and silver—The story of Hazarat Mahal*. New Delhi: Full Circle Publishing.

Pollard, S. (1968). *The genesis of modern management*. London: Penguin.

Putnam, R. (2000). *Bowling action: The collapse and revival of American community*. New York: Simon and Schuster.

Raff, D. M. G., & Summers, L. H. (1987). Did Henery Ford pay efficiency wages. *Journal of Labour Economics, 5*(4, II), 557–586.

Roy, T. (1993). *Artisans and industrialization—Indian weaving in twentieth century*. New Delhi: Oxford University Press.

Robins, N. (2006). *The corporation that changed the world—How the East India Company shaped modern multinationals*. Hyderabad: Orient Longman.

Robinson, F. P. (1912). *The trade of East India Company form C.1709–C.1813*. Cambridge: Cambridge University Press.

Roe, M. J. (2002). The political foundations of separating ownership from control. In J. A. Mc Cathery, P. Moreland, T. R. Makero, & L. Rennebug (Eds.), *Corporate governance, regimes, convergence and diversity* (pp. 113–140). New York: Oxford University Press.

Salop, S. C. (1979). Strategic entry deterrence. *American Economic Review (Paper and proceedings),69*, 305–320.

Scott, W. R. (1909–1912). *The constitution and finance of the English and Irish joint-stock companies to 1720* (Vols. I–III). Cambridge: Cambridge University Press.

Seth, V. K. (1998). Economic reforms and capacity utilization in Indian industry. *Productivity, 39*(3), 391–398.

Seth, V. K. (2014). Debate on de-industrialization revisited: The process of decline of traditional flexible manufacturing. *Global Business Review, 15*(3), 597–610.

Shahj, K. T. (1948). *National planning principles and administration*. Bombay: Vora.

Smith, A. (1776, 2003). Introduction. In A. B. Krueger (Ed.), *The wealth of nations* (Bantam classic). New York: Bantam Books.

Stern, P. (2004). *One body corporate and politic—The growth of East India Company state in the late seventeenth century*. Unpublished PhD thesis, Columbia University, Columbia.

Stiglitz, J. E. (1984). *Theory of wage rigidities*. Working paper no 1442, NBER. Cambridge, MA.

Stiglitz, J. E. (1988). Equilibrium wage distribution. *Economic Journal, 95*(3), 599–618.

Times of India. (2010). East India company is back with Indian owner. *Times of India*, August 16.

Toynbee, A. (1958). *Industrial revolution*. Boston: Becon Press.

Varshney, A. (1999). *Democracy, development and country side urban and rural struggle in India*. New York: Cambridge University Press.

Weber, M. (1958). *The Protestant ethic and the spirit of capitalism*. New York: Charles Scribner's Sons.

CHAPTER 6

Modern Manufacturing and the British Raj

After the invention of the steam engine, which paved the way for the birth of the Industrial Revolution, British society, polity and economy also underwent revolutionary transformation. The changes transformed its traditional flexible manufacturing into modern manufacturing, and Britain emerged as the workshop of the world. This process has been described in greater detail in Chap. 3. The Industrial Revolution, for the first time in human history, removed constraints that had prohibited the possibilities of realization of human potential. It gave rise to a new period, when revolutionary changes became the norm and change became a constant. The importance of this event for our narrative is that from the womb of these historical changes, a new child was born. This child has been variously described as manufacturer, industrialist, entrepreneur or capitalist, depending upon which aspect of his biography one intends to highlight.

The manufacturers, as they increased in numerical strength, gained prominence in the society, polity and economy of Britain. This was inevitable because manufacturing provided possibilities to create unlimited amounts of wealth. The same was not true in agrarian economies, where owners of land could create wealth up to the extent the intrinsic quality of soil, or its fertility, permitted. Moreover, returns on land were 'diminishing' in nature. The emerging prominence of manufacturers as creators of wealth and employment opportunities resulted in the birth of new power elite in Britain. At the same time, Adam Smith's book, *The Wealth of Nations*, provided ideological support to the new power elite. The new ideology, the 'classical political economics', replaced the dominance of physiocracy

V.K. Seth, *The Story of Indian Manufacturing*,
DOI 10.1007/978-981-10-5574-4_6

(which protected the interest of peasants) and mercantilism (which provided ideological support for the activities of merchants and the existence of companies like the East India Company). This shift in the ideological premises of policymakers in Britain is narrated in Chap. 4 of the book.

Classical political economics became the new testament of the new elite, which preached the appropriateness of policies based on laissez faire, private property right and free trade. The new elite, through its influence on policymaking, began to challenge the premise that provided justification for the rule of the company of merchants on the Indian sub-continent. The new elite also challenged the monopoly of trade with the East, given to the East India Company by the British Crown. Influenced by the ideology of manufacturers, policymakers in Britain began to impose more stringent regulation on the East India Company. The process began with the Regulation Act of c.1778. It was followed by Pitt's India Act c.1778, and the Charter Act c.1793. Each successive regulation reduced the extent of the autonomy of the East India Company. In c.1813, the new political force succeeded in removing the monopoly of trade with the Indian sub-continent given to the company by the British Crown. Finally, in c.1833, the company was stripped off all its commercial activities. This process is described in Chap. 5. Therefore, c.1813 can be considered an important date, since this is when the paramountcy of the interest of British manufacturers began to influence policymaking in Britain.

A similar process against the rule of the company was also unfolding in the Indian sub-continent. The rule of the company had affected the interests of several groups and strata of the Indian sub-continent. The collective discontent of these diverse sections of society was expressed in the form of the Sepoy Mutiny (as described by imperialist historians), or first war of independence (as projected by nationalist historians). The revolt began when the troops of the 19th native cavalry deserted and marched towards Delhi on the night of May 10, 1857. Several attempts have been made to analyse this episode by different historians (imperialist, nationalist, Marxist and rationalist). The interest in this episode was revived recently with the appearance of a well-researched, well-documented and brilliantly written book by William Dalrymple, *The Last Mughal.*[1] Several other books also appeared in c.2007 (some old reprints and some new contributions), when India celebrated the 150th anniversary of the episode. We are not going to digress to analyse different dimensions of the revolt, because the anatomy of this event has already been covered by several historians.[2,3]

The most important consequence of the first war of independence for the story of Indian manufacturing was that the rule of the company was abolished and the Indian sub-continent came directly under the rule of the British Crown through the Better Government of India Act c.1858. The end of the rule of the company was the beginning of the British Raj. The transfer of the rule of the East India Company to the rule of the British Crown and Parliament uncovered the reality to the masses in the Indian sub-continent that from then on, India would be directly ruled by the British Raj and not by any of its proxies. The transfer of all the territories ruled by the East India Company to the British government created a unitary political entity called India, which could be transferred to any new political entity, en bloc, in the future. In c.1876, the British government organized the Delhi Darbar to revive the Mughal tradition and display of imperial splendour. Around four hundred princes were invited to participate in the event. The Delhi Darbar was organized to emphasize the new unity under the British Raj, which had replaced the centrality of the Mughal Empire. To this historical event Sir Dinkar Rao, Chief Minister of Maharaja of Gwalior described, 'India has been until now a vast heap of stones, some of them small. Now from roof to basement, each stone of it is in its right place'.[4] Direct rule of the British Crown also meant that henceforth, the interest of manufacturers of Britain would directly influence the policies of British government towards the Indian economy in general and towards Indian manufacturing in particular. The buffer created by the existence of the East India Company would no longer be able to stop this from happening.

In this chapter, I attempt to analyse the impact of the transfer of power to govern the Indian sub-continent, from the East India Company to the British Raj, on Indian manufacturing, which, in fact, amounts to an analysis of the consequences of shifts in paramountcy of the interest of merchants to the interest of manufacturers in the policymaking of the British government. To understand the impact of this historical fact, this chapter is divided into three sections. Section 6.1 describes the changes that occurred in the pattern of governance due to transfer of power directly to the British Raj. Section 6.2 deals with the process of emergence of modern manufacturing in India.

6.1 Changes in the Pattern of Governance

Before c.1857, the administration of the Indian sub-continent was conducted by the Court of Directors of the East India Company, which was elected by the shareholders of the company. The court of directors was answerable to the board of control, a body set up by the British government in c.1784. The court of directors appointed a hierarchy of examiners, writers and merchandisers to run the company's affairs. Apart from officials working in the London office of the company, several officials who were working in different parts of the Indian sub-continent were also appointed. The highest position in the administration of the East India Company, in the Indian sub-continent, was enjoyed by the governor-general, whose office was located at Calcutta (Kolkata), followed by the governors of each of the presidencies, Bombay (Mumbai) and Madras (Chennai).

Subsequent to the transfer of power, the court of directors and the board of control were abolished and replaced by the office of the secretary of state along with his advisory board. The office of secretary of state was occupied by the person who enjoyed the confidence of the ruling party in the British Parliament. The office of the governor-general was replaced by the office of viceroy. Similarly, examiners and writers were replaced by permanent secretaries. In new circumstances, the policies regarding the Indian sub-continent were largely influenced by the shifting fortunes of the political parties in the British Parliament. Prior to the transfer of power, in the kind of governance structure that existed, the company's administration was quite authoritarian, devoid of any external interference. In the new structure of governance, the influence of the ruling political party in the British Parliament became paramount in policymaking. The involvement of British Parliament in the administration of the Indian sub-continent provided scope for the influence of pressure groups that existed in British political structure at that point in time, such as the textile manufacturers of Manchester and Lancashire, jute manufacturers of Dundee, salt manufacturers of Cheshire and traders of opium and indigo.

The civil servants of the East India Company were trained at the East India College, which was located at Haileybury, Hertfordshire. This college was established in c.1805 and civil servants of the company were especially initiated in the ideas contained in the writings of political economy in the East India College. Thomas Malthus held the chair on political economy of the college between c.1805 and c.1834, Richard Jones between c.1834 and c.1855, and James Stephen between c.1855 and c.1858. The college was closed in c.1858. Civil servants were selected through a competitive exam. The civil servants 'were trained to rule not to serve. Though by ruling they served'.[5] These officials were not expected to be great scholars or intellectuals, but were expected to have the ability to collect data and argue convincingly based on these data. Statistics became their obsession, perhaps because it was known at that time as 'political arithmetic'. However, the exception to this stereotype was some of the members of the Indian Civil Services (ICS). Scholars of Indian history and economy must be familiar with the brilliant writings of ICS officers like Malcolm Darling, William Todd and W.W. Hunter.[6]

All the candidates who were selected to join the ICS had to study classical political economy as a compulsory subject. They had to read all the important texts on political economy like McCulloch's edition of the *Wealth of Nations,* and Ricardo's *Principles of Political Economy and Taxation.*[7] It was believed by policymakers that the knowledge of political economy would equip the civil servants to find appropriate answer to questions relating to policy formulation. Most civil servants, who were selected through competitive examination, generally came from the British middle classes. Principles of political economy had special appeal to the mind-set of middle classes of Britain because it provided importance to competition and merit rather than to family and kinship ties. The value system based on the premises of political economy, in fact, had replaced the acceptance of tenants of orthodox Christianity as a dominant value system amongst British middle classes.

After the birth of political economy, policymakers in Britain began to depend on the theoretical premises implicit in the text of political economy. Despite the fact that there existed several areas of disagreement amongst classical political economists, there existed consensus amongst them on wide areas. All of them agreed that policies made by the government should not affect the working of free markets. In fact, they advocated that government policies should facilitate the efficient working of the market economy. To achieve this objective, most policymakers of the British Raj accepted that there should be (i) free inter-regional and international trade, (ii) interference of the government in only those areas where markets fail, and (iii) rule of law and regulations to enforce contracts.[8]

The British Raj intended to erase the image of the empire created by the East India Company as an agency of transfer of resources from colony to Britain. Therefore, unlike the Mughal state, which did not legislate to give freedom to a conventional legal framework that existed in the form of a community-based redressal system.[9] The British Raj believed in the legislative state. The British policymakers thought that the Indian society was segmented into different religions and was also hierarchical due to the prevalence of caste-based hierarchy. Therefore, the British Raj wanted to project itself as an institution that provides impartial justice.

This was necessary to project because the government during the British Raj wanted to establish that the British rule is necessary for India, otherwise the fragmented and hierarchical society of India will get translated into fragmented polity. However, it was impossible to provide equality in justice when the expatriate British population was politically unequal. In practice, the legal structure placed British citizens above law and protected their acts of omission and commission. The Regulation Act of C-1773 laid the foundation for the establishment of the supreme court of India. It was established essentially to maintain the principle of extra-territoriality to all British subjects in India, to project that British citizens, wherever they live in the colonies, were subject to only British laws. The premise on which this principle was based suggested that irrespective of the country where British are settled as colonizers, its citizens would be tried by English courts and presided over by English judges, according to English laws and argued by English advocates.

The first example of this extra-territorial principle was observed immediately after the establishment of the supreme court in India with the trial of Nand Kumar who was diwan of Nawab of Bengal. In March c.1775, he filed a case against Warren Hastings, in which he charged him of taking a bribe of £35,000 in lieu of help he provided for appointment in the court of the Nawab. Instead, Warren Hastings sued him for forging the document, which was proof that he had taken the bribe. The first chief justice of the supreme court, who was a close friend of Hastings, Elijah Impay, gave Nand Kumar the death sentence on the charges of forgery. The forging of a document was not such a serious crime to be punished by the death sentence. This was described at that time as 'judicial murder'. Bal Gangadhar Tilak appropriately described the prevailing reality when he wrote that 'the Goddess of British justice though blind, is able to distinguish between black and white'.[10] Apart from providing justice, legislations were also required for guaranteeing efficient working of the market. This necessitated the legislations regarding property rights and enforcement of contracts.[11]

By c.1857, British manufacturing had been enjoying competitive advantage, emanating from several inventions and innovations, which facilitated the birth of the Industrial Revolution. These advantages had eliminated the fear of competition from the Indian traditional flexible manufacturing. In fact, most of the traditional manufacturing had experienced a decline, as has already been discussed in detail in the section on de-industrialization in Chap. 4. Therefore, for the British policymakers, ideology of free trade advocated by classical political economists was most appropriate. The rule of the East India Company was perceived by British manufacturers as an institution that was protecting the interest of Indian traditional flexible manufacturing. After all, the East India Company had earned its profits from the exports of products made by Indian traditional manufacturing. Moreover, its monopoly of trade with India denied them access to Indian markets. In the aftermath of the East India Company, such fears also disappeared. Therefore, British manufacturers wanted the policy of free trade to have free access to Indian markets.

The policy of free trade was advocated by British manufacturers, despite knowing that their own growth was made possible under protectionist policies of the British government. Not only were their protectionist policies protected by wars against competing nations, but so were their interests. In new circumstances, protection was helping the landed gentry of Britain rather than manufacturing because protection increased the price of finished goods due to higher costs of imported raw materials. Imposing tariffs on the imported food grains increased prices of wage goods, which necessitated the payment of higher wages. Therefore, the growth of modern manufacturing in Britain is considered as the first historical example of 'infant industry' arguments.[12]

The principles of classical political economy provided theoretical support to the interest of British manufacturers. The institutional structure that these principles justified consisted of a legal framework to protect private property rights and enforcement of contracts. These principles also supported the policies of free trade and freedom to enterprise. Classical political economists believed that if such institutions do not exist in any country, and existing value systems and habits of thoughts of the society are constraints in evolving these institutions, government should create them through legislations. Classical political economists laid more faith on the working of the 'invisible hand' of the market, which gives freedom to economic agents to make choices, rather than on 'visible hand' of the state.

Accepting the logic inherent in the classical political economy, the British Raj followed the policy of free trade in India, while all other independent nation states of the world were imposing tariff walls to protect their infant manufacturing enterprises. Bismarck raised tariffs in Germany in c.1879, and tariffs were raised in France c.1881, Russia c.1882, America c.1890 and Canada c.1897.[13] The British Raj in India followed the policy of non-intervention by state. The absence of the role of state in providing protection to local enterprises has been described as 'night watchman role of the state';[14] the objective being to protect the interest of Britain manufacturers in disguise.[15] It is also an historical fact that while imports from Britain were without tariffs, exports from India were subject to tariffs. This is how the colonial government was carrying out free trade, which has been described as 'one-way free trade'.[16]

Despite the fact that during the British Raj, the role of state was insignificant, its indirect influences on the economy were quite significant. Indirect role of the state emerged as a consequence of public spending and through the mode of earning revenue. Due to the insignificant direct role of state, it was believed that if modern manufacturing did not emerge rapidly during the British Raj, then its reasons should be located in the lack of opportunities or constraints on investment decisions. This is believed because 'during the century and half of British rule, the market was given its head. British India was the great social experiment in letting interest and market forces do virtually everything'.[17] Hence, scholars who believe in the argument attribute the reasons for delay in the emergence of modern manufacturing in India to endogenous factors that restricted the efficient working of the market. These restrictions existed on the supply side (like lack of means of transportation and communication, economic and social infrastructure, availability of skilled manpower and managerial talent), and demand side (like the fragmented markets and inequalities of income, which did not provide necessary conditions for the production of standardized goods on a mass scale).

The discrimination inherent in the policymaking became quite obvious when British government intended to develop railways in India. The railways were introduced in India during the rule of the East India Company in c.1845, when the first railway line was constructed between Bombay and Thane. However, the railways were strategically used to protect colonial interests. For example, the revolt of c.1857 prompted the British Raj to develop railways on a massive scale for the purposes of defence, so that the movement of troops, as well as military hardware to locations where military operations were required, could be facilitated. Moreover, by c.1857, traditional flexible manufacturing had experienced decline and India had emerged as an important supplier of raw materials for British manufacturing enterprises. Railways had to play an important role to establish connectivity between the centres of production of raw materials and port cities. Initially railway lines were constructed to link important port cities with hinterlands, which supplied important raw materials, such as Bombay city in c.1853, Calcutta in c.1854 and Madras in c.1856. In addition to these strategic needs, railways helped distribute British manufactured goods to different markets in India. Hence, railways assumed strategic importance for the British Raj to achieve its political as well as commercial interest.[18]

To maximize the strategic advantage of developing railways, British government intended to expand network of railways across the Indian sub-continent in the shortest possible time. This necessitated encouragement of private investment in railways. To encourage private investment from Britain, the British government evolved the policy of guaranteeing a 5 per cent rate of return to private railway companies. In case a private company failed to achieve this 5 per cent rate of return on its investment, the government promised to compensate the railway company. This policy flooded the railway companies with requisite finance from the British investing public. However, a similar advantage was not extended to the investing public in India. The guarantee of return scheme gave to the British government the right to purchase the railway lines after twenty-five years had lapsed. This was perhaps the first example of Private Public Participation (PPP). Through this mode of organization of investment, by the 1920s, the government became the owner of 67 per cent of total railway mileage.[19]

As a consequence of the scheme, several private railway companies were established that developed their own networks of railway lines, such as East India Railway Company, Great Indian Peninsular Railway, Sindh Punjab and Delhi Railway, and North Western State Railway. Apart from these big railway companies, some smaller railway companies also emerged that provided railway services to smaller areas, like Rajputana Malwa Railway, East Coast Railway, Bengal Nagpur Railway and Madras Railway. Along with these private railway companies, the government of Britain also developed a government-owned railway network to earn revenue. Due to all these efforts, by the beginning of World War I, the aggregate distance covered by the entire railway network, both private as well as public, consisted of around 40,000 km.[20]

While a large part of investment for the railway was raised in Britain with guaranteed returns, most of the requirements of the railway were procured from British suppliers. Therefore, a large part of investable funds raised from Britain flowed back to Britain by way of payments for imported locomotives, rolling stock, track materials, sleepers and wages (including salaries of British managers and skilled workers). To buy the equipment for railways, the government floated tenders only in Britain.[21] Due to these faulty policies, even the development of railways in the Indian sub-continent by the British government did not facilitate the development of its economy; it largely served the interest of British manufacturing and administration of the economy. The benefits that occurred to the Indian economy were incidental. One of most important unintended consequences of the railway network developed by the British administration was on the biography of Gandhi. It is while travelling in a train in South Africa (also developed by the British), that Gandhi learnt important lessons about the value of human dignity. He travelled extensively across different parts of India using the Indian railway after coming back from South Africa in c.1905. These travels transformed Mohandas Karamchand Gandhi into Mahatma Gandhi, and provided added strength to his conviction that India would be best governed by Indians rather than by the British Raj. Therefore, he initiated the Quit India Movement in October 1942.

Indian railways also charged discriminating freight rates. From port cities to inland movement of goods, tariffs were less in comparison to inter-city freight rates. This policy subsidized movement of goods meant for international trade, but increased the costs of movement of goods meant for inter-regional trade. The nationalist leaders provided an important example to make the British government aware of this anomaly. For instance, transporting of imported matches from Bombay port to Delhi cost the same in tariff as transporting them from Ahmedabad to Delhi. This, despite the fact that the distance between Delhi and Ahmedabad was 483 km less.[22] Due to its emphasis on facilitating exports of raw materials and movement of troops, the Indian railway network did not revitalize the Indian economy.

Railways might have emerged as a leading sector of the Indian economy if British government had initiated a policy for buying of railway equipment from an India-based supplier. This would have provided immense scope for the growth of Indian manufacturing industries that have strong technological linkages with railways, such as the iron and steel industry, engineering industries, coal mining, and manufacturing of railway wagons and signalling equipment. If the British government had met the requirement of railway suppliers based in the Indian sub-continent, it would have provided an incentive to Indian entrepreneurs to establish these industries, or the policy might have forced foreign manufactures to set up operations in India, in which case it would have provided learning opportunities to Indian entrepreneurs and workers.

However, circumstances changed significantly during World War I. The wartime needs forced British policymakers to deviate from the policy of laissez faire as well as free trade. The departure occurred through the appointment of Indian Industrial Commission in c.1916, followed by Indian Fiscal Commission in c.1922, and later by Montagu-Chelmsford reforms. There exists an alternative strand of thought on the consequences of these events on policymaking in India. Economic historians like Bagchi, Chatterji, Ray and Tomilson believe that these events did not significantly alter the stance of policymaking in India.[23] On the contrary, historians like Dewey believe that through these legislations and reforms, a certain amount of sovereignty was transferred to policymakers in the Indian sub-continent, which opened scope for bargaining by different pressure groups that were functional in India at that point in time.[24]

A short stint by the Labour Party in the British government between c.1929 and c.1931 slightly benefited the Indian interest because the Labour Party was willing to accommodate the interest of Indian manufacturers. The changes that occurred in the global economic scenario during the Great Depression of the 1930s also compelled the British government to deviate from the policy of laissez faire and free trade. The Great Depression began to spread from the centre of the world market to countries that were on its periphery (colonies) like a communicable disease. Once the countries on the periphery caught this virus, the disease became more persistent because virus affects weaker economies for a longer period. By the 1930s, the Indian manufacturing sector, though not very large, had developed needs of its own for its continuous survival. The British government, which had given up the idea of free trade to protect the interest of its own manufacturing, also introduced protective tariffs in India. Therefore, those manufacturing industries that produced largely for the domestic market or produced import-substituting product benefited a lot. The tariffs imposed during the period of the Great Depression helped the growth of cotton textiles, sugar and steel manufacturing.[25] Moreover, British government also gave up the policy of gold standards in c.1931. This helped in stabilizing the exchange rate of the Indian rupee.

The end of the Depression almost coincided with the outbreak of World War II in c.1939. This time, cessation of imports was almost totally due to the disruption of sea routes. The entry of Japan brought the war to India's doorstep. India became an important base for military supplies including arms and ammunition on behalf of the allied forces. The ordnance factories in India were producing 700 types of munition.[26] To force Indian manufacturing enterprises into meeting the wartime requirements of the British government, the government began to interfere in the working of Indian manufacturing with the Defence of India Rules, which were promulgated in c.1939. These rules gave far-reaching power to the state to influence the growth of manufacturing enterprises in India. For instance, Rule 81 of the Defence of India Rules gave authority to the government to regulate or prohibit the production of certain articles or things of any description. Similarly, Rule 84 provided to government the power to prohibit imports or exports to control the purchase of foreign exchange. The British government used the provisions given in the Defence of India Rules to regulate Indian manufacturing to achieve its wartime objectives.[27]

During the British Raj, an important innovation occurred in India in the organization and management of manufacturing enterprises. This innovation was the evolution of the managing agency system. As mentioned in Chap. 4, to accommodate the interest of British manufacturers, the East India Company had permitted the entry of free merchants who acted as agents of British manufacturers to supply products produced by the British mills in the Indian markets. These agents formed agency houses, which were partnership companies founded by free merchants, ex-civil servants and military officials of the company under the patronage of the administration of the East India Company. These companies were generally financed with the earnings of the company officials, from their private trade or their savings.

The managing agency system came into existence in the Indian sub-continent when an agency house first promoted a manufacturing enterprise as a joint-stock company and then acquired the right to manage it. This kind of event, first of all, occurred in c.1836, when one of the agency houses, Carr Tagore & Company, founded by a partnership between William Carr and Dwarkanath Tagore (grandfather of Rabindranath Tagore), promoted Calcutta Tugg Association and assumed its management. At the same time, Parry & Company became managing agents of Proto Nova Steel and Iron Company.[28] The managing agency system grew out of the needs of prevailing circumstances, when there existed shortages of managerial and technical skills. It facilitated the financing of manufacturing projects through public investment, through share capital, and protected at the same time the interests of promoters. To protect the interest of promoters, it gave them right to manage the enterprise as managing agents who earned commission as a fixed per cent of the total sales of the enterprise. The managing agency system created a unique pattern of organization of business enterprises in India where private interest managed public funds. Some of the important managing agency houses of that period were Andrew Yule & Company, Duncan Brothers, Williamson Magor & Company, Birds & Company, Martin Burn & Company, Balmer Lawrie & Company, Shaw Wallace & Company Limited, Parry & Company, Mcleod & Company and Rolli Brothers.[29]

6.2 Emergence of Modern Manufacturing

As mentioned, there existed implicit as well explicit discrimination in British policies regarding the emergence of modern manufacturing, especially during the early years of the British Raj. Despite these discriminations, a modern manufacturing sector did emerge in India. This happened because historical circumstances provided opportunities to the Indian entrepreneurs to accumulate capital that was necessary to invest in modern manufacturing. Circumstances also provided them with the availability of modern technology through their interaction with the British manufacturers. The existence of these two important preconditions, which are necessary for the growth of modern manufacturing, helped Indian entrepreneurs to start modern manufacturing enterprises. Initially, they initiated enterprises in two textile manufacturing industries—cotton and jute. In both the manufacturing industries, India had competitive advantage as it had long traditions of craft-based production in these industries, and also had local availability of raw materials.

These two textile manufacturing industries took root at two different locations. The cotton textile manufacturing emerged in the western part of the Indian sub-continent at Bombay, whereas the jute manufacturing industry took root in the eastern part of the sub-continent, at Calcutta. These two locations were separated from each other by a distance of about 1600 kilometers. Prior to the emergence of a railway network and means of communication, there did not exist an integrated market economy in India. In fact it consisted of several fragmented markets unified by only single currency and uniform laws imposed by the British Raj. Therefore, we can assume that, at that point in time, eastern and western parts of the continent existed as separate regional economies.

Bombay had emerged as an important centre of maritime trade after the decline of Surat in the western part of the Indian sub-continent (decline of

Surat has been explained in an earlier chapter of the book). In c.1818, when the Maratha Empire was annexed by the East India Company, Bombay was just a port city. Bombay assumed importance after the abolition of monopoly of trade given to the East India Company in c.1833 and opening up of the market in China by British military in c.1842. These two historical events provided opportunity to Indian merchants who had migrated to Bombay from Gujarat. Some of the communities of merchants that migrated to Bombay were Parsis, Bhatia merchants, Bohras, Khojas, Memons and Marwaris. The most important articles for export from Bombay were cotton and opium. Opium was primarily exported to China and cotton was exported to the textile mills of Manchester and Lancashire. Subsequent to the opening of the Chinese market for opium, Bombay merchants became drug barons of the nineteenth century and Bombay emerged as Medellin.[30] This happened because the opium of Malwa was also routed through Bombay to China.[31]

Calcutta, at that point in time, was the capital of the British Raj until c.1911. It had the largest concentration of Europeans and European merchants. In addition, Calcutta had the largest concentration of agency houses floated by British merchants, who enjoyed a strong link with British administration as well as with British manufacturers. Calcutta also attracted Indian business communities; the Marwari merchants of Calcutta especially had earned substantial wealth from the exports of raw silk, jute, sugar and indigo, and by selling the manufactured products of British manufacturing companies in Indian markets. Moreover, coal, which had emerged as a prime mover of modern machines, was also locally available.

The British government gave freedom to British manufacturers of textile machinery to export it to other countries in c.1848. Until then, the export of textile machinery and equipment to other countries had been prohibited to maintain the monopoly of textile manufacturers of Britain. The British government was forced to liberalize exports of textile machinery because textile machinery manufacturers of Germany, France and Belgium had begun to export textile machinery by this time.[32] Following the liberalization of trade in textile machinery, standard equipment producers of Britain, like Platt Brothers, Oldham and John Hetherington & Sons, began to supply, to Indian promoters of textile mills, turnkey solutions, which included supply of plants, equipment, skilled persons and managers.[33] Availability of capital and technology for manufacturing cotton textiles provided objective conditions for the Indian merchants of Bombay to establish modern textile mills.

Trading in jute was important since the early days of the East India Company when it was used for manufacturing cordage and ropes. Until the 1830s, Bengal had a monopoly of handloom manufacturing of jute textiles. In c.1838, the Dundee mills in Ireland were successful in transforming the jute fibre into yarn with the help of a modified form of power loom. This led to an increase in the demand of Indian jute. Indian jute achieved near monopoly position during the Crimean War, which cut off regular supply of Russian hemp, which was an important substitute for Indian jute.[34] This boom provided considerable earnings to the agency houses of Calcutta that were involved in the exports of jute. These agency houses had strong linkages with the Dundee mills due to their contacts with them. They were able to acquire knowledge about jute textile machinery and equipment necessary to establish jute textile mills.[35] Due to these advantages, the agency houses of Calcutta took the leadership in promoting jute textile manufacturing in Calcutta.

The simultaneous growth of two textile manufacturing industries, at these two separate locations, has been described by scholars as '*dual dominance*', because these two locations continued to enjoy significant position in the manufacturing map of the Indian sub-continent.[36] These two different manufacturing industries, at two separate locations, were example of two different modes of growth of manufacturing where each mode represented two interrelated but antagonistic elements. In the case of cotton textile manufacturing, entrepreneurship came from Indian merchants, who had made riches from the trade in cotton and opium. In the case of jute textiles, the leadership was provided by ex-civil servants of the East India Company and Calcutta-based agency houses managed by British citizens. Due to this reason, investments in the jute textile industry were partly financed by British capital (i.e., sterling capital) and savings of British civil servants.[37] On the other hand, investments in cotton textile manufacturing came from indigenous accumulation of capital by Indian merchants. The attitudes of British policymakers were also different in respect to these two manufacturing industries. Cotton textile manufacturing was slowly emerging as an important competitor to Manchester and Lancashire cotton textile mills. On the contrary, jute textile mills were owned by expatriate British and were engaged in the production of a product that was emerging as a necessary complement to British manufacturing industries.

In the manufacturing of cotton textiles, earlier attempts to start modern mills were made by British and other European entrepreneurs. However, due to wrong location and financial problems, these enterprises had mixed success. The first cotton textile mill, called Bowreah Mill, was opened by British entrepreneurs near Calcutta in c.1818, by Kettlewell Butler & Company. The company also managed two other cotton textile mills in Bengal, the Durban Mill and New Ring Mill. In the 1830s, another spinning and weaving mill was functioning in Fort Gloucester about fifteen miles from Calcutta. This factory had employed five steam engines to run its machine. Steam engines were operated by bringing coal from Burdwan. Two cotton textile mills were also established by French entrepreneurs in Pondicherry (now called Puducherry) in c.1831 and c.1838, respectively. Most of the equipment and machinery for these cotton textile mills were imported from France, apart from the steam engines, which were imported from Britain,. France was exporting blue cloth dyed with indigo to Senegal, where it was exchanged for gum Arabic, which was essential input for the French clothing industry. French mills had taken the services of a master weaver, Thomas Geodefrog, to improve the skills of local weavers, and services of a chemist, Michael Gonfreville, to improve the quality of indigo-dying techniques. In c.1829, the factory management brought one mechanic, two assistant mechanics, a turner for woods and metals and a fitter who was trained in making combs and shuttles for weaving machines. These craftsmen were brought to India to train local artisans and to establish a model workshop. By c.1832, most of the French craftsmen were replaced by Indian workers.[38]

A British entrepreneur, James Landen, started a cotton textile mill at Broach in Gujarat in c.1853. Landen had been appointed by the British government in c.1840 to help in improving the quality of Indian cotton. He was involved in experiments to cultivate long-staple American cotton in India. He was also instrumental in introducing American saw gins in Gujarat's cotton growing areas. The cotton cleaned with the help of this equipment was more acceptable in British mills.

The first Indian who thought about starting a modern cotton textile manufacturing mill was Ranchhodlal Chhotalal, who was from Gujarat. He worked in British administration and occupied the position of assistant superintendent of customs (the highest position that could be occupied by an Indian at that point in time). However, he was removed from this position on corruption charges. Being an important bureaucrat, he had contacts with several British officials. He was in contact with Captain Fulljames and James Landen. Fulljames had collected plenty of information regarding textile equipment suppliers of Britain while helping a group of

Gujarati merchants who intended to establish a spinning mill at Surat, called Surat Cotton Spinning Association. Landen had already opened a cotton textile mill at Broach and was encouraging local merchants to start cotton ginning factories to supply clean cotton to his mill. Encouraged by Landen, Ranchodlal Chotelal started his cotton gin factory at Broach and planned to establish a cotton textile manufacturing mill at Ahmedabad. However, due to reluctance of native Gujarati merchants to bear the risk of entering into cotton textile manufacturing, he could not start the mill at Ahmedabad. If he had been successful in raising sufficient capital to start the mill at Ahmedabad, Ahmedabad could have been the birth place of cotton textile manufacturing in India.[39]

Cowasjee Davar floated the first successful steam-powered cotton textile mill at Tardeo in Bombay, called Bombay Weaving and Spinning Company on July 7, c.1854. The equipment for this mill was supplied by John Hetherington & Sons of Manchester and skilled manpower was also sourced from Britain. Sir William Fairbairn, an architect from Britain, designed the building of this factory. The team of British technicians, consisting of Whitehead, Lambert, Newton and Beardsall, acted as erectors. Grean Halm was appointed as carding master and Knott as spinning master.[40] Maneckji Petit established another cotton textile manufacturing mill, the Oriental Spinning and Weaving Mill also in c.1854, which began production in c.1858. Opening of this mill was followed by starting of Victoria Mill in c.1860 and in the same year, Mangaldas Nathubhai promoted Bombay United Spinning and Weaving Company, and Dinshaw Petit established Maneckji Petit Mill. In the month of August c.1860, Bomanji Wadia opened Bomanji Hormusjee Spinning and Weaving Mill, and in September c.1860, Merwanji Bhavnagar and Kapadia established New Great Eastern Spinning and Weaving Mill. Subsequently, Maneckji Petit started Royal Mills and Petit Mill during c.1861–1862.[41]

After great struggle, Chhotalal was finally able to float a cotton textile mill at Ahmedabad. After the success of several cotton textile mills at Bombay, Ahmedabad merchants now began to believe that successful cotton textile mills could be established. By c.1858, Ranchhodlal renewed his efforts to open a textile mill at Ahmedabad. This time his efforts received positive response from local merchants. He ordered the equipment for the proposed mill from Britain. However, the first shipment of textile machinery was lost due to a storm at sea and the British engineers who had to establish the factory at Ahmedabad died of cholera. A replacement cargo of equipment was claimed under the provision of insurance. The equipment finally arrived at a sea port near Ahmedabad, the Cambay port. In the absence of railway between Cambay and Ahmedabad, textile machinery was brought to Ahmedabad on bullock carts, covering distances of more than 100 kilometers.[42] The establishment of this mill at Ahmedabad led to the promotion of several cotton textile mills at Ahmedabad. Subsequently, Ahmadabad emerged as an important rival to Bombay-based cotton textile mills.

On April 12, 1861, the American Civil War began, which suspended the supply of American cotton to British cotton textile mills. This led to massive demand for Indian cotton by British mills. The exports of cotton from India to Britain were valued at £148 lakhs in c.1862–1863, compared to £93 lakhs in the preceding year.[43] This incident created a strong linkage between British Manchester and the Indian Manchester (Bombay). Despite the fact that the Bombay merchant earned substantial profits during the cotton boom, they lost most of the profit earned on speculation rather than on investment. On May 2, 1865, the American Civil War came to an end, which resulted in a sharp fall in the price of Indian cotton. This led to a serious financial crisis in Bombay. The expansion in investments in cotton textile mills stopped, which were renewed only after the Indian business community regained confidence.

Between c.1873 and c.1874, twenty cotton textile mills were added in Bombay. The concentration of cotton textile mills in Bombay provided preconditions for mill owners with common problems to form in c.1875 the Bombay Mill Owners Association. Some of the important mills that were established during this period were promoted by a highly diversified group of entrepreneurs. In c.1870, Morarji Gokuldas converted his jute and wool unit into a cotton textile unit and named it Morarji Gokuldas Mill. With this mill, Kutchi Bhatia merchants entered cotton textile manufacturing. In c.1874, another Bhatia merchant, Maneckji Khatau, opened Khatau Mills. These new mills ended the monopoly of Parsi merchants in cotton textile manufacturing in Bombay. Thereafter, Albert Sassoon founded Sassoon Spinning and Weaving Company at Bombay in c.1874 and c.1893. Kowasji Naik promoted Alexandra Mills in 1892. This was subsequently purchased by Nusserwanji Tata in c.1893. James Greaves and George Cotton, two British entrepreneurs, established Greaves Cotton and Company in c.1879; and Charles Killick and Presto Nixon launched Kohinoor Mills. The managing agency house of Bradburry and Brady opened several cotton textile mills at Bombay. Around the same time, Nowroj Nusserwanji Wadia established Bombay Dyeing and Century Mills. Until the1890s,

Bombay remained the most favoured location for cotton textile manufacturers. At that point in time, cotton manufacturers of Bombay were primarily exporting yarn to the markets of China and Japan. The location of cotton textiles near Bombay, being an export-oriented industry, was highly appropriate. Even the humid climate of Bombay was suitable for the growth of cotton textile manufacturing. And the manufacturers at Bombay had captured a significant share in the Chinese and Japanese markets. The Indian twist yarn had emerged as such a phenomenal success that the Manchester Chamber of Commerce conducted an enquiry regarding its success in the markets of Southeast Asia.[44]

The dominance of Indian cotton yarn in the markets of China and Japan was short-lived. During c.1860s, the spinning of cotton by machines was introduced in Japan. The spread of modern technology of spinning reduced the demand of Indian yarn in the Japanese market.[45] In c.1894–1895, Japan defeated China in the war, and the Chinese markets were captured by Japanese mills. Until c.1893, Indian cotton textile mills were supplying 96 per cent of the total demand of Chinese markets. However, after the defeat of China, export of Japanese cotton yarn increased significantly to China, and Indian exports registered a dramatic fall.[46]

These new developments created the necessity for a structural adjustment of the Indian cotton textile industry. It began with targeting the Indian handloom weavers for supplying them yarn. Handloom weaving was still continuing in India at several centres. It also necessitated vertical integration of cotton spinning companies from only spinning yarn to weaving of cloth, also to capture the domestic market of cloth. This led to the emergence of integrated mills. This transformation made Indian cotton textile manufacturing inward looking. At the same time, cotton textile manufacturers realized that to maintain their competitive advantage they had to spread to different cotton growing areas of the Indian sub-continent. All these factors slowly led to the spread of manufacturing of cotton textiles from Bombay to other parts of India. This change occurred because, by now, the cotton textile manufacturing in India had moved from an import-substituting industry to an export-oriented one.

It has already been described how Ranchhodlal Chhotalal made a pioneering effort to establish the first cotton textile manufacturing mill at Ahmedabad in c.1858. Ahmedabad had a long tradition in manufacturing cotton, silk, wool, and silver and gold brocade work. It was also an important market of indigo, which was produced at Surkhejee, a place that is now part of the Ahmedabad district. Despite these advantages, it took several years for another Gujarati merchant to promote the next textile mill in the town. In c.1876 Chotelal started another cotton textile mill in Ahmedabad. This mill was followed by Mansukhbhai Bhagubhai, a Gujarati Jain entrepreneur, who set up the Gujarat Spinning and Weaving Company in c.1877. Another Gujarati Jain entrepreneur, Karamchand Premchand, launched Ahmedabad Manufacturing and Calico Mills in c.1880. During the decade of the 1890s, several entrepreneurs belonging to the Vaishnav Vania community of Ahmedabad established a number of cotton textile mills. The famous cotton textile baron of Gujarat, Mafatlal Gagalbhai, established his first cotton textile mill, the Ahmedabad Shorrock Spinning and Weaving Mill in c.1905. He took the manufacturing of cotton textiles to other towns of Gujarat, when he established a mill at Nadiad in c.1912 and at Surat in c.1916, to take advantage of incentives given by the Maharaja of Baroda. Consequently, through takeovers of different mills, he established the first integrated empire of cotton textile manufacturing from gin to retailing of cloth. The family of Kasturbhai Lalbhai promoted Calico Mills at Ahmedabad, followed by Kasturbhai mills. This family also established Arvind Mills in c.1931.[47]

The expansion of the cotton textile manufacturing in Ahmedabad created a rival cotton textile manufacturing centre to the Bombay cotton textile mills. The mill owners of the city also established their own mill owners' association called Ahmedabad Mill Owners Association in c.1891. The cotton textile mills of Ahmedabad remained preoccupied with production of mill yarn for a considerable period of time, like the Bombay mills. Unlike Bombay mills, which depended on the export of yarn to Japanese and Chinese markets, the Ahmedabad mills developed a symbiotic relationship with local centres of handloom weaving by supplying them machine-spun yarn.[48] It is interesting to note that in Ahmedabad close linkages emerged between promoters of the modern cotton textile mills, financiers, moneylenders, traders of cotton and traditional flexible manufacturing. The existence of these linkages provided a distinct competitive advantage to Ahmedabad-based cotton textile mills. Ahmedabad also had availability of skilled manpower from the established centres of spinning and weaving communities.

Contrary to the advantages of Ahmedabad cotton textile mills, the Bombay mills were hiring their workforce from distant locations like Southern Konkan districts of Bombay Presidency and eastern districts of the United Provinces through middlemen known as '*jobbers*'. This system reduced the management's control on workers. Therefore, these workers using their regional affinities evolved regional solidarity to assert their demands on mill owners. The data on industrial disputes between mill owners and workers shows that the manufacturers of cotton textile mills of Bombay experienced eight general strikes over a one-month duration between c.1919 and c.1940. One of the strikes lasted for more than one-and-a-half years in c.1928–1929.[49] Another problem Indian cotton textile manufacturing faced was that these mills depended heavily on male, displaced, handloom weavers, who were available after the decline of traditional flexible manufacturing of cotton textiles. Such workers were prone to form unions. This structure was quite at variance with the experience of cotton textile manufacturing mills of Britain and Japan. Here the mills owners depended on women and child labour.

Jamsetji Tata took the manufacturing of cotton textiles to Nagpur, where he established the Central Indian Spinning and Weaving Mills in c.1874. The mill began manufacturing cotton textiles on January 1, c.1877; the date of Queen Victoria's coronation in London. To commemorate the event, the factory was named Empress Mill. For the first time in India, Tata used ring spindles, which were developed in the United States and as yet were not very popular, even in America. Nagpur city was in close proximity to the cotton-growing areas and was also quite near to coal mines of eastern India. Nagpur was also an important centre of handloom production and served as an important market of yarn. It had also emerged as an important railway junction (terminus) of the Great Indian Peninsular Railway, which provided access to all the important markets of the sub-continent. After the opening of the textile mill by Tata, Nagpur attracted several other cotton textile mills. Sarup Chand Hukumchand introduced manufacturing of cotton textiles at Indore.

Before the Sepoy Mutiny or the first war of independence in c.1857, the British administration started a cantonment at Kanpur. This cantonment played an important role during the revolt. Here, the British army established the Central Ordinance Depot (COD). The main work of the COD was to procure supplies for the army. In c.1861, British citizens Hugh Maxwell and David Begg formed Begg Maxwell & Co. and promoted Elegin Mills, which was the first cotton textile mill at Kanpur. Begg Maxwell & Co. later became Begg Sutherland & Company. Elgin Mills produced textiles and tents for the army. Later on, three expatriate British entrepreneurs floated Cawnpore Woollen Mills and Army Cloth Factory. These mills were converted into joint-stock companies in c.1882. It is interesting to note that most of the investment in the cotton textile mills at Kanpur was made by British entrepreneurs, while the local merchant community remained preoccupied by trade. The only Indian entrepreneur who started a cotton textile mill at Kanpur was Kamlapat Singhania.

Soon, cotton textile manufacturing spread to other towns of northern India. On May 1, 1889, the first cotton textile mill was set up in Delhi by Lala Shri Ram and was named Delhi Cloth Mill (DCM). During World War I, Shri Ram floated another separate unit to manufacture tents. For this, DCM entered into a partnership with a government contractor who had strong liaison with the British administration. After c.1911, when capital of India was shifted from Calcutta to Delhi, there was substantial growth in the demand for tents because at that time some of the government offices were operating in tents, and tents also provided residential accommodation to some of the government officials.[50] Lala Shri Ram's family had migrated to Delhi from the Jhind district of Punjab (now Haryana). In c.1935, Lala Shri Ram promoted another cotton textile mill at Lyallpur in Punjab in c.1916 called, Lyallpur Cotton Mill. After the establishment of DCM, G.D. Birla promoted a cotton textile company at Delhi, the Birla Mills. Birla also located one of his cotton textile mills, Jiyajirao Cotton Mills, at Gwalior in c.1919. In the 1940s, one of the entrepreneurs of Punjab, Karmachand Thapar, situated Jagatjit Cotton Textile Mills (JCT) at Phagwara in the district of Kapurthala of Punjab. At the same time a British entrepreneur, E. Dyar, established the first woollen textile mill, Egerton Woollen Mills at Dhariwal in Punjab, which later became popular as Dhariwal Mills.[51]

At approximately the same time, possibilities of setting up cotton textile mills in the cotton-growing areas of southern India, were also explored. Coimbatore region, which was part of Madras Presidency's Kangaund area, was bestowed with large areas of black soil, which is ideal for the cultivation of cotton. Farmers in this region were cultivating the short-staple local variety of cotton. This variety was suitable for handloom weaving, but not for modern textile mills. James Stance, a British entrepreneur, started the Coimbatore Spinning and Weaving Company in c.1888. He was basically a coffee planter from the Nilgiris, who had come to Coimbatore to establish a curing factory in c.1861. In the Tuticorin area at Valkarts, a British managing agency house, A. & E. Harvey, established the first gin factory in c.1880. A. & E. Harvey also started Tinnevelley Mills in c.1889, Coral Mills in c.1887, Madura Mills in c.1889 and Coimbatore Spinning and Weaving Company in c.1888.[52] The economic transformation of the region occurred in c.1904–1905, when a new variety of cotton called *Cambodia* was introduced here for cultivation. This variety improved the quality of cotton produced here and increased its demand for export to Britain. The name Cambodia was given because it was a variety of long-staple American cotton that was introduced in India after experimentation in Indo-China (Cambodia). Due to its demand, the new variety of cotton rapidly replaced the cultivation of the traditional variety of cotton. As a consequence, Tripur, which was located at close proximity to Coimbatore, emerged as an important ginning and trading centre of cotton.[53]

James Stance was encouraging local rich farmers and traders to start cotton gins to supply cotton to his mill. One local farmer, Kuppuswamy Naidu, set up a small gin unit in c.1905 and promoted Lakshmi Mills in c.1910. However, his dream of becoming the first successful pioneer in the introduction of cotton textile mills in the region experienced a setback when he incurred huge financial losses in the speculation of cotton. The first successful cotton textile mill in this region was started by Peelamedu Samanaidu, Govindaswamy Naidu (P.S.G. Naidu), who opened Peelamedu Sri Ranga Vilas Gin Spinning and Weaving Company in Coimbatore in c.1922. K.K. Naidu, son-in-law of P.S.G. Naidu, promoted Sri Balasubramania Mills in c.1934. Two British managing agency houses, the houses of A. & E. Harvey and Binny's, together promoted Buckingham Mill and Carnatic Mills in c.1940. Following the lead from earlier attempts to float cotton textile mills in Coimbatore, several entrepreneurs of the Naidu community started cotton textile mills here. K.K. Naidu and Venkapatty Naidu started Balasubramani and Kasturi Mills; S.N. Rangaswamy Naidu established Sri Ramakrishna Mills Coimbatore Ltd., and N. Vellapan Naidu opened Sri Ram Narayan Mills. These Naidu entrepreneurs transformed Coimbatore into the Manchester of the south.[54]

The entrepreneurial adventures of the Naidu community of Coimbatore also encouraged another peasant community of Gounders of the Kongunad region to enter into the manufacturing of textiles. The entrepreneurs belonging to the gounder community started only two textile mills during the period of British Raj. V.C. Vellingiri Gaunder and P. Somasundram Chetty established Kaleeswar Mills, and V.C. Vellingiri Gaunder started Granabika Mills.[55]

As has already been mentioned, Tripur, which is located near Coimbatore, became an important centre of production of the cotton knitwear industry after the introduction of Cambodia cotton. Knitting of yarn is different from weaving of yarn and needs different kinds of equipment. The foundation of the knitwear industry at Tripur was laid by three pioneering entrepreneurs, M. Abdul Rguman Sahib, who founded Azad Knitting Company in c.1926; Pethi Chetty who started Rajendera Knitting; and S.A. Khadar who promoted Khadar Knitting. Tripur became the ideal location for knitwear manufacturing for many reasons. The availability of yarn from cotton textile mills of the region, cheap labour and availability of hard water from the local Noyyal River. Tripur began with the production of *banians* (vests), but in due course, it diversified into the manufacturing of several kinds of knitwear.[56] Slowly, Tripur became an important cluster of flexible manufacturing units producing knitwear, and established technological linkages with another important centre of manufacturing of knitwear and knitwear equipment, Ludhiana of Punjab. The Ramgarhias, the community of Sikh artisans, were making equipment for knitwear manufacturing units, which were used by local knitwear manufacturers of Ludhiana and also by the companies located at Tripur. Ramgarhia-owned units like G.S.L. Textiles, started by Ghansham Singh Lotey, and Swarup Mechanical works of Narpal Singh, provided carding machines, draw frames, shearing machines and sewing machines.[57]

The emergence of modern cotton textile mills in India, and their rapid expansion, invited the hostility of British textile manufacturers. To protect their interest, they influenced the British government to impose discriminating tariffs on Indian cotton textile manufacturers. The process of discrimination began when Lord Lytton removed custom duties on English textiles imported into India.[58] This was an important victory for British cotton textile manufacturers. However, cotton textile mills of Britain were not satisfied with this achievement. They expected much more. In c.1890, Manchester secured another victory when the government imposed an excise duty on Indian cotton textile manufacturing.[59] Imposition of an excise duty further reduced the competitive advantage of Indian cotton textiles. In this way, the Indian cotton textile industry is an important example that explains the hostility between the Indian mill owners and the cotton textile mill owners of Britain.

Despite the existence of several positive conditions for investment in jute textile manufacturing, Bengali merchants did not take initiative to invest in the industry. This happened despite Bengal having an early advantage of modern education. However, the increasing profitability of Dundee Mills provided motivation to Mr George Auckland to start the processing of jute in a country where raw material was available in abundance. He set up the first jute textile mill, called Auckland & Company, at Rishra, near Serampur in c.1854. Auckland was a Scot, who was an ex-officer of the British navy and had experience working on coffee plantations in Ceylon (Sri Lanka). Thereafter, another Scot, Mr George Henderson started the Borneo Company in c.1859.[60] Between c.1862 and c.1865, three more jute textile mills were added.

The first jute textile mill, which was a sterling company registered at the London Stock Exchange, was set up by a British entrepreneur in c.1872. The company was named Champadany Jute Mill. The following year, in c.1873, Thomas Duff promoted another sterling company called Sammuggur Jute Company at Hoogly. During the same year, a Scottish mercantile house of Calcutta, Kettle Buller & Company, started Fort Gloster Jute Mill. Richard Macallister, another British entrepreneur, started Oriental Jute Company in c.1875. In c.1876, other important managing agency houses of Calcutta, Jardine Skinner & Company and Gillander Arbuthnot & Co. Ltd., established jute textile mills in c.1883. After this pioneering effort, different managing agency houses of Calcutta set up several jute textile manufacturing companies. The promoters of these jute textile mills used their Scottish relationships to enter the markets of America, New Zealand and Australia. By this time, Calcutta-based jute textile mills had cornered a substantial proportion of markets of coarse bags in America, Australia and New Zeeland. Despite the fact that the Indian jute textile mills had entered several markets, the excessive investment in the creation of new capacity resulted in the underutilization of capacity. This stopped the expansion of investment in the jute textile industry during c.1884–1895. The problem of excess capacity was further aggravated by the occurrence of famines in the jute-growing areas of Bengal between c.1895 and c.1900.

As the jute textile mills were able to overcome the problems created by the two aforementioned crises, most of the managing agency houses of the eastern part of India again began to invest in jute textile manufacturing. Managing agency houses like Andrew Yule & Company, Shaw Wallace & Company, James Finlay & Company, Bird & Company, Rolli Brothers, Mcleod & Company and Graham & Company opened several jute textile manufacturing mills. As a consequence, jute textile became an important export from India. Up to World War I, there existed total dominance of European managing agency houses in the manufacturing of jute textiles, and the entire manufacturing of jute textiles remained localized in Bengal. Bengal had advantage in terms of availability of raw material and the Calcutta port provided a suitable location for the export-oriented industry. Even the fuel (coal), which was necessary to run jute textile mills, was available in close proximity.

The dominance of British managing agency houses in the manufacturing of jute textiles emerged because of their complete monopoly in shipping as well as river transport network of Eastern India. Railways, another important mode of transportation, were also under British control. Railway companies like Eastern Bengal State Railway, the Assam Bengal Railway, and the East India Railway were owned by British citizens.[61] Due to their monopoly on all the modes of transportation, one finds that at that time, the offices of all the important managing agency houses existed in all the jute markets of Bengal.[62] Even the banking sector in Eastern India was totally under British control. Hence, having complete monopoly over finance, transportation, trading and markets in India and abroad provided distinct advantages to British managing agency houses. These advantages reduced the chances of the emergence of native entrepreneurship in jute textiles.[63]

Contrary to the pattern of ownership of jute mills, which were largely owned by British managing agency houses, Marwari traders of Calcutta were quite important players in the procurement of raw jute required by Calcutta mills, as well as for exports. The Marwari merchants were owners of most of the baling houses of Bengal. Marwari trading houses had been migrating from Rajasthan to Bengal since the sixteenth century. The beginning of ascendancy of the East India Company after c.1757 provided new business opportunities to the Marwari traders, who began to play an important role in the economy of Bengal as brokers, agents, sub-contractors and who got involved in helping British agency houses. The numerical strength of Marwari immigrant community houses had increased significantly and the community had founded the Marwari Panchayat in Calcutta in c.1828. The names of some of the important Marwari trading houses were included in the commercial directory of Calcutta city published in c.1860–1863. By c.1903,[64] the Calcutta Baled Jute Association, a body affiliated with the Bengal Chamber of Commerce, had 45 Marwari firms as members, out of a total membership of 133. However, by the end of world War I in c.1918, membership of Marwari firms had crossed 63 per cent of the total membership of the association. Some of the important Marwari members of the association were Shri G.D. Birla, Saroopchand Hukumchand, Surajmal Nagarmull and Badridas Goenka.

It is quite obvious from these facts that by the end of World War I, Marwari merchants had accumulated sufficient financial resources. They were familiar with all operations involved in jute textile manufacturing. Moreover, the method of manufacturing gunny was quite simple. Marwari trading houses had well-established links with British agency houses that owned most of the jute textile manufacturing mills. For instance, Goenkas were associated with Rolli Brothers, Onkar Nath Jatin with Andrew Yule & Co., Surajmull Jhunjhunwala with Graham & Co., Luxmi Narayan with Mcleod & Company and Tarachand Ghanshyamdas with the managing agency house of Shaw Wallace & Company. Given their place in the overall jute economy of Bengal and prevailing historical circumstances, Marwari merchants began to enter into the manufacturing of jute textiles. For this, these merchants followed two different strategies.[65] The first strategy that was used by the Marwari trading houses was purchasing substantial shares of the companies owned by the British managing agency houses. These managing agency houses had to transfer a large number of shares of their companies to Marwari merchants because they were experiencing shortage of funds to meet the requirement of working capital as well as for investment in fixed capital. British managing agencies borrowed funds from Marwari trading houses against the collateral of blocks of shares of jute mills. This system of getting funds at that point in time was known as '*fatka*'. This gave opportunity to Marwari merchants to become members of the board of directors of the company. Slowly, when the Marwari merchants had accumulated substantial holdings, they took over the companies. The strategy of entering in the manufacturing of jute textile through '*fatka*' was used successfully by the Marwari trading houses of Bajorias, Bangurs, Goenkas, Kanorias, Kedias and Jalans. For instance, Bangurs took over Kettle Buller & Co, Kedias took over Anderson Wright & Company, Bajorias took over Begg Dunlop & Company, Keshav Prasad Goenka took over Duncan, and Radha Kissen Kanoria took over the management of Mcleod & Company.[66]

The second strategy that Marwari merchants adopted to enter in the manufacturing of jute textile was to start entirely new mills. Following this strategy, in c.1922, G.D. Birla and Sarupchand Hukumchand floated wholly Indian-owned and Indian-managed jute mills. Afterwards, during the 1920s, several new jute textile manufacturing mills were started by Marwari merchants—six mills around Calcutta and one by Kamlapat Singhania in Kanpur. The trading house of Surajmull Nagarmull also opened new jute textile mills. As a consequence of these alternative strategies adopted by Marwari trading houses, by the end of World War II, there were 22 jute textile mills wholly owned by Indians business houses. In 62 mills owned by the British managing agency houses, there were Marwari directors, and five jute textile mills owned by British managing agency houses had been taken over by Marwari trading houses.[67]

Manufacturing of iron and steel was another important industry that also originated during the rule of the East India Company, but assumed importance during the British Raj. The art of smelting iron was known to the people of the Indian sub-continent for ages. It was produced at several locations in the Indian sub-continent. It is believed by historians of technology that India was the pioneer in the manufacturing of carbon steel (wootz iron). The iron produced in India at that time was of good quality. European experts have approved the quality of Indian iron in scientific journals, published by their professional associations like Royal Society.[68] The famous scientist Michael Faraday was quite fascinated by the quality of India's wootz iron, and conducted several experiments to know its qualities.[69] In spite of being one of the oldest manufacturers of iron, modern iron and steel manufacturing began quite late in India (for details regarding the history of iron and steel manufacturing during the stage of traditional flexible manufacturing see Chap. 2 of the book).

During the rule of the East India Company, several attempts were made to manufacture iron on European lines. In c.1762, an iron foundry was established in Begpore called Begpore Iron Foundry. This foundry was promoted under the supervision of Mr Andrew Duncan, who had acquired experience for starting an iron foundry in Russia.[70] But this enterprise did not succeed due to availability of cheaper imported iron, which was imported as deadweight cargo on ships coming from Britain. Due to this reason, it was felt that Indian iron did not have a competitive advantage over imported iron. Therefore, the East India Company never encouraged its production in India.

In c.1825, Josiah Marshall Heath of Madras Civil Services filed a proposal to the directors of the East India Company to set up an iron smelting mill in Madras Presidency, using European techniques of production. Due to the negative attitude of the officials of the East India Company, his request did not receive any response. He repeated his request again in c.1829, along with another request to grant his proposed iron mill monopoly to produce iron in India. To support his demand for monopoly right, he provided the example of monopoly right given to British entrepreneurs who established the first iron foundry in Britain during the rule of King James. This time his request was accepted, and he was granted permission to set up an iron foundry. He was also given exclusive privileges and an interest-free loan by the East India Company in c.1830. The factory was established in Proto Novo in the south Arcot district of Madras Presidency 125 miles away from the city of Madras. The plant was built to use charcoal for smelting iron and used bullocks for the same. The iron ore for the plant was brought from Salem by boats, through the river Cauvery. This river was navigable for only six months of the year. The ore at Salem was of good quality.

Heath originally established two large furnaces, and afterward added two cupola furnaces and two revelatory furnaces. The mill used puddling and rolling methods to produce wrought iron. Heath had taken the services of two expert British ironmasters to set up and maintain the plant. However, soon Indian workers learnt all the operations of the plant and they replaced the expatriate workers. Thereafter, Proto Nov Mills never experienced a shortage of skilled workers. The iron produced at the Proto Nov plant was of good quality, and soon the plant began to export iron to Britain. In c.1839, the iron produced by that mill helped the British Royal Arsenal, located at Woolwich, to substitute it with imported iron from Sweden.[70] The iron produced at this plant was also used in the construction of Britannia Tubular Bridge over Menai Strait.[71]

Despite the fact that the Proto Nova mill produced good quality iron, it experienced frequent financial problems to not only expand its scale of operation, but even to run its existing facilities. This was happening because by this time, European ironmaking technology was experiencing a revolutionary change due to its shift from using charcoal to coal. Therefore, it became impossible for it to compete in the British market. Due to its structural as well as financial problems, the Proto Nova plant was taken over by Proto Nova Steel and Iron Company. Even the new company could not cope with its financial problems. Finally, in c.1853, the Proto Nova Steel and Iron Company was taken over by the East India Company. The plant met its ultimate fate when it went into liquidation in c.1874. It is believed that when Heath began to construct his factory at Proto Nova, one of his close British friends had advised him to shift his factory to the Burdwan district of Bengal Presidency, which had abundant supply of coal.

Similar attempts were made by Jessop & Company, when it set up an iron smelting factory near Calcutta in c.1839. In c.1855, Mackay & Company started Birbhum Iron Works, and in the same year, iron foundries were established at Kumaon, Indore and Chanda. These units also did not survive for long.[72] In c.1874, Bonker Iron Works was set up at Kulti. Soon this plant also experienced financial problems, and in c.1882, it was taken over by the government and resold to a company, Bengal Iron and Steel Works, which was registered in Britain. Bengal Iron and Steel Works was able to get support from the government when it agreed to buy back 10,000 tons of pig iron, every year for ten years, at a price fixed at 5 per cent below the price of imported steel. The government had begun to support iron and steel manufacturing in India because by this time Belgian and German steel was replacing imported steel from Britain.[73]

While one plant after the other was meeting with failure in the Indian sub-continent, the European iron and steel manufacturing was experiencing revolutionary transformation with the introduction of the puddling furnace devised by Henry Cox in c.1783, and by c.1830, coal had replaced charcoal as the main fuel. In c.1856, Henry Bessemer evolved a new method to convert molten pig iron in steel known as the Bessemer converter. In c.1869, Siemens Martin devised the open hearth furnace.[74] These changes in the manufacturing of iron and steel altered the characteristics of iron and steel making. The new combination of raw materials required to produce iron and steel was coal, limestone, iron ore and water. These new inputs influenced the location pattern of iron and steel manufacturing from being near woods to near coal and iron ore mines. Since the new materials were bulky as well as *weight losing*,[75] and accounted for around 80 per cent of the total cost of manufacturing of iron and steel, their geographical location began to determine the location of its manufacturing.[76] Hence, the most important determinant of location of steel plants became the objective of minimizing transport costs, when two important materials coal and iron ore are located at different places, separated from each other. For instance, in the Pillsbury steel plant, ore moves towards coal. In the steel plant located at Lake Superior, coal goes to ore, and at Lake Erie, the plant is located at the meeting place of the two.[77]

The technological changes experienced by European iron and steel manufacturing also influenced the manufacturing of iron and steel in India. Decisions regarding appropriate location became more decisive in determining the success of new steel manufacturing plants. Due to these important factors, the next steel mill established by Jamsetji Nusserwanji Tata took several years between conception and building of the plant. Tata had to spend several years, as well as substantial financial resources, to identify the appropriate location of his proposed steel mill and to shortlist the services of a consultant to establish the new plant.

The idea behind entering the business of manufacturing iron and steel originated in the mind of Tata when he read about the presence of iron ore and coal deposits in Chanda district of Central Provinces (present-day Madhya Pradesh), prepared by a German expert, Ritter Von Schwarz. This report appeared in c.1882. After reading the report, Tata visited Chanda district and met the chief commissioner of the district. He discussed with him his intention of setting up a steel manufacturing plant in Chanda. He even promised that in case he decided to set up a steel mill in Chanda, his company would take the responsibility of constructions the railway line between Wardha junction and Warora, where coal deposits were located. However, Tata did not receive positive response from British authorities because at that point in time, British government was not allowing private sector companies to exploit natural resources of the country.

In c.1899, under Governor-General Lord Curzon, the British government formulated its policy regarding prospecting of minerals. Under the new policy regime, the government was prepared to grant permission to prospect minerals to private companies by giving mines on lease. Moreover, Curzon had also begun the practice of buying steel form India, when the government entered into contract with Bengal Iron and Steel Works to buy steel form it.[78] These policy changes were accompanied by an encouraging report prepared by Mr. R.H. Mahon on the manufacturing of iron and steel in India. In his report, Mahon had suggested Bengal as an appropriate location for the manufacturing of iron and steel. His suggestion was based on the fact that coal available in the mines of Bengal Presidency could be converted into coke, which had emerged as an important fuel for blast furnaces. These intertwined events again motivated Jamsetji Tata to enter into the manufacturing of iron and steel.

In order to realize his dream of setting up a company to manufacture steel, Tata visited London and met George Hamilton, who was at that time, secretary of state of India. Hamilton promised to provide all necessary support in starting a steel mill. After receiving this positive response, Tata went to the United States in search of the best possible technical support to set up a steel plant. After travelling to several centres of steel manufacturing in America, he identified Kennedy Sahlin & Company of Pittsburgh to help him achieve his dream. The company consisted of two partners, Julian Kennedy and Alexander Sahlin who advised Tata to spend more time in identifying the best location for the plant before they commenced the work of building it. For this purpose, they suggested the name of Charles Page Perin who sent his partner C.M. Weld in c.1903 to help Tata to find a suitable location.

C.M. Weld, along with the son of Tata, Dorabji Tata, and his nephew Shapurji Saklatvala, visited Chanda mines. Weld rejected Chanda as a suitable location for the proposed plant because coal deposits of Chanda district were not of good quality, and iron ore deposits were too small to last for a longer period. Tata did not give up his dream and asked Weld to explore other possible places in India. Soon Weld's attention was drawn to Durg district of Central Provinces, where a geological survey had indicated the availability of good-quality iron ore. This report was prepared by P.N. Bose in c.1898. The team sent by Tata found that the hills of Dhali and Rajhara were indeed endowed with good-quality iron ore, but this place did not have regular supply of water. At this point in time, P.N. Bose wrote a letter to Tata, in which he mentioned the discovery of good-quality iron ore near Jharia and Ranigang coal deposits. Since the place was near to Calcutta, Weld, his senior partner Perin, and Bose reached Gorumahisani Hills, located in the princely state of Mayurbhanj. Here, Perin and Weld identified a place called Sini, which was quite close to sources of three important raw materials—iron ore, coal and limestone. The selection of this site also received support from the government, when John Hewitt of the department of commerce and industry promised to link Gorumahisani by railway line with the Bengal-Nagpur line. Tata's proposed site also received support from Maharaja of Mayurbhanj because he expected that the new plant would help the entire princely state to get railway connectivity. However, Sini was also rejected by the team, because the place had a shortage of water.

The team of experts began its search for an alternative location where it could identity a source of adequate supply of water. For this purpose, Weld and his team travelled on horses until they reached near the village of Sakchi where two rivers, Subarnarekha and Korkai, met. The village was very near to the railway station of Kalimati. Thus, Sakchi became the final choice for locating the steel plant. After identifying the appropriate location for the proposed mill, Tata formally appointed Kennedy Sahlin & Company for building the steel mill.

After finalizing technical details of the plant, the next problem that Tata faced was to raise necessary financial resources for the project. Tata initially tried to raise capital from the London financial market, but he did not get support from the key players. However, a group of London financiers, who had shown some interest in the project, sent Charles Stoddart, who was chairman of the Parkgate Iron and Steel Company, along with William Selkirk, a mining engineer, to India. Their purpose was to assess the potential of Tata's project. Despite the fact that both the experts gave encouraging report about the project, London financiers did not provide finances.

This was the period when the Swadeshi movement had emerged in Bengal as a consequence of Lord Curzon's policy of dividing the Bengal Presidency (c.1905–1906). Nationalist leaders viewed the partition of Bengal as a strategy to create a wedge between Hindus and Muslims of the presidency. One of the important methods of protest adopted by the nationalist leader was to boycott the use of imported goods and buy only goods produced in India. This movement had encouraged several entrepreneurs of Bengal to set up new enterprises. Under such historical circumstances, Tata also thought to raise funds in India. On August 26, c.1907, Tata Iron and Steel Company was registered in Bombay, and within three weeks, the entire initial capital needed by the company, £23 million, was fully subscribed. The construction of the plant began in c.1908, and in the winter of c.1911, the first piece of iron was made at the plant.

Tata's project received government support also, because by c.1900 continental steel had made inroads in the Indian market and had substantially replaced British steel from most of the private sector needs. Therefore, British policymakers had realized that Tata steel would act as substitute for continental steel, especially imported from Germany and Belgium.[79] Under these circumstances, Tata's plant received government support, when it guaranteed the purchase of 20,000 tons of steel rails for the next ten years at the price that was fixed equal to the price of imported steel. However, outbreak of World War I in c.1914 forced the government to buy the entire output of steel produced by Tata's plant. In the process, Tata's mill evolved a close symbiotic relationship with the British government's war effort. During the period of World War I, Tata Iron and Steel Company supplied to the British government 1500 miles of rails made of 3 lakh tons of steel for the construction of railway networks in Mesopotamia, Egypt, Salonica and East Africa.[80]

Subsequently, after World War I, Tata's steel plant expanded its manufacturing capacity manifold. It also made efforts to produce different variety of steels for specific products. To achieve its objective, Tata's steel plant set up a research and development laboratory in c.1932. This laboratory helped in developing several new kinds of steel products. The centre developed high-tensile alloy steel (TISC ROM), which was used in the construction of the Howrah Bridge in Calcutta. It also developed a corrosion-resistant steel (TISCOR), which was used in the manufacture of steel coaches for Indian railways.[81] The most important innovation of Tata's research laboratory was the steel that was used for the manufacture of armoured cars, called Tatanagars, for the army, which were used in World War II in North African war theatre.[82]

In c.1918, another company was founded to manufacture iron and steel in India at Burnpur in Bengal called Indian Iron and Steel Company (IISCO), which was associated with the Bengal Iron and Steel Works through the managing agency firm, Martin Burn & Company. In c.1936, Bengal Iron and Steel Works was merged with IISCO to form a new company—Steel Corporation of Bengal. During the same period, a small iron works was also established by Maharaja of Mysore in c.1923. The name of this company was Mysore Iron and Steel Works.[83]

Sugar has been one of the important consumer products of India, which is also considered the house of sugarcane, as has already been described in Chap. 2 of the book. The traditional flexible manufacturing of the Indian sub-continent mainly produced either *gur* or *khandsari*. These two products were able to meet the entire domestic demand of sugar and also for export. Manufacturing of white crystal sugar originated in India, initially, during the rule of the East India Company. At that time, British expatriates managing agency houses were making white crystal sugar in the West Indies. The West Indies-based sugar mills were using sugarcane cultivated as a plantation crop with the help of slaves. This sugar was much cheaper in comparison to sugar produced in India, which was produced form the sugarcane cultivated by small farmers. Therefore, British markets were totally dependent on sugar imports from the West Indies. The requirements of white sugar for the Indian market were met by imports of white sugar from Java, which at that time was a Dutch colony. The beginning of manufacturing of white sugar in India can be attributed to the killing of a very large part of the white expatriate population in a rebellion that occurred in c.1791 in Hayati and San Domingo. During this rebellion, several manufacturers and exporters of white sugar were also killed. This historical event disrupted the supply of white sugar in Britain because of which the prices of sugar increased substantially.[84] In c.1834, British government removed slavery in the West Indies. Liberation of slaves from the sugarcane plantations in the West Indies altered the availability of workers. Therefore, cost of cultivation increased substantially. The emerging gap in the supply of sugar in Britain encouraged the East India Company to promote manufacturing of sugar in the Indian sub-continent.

As white sugar was meant for export only to Britain, initial investments in the manufacturing of sugar were made by the British managing agency houses. Binny & Company set up a sugar manufacturing mill in Madras Presidency, and Parry & Company established its first sugar mill and distillery at Nallikupam in the district of Cuddulore in c.1845. At that time, only one other sugar mill, called Rosa Sugar Mill, was working at Shahajanpur in the United Provinces (present-day Uttar Pradesh). After a gap of several years, another managing agency house, Begg Sutherland & Company, established Kanpur Sugar Mill in c.1884, which began to refine *gur* by a new method, which did not use bone charcoal. This new method was used to encourage the consumption of white sugar by the Hindu population of the country. Subsequently, Begg Sutherland & Company promoted several sugar mills in different parts of United Provinces and Bihar—in district Saran, in c.1905, by the name of Champaran Sugar Mill; Samastipur Sugar Mill in district Samastipur in c.1906; and Ryan Sugar Mill in district Darbhanga of Bihar in c.1913. Before the beginning of the Great Depression, many more managing agency houses also entered in the business of manufacturing of sugar. For instance, Octavious & Company had started two sugar mills, and James Finlay & Company and Marshal & Company had also established two sugar mills each.[85,86]

However, when supplies of sugar resumed from the West Indies, these managing agency houses that had established sugar mills in India primarily to export sugar to Britain, were highly disappointed because they could not compete with the sugar manufacturers of the West Indies and other British colonies because in these locations, mills were getting sugarcane cultivated as a plantation crop. Moreover, British government had imposed discriminating tariffs, which discriminated against the Indian manufacturers of sugar.

In order to plead their case and make the British government aware of their problems, all the managing agency houses that were engaged in the manufacturing of sugar in India sent a petition to the British Parliament in c.1931. At the same time, British officials in India were also worried about the increasing imports of sugar from Java. Due to these reasons, the British government provided, for the first time, protection to Indian sugar manufacturers under the Indian Sugar (Protection) Bill of c.1932. The government also appointed a sugar technologist in India in c.1933 from the Imperial Council of Agriculture. The Sugar Bureau of Britain, at the same time, opened a department of sugar technology in the Harcourt Butler Institute of Technology at Kanpur. Around the same time, the British government constructed the Sarda Canal in the United Provinces (UP). The Tariff Commission submitted its report in c.1930 and suggested imposing a tariff of Rs 6 per quintal on imported sugar (this rate was equal to 120 per cent *ad valorem* on the existing prices of imported sugar). In March c.1931, the duty was further raised to Rs 7 per quintal (which was about 190 per cent *ad valorem* on the existing prices of imported sugar). In September 1931, the duty was further raised to Rs 9 per quintal. These protective rates of tariff, especially at that point in time when the European economies were experiencing recession, gave enormous advantage to the sugar manufacturers of India. This was also the time when plants and equipment, and consultants, were available at lower prices in the European markets.

Sugar manufacturing, by its very nature, is characterized by raw material orientation, because the basic input of sugarcane is weight losing. Therefore, transporting sugar is cheaper than transporting sugarcane. Moreover, after harvesting, sugarcane should be crushed within twenty-four hours, to maximize the production of sugar. Due to these reasons, spread of sugar manufacturing in India followed the geographical location of sugarcane cultivation. This suggests why sugar manufacturing remained localized in United Provinces, especially the regions irrigated by the Sarda Canal, Sangli and Ahmednagar in Maharashtra, Bihar, Parts of Andhra Pradesh and Madras Presidency. Moreover, the Coimbatore agricultural station started by the government also helped in improving the quality of sugarcane in India.

The changed favourable conditions motivated several Indian entrepreneurs to enter into the business of manufacturing sugar. It is important to mention, soon after the Great Depression of the 1930s, that Indian business houses became dominant players in sugar manufacturing. Jamlal Bajaj entered in the manufacturing of sugar in c.1932, when he started Hindustan Sugar Mill at Gologokarnath in UP; Karam Chand Thaper established Deoria Sugar Mill in c.1935 and Sri Sita Ram Sugar Mill in c.1830 at Baitalpur in the Gorakhpur district of UP. He also set up Jamuna Nagar Sugar Mill at Jamuna Nagar in District Ambala of Punjab. Gokulchand Narang started the Basti Sugar Mill in c.1927. Subsequently, he floated seven other sugar mills and emerged as the sugar baron of India. At the same time, Ramkrishan Dalmia opened five sugar mills in Bihar under the name Rohtas Mills. Gujarmal Modi opened Modi Nagar Sugar Mill in c.1933 at Begamabad in UP near Delhi. Soon, manufacturing of sugar also spread to Maharashtra. Walchand Hirachand opened Ravalgaon Sugar Mill, Karanishi bhai Jethabhai Somaiya started Godavari Sugar Mill, Radha Krishan Ruia established Kohlapur Sugar Mill, and V.S. Apte floated the Phatan Sugar Mill. The most important characteristic of all these sugar mills was that all of them had their own captive sugarcane farms.[87]

A similar trend also emerged in the sugarcane-growing areas of Southern India. For quite some time, Parry & Company had a monopoly of manufacturing sugar in Southern India, which was soon broken by native entrepreneurs such as Yardagedda Sriram Prasad, known as Raja of

Chellapali, in the Krishna district, who established his first sugar mill in c.1934, followed by another in Rayagada. Despite the fact that the manufacturing of sugar dispersed to several locations in the Indian sub-continent during this period, UP, Bihar and Maharashtra remained more important. These facts clearly show that the imposition of tariffs on imported sugar in India initiated the process of import substitution in the production of sugar. This process was so rapid that the economy, which used to meet most of its requirement of sugar through imports, became self-sufficient by c.1942–1943. In fact, it developed capability to export sugar.

Manufacturing of paper also began in India during the British rule. Most companies that were manufacturing paper in India before World War I were started by managing agency companies managed by British expatriate entrepreneurs. This is because the technology to manufacture paper was only available abroad. Moreover, most of the important inputs required, such as paper pulp, Sabai grass pulp, caustic soda, and chloride, for the manufacturing of paper were also imported. The managing agency houses of Helger & Co. and Balmer Lawries had opened Titagarh Paper Mill and Bengal Paper Mill before World War I. With this, native entrepreneurs were also encouraged to start small paper mills like Upper India Paper Mill and Deccan Paper Mill. Despite increased demand and profitability in the manufacturing of paper, the industry did not attract investment during World War I. This was because the war had disrupted trade flows, making it difficult to import plants and machinery, as well as other important inputs that were necessary for the manufacturing of paper. During the war, only one new paper mill was established by Andrew Yule & Company in c.1916.

Boost to invest in the manufacturing of paper occurred during the 1930s when economies of Europe and America were experiencing the Great Depression. This was the time when plants and equipment and important raw materials were available at substantially low prices. Moreover, by this time there had emerged possibilities of using substitute local raw materials. Successful experiments had been conducted to use bamboo pulp for making paper. In the development of this technology, Forest Research Institute located at Dehradun played an important role. Tata began to manufacture caustic soda and soda ash in c.1939 at the Mithapur plant near Dwarka in Gujarat. This plant was opened in Gujarat after considerable help from the Maharaja of Baroda. During the same period, several units were opened to produce chemicals that were required by paper manufacturers. In c.1937, Kasturbhai Lalbhai had opened Mettur Chemical Industrial Corporation. The Imperial Chemical Industries set up their plant in 1941, followed by the plant established by Dalmias. Around the same time, Shri Ram Group established Delhi Chemical Works to manufacture chemicals.[88] Once the technological requirements necessary to open paper mills were largely met by indigenous resources, investment in the manufacturing of paper by native entrepreneurs increased. Birlas established Oriental Paper Mill at Sambalpur in Orissa; Thapars' established Shri Gopal Paper Mill at Jagadhari in Punjab; and Bajorias started Star Paper Mill at Saharanpur in the UP.[89] Maharaja of Mysore opened Mysore Paper Mill, Nizam of Hyderabad started Sirpur Paper Mill, and Harvey & Company set up Punar Paper Mill.[90]

Manufacturing of Portland cement in India did not start for quite some time due to lack of threshold demand, which would have motivated entrepreneurs to venture into its manufacturing. Due to lack of demand, the first cement manufacturing mill, the South Indian Industries Ltd., which was established in c.1904, could not sustain itself. Almost eight years later in c.1912, Tata sons started the Indian Cement Company at Porbandar near Kathiawar in Gujarat and Sahabad Cement Company in c.1915. Immediately afterwards, managing agency houses of McDonald & Company started Katni Cement Company and Killick Nixon & Company opened Bundi Portland Cement Company.

As information about the availability of basic inputs required to manufacture Portland cement plants in Southern India spread, it encouraged several entrepreneurs to set up cement plants. Kistna Cement Company came into existence in c.1920, followed by Coimbatore Cement Company at Madkurai in Coimbatore district in the same year. In c.1922, Bezwada Cement Company was started in Vijaywada in Andhra Pradesh, and in c.1936, Andhra Cement Company was established.[91]

It is important to mention that cement industry, since its inception, continuously experienced the existence of excess capacity. Moreover, this industry also suffered from intense price competition amongst different companies that were involved in the manufacturing of cement. This occurred because of lack of product differentiation amongst different cement producers. These two objective conditions forced F.E. Dinshaw to initiate the merger of ten out of twelve cement manufacturing companies that existed in c.1936. The ten companies that merged together formed Associated Cement Companies (ACC) in c.1936. The merger of these companies provided respite to these units from price competition. However, this advantage was short lived. In c.1938, Dalmia entered into the manufacturing of cement. The companies started by Dalmia gave stiff competition to the ACC group of cement companies. Subsequently, Dalmia also became part of the cartel in c.1941.[92]

The Indian manufacturing sector also attracted foreign investment during this period. The motivations of foreign investors were different to enter the Indian sub-continent. One set of colonial firms entered into the Indian sub-continent with the intention of monopolizing export-oriented manufacturing industries. These firms heavily invested in jute textiles, tea and manufacturing of sugar. Another set of foreign firms that entered into the Indian markets were those firms that were forced to open their establishments due to changes in the stores policy of the government. Just before World War I, the government began to buy stores through tenders, filed by only by those enterprises that were located in India. To take advantage of the change in the government policy, those foreign firms that were supplying their products to the government opened their establishments in India. Companies like John Thompson, which entered in India in c.1919, Harry Williams in c.1921, Alfred Herbert in c.1919 and Guest Kean Williams in c.1931, came to India to take the advantage of changes in the stores policy of the government.

After the 1920s, when government began to impose tariffs on imported manufactured products, those foreign companies, which viewed India as an important market, also opened their subsidiaries in India to avoid tariff restriction. These companies were either exporting consumer goods or industrial inputs, which were essential for the Indian manufacturing firms. Siemens, a German multinational firm, opened its subsidiary in India in 1923. Dunlop Rubber Company opened its subsidiary, Dunlop Rubber (India) Ltd, in c.1926; Braithwaitt Company established its subsidiary unit in 1930; Jenson and Nicholson Ltd. in 1922; ICI in 1923; Goodlass Walls Ltd. in 1933; International General Electrical Ltd. in 1931; Metal Box in 1933; Crompton and Parkinson opened their subsidiary Greaves and Greaves Cotton in 1934; and Union Carbide in 1934. All these companies were making products that were needed by other companies.

Another set of foreign companies established their subsidiaries in India in the manufacturing of consumer goods to overcome the tariff barriers imposed by the government. Swedish Match Company, which was the largest supplier of safety matches in India, opened its subsidiary Western India Match Company (WIMCO) in c.1923 at Bombay, and Assam Match Company Ltd., in c.1926. Associated Biscuit Manufacturing Company of Britain started its subsidiary, Britannia Biscuit Company, in c.1924. Colgate–Palmolive Company came to India in c.1930. Unilever established its subsidiary Lever Brothers in 1931 and Hindustan Vanaspati in c.1931. Bata Shoe Company of Czechoslovakia opened its subsidiary Bata India Ltd. in 1931. Philips Company set up its subsidiary Philips India Ltd. in c.1931. Imperial Tobacco Company and Godfrey Philips entered India in c.1908 and c.1936, respectively.

It is important to mention here that these foreign firms, which entered in the manufacturing sector in India during this period, adversely affected the growth of some of the companies that were started by Indian entrepreneurs. Those enterprises that were established by Bengali entrepreneurs in response to the Swadeshi Movement that emerged after the partition of Bengal in c.1905 suffered the most.

The partition of Bengal in c.1905 by Viceroy Lord Curzon gave rise to nationalist feelings amongst the Bengali elite and motivated them to enter into manufacturing of high-end technology products. Two Bengali entrepreneurs established Bande Match Factory and Oriental Match Factory in Calcutta in c.1906. Profull Chand Roy, who had received a doctorate degree from the University of Edinburgh, along with Jagdish Bose, opened Bengal Chemicals and Pharmaceutical Company. Similarly, Kiron Shankar Roy, who completed his education at Oxford University, promoted Bengal Lamps. Two prominent families of Mitra and Dasgupta set up Calcutta Chemical Company to manufacture soap, toothpaste and herbal cosmetics, and Bengal Waterproof Company to manufacture *Duckback* brand of raincoats and gumboots. At the same time, Mandira Chandra Nandy, the Maharaja of Kasimbazar, started Bengal Pottery, and Kolay & Sons, a company floated by another Bengal entrepreneur, began to manufacture biscuits. All these companies were owned by families that had limited financial resources. Moreover, all of them were manufacturing products that were produced by foreign companies that entered the Indian market after the 1920s. Therefore, most of them were not able to sustain themselves for a longer period.[93]

Manufacturing of matches in India also began before World War I. There existed several units for manufacturing safety matches in different parts of the Indian sub-continent. Amrit Match Factory was started in c.1894 at Kota in Bilaspur; Islam Match Factory was functioning in Gujarat since c.1895. Two Bengali entrepreneurs had opened Bande Match Factory and Oriental Match Factory in c.1906 in Calcutta. During the 1920s when the government imposed tariffs on imported safety matches, it encouraged the process of import substitution in the manufacturing of safety matches. This process led to the opening of Esavi India Match manufacturing Company in

Calcutta in c.1923. In c.1925, Mahalakshmi Match Factory began to produce safety matches at Lahore, and in c.1926, Bareilly Match Factory was started in the United Provinces.[94]

Despite the fact that there existed several safety match factories in India at that time, their combined production capacity was not sufficient to meet the domestic consumption needs. Due to this reason, India was importing large consignment of matches form a Swedish company called Swedish Match Company (which was described previously). Imposition of tariffs in the 1920s reduced the volume of imports. This motivated Swedish Match Factory to open its subsidiaries in India. One of the subsidiary units was located at Ambarnath near Bombay and another at Calcutta.[95] The entry of this company affected the future growth of safety match manufacturing in India.

Two young persons from Sivakasi, P. Ayya Naidu and his cousin A. Shanmuga, learnt the method of making safety matches while working in Bande Match Factory at Calcutta. They worked in the factory for six months during c.1922. On their return, they established an enterprise to manufacture safety matches in Sivakasi. However, soon they realized that the imported technology and imported inputs would not make the manufacturing of safety matches successful. Therefore, they adopted a labour-intensive, handmade method of making matches, using local cheap labour. They also used wood veneers obtained from local trees to make matches. They began to produce handmade safety matches using the brand name *Bengal Light*. Their pioneering effort created a cluster of several units for producing handmade matches in Sivakasi. These two cousins also pioneered the manufacturing of fireworks at Sivakasi. P. Ayya Naidu opened Ayyan Fireworks and A. Shanmugam Naidu established Kaliswari Fireworks in c.1926. They were selling their fireworks using Peacock and Cock brands respectively. After their pioneering effort, there emerged several units that started manufacturing fireworks.[96] Now Sivakasi is known as the fireworks capital of India.

The imposition of tariffs provided opportunities to several entrepreneurs to start manufacturing units for consumer as well as producer goods, because the market was starving for manufactured goods. Therefore, this period can be considered as the period of great opportunity. Few Indian entrepreneurs responded to take advantage of prevailing circumstances and helped in creating a highly diversified manufacturing base in India.

The first glass manufacturing company was started by Antaji Damodar Kale and Ishwar Das Varshnei in c.1905 at Talegaon near Pune, known as

Paisa Fund Glass Work. Ishwar Das Varshnei had studied engineering at the Massachusetts Institute of Technology. This enterprise was promoted under the patronage of Bal Gangadhar Tilak, and was established from a public contribution of one paisa each. The main objective of setting of this unit was to train people in the methods of glassmaking.[97,98] One of the apprentices of this company, Prabhakar Ogale, started Ogale Glass Works at Satara in c.1919. It is not surprising that in post-independent India, Maharashtra emerged as an important place for the birth of community-owned cooperative sugar mills.

Laxman Rao Kirloskar, who acquired technical skills while working as mechanical drawing instructor at the Victoria Jubilee Technical Institute located at Bombay, became the pioneer of several enterprises. He began to manufacture iron ploughs in a small factory in the princely state of Aundh in district Satara of Maharashtra. The son of the Raja, Balasaheb Pant Pratinidhi, donated a plot of land to him. This piece of land subsequently became the location of Kirloskar enterprises. He diversified his unit into the manufacturing of other agricultural implements also like chaff-cutters, windlasses and pulleys to lift water from wells. In 1920, the Kirloskar Brothers Company was incorporated. By c.1926, the company started manufacturing diesel and centrifugal pump sets. In c.1941, Mysore Kirloskar Ltd. was opened to manufacture machine tools at Harihar in district Davangere under the patronage of Dewan of Maharaja of Mysore, Sri Visvesvaraya. Laxman Rao Kirloskar's son did his engineering from MIT. He developed a specific type of sugarcane crusher, which the company began to manufacture under the brand name *kamal*. Subsequently, Kirloskar group entered into the manufacturing of electric motors, generators and transformers. In c.1946, Kirloskar Oil Engine Ltd. was founded.

Ardeshir Burjoria Godrej founded the Godrej & Boyce manufacturing company in c.1897 in Bombay, to make security locks with the help of modern machines. He also started Godrej Soap Company in c.1919 to manufacture soaps.[99] Alambic Chemical works was established in c.1907 by a professor of chemistry of Baroda College, his student Tribhuvandas Kalyandas and a rich landowner Mr. B.D. Amin in Baroda. The company experienced several problems, partly financial and partly emanating from government policies of that time of Maharaja of Baroda, and British administration.[100] This company survived by producing country liquor manufactured from the *Mahua flower*.[101,102] Kasturbhai Lalbhai group, which was mentioned while discussing textile manufacturing in India, opened Atul Chemical and Atul Glass Factory in Gujarat in c.1937.

Around the same time in c.1938, two Danish engineers, Henning Holk Larsen and Soren Kristin Toubro, founded the Larsen and Toubro Company in Bombay. These two Danish citizens had come to India as representatives of a Danish company, F.L. Schmidt & Company, which was supplying equipment for cement manufacturing plants. Visualizing scope for business opportunities, they stayed back in India. In c.1946, they promoted Indian Crown Cork Company Ltd. to manufacture crown corks for bottles, and Alu Capsules Ltd. to manufacture aluminium bottle caps.

Walchand Hirachand group of companies, which were engaged in the manufacturing of sugar, decided to make aircrafts in India. This decision was taken in a chance meeting between Walchand and William D. Pawley, head of the Harlow Aircraft Manufacturing Company of the United States. Walchand entered into a contract with Harlow Aircraft Manufacturing Company in 1942 to manufacture their aircraft in India. Walchand approached the British government and his proposal was promptly accepted because, during war, the British government was not able to provide air defence cover to the Indian sub-continent. Therefore, they permitted the government machinery in India to make arrangements for aircraft needed to meet the defence needs of India. After obtaining permission from the government, Walchand, along with Dharamsey Mulji Khatau and Tulsidas Kilachand, formed a managing agency company, Walchand-Khatau-Kilachand. The company also received patronage from the Maharaja of Mysore. They promoted Hindustan Aircraft Company Ltd. in c.1940 at Bangalore. The company began to assemble Hawk P36S and Harlow 5S aircrafts in India. The company immediately received a bulk order from the government to supply seventy-four long-range Vultee Attack bombers, forty-eight Curtis hawk fighters, and thirty Harlow trainers to be delivered by c.1942. In August c.1941, the company delivered its first Hawk to the government. However, after the attack on Pearl Harbor and the attack on Rangoon in c.1942 by the Japanese, the government got the excuse to nationalize the company, stating that it would become an important target if World War II spread to the Indian sub-continent.

As industrialization progressed, there emerged a demand for producer goods also. Realizing that P.N. Dutta & Company of Calcutta started the production of machine tools in c.1930, Birla also entered in the making of textile machinery and founded Textile Machinery Corporation (TEXMACO) at Calcutta and Central India Machinery Manufacturing Company (CIMMCO) in Bharatpur in c.1940. Again, to meet the wartime production needs, TEXMACO's plant was taken over by the government. In c.1943, the government allowed the company to move a certain amount of equipment of the company from Calcutta to its textile plant at Gwalior. Here the company began to manufacture looms. Only in c.1945 did TEXMACO receive its Calcutta factory back from the government, and in c.1946 it produced its first ring frame.[103] Even before the attempts of Birla to manufacture textile machinery, Alcon Ashdown & Company and Jost Engineering Company were engaged in the manufacturing of equipment for textile mills. Similarly, the expansion in sugar production led to increases in the demand for plants and equipment required to establish sugar mills. This was met by Butler & Company located at Muzaffarpur and Joran Engineering Company located at Saran district of Bihar. In the meantime, the United States also became an important supplier of equipment to Indian manufacturing, using lend-lease delivery mechanism.[104] It is estimated that the total cost of equipment supplied to Indian manufacturing enterprises during World War II was $913 million.[105] It is also essential to mention here that when the Tatas began to establish Tata Iron and Steel Company, they took consultancy from the most renowned consultancy firm located at Pittsburgh founded by Alexander Sahlin and Julian Kennedy called Kennedy, Sahlin & Company.[106]

It is also interesting to mention that, soon after the beginning of the British Raj in India (approximately after two decades), Britain was losing its status as the 'workshop of the world'. Modern manufacturing spread to other countries of Europe (like Germany) and America. Especially, after the end of the nineteenth century, German and American manufacturing began to give stiff competition to British manufacturers. After the unification of Germany by Otto Von Bismarck in c.1871, modern manufacturing grew quite rapidly. Between c.1870 and c.1900, Germany not only established manufacturing enterprises in the mainstream manufacturing products like textiles, it pioneered the manufacturing of entirely new products, which assumed importance in the twentieth century. Germany evolved these new products using coal, which was available in abundance in the country. These products were synthetic products, like chemical dyes, synthetic fibres and fertilizers. Germany also introduced new pharmaceutical products, based on inorganic chemistry, like aspirin, antibiotics and chemicals used in photography. These products were developed by two German firms, Bayer and Hochst. Germany also became a leader in the manufacturing of electrical goods after the development of electric dynamo by Ernst Warner Von

Siemens. The production of automobiles was also perfected by Germany during this period.[107]

Spread of modern manufacturing in the United States began through transfer of technology from Britain and other European countries, which was facilitated by the immigrant population. One of the famous immigrants from Britain, who made significant contributions in introducing modern manufacturing in the United States, was Samuel Slater. He had acquired knowledge of establishing textile mills while working in the cotton textile mill established by Richard Arkwright. In c.1793, he established that first cotton textile mill at Rawtucket in Rhode Island. He formed Samuel Slater & Company in c.1798 and opened several textile mills. In c.1814, Francis Cabot Lowell founded the Boston Manufacturing Company at Waltham in Massachusetts.[108]

The competitive advantage of United States manufacturing over British manufacturing began when Eli Whitney, a gun-making factory in New Haven, Connecticut started using assembly-line method of production to produce standardized products from interchangeable parts. This became a unique contribution of American manufacturing. The same process was used by Samuel Colts in gun making, by Cyrus McCormick in manufacturing of reapers, by Isaac Singer to manufacture sewing machines, and Henry Ford to manufacture automobiles. The use of a running assembly line to produce standardized products, in large volumes, using interchangeable parts in vertically integrated large manufacturing enterprises, emerged as an American system of organization of production. This was accompanied by raising of large volumes of capital, using the joint-stock form of ownership of companies run with the help of a hierarchy of professional managers. These became an important characteristic of US manufacturing. These unique manufacturing practices are now popularly known as *Fordism*.[109]

The American automobile manufacturing companies, after consolidating their market shares in Europe and America, began to establish their manufacturing facilities in India. In c.1928, General Motors Company established its assembly unit at Bombay, where it began to assemble its brand of trucks and cars with the help of imported knock-down kits. The Ford Motor Company established its assembling units at Madras, Bombay and Calcutta in c.1931. In these facilities, it assembled its own brand of automobiles. Around the same time, Addison & Company had also established its assembly unit in Madras in c.1936. At this facility, it assembled trucks and cars.

During this period, the Indian army was the largest buyer of automobiles, followed by the British administration. At that point in time, the army was buying automobiles only from Ford Motor Company and General Motors. The army did not want to buy automobiles from other companies because it would have resulted in buying a variety of spare parts, which would have increased inventory costs. On the other hand, American automobile companies were not interested in entering into collaboration with Indian entrepreneurs, as they had already cornered a substantial part of the market. In fact, they actively lobbied with authorities to create entry barriers for new entrepreneurs. Due to these reasons, Indian entrepreneurs, who intended to enter into the manufacturing of automobiles, faced stiff entry barriers.

Walchand Hirachand dreamed about manufacturing automobiles in c.1935. In the same year, when the Congress Party formed the provincial government, Walchand was able to negotiate an interest-free loan. However, by the time the idea of the project took concrete shape, World War II began in c.1939. In the wake of the non-cooperation movement, initiated by the Congress Party, the provincial government led by the Congress resigned. Walchand approached the British administration of the Bombay Presidency to honour the agreement that he had entered with the elected government. British administration refused to honour the agreement. This development forced Walchand to approach Maharaja of Mysore, who had helped him earlier in starting Hindustan Aircraft Company at Bangalore. However, under the influence of automobile lobby of American automobile manufacturers, Maharaja did not help Walchand. Despite the fact that he had received positive response from Maharaja of Baroda, he decided to promote the company through his own effort. Therefore, under the same managing agency company, Walchand-Khatu-Kilachand, which he formed to launch Hindustan Aircraft Company, he promoted Premier Automobiles in c.1944 and located his company at Kurla near Bombay. He entered into contract with Chrysler Corporation to assemble Plymouth cars and with Dodge Company to assemble their trucks. At the same time, Birla was also exploring the possibilities to enter into the manufacturing of automobiles. Birla took advantage of the interest shown by Maharaja of Baroda in helping to establish automobile manufacturing company at Baroda. Birla promoted Hindustan Motors in c.1942 and entered into contract with Nuffield Organization to assemble Morris ten cars and Studebaker trucks from knock-down kits.[110]

Mahindra and Mahindra, the company that has expanded manifold since independence, originated in Ludhiana in 1945 as Mahindra and Mohammed Company, to assemble automobiles. The company was promoted by Keshub Mahindra and Malik Ghulam Mohammed as M&M Company. After partition of the country, Mohammed quit his business in India and migrated to Pakistan. He became Pakistan's first finance minister, and later on governor-general in c.1951. Strangely, the company retained its original name M&M Company after independence also. However, now it is known as Mahindra and Mahindra Company.[111]

The account given here regarding the growth of Indian manufacturing shows that by c.1940s, India had emerged as the most industrialized amongst all the colonies of the British Raj. This happened because there had emerged a national market for a number of goods. Some of the markets also got integrated due to safety and security in interregional travel and transportation. British Raj also established rule of law and created legislations to enforce contracts and protect private property rights. These conditions, accompanied by competitive advantage of Indian manufacturing in the production of certain goods, provided conditions for the emergence of modern manufacturing in India. During this period, a substantial part of income and employment in modern manufacturing was generated by cotton and jute textile manufacturing enterprises. As has been described, modern manufacturing enterprises also emerged in the production of sugar, cement, paper, steel, safety matches, chemicals and some general engineering products. This shows that India at that time had evolved a well-diversified manufacturing base. Modern manufacturing had also spread to several locations in the Indian sub-continent, but it remained mostly localized at Bombay and Bengal presidencies. Madras Presidency has emerged as an important location of modern small enterprises.

Through the growth of modern manufacturing, the Indian economy had achieved the stage of primary import substitution, whereby it had developed capability to process most of the primary raw materials that it had been exporting earlier. Despite all these achievements, the growth of the manufacturing sector was not sufficient to transform a continental-size, predominantly agrarian economy, into a manufacturing nation of the world. On the eve of independence, only 9.6 per cent of its working population was engaged in various forms of manufacturing activities. These forms consisted of modern large, medium and small enterprises. In certain segments, traditional flexible manufacturing continued to exist and in some others, traditional manufacturing assumed different patterns of organization of production. Amongst all these forms of manufacturing enterprises, the share of workforce engaged in large, modern manufacturing enterprises was just 1.8 per cent of the total working population, and share of all other forms of manufacturing was 7.8 per cent.[112] In terms of their contribution to national income (which shows their productivity), the aggregate contribution of the entire manufacturing sector was 17 per cent in c.1946. This shows that despite the fact only 9.6 per cent of the working population was employed in the manufacturing sector, it contributed 17 per cent to the national income of the country. This clearly reveals that productivity per worker was higher in the manufacturing sector compared to productivity of workers in the other sectors of the economy. However, when we see disaggregated pictures of the manufacturing sector, we find that despite the share of modern, large manufacturing enterprises in employment was just 1.8 per cent whereas the shares of small enterprises in employment was 7.8 per cent, their contribution to national income was 6 per cent.[113,114]

The existence of duality in the structure of the Indian manufacturing sector was not unique to India. It is described in Chap. 3 of the book that even in Britain's modern manufacturing, there evolved a symbiotic relationship between its preexisting forms of traditional flexible manufacturing. The symbiosis between the two provided to modern manufacturing some of the flexibilities that are built into the organizational structure of traditional flexible manufacturing. However, duality in the structure of Indian manufacturing emerged partly as a consequence of it being a colony of Britain and partly due to different supply-side and demand-side factors, which affected the composition of Indian manufacturing. On the supply side, these factors were lower rates of accumulation of capital accompanied by underdeveloped capital market. This made capital costly and scarce. Shortages of skilled manpower and managerial talents necessitated the use of expatriate workers. At the same time, there existed excess supply of unskilled labour and a labour force trained in traditional skills. The impact of these supply-side factors was complemented by the narrow size of the market for standardized products, which were produced by modern, large-scale manufacturing enterprises. The limited size of the market for standardized products existed due to widespread poverty amongst the Indian population.

During the period of British Raj, land revenue was the major source of financing government expenditure. To meet the requirements of mounting government expenditure, British government had imposed on the peasantry high rates of land revenue. This left very little savings with the peasants, either to invest in improving methods of cultivation or to consume products produced by the manufacturing sector. Therefore, backward agriculture and poor peasantry sustained the continuity of traditional, as well as modern, small-scale enterprises. On the other hand, there also existed traditional crafts-based manufacturing, which produced high-quality products with better designs and craftsmanship that were patronized by rich consumers, and also enjoyed demand for exports. These products also reduced the demand for standardized products produced by large modern enterprises. Indian markets were also fragmented along caste lines, religious affiliations, and regional diversities in customs, tastes, and climates. The fragmented markets reduced the demand for standardized goods and provided conditions for the existence of *niche markets*, which could sustain traditional, as well as modern small manufacturing enterprises.

High cost of capital and availability of labour at less than subsistence level of wages provided competitive advantage for the modern as well as traditional small scale enterprises. The factor prices that existed at that time as a consequence of supply of factors of production provided unique advantage to small manufacturing units that used modern methods of production. The existence of a railway network had facilitated the mobility of labour, including the migration of craftsmen from rural areas to new towns, trading centres and near the centres of modern manufacturing. This process became more pronounced in the case of weavers who migrated from different parts of India during the 1940s to some of the towns of Western India. The towns that attracted these migrant weavers were Sholapur, Malegaon, Bhiwandi, Burhanpur and Surat. Some of these weavers purchased second-hand power looms, which were being discarded by modern textile mills. These weavers not only used their own labour, but also used available cheap labour and laid the foundation of small textile enterprises using power looms. This sector is now popularly known as the 'power loom sector'.

The preceding description about the structure and composition of the Indian manufacturing sector challenges the received understanding regarding the expansion of modern manufacturing. The received understanding is based on the premise that modern manufacturing grows by substituting fixed capital for working capital and labour. Instead of this process, in India, traditional flexible manufacturing as well as modern and traditional small units survived by retaining a competitive advantage by following different strategies. For instance, some of these enterprises used better division of labour to improve their efficiency. Some others replaced family labour with wage labour, which was more efficient than family labour. These enterprises also organized production in cluster of units using division of labour across firms, developing inter-firm networks and relations to evolve *flexible specialization.*[115] These firms survived by maximizing '*external economies*' rather than '*economies of scale*'. One can give an example of clustering of lock-making firms and firms manufacturing scissors in Aligarh and Meerut, units manufacturing handmade safety matches, *bidi* makers, firework manufacturers at Sivakasi, cluster of firms manufacturing knitwear at Coimbatore and Ludhiana, brassware manufacturing at Moradabad, carpet-weaving at Kashmir, Murshidabad, and Agra. Following same strategies, some of the centres of traditional flexible manufacturing have also survived at different locations. Economists and business historians are generally familiar with only two forms of organization of production, market and hierarchies (i.e., stand-alone, small, single-proprietor owned company or vertically integrated firm run with the help of hierarchy of managers). The driving force in both forms of organization is *competition*. The existence of inter-firm network rests on the proposition that there exists a certain amount of cooperation between firms. This cooperation emerges either due to prevalence of trust and goodwill amongst them, or due to the possibilities of interaction between them.[116] Therefore, the motive for forming of inter-firm network lies beyond market and hierarchy. In the inter-firm networks, units that form a network enjoy more autonomy in respect of ownership, control, and decision-making. Hence, inter-firm networks emerge between independent and loosely held firms. Inter-firm networks provide advantage to its members in terms of sharing of information, knowledge and strategic resources, like in Japan where it exists as organization of firms in *Kigyo Shudan*. Sometimes, inter-firm cooperation can also emerge amongst a group of firms, where they are linked with each other through supply of inputs or as buyers of output. This format is like the organization of *Keirestu* in Japan.[117]

The analysis given suggests that the expansion of modern manufacturing from Britain to the already inhabited regions, where traditional manufacturing already existed, created hybrid structures, part developed as an enclave and the other part maintaining its old characteristics. Indian manufacturing, at the time of independence, exactly manifested the phenomenon of dualistic existence. The idea of dualism rests on the premise that there exist concurrently two levels of technology, two different kinds of economies, two different kinds of cultures. These dualistic structures have been variously conceptualized in the literature of economic development viz.

capitalist and non-capitalist sectors,[118] modern and traditional sectors,[119] or industrial and pre-industrial sectors.[120,121] In these conceptual works on dualism, modern sector, capitalist sector, or modern manufacturing seems to have emerged as a consequence of colonial inheritance. It was the dualistic structure of the Indian manufacturing sector that postponed the birth of industrial revolution in India because it prevented the transformation of the economy into a manufacturing one. The existence of this dualistic structure posed the most important challenge to policymakers in post-independence India.

Notes

1. See Dalrymple (2006).
2. For literature on c.1857 revolt, see David (2002), Mukherjee (1988 [2007]), Joshi (ed) (2007) and Hibbert (1978).
3. See Stein (1998), p. 239 and Bayly (1988a, b).
4. See James (1997 [2003]), p. 316.
5. See James (1997 [2003]), p. 308.
6. To date, scholars working on the history of the period refer to the writings of Darling (1926), Tod (1829) and Hunter (1888 [1974]).
7. For this, read Ambirajan (1987), p. 13.
8. Bhattacharya (1965) and Hoznettey (1972).
9. See Chap. 2 on the Mughal Empire and traditional flexible manufacturing in this book.
10. For the trial of Nandkumar see Beveridge (1886) who used the term; judical murder and the quotation of Tilak has been taken form Kolsky (2010), p. 4; and the book also provides several instances when the legal system evolved by the British Raj ignored the act of violence of British citizens against the native population. For details regarding these aspects of British judicial system in India. See Kolsky (2003) and Singh (1998).
11. See Mukherjee (2010) and Cassel (2010).
12. See for this, Dutt (1956), p. 103.
13. See New Cambridge Modern History, vol. IX pp. 56–57.
14. See Morris (1982), pp. 553–676.
15. See Rajat Kanta Ray (1982), p. 66.
16. Morris (1987), fn. 4, p. 154.
17. See Macpherson (1955), p. 180.
18. See Morris and Dudley (1975).
19. See Tomilson (1981, 1993) and Sarkar (1983).
20. See Hurd (1982), p. 758 and writings of Thorner (1951) and Astha (1976).

21. See Lebman (1965).
22. See Rungta (1970).
23. For details regarding the ideas of these economic historians, see Tomilson (1979), Bagchi (1972) and Chatterji (1983).
24. For the arguments given by Dewey, see Dewey (1988).
25. Rothermund (1992) and Latham (1981).
26. See Seth (1987), Chap. II.
27. For Defence of India Rules, see Government of India (1939) and Mohan (1992).
28. See King (1969).
29. For details regarding the working of managing agency system, see Nigam (1957), Basu (1958), Singh (1966) and Sethia (1996).
30. Maikovits (2008), p. 130.
31. Kosambi (1986) and Hunter (1888 [1974]).
32. For these facts, see Tripathi (2004).
33. See Gandhi (1930) and Morris (1965), p. 12.
34. The Crimean War began in c.1854 and continued until c.1856; see Williams (1975), p. 224.
35. See Hunter (1888 [1974]).
36. Hunter (1888 [1974]) observed in the early 1880s that Calcutta and Bombay emerged as two nodal centres of trade, without parallel in any other century. For how these two port cities enjoyed double dominance in manufacturing, see Markovits (1985).
37. See Bagchi (1972).
38. For these important historical facts, see Roberts (1960), Roy (1999), Riello and Parthasarathi (eds) (2009), Riello and Ray (eds) (2009) and Parthasarathi (2011).
39. See Tripathi and Mehta (1990a, b).
40. For details regarding these facts, see Gandhi (1930), p. 52.
41. See Mehta (1954).
42. For these historical facts, see Mehta (1954) and Tripathi and Mehta (1990a, b).
43. Se Mehta (1954), pp. 27–29.
44. See Gadgil (1924), p. 76.
45. See Koh (1966), p. 23.
46. See Koh (1966), p. 14.
47. Kasturbhai Lalbhai's family traces its long lineage of being merchants and bankers. Their famous ancestor, Shantidas, was jeweller to the Mughal Emperor Akbar. Another ancestor saved the city of Ahmedabad from the plunder by the Martha Peshwa army by paying them a huge tribute. This act earned him the title of Nagar Seth.

48. The process of integration of traditional flexible manufacturing and modern textile mills has been described in greater detail by Ray (1982).
49. See Bagchi (1972), p. 143 and Chandavarkar (1994), Chaps. I, VI, and VII.
50. For the details regarding the birth of Delhi Cloth Mill, see Arun Joshi (1955), p. 58.
51. Ray (1979), pp. 281–284.
52. See Mehta (1954).
53. For the transformation of Coimbatore region, see Baker (1984).
54. For the contribution of Naidu entrepreneurs, see Mahadevan (1984, 1992, 1999) and Damodaran (2008).
55. Damodaran (2008), p. 153.
56. For the growth of Tripur as an important centre of knitwear manufacturing, see Vijayabhaskar (2001) and Chari (2004).
57. For details regarding Ramgharia community, see Damodaran (2008), p. 2.
58. The removal of import duties on British imported cotton textiles was embodied in the Notification, Sea custom No. 43, Calcutta March 18, 1878.
59. For discrimination against Indian cotton textile manufacturing firms, see Mehta (1954) and Morris (1965).
60. For details, see Buchanan (1966) and Wallace (1928).
61. See Wallace (1928), Bagchi (1972) and Buchanan (1966), Investor's India year Book of relevant year and the section on railway of the chapter.
62. See District Gazetteers of Eastern Bengal, Dacca, Pabna and Mymensingh.
63. See Gadgil (1924), Chap. 15, and Levkovsky (1965).
64. See Tripathi (2011).
65. For understanding alternative strategies that were followed by Marwari trading houses of Calcutta to enter into the manufacturing of jute textiles, see Lokanathan (1935).
66. For details regarding the use of fatka by Marwari merchants, see Goswami (2011), pp. 235–257.
67. For details regarding the ownership pattern of jute textile mills by the end of World War II, see Inverters India Year Book (1942), Goswami (1982, 1985, 2011).
68. See for details, Heath (1839).
69. For this fact, see Biswas (2001), pp. 120–122.
70. See, Parthasarathi (2001), p. 237.
71. See for this fact, Parthasarathi (2001), p. 237.
72. For this important fact, see Bhardwaj (1982), pp. 222–237.
73. See, Leidholmes (1972).
74. See for details, Rungta (1970).

75. For details regarding inventions that occurred in the manufacturing of iron and steel in different parts of Europe, see Singer et al. (eds) (1958), vol. IV.
76. The term weight-losing materials was used by Alfred Weber to describe those materials that lose their weight during processing, e.g. coal loses 100 per cent weight in the manufacturing of steel because its weight is not part of the final product. Similarly, when sugarcane in used in the manufacturing of sugar, the weight of 100 kilograms of sugarcane helps in producing only 10 kilograms of sugar; see for details Weber (1929) and Seth (1987), Chap. IV.
77. See Estall and Buchanan (1946), p. 172.
78. See Seth (1987), p. 42.
79. See Rungta (1970).
80. See Bagchi (1972) and Ray (1982), p. 82.
81. For these facts, see Mukherjee (2008), p. 90.
82. Piramal (1998).
83. For Tata's story in steel manufacturing, see Lala (1981), Harris (1954), Mukherjee (2008).
84. For details regarding the Indian iron and steel manufacturing industry, see Bagchi (1972) Johnson (1966), Harris (1954) and Mukherjee (2008).
85. See Gandhi (1934), pp. 2–8.
86. For details regarding these events, see Ray (1982), pp. 142–144.
87. See Dey (1933), pp. 238–282.
88. For these facts, see Damodaran (2008).
89. Bagchi (1972), pp. 406–409, Piramal (1998).
90. See Mehta (1991), pp. 233–248.
91. See Mehta (1991), pp. 189 and Ray (1982), p. 132.
92. See Bagchi (1972), pp. 353–358, Ahuja (1992).
93. See for details, Rosen (1958), pp. 20–23.
94. For details regarding these companies, see Ray (1982).
95. See Ray (1982).
96. See Kidron (1965), Tomlison (1978) and Bagchi (1972).
97. See Damodaran (2008), pp. 138–142.
98. See Damodaran (2008) and Mazumdar (2012), p. 106.
99. See Tripathi and Mehta (1990a, b) and Tripathi (2004), p. 157, 186, 234 and 293.
100. Kranjia (1997), Vol. I and Vol. II.
101. For details regarding these problems faced by Alembic Chemicals Ltd., see Tripathi (1987), pp. 293–295.
102. See Tripathi and Mehta (1990a, b), Tripathi (2004) and Desouza (1961).
103. See Khanolkar (1969) and Piramal (1998).
104. Piramal (1998), Piramal and Herdeck (1986) and Tripathi and Mehta (1990a, b).

105. See Thorner (1980), p. IX.
106. See Shrikov (1973) and Hall (1955).
107. See Lala (2007), p. 19.
108. See Clapham (1968), Henderson (1975) and Maddison (1991).
109. See Lichet (1995).
110. For the American method of organization of production, see Chandler (1977).
111. Fateh (2012).
112. Government of India (1952).
113. Roy (1999).
114. See for these processes, Piore and Sable (1984) and Seth (2003).
115. For the concept of flexible specialization, see Amin (1989, 1995), Stroper and Christopherson (1987), Christoplherson and Storper (1989) and Seth (2000).
116. For understanding regarding the role of trust and goodwill in forming inter-firm networks, see Dore (1986).
117. For the existence of inter-firm networks based on scope of interaction, see Bordien (2005) and Davis (2003).
118. For the knowledge regarding the contemporary forms of networks of firms, see Fruin (1983, 2008).
119. The Conceptualization of dualism into capitalist and subsistence sector was evolved by Lewis (1954).
120. For the idea of dualism based on modern and traditional sector, see Rostow (1963). For alternative models of dualism, see Jorgenson (1961).
121. The conceptualization of dualism sector, see Jorgenson (1961) pp. 309–334.

Bibliography

Ahuja, S. P. (1992). *Paper industry in India*. New Delhi: The Institute of Economic and Market Research.

Ambirajan, S. (1987). *Classical political economy and British policy in India*. Cambridge: Cambridge University Press.

Amin, A. (1989). Flexible specialization and small firms in Italy—Myth and reality. *Antipode, 21*(1), 13–24.

Amin, A. (Ed.). (1995). *Post fordism—A reader*. Oxford: Blackwell.

Astha, M. C. (1976). *Railway rates and fares in India 1849–1922—A historical cum analytical study*. PhD thesis, Department of Economics, Delhi School of Economics, New Delhi.

Bagchi, A. K. (1972). *Private investment in India 1900–1939*. Cambridge: Cambridge University Press.

Baker, C. J. (1984). *An Indian rural economy 1880–1950: The Tamil Nadu country side*. New Delhi: Oxford University Press.

Basu, S. K. (1958). *The managing agency system*. Calcutta: World Press.

Bayly, C. A. (1988a). *The British empire and the world 1780–1830*. Oxford: Oxford University Press.

Bayly, C. A. (1988b). *The new Cambridge history of India: The Indian society and the making of British empire* (Vol. II, p. 1). Cambridge, UK: Cambridge University Press.

Beveridge, H. (1886). *The trial of Maharaja Nand Kumar—A narrative of judicial murder*. Calcutta: Spock&Company/Thacker.

Bhardwaj, H. C. (1982). Development of iron and steel technology in India during 18th and 19th century. *Indian Journal of History of Science and Technology, 117* (2), 222–237.

Bhattacharya, S. (1965). Laissez Faire in India. *Indian Economic and Social History Review, 2*, 1–22.

Biswas, A. K. (2001). *Minerals and metals in pre-modern India*. New Delhi: Oxford University Press.

Bordien, P. (2005). Principles of an economic anthropology. In N. J. Smelser & R. Swedberg (Eds.), *Handbook of economic sociology* (pp. 75–89). Princeton: Princeton University Press.

Buchanan, D. H. (1966). *The development of capitalist enterprises in India*. London: Frank Cass.

Cassels, N. G. (2010). *Social legislation of the East India Company: Public justice versus public instruction*. New Delhi: Sage.

Chandavarkar, R. (1994). *The origins of industrial capitalism in India. Business strategies and the working class in Bombay 1900–1940*. Cambridge, UK: Cambridge University Press.

Chandler, A. (1977). *The visible hand; the managerial revolution in American business*. Cambridge, MA: Harvard University Press.

Chari, S. (2004). *Fraternal capital—Peasant, workers, self-made men—Globalization in provincial India*. Ranikhet: Permanent Black.

Chatterji, M. (1983). The political economy of 'discriminating protection', the case of textiles in the 1920s. *Indian Economic and Social History Review, 20*, 239–276.

Christoplherson, S., & Storper, M. (1989). New forms of labour organizations and production policies of flexibility in specialized industry. *Industrial and Labour Relaitons Review, 42*(3), 331–347.

Clapham, J. F. (1968). *Economic development of France and Germany 1815–1914*. Cambridge, MA: Cambridge University Press.

Dalrymple, W. (2006). *Last Mughal—The fall of the dynasty, Delhi 1857*. New Delhi: Penguin Viking.

Damodaran, H. (2008). *India new capitalist, caste, business and industry in modern nation*. Ranikhet: Permanent Black.

Darling, M. (1926). *Punjab peasantry in prosperity and debt*. London: Oxford University Press.

David, S. (2002). *The Indian mutiny*. New Delhi: Penguin Books.

Davis, G. F. (2003). Firms and environment. In N. J. Smelser & R. Swedberg (Eds.), *Handbook of economic sociology* (pp. 478–502). Princeton: Princeton University Press.

Desouza, J. D. (1961). *History of chemical industry in India*. Bombay: Technical Press.

Dewey, C. (Ed.). (1988). *Arrested development in India—The historical dimension*. New Delhi: Manohar Publication.

Dey, H. L. (1933). *The Indian tariff problem in relation to industry and taxation*. London: George Allen and Unwin.

Dore, R. (1986). *Flexible regidities*. Standford: Standford University Press.

Dutt, R. C. (1956). *The economic history of India under early British rule* (p. 103). London: Rentledge and Kegan Paul.

Estall, R. G., & Buchanan, O. R. (1946). *Industrial activity and economic geography*. London: Hutchisan University Press.

Fateh, S. (2012, June 14). Tracing the origins of Pak's doom and gloom. *Times of India*.

Fruin, W. M. (1983). *Kikkoman—Company, clan and community*. Cambridge, MA: Harvard University Press.

Fruin, W. M. (2008). Business groups and inter-firm networks. In G. Jones & J. Zeitlin (Eds.), *The Oxford handbook of business history* (pp. 244–268). New York: Oxford University Press.

Gadgil, D. R. (1924). *The industrial evolution of India in recent times: 1860–1939*. Oxford University Press.

Gandhi, M. P. (1930). *The Indian cotton textile industry*. Bombay.

Gandhi, M. P. (1934). *Indian sugar: Its past, present and future*. Calcutta: Government of India Press.

Goswami, O. (1982). Collaboration and conflict—European and Indian capitalists in jute economy of Bengal 1919–1939. *Indian Economic and Social History Review, 19*(2), 141–179.

Goswami, O. (1985). Then came Marwaris: Some aspect of changes in the pattern in industrial control in Eastern India. *The Indian Economics and Social History Review, 22*(3), 225–249.

Goswami, O. (2011). From bazar to industry. In M. M. Kudaisya (Ed.), *Anthology of business history* (pp. 235–257). New Delhi: Oxford University Press.

Government of India. (1939). *Defense of India act*. Shimla: Government of India Press.

Government on India. (1952). *Large industrial establishments in India in 1947*. New Delhi: Government of India Press.
Hall, H. D. (1955). *North American supply*. Oxford: Oxford University Press.
Harris, F. R. (1954). *Jamsatji Nussarwanji Tata – A chronicle of life*. Bombay: Asia Publishing House.
Heath, J. M. (1839). Indian iron and steel industry. *Journal of Royal Asiatic Society, 5*(3), 390–399.
Henderson, W. O. (1975). *Rise of German industrial power, 1834–1914*. Berkeley: University of Berkley.
Hibbert, C. (1978). *The great mutiny: India 1857*. London: Allen Lane.
Hozenetty, P. (1972). *Imperialism and free trade, Lancashire and Indian in mid-ninetieth century*. Manchester: University of Manchest Press.
Hunter, W.W. (1888 [1974]). Statistical accounts of Bengal, Vol. IX, Concept Publishing, New Delhi.
Hurd, J. M. (1982). Railways. In D. Kumar (Ed.), *The Cambridge economic history of India* (Vol. II, pp. 737–761). Cambridge: Cambridge University Press.
James, L. (1997). *Raj: The making of British India*. London: Abacus.
Johnson, W. A. (1966). *The steel industry of India*. Cambridge, MA: Cambridge University Press.
Jorgenson, D. W. (1961). The development of dual economy. *Economic Journal, 71* (3), 309–334.
Joshi, A. (1955). *Lala Shri ram—A study in entrepreneurship and industrial management*. London/Delhi: Orient Longman.
Joshi, P. C. (2007). *Rebellion 1857*. New Delhi: National Book Trust of India.
Khanolkar, G. D. (1969). *Walchand Hirachand: The man his times and achievements*. Bombay: Walchand and Company.
Kidron, M. (1965). *Foreign private investment in India*. London: Oxford University Press.
King, B. (1969). The origin of managing agency system in India. *Journal of Asian Studies, 26*, 120–130.
Koh, S. J. (1966). *Stages of industrial development in Asia – A Comparative history of cotton textiles industry in Japan, India, China and Korea*. Pennsylvania: University of Pennsylvania Press.
Kolsky, E. (2003). Confidication and rule of colonial difference: Criminal procedures in British India. *Law and History Review, 23*(3), 631–684.
Kolsky, E. (2010). *Colonial justice in British India: White violence and the rule of the law*. New Delhi: Cambridge University Press.
Kosambi, M. (1986). *Bombay in transition – The growth and social ecology of a colonial city, 1880–1980*. Stockholm: Almqvist and Wicksell.
Kranjia, B. K. (1997). *Godrez—A hundred years* (Vol. I and II). New Delhi: Viking.
Lala, R. M. (1981). *The generation of wealth—The Tata story*. Bombay: IBH Publishing.

Lala, R. (2007). *Romance of Tata Steel.* New Delhi: Penguin Viking.
Latham, A. J. H. (1981). *The depression and the developing world 1914–1939.* London: Croom Helm.
Lebman, F. (1965). Great Britain and the supply of railway locomotives to India—A case study of economic imperialism. *Indian Economic and Social History Review, 2*(4), 297–306.
Leidholmes, C. (1972). *Indian iron and steel industry—An analysis of comparative advantage.* East Lansing: Michigan State University.
Levkovsky, A. I. (1965). *Capitalism in India.* Bombay: Popular Prakashan.
Lewis, W. A. (1954). Economic development with unlimited supply of labor. *Journal of Manchester School of Economics and Social Studies, 22,* 139–192.
Lichet, W. (1995). *Industrialization of America: The nineteenth century.* Baltimore: John Hopkins Press.
Lokanathan, P. S. (1935 [1965]). *Industrial organization of India.* London: Orient Longman.
Macpherson, W. J. (1955). Investment in Indian railways c. 1845–1875. *Economic History Review, VIII,* 177–187.
Maddison, A. (1991). *Dynamic forces in capitalist development.* Oxford: Oxford University Press.
Mahadevan, R. (1984). Entrepreneurship and business community in Colonial Madras c.1902–1929—Some preliminary observations. In D. Tripathi (Ed.), *Business communities of India—A historical perspective* (pp. 325–239). New Delhi: Manohar Publishing.
Mahadevan, R. (1992). Pattern of industrial control in colonial madras: Some critical observations on relative position of Indian and foreign capital 1930–1930. In A. Ghosh (Ed.), *Indian industrialization structure and policy issues* (pp. 336–364). New Delhi: Oxford University Press.
Maikovits, C. (2008). *Merchants, traders, entrepreneurs–Indian business in colonial era.* Renikhet: Permanent Block.
Markovits, C. (1985). *Indian business and national politics, 1930–1939.* Cambridge: Cambridge University Press.
Mazumdar, S. K. (2012). *India's late, late industrial revolution—Democratization of entrepreneurship.* New Delhi: Cambridge University Press.
Mehta, S. D. (1954). *Cotton mills of India 1854–1954.* Bombay: Textile Association.
Mehta, M. (1991). *Indian merchants and entrepreneurship in historical perspective.* New Delhi: Academic Foundation.
Mohan, R. (1992). Industrial policy and control. In B. Jalan (Ed.), *Indian economy, problems and prospects* (pp. 85–104). New Delhi: Vikings.
Morris, M. D. (1965). Towards to re-interpretation of nineteenth century Indian economic history. *Journal of Economic History, 23*(4), 510–540.

Morris, D. M. (1982). Growth of large scale industry in India to 1947. In D. Kumar (Ed.), *The Cambridge economics history of India c.1757–c.1970* (Vol. II, pp. 553–676). Hyderabad: Cambridge University Press.

Morris, D. M. (1987). Indian industry and business in the age of laissez faire. In D. Tripathi (Ed.), *State and business in India* (pp. 123–156). New Delhi: Manohar Publication.

Morris, D. M., & Dudley, C. B. (1975). Selected railway statistics for Indian sub-continent (India, Pakistan, Bangladesh) c.1853–1946/7. *Artha Vijnana, 17*(3), 185–298.

Mukherjee, R. (1988 [2007]). *Spectre of violence: The 1857 Kanpur massacres.* New Delhi: Penguin Books.

Mukherjee, R. (2008). *A century of trust—A story of Tata Steel.* New Delhi: Penguin Books.

Mukherjee, M. (2010). *India in the shadow of empire: A legal and political history 1774–1950.* New Delhi: Oxford University Press.

Nigam, R. K. (1957). *Managing agency system in India: Ministry of Commerce.* New Delhi: Government of India.

Parthasarathi, P. (2001). *The transition to colonial economy weavers, merchants and kings in South India 1720–1800.* Cambridge: Cambridge University Press.

Parthasarathi, P. (2011). *Why Europe grew rich and Asia did not—Global divergence 1600–1850.* Cambridge: Cambridge University Press.

Piore, M. J., & Sable, C. F. (1984). *The second industrial divide.* New York: Basic Books.

Piramal, G. (1998). *Business legends: Indepth profiles of India's most daring and innovative entrepreneurs.* New Delhi: Penguin Books.

Piramal, G., & Herdeck, M. (1986). *India's industrialist.* Bombay: India Book House Distributors.

Ray, R. K. (1979). *Industrialization in India: Growth and conflict in private corporate sector, 1914–1947.* New Delhi: Oxford University Press.

Ray, R. K. (1982). Pedhis and Mills—The historical integration of formal and informal sectors of the economy in Ahmedabad. *Indian Eocnomic and Social History Review, 19*(3 and 4), 239–265.

Roy, T. (1999). *Traditional industry in the economy of colonial India.* Cambridge: Cambridge University Press.

Riello, G., & Parthasarathi, P. (Eds.). (2009). *The spinning world—A global history of textiles 1200–1850.* Oxford: Oxford University Press.

Reillo, G., & Ray, T. (Eds.). (2009). *How India clothed the world; Cotton textiles and the Indian Ocean 1500–1850.* Leiden: Brill.

Roberts, E. (1960). *Embassy to the eastern courts – Cochin, China, Saim, and Muscat in the US sloop of war during 1832–34.* Wellington: Scholarly Recourses Limited.

Rosen, G. (1958). *Industrial change in India.* Glencoe: Jllionais.

Rostow, W. W. (Ed.). (1963). *The take-off into self sustained growth*. London: Macmillan.

Rothermund, D. (1992). *India in the great depression, 1929–1939*. New Delhi: Manohar.

Rungta, R. S. (1970). *The rise of business corporation in India, 1851–1900*. Cambridge, MA: Cambridge University Press.

Sarkar, S. (1983). *Modern India 1885–1947*. New Delhi: Macmillan.

Seth, V. K. (1987). *Industrialization in India spatial perspective*. New Delhi: Common Wealth Publication.

Seth, V. K. (2000). Technology and labour in retrospect and prospect. *Managment and Change, 4*(1), 61–90.

Seth, V. K. (2003). The beginning and the end of primitive flexible manufacturing system. *Journal of Management Research, 3*(2), 73–86.

Sethia, T. (1996). The rise of jute manufacturing industry in colonial India: A global perspective. *Journal of World History, 7*(1), 35–56.

Shrikov, G. K. (1973). *Industrialization in India*. Moscow: Progress Publishing House.

Singer, Chargles et al. (Eds.). (1958). *History of technology* (Vols. II–IV). Oxford: Clarendon Press.

Singh, B. (1966). *European agency houses in Bengal*. Calcutta: Firma Mukhopadhyay.

Singh, R. (1998). *A despotism of law: Crime and justice in early colonial India*. New Delhi: Oxford University Press.

Stein, B. (1998). *A history of India*. New Delhi: Oxford University Press.

Stroper, M., & Christopherson, S. (1987). Flexible specialization and regional agglomeration. *Annal of the Association of American Geographers, 77*(1), 25–48.

Thorner, D. (1951). Great Britain and development of Indian Railway. *Journal of Economic History, XI*, 326–350.

Thorner, D. (1980). *Shaping of modern India*. Bombay: Sameeksha Trust.

Tod, J. (1829 [1993]). *Annels and antiquities of Rajasthan* (W. Crooke, Ed.), reprint Bibliolithic, vols 1 & 2. New Delhi.

Tomilson, B. P. (1978). Foreign private investment in India 1920–1960. *Modern Asian Studies, 12*(4), 655–677.

Tomilson, B. R. (1979). *Political economy of the raj*. London: Oxford University Press.

Tomilson, B. R. (1981). Colonial firms and decline of colonialism in eastern India 1914–1947. *Modern Asian Studies, 15*(3), 455–486.

Tomilson, B. R. (1993). *The economy of modern India 1860–1970*. Cambridge: Cambridge University Press.

Tripathi, D. (2004). *Oxford history of Indian business*. New Delhi: Oxford University Press.

Tripathi, D. (1987). Business and princely states—The case of Baroda. In D. Tripathi (Ed.), *State and business in India—A historical perspective* (pp. 274–301). New Delhi: Manohar.

Tripathi, D. (2011). Marwaris from community to class. In M. M. Kudaisya (Ed.), *The Oxford India anthology of business history* (pp. 133–142). New Delhi: Oxford University Press.

Tripathi, D., & Mehta, M. (1990a). The Nagarsheth of Ahmedabad: The history of urban institutions in a Gujarat city. In S. Chandra (Ed.), *Essays in medieval Indian economic history*. New Delhi: Munshi Ram Manoharlal.

Tripathi, D., & Mehta, M. (1990b). *Business houses of Western India—A study of Entreprenerurial response, 1850–1956*. New Delhi: Manohar.

Vijayabhaskar, M. (2001). *Industrial formation under conditions of flexible accumulation—The case of global knitwear node, in Southern India*. Unpublished PhD thesis, Centre of Development Studies, Thiruvananthpuram (Mimio).

Wallace, D. R. (1928). *Romance of jute—A short history of the Calcutta jute mill industry 1855–1927*. London: W. Thacker.

Weber, A. (1929). *Theory of location of industry* (trans: Freidrich). Chicago: Chicago University Press.

Williams, N. (1975). *Chronology of the modern world; 1763–1965*. Middlesex: Renguin Books.

CHAPTER 7

Lessons Learnt from *A Story of Indian Manufacturing*

Understanding of the process of economic development in historical perspective reveals that only those economies that have achieved industrial revolution have been able to maintain high rates of growth, sustained over a longer period. It is because revolution in the manufacturing sector brings higher growth rates in all other sectors of the economy. However, scholars who have been studying the pattern of economic development of the Indian economy have observed that under the compulsions of its socio-economic and political circumstances, it has deviated from the historically known path of development. It has followed a path of economic development in which an economy skips the stage where the manufacturing sector dominates in terms of its contribution to income and employment. This has happened because the Indian economy has missed industrial revolution. Accepting that it is generally the 'past' of an economy that determines its present, we began to explore the roots of this historical exception in *A Story of Indian Manufacturing*.

Our explorations revealed that the Indian economy had become an agrarian economy much earlier than other economies of the world. The crafts-based methods of manufacturing products were used for ages. Despite the fact that the division of labour had occurred between cultivators and craftsmen, their functions remained embedded in the village economy. The division of labour between different occupational groups was organized on caste lines.

It is also a historical fact that, despite the existence of division of labour for a considerable period, craftsmen were generally producing manufactured

V.K. Seth, *The Story of Indian Manufacturing*,
DOI 10.1007/978-981-10-5574-4_7

products to meet the needs of the local village economy. At this point in time, production of manufactured products was not meant for market-mediated exchange. This phenomenon existed due to the prevalence of a unique relationship between different occupational groups. Created by a distinct social institution called the *jajmani* system. The *jajmani* system was based on a 'gift exchange' relationship amongst groups of people pursuing different occupations in the village economy. The consequence of this form of organization of production and exchange was that the manufacturing remained an 'integral part of the self-sustained village economy'. Along with the craftsmen who were part of the village economy, there always existed a number of craftsmen who produced luxury handicrafts for rulers, nobles and gentry living in urban areas of the Indian sub-continent, and also for those living in other parts of the world.

The equilibrium in the economy and society, which had emerged as a consequence of integration between manufacturing and agriculture, came to an end when the autonomy of the village economy was slowly eroded by several exogenous forces. This process began with the emergence of a new set of rulers, who began to collect land revenue in cash. This led to the monetization of the village economy. This process was also accompanied by increasing urbanization, which necessitated the migration of craftsmen to urban areas. On one hand, these new circumstances created larger markets of manufactured goods, and on the other hand, they established market-mediated exchange between different occupational groups and their consumers.

The process of transformation of the economy of the Indian sub-continent from production for distribution amongst households based on customary payments, to an economy where a market-mediated exchange relationship emerged, occurred around c.1500. The process was galvanized when two intertwined forces of globalization impacted the economies of the Indian sub-continent. One was the entry of European merchants through a sea route, after the discovery of a sea route to the Indian sub-continent by Vasco da Gama in c.1498. The other was the beginning of the Mughal rule in c.1526, with the entry of Mughals through a land route. These two concurrent events set in motion necessary preconditions for the emergence of a new form of organization of production of manufactured products. This new form developed the capability of the Indian manufacturers to supply crafts-based manufactured products to mass markets, and came to be known as traditional flexible manufacturing.

These two interconnected episodes of globalization created new socio-political and economic forces, which conditioned the pattern of growth of manufacturing in the Indian sub-continent. The Mughal Empire created an integrated economy of continental size, and arrival of European trading companies expanded the commercial horizon of the economy of the Indian sub-continent to European economies. The increasing size of domestic market and integration of the Indian economy with a global trading system provided favourable conditions for the growth of traditional flexible manufacturing. To meet the requirements of the growing market, there was a need to expand the scale of operation of production of manufactured products because growing demand for manufactured goods could not be met by household workshops of craftsmen, as these did not have the resources to expand the scale of operation. These prevailing conditions necessitated structural transformation in the organization of production in manufacturing.

This transformation was initiated by merchants who had substantial financial resources. To meet the growing demand of manufactured goods, merchants evolved networks of several craftsmen through the putting-out system. This not only provided to the craftsmen necessary working capital and their subsistence needs, but also took the responsibility of marketing and selling of their products.

The traditional flexible manufacturing form of organization of production emerged as a consequence of loss of independence of craftsmen to merchants, who supplied them raw materials and working capital along with their subsistence needs in lieu of controlling the marketing of the final output of manufactured goods. During Mughal rule, this form of organization of production was known as the *dadni* system. This system of production of manufactured goods created a dispersed household workshop economy, based on vertically disintegrated inter-industry and intra-industry specialization, localized in clusters in specific areas. Such clusters of household workshops had emerged in different parts of the Indian sub-continent during Mughal rule. Each cluster specialized in the production of specific manufactured goods. These clusters have emerged in locations such as Dacca, Patna, Murshidabad, Kasimbazar, Sonargaon, Assam, Balasore and Banaras in Eastern India; Surat, Baroda, Ahmedabad and Jaipur in the Western India; Agra, Aligarh, Gwalior, Moradabad and Lucknow in Central India, Amritsar, Lahore and Ludhiana in northern India; and the Coromandal and Malabar Coasts in Southern India.

This book has told us that the historical circumstances that were prevalent in India during Mughal rule transformed Indian manufacturing. The transformation was made possible because the Mughal Empire facilitated this changeover by establishing law and order, integrating fragmented markets, monetizing the economy, increasing the volume of market-mediated exchange of goods and services, and introducing single currency. Increases in the urban population, urban culture through lifestyle consumption needs of nobles and elites, existence of several communities of merchants, and emerging linkage of the domestic economy with the rest of the world, were some other factors. These forces collectively helped in the spread of traditional flexible manufacturing. This form of manufacturing evolved through the formation of a nexus between craftsmen and merchants. The nexus provided capability to Indian manufacturers to supply quality manufactured goods to not only domestic markets but also to meet the needs of markets scattered around the world.

Though Mughal emperors were interested in promoting the growth of traditional flexible manufacturing, there were variations in the conduct of *subedars* of different *subas*. There were instances when these local governors indulged in extortion, coercion and intimidation of merchants to amass private wealth. The existence of absolute monarchy provided scope for this kind of opportunistic behaviour. This was at odds with the prevailing situation in Britain at that time. In Britain, the power of absolute monarchy was made limited by the emergence of Parliament. Limited monarchy had facilitated the birth of the rule of the law, which protected private property rights of individuals.

The slow and systematic decline in the authority of the Mughal state, after the death of Aurangzeb, resulted in changes in the conditions that facilitated the continuity of traditional flexible manufacturing. The markets again were fragmented with fragmentation of polity. This led to the emergence of inter-regional trade barriers accompanied by re-emergence of multiple currencies. Due to these changes in the circumstances, some scholars believed that the economy of the Indian sub-continent experienced decline of traditional flexible manufacturing. This is also viewed as the beginning of the process of de-industrialization.

The research conducted by scholars on specific regions that enjoyed autonomy after the death of Aurangzeb, suggest that the preconditions that are necessary for the existence of traditional flexible manufacturing continued to prevail in several autonomous regions. Therefore, the process of de-industrialization did not begin with the death of Aurangzeb. Traditional flexible manufacturing continued to exist in regions like Hyderabad, Punjab, Rajasthan, Oudh and southern parts of India. New evidence suggests that the decline of the Mughal Empire did not result in decline of traditional flexible manufacturing. This also happened because traditional flexible manufacturing is just only one form of organization of production. The handicrafts-based mode of production of manufacturing goods can be embedded in several kinds of organizational formats, namely, the standalone family workshop catering to local needs, and the household workshop part of a craft guild. Hence, associating de-industrialization with the decline of the Mughal Empire needed re-assessment. Our story revealed that scholars generally do not realize that the decline of the Mughal Empire (which began in c.1707) was accompanied by a slow and steady rise of the rule of the East India Company after the Battle of Plassey in c.1757. Historians have established that India remained one of the largest exporters of manufactured goods up to c.1750. The post-1750 period coincides with the period of political ascendency of the East India Company. Therefore, in our story, we explored the role of the East India Company in affecting the pattern of growth of traditional flexible manufacturing.

Our analysis revealed that to know the role of the East India Company in affecting the growth of traditional flexible manufacturing, it is essential to understand the process of transformation of British traditional manufacturing into modern manufacturing. The emergence of industrial revolution in Britain significantly altered the nature of the relationship between the East India Company and the traditional flexible manufacturing of the Indian sub-continent. The emergence of a new class of manufacturers or industrialists in Britain slowly eroded the paramountcy of the merchants in the British society and polity. This change in the nature of the dominant class altered the political economy of policy formulation in Britain. Due to these changes, the East India Company was entrapped between the interests of merchants and the interests of industrialists. Therefore, we realized the need for evaluating the role of the British Industrial Revolution in affecting the pattern of growth of traditional flexible manufacturing of the Indian sub-continent.

Similar to the prevalence of traditional flexible manufacturing in the Indian sub-continent, Britain also had a crafts-based traditional manufacturing industry around the same time. In European history, this stage of manufacturing defined as a stage of 'proto-industry'. The characteristics of the European proto-industry and traditional flexible manufacturing that existed in the Indian sub-continent were the same. However, unlike the dominance of production of cotton textiles, British proto-industry specialized in the production of woollen textiles. This form of organization of production remained dominant in Britain because it provided several advantages to small household workshops available in the form of economies in costs, as a consequence of division of labour. It also helped in reducing cost of production resulting from economies of scale and scope to the merchants, who were organizing networks of craftsmen into the putting-out system. The expansion of this form of organization of production was facilitated by the existence of demographic dividend, which occurred due to demographic trends in the British economy. Moreover, this form of organization of production of crafts-based products was compatible with the behaviour of household economy of the family workshops.

Since traditional flexible manufacturing of Britain provided different kinds of benefits to British manufacturers, these benefits ensured its continuity for a considerable period. However, this form of organization of production of handicrafts contained seeds of its own transformation into modern manufacturing. The traditional flexible manufacturing helped in accumulation of capital in the hands of merchants, which is an important precondition for the birth of modern manufacturing. Moreover, there also existed an 'agency problem' between merchants and craftsmen. The prevalence of the agency problem forced merchants to organize the production process in such a way that ownership and control were integrated. This could emerge only under the condition where merchants invested in the equipment and craftsmen worked as workers under merchant supervision. The other important condition, which was essential for the birth of integration between ownership and control was that the new organization should have the capability to outcompete low-wage, low-capital intensive, household workshops of traditional flexible manufacturing.

This was made possible when there sprang up, within a short duration, a series of inventions regarding processing of cotton. These inventions became more useful with the invention of the steam engine. The competitive advantage was created when a source of inanimate power was harnessed with the help of the steam engine and was used to run tools. The integration of tools and the source of inanimate power gave birth to 'machines'. This technological integration changed the meaning of the word manufacturing from 'made of hand' to 'made of machine'. The new paradigm of manufacturing that emerged as a consequence of integration of inanimate sources of power with tools transformed the nature and characteristic of British manufacturing. This process of transformation is described as the beginning of the Industrial Revolution in Britain.

Introduction of machines created a new class of people, who specialized in the manufacturing of machines and equipment. Some of these machines were general-purpose machines and some of them were dedicated to help manufacturing in specific industries. In new circumstances, some of the machine and equipment producers themselves became manufacturers of products, and the availability of machines provided opportunities to merchants to buy these machines to establish enterprises. Ownership of capital became an important characteristic of new manufacturing, because it facilitated the emergence of the factory form of organization of production. The birth of factories forced craftsmen, whose household workshops could not compete with factories, to come out of their family-owned workshops to work in factories. However, to establish factories or mills, entrepreneurs had to invest a substantial proportion of their circulating capital into fixed capital. Therefore, in order to achieve convergence between ownership and control, entrepreneurs had to become slaves of their own investments in fixed assets.

The most pertinent lesson that we learnt from the story was that the industrial revolution of Britain—instead of transforming its manufacturing of woollen textiles in which Britain had competitive advantage—transformed manufacturing of cotton textiles. Why did Britain transform manufacturing of cotton, where it did not have competitive advantage? It is while seeking answer to this question that we realized that there exists a very strong connection between the story of Indian manufacturing and the British Industrial Revolution.

The manufacturing of cotton textiles in Britain began with its interaction with the Indian sub-continent. Britain began earning bullion through trade in Indian cotton textiles. Looking at the advantage of fostering the manufacturing of cotton textiles, British manufacturers initiated the process of import substitution of Indian cotton textiles. The process began in the sixteenth century in Britain with the manufacturing of a variety of cotton textiles called 'fustians' in the East Anglia region of Britain. Fustians were manufactured by combining linen and cotton. Their production spread to various regions of Britain. The expanding home market for cotton textiles along with the possibilities of exporting it to other parts of Europe initiated this demand-led process of import substitution. Subsequently, most of the technological changes that occurred in the manufacturing of cotton textiles cannot be explained without accepting the fact that the process of industrial revolution of Britain began with the emergence of its trading relationship of with India.

The process of import substitution received further impetus by the introduction of cotton plantations in southern parts of America by people who had migrated to America from Britain. In these cotton plantations, they were cultivating cotton with the help of slaves brought from the African continent. This made cotton available at much cheaper prices to the British manufacturers of cotton textiles than the cotton available to Indian manufacturers of cotton textiles. These intertwined processers of technological changes in the manufacturing of cotton textiles, along with the availability of cheaper cotton, shifted the competitive advantage of manufacturing cotton textiles from the Indian sub-continent to Britain. British machine-made cotton textiles began to replace Indian cotton textiles not only in the global markets but also even in the Indian markets. This episode shifted advantage of manufacturing from India to Britain. This happened because it led to decline of Indian traditional flexible manufacturing. Cotton textile manufacturing was one of the most important manufacturing industries of the Indian sub-continent, which had provided income and employment to a very large proportion of the working population. This episode has been described by Indian nationalists as well as by Marxian historians as the process of de-industrialization.

At that point in time, when Britain was experiencing the emergence of modern manufacturing in the Indian sub-continent, the authority of the Mughal Empire was slowly declining after the death of Aurangzeb. This led to fragmentation of polity into several regional centres of power, competing with each other to fill the political space vacated by the Mughal state. After the Battle of Plassey (c. 1757), the East India Company began to fill the empty space created by the decline of the Mughal Empire. The Battle of Buxar (c. 1764), which led to granting of diwani right to the East India Company on Bengal Suba changed the character of the East India Company. The arrangement laid the foundation of transformation of the trading company into a territorial empire. Thus, our story told us that the period during which the Indian sub-continent witnessed the territorial expansion of the East India Company coincided with the period when Britain was experiencing industrial revolution. This historical coincidence resulted in a situation where the interest of Indian traditional flexible manufacturing got entangled with the basic conflict between the merchants and manufacturers of Britain.

The transformation of the East India Company from a trading company to an agency of the British Empire created new objective conditions, which led to restructuring of the relationship between the company and the economy of the Indian sub-continent in general, and Indian flexible manufacturing in particular. In the new relationship, the most important macro-economic consequence was that the East India Company began to use income that it earned from land revenue on financing its purchases of products manufactured by the Indian traditional manufacturing for exports without making payments in bullion. This form of trading arrangement has been described as 'unrequited trade'. The bullion earned through foreign trade was important for the functioning of the economy of the continent because it provided supply of metal for minting coins. Therefore, new arrangements created by the East India Company resulted in shortage of currency, which caused virtual stagnation of business activities in the sub-continent.

The story also revealed that once the East India Company's political ascendency began in India, it used its political power to consolidate its monopoly over centres of traditional flexible manufacturing to get unrestricted access to their best products. To achieve this goal, the East India Company used market as well as non-market strategies to establish its paramountcy in trade of products produced by traditional flexible manufacturing. This strategy necessitated that it had to marginalize the role of Indian merchants from the process of production to break the nexus that existed between Indian merchants and craftsmen. It was this nexus that had provided to the Indian manufacturing sector the capability to supply unlimited volume of quality, manufactured goods; not only to meet domestic needs, but also to meet their global demand. The strategies followed by the East India Company to break this nexus intended to create dependence of Indian craftsmen on the agents of the company to sell their products. This strategy led to the decline of traditional flexible manufacturing, which has been described by historians as the process of de-industrialization. This process took the economy of the Indian sub-continent one stage backward from industrial revolution. Therefore, the analysis of the process of de-industrialization amounts to the understanding of factors that destroyed the seeds that are essential to give birth to industrial revolution.

To break the nexus between Indian traders and craftsmen, the East India Company declared its direct monopoly of trade in certain products. The process began with assumption of monopoly of trade in saltpetre, followed by salt and opium. In such products, the company eliminated the role of Indian merchants and became the sole buyer or monopsonist. Since the company became the sole buyer, it acquired the capability to manipulate the prices of these products. To reduce the role of Indian manufacturers in the production of handicraft products, the East India Company reduced its dependence on Indian merchants for procurement of merchandise. It instead appointed its own *gomashtas* (agents) for the purposes of procurement. Finally, the East India Company banned the *Dadni* system in c.1770, which had given competitive advantage to Indian traditional manufacturing to sell its products globally. Moreover, under the pressure of British manufacturers, who wanted to sell their products produced by their modern manufacturing sector, the East India Company gave them access to Indian markets. This happened when the East India Company was forced to allow the entry of free British merchants, who acted as agents of British manufacturing companies in India. These agents formed agency houses, which in due course of time, became wealthier and developed their own network of influence in the Indian market due to the patronage given to them by the officials of the East India Company. These agency houses also reduced the significance of Indian merchants.

The East India Company, along with marginalizing the role of Indian merchants, also imposed several kinds of regulations on the Indian craftsmen. Like indenture regulation (*khatbandi*), which helped in binding Indian craftsmen exclusively to the agents of the company. Through this regulation, craftsmen were prohibited from selling their products to the Indian merchants. This regulation converted Indian craftsmen into bonded labourers of the company. The East India Company also introduced a *motarfa* tax on weavers, which was levied on the basis of number of looms employed in the household workshop of the weaver. The company also established its own centres of manufacturing, where craftsmen were forced to work at lower wages and for longer hours. These regulations provided immense bargaining power to the officials of the East India Company in terms of buying better quality products at cheaper prices and hiring Indian craftsmen at lower wages.

The aforementioned strategies and regulations used by the East India Company created a disconnect between Indian merchants and craftsmen. The disengagement of merchants in the process of production of crafts-based traditional flexible manufacturing exposed craftsmen to several kinds of risks and uncertainties. This occurred due to altered circumstances when craftsmen lost the support of merchants in the organization of production.

Traditional flexible manufacturing, which was engaged in the production of luxury handicrafts, also experienced significant decline because of the emergence of the East India Company as the ruler of India. The East India Company replaced the hierarchy of *mansabdars,* nobles and Mughal aristocracy with salaried administrators. The elites created by the Mughal Empire supported the demand for luxury handicrafts; the new elites were more adept to the use of manufactured products produced by the British manufacturers. Even regional rulers, who occupied positions of power in the successor states of the Mughal Empire, intended to differentiate themselves from the Mughal aristocracy. To these new power elites, the availability of British and European manufactured goods provided alternative choice. This practice further reduced the demand for products manufactured by traditional flexible manufacturing.

The story revealed to us an important aspect of the East India Company, which has not been analysed by scholars in greater detail. We learnt that the opportunistic behaviour of its management of the East India Company was embodied in the DNA of the company. The East India Company was perhaps the first joint-stock company of the world. It had a continuous existence as a commercial corporation from c.1600 to c.1833, and from c.1833 to c.1857 as an agency of the British Empire. The company was formed by royal charter granted by Queen Elizabeth I. The chartered status given to companies gave them monopoly of trade with specific territories. The justification for giving chartered status to companies was based on the premise that the opening of trade in new markets and products is as important as innovation and therefore, such companies should enjoy monopoly right for certain periods of time. Therefore, granting of royal charter was considered to be equivalent to granting intellectual property right. Moreover, granting of royal charter by the Crown gave the company quasi-sovereign rights, like the right to mint its own coins, exercise law and order and provide justice in its own overseas settlements. These quasi-sovereign rights made it possible for the East India Company to occupy territories in India, despite being a trading company.

The East India Company was a joint-stock multinational company. It was a unique innovation in organization of business at that point in time. The joint-stock form of business organization gave it capability to pool resources from thousands of stockholders as well as source talent from thousands of inspiring persons to run its management, without relying on hereditary managers. However, being a joint-stock form of business organization, it was perhaps the first company that experienced separation between ownership and control. To understand the consequences of this separation we studied the nature of corporate governance in the case of the East India Company. One of the important consequences of separation between ownership and control is that it results in the 'principal-agent problem'.

In the case of the East India Company, there existed multiple layers of management. For instance, the management was based at the head office of the company in London. It had managers working at different presidency towns and field managers working at different markets in the Indian sub-continent. The multiple layers of management created several kinds of principle-agent problems. The most serious was between the management and staff located at the head office of the company at London and managers and employees working in the Indian sub-continent. The degree of seriousness of the principal-agent problem can be appreciated if one realizes the quantum of information asymmetry, which existed because of spatial distance that separated these two sets of company managers. The enormity of the information asymmetry had also increased between these two separate managements because company employees working in the Indian sub-continent were also permitted to conduct business on their own account. Therefore, the story also dealt with the principal-agent problem that existed between the management of the company and its overseas employees. This necessitated the analysis of the organizational structure of the East India Company and the nature of the contract between the company and its overseas employees.

The story revealed that given the historical circumstances of the period, the East India Company had devised an appropriate contract between the company and its overseas employees. The contract consisted of paying low wages (to minimize cost to the company), permission to conduct private trade to supplement low wages as a 'carrot' and threat of dismissal as a 'stick'. Despite the formulation of an appropriate contract, the story told us about several instances of opportunistic behaviour on the part of overseas employees, which compromised the quality of corporate governance in the East India Company.

To minimize the chances of opportunistic behaviour on the part of company officials working in the Indian sub-continent, British Parliament evolved several regulations. However, at that point in time, the British polity was struggling to arrive at a new balance of power between the Crown and the Parliament. In these circumstances, to overcome the threats imposed by the Parliament on its autonomy, the company evolved nexus with the Crown. Over time, several executives of the company got involved in acts of financial irregularities in order to enhance their private wealth. These executives came back to Britain after accumulating substantial amounts of wealth. They used the power of their wealth to enter into Parliament. The top executives of the company, like Clive and Warren Hastings, were reprimanded by the British Parliament on account of serious charges levied against them. Despite these attempts made by the British

Parliament and by the civil society, none of the executives of the company was punished. This happened because these executives were considered national heroes who had helped in laying the foundation of British Empire in the Indian sub-continent. British authorities ignored them as minor irritants in the process of empire building. It is also true that, at that point in time, there was absence of necessary legal institutions and provisions to regulate the opportunistic behaviour of the executives of the East India Company. The inability of the British Parliament to regulate the behaviour of the overseas executives of the East India Company reflected the character of British democracy of the period, when the electorate consisted of a very narrow social base that did not reflect the ethos of common British citizens.

The story made it evident that the chartered status of the East India Company provided it ample scope to fight wars and participate in local conflicts and occupy territories in the Indian sub-continent. It was the chartered status of the company that facilitated it to gain political power in the Indian sub-continent. The company used its political power to break the nexus between Indian merchants and craftsmen, which was the basis of the emergence of traditional flexible manufacturing. Thus, the story revealed that the East India Company was uniquely placed at that time, due to prevailing historical circumstances, as well as due to its organizational structure to emerge as a ruler of a vast part of the Indian sub-continent and cause decline in its traditional flexible manufacturing.

The story also made it clear that while the East India Company was expanding its political control over the Indian sub-continent, the polity, society and economy of Britain was undergoing structural transformation, after the Industrial Revolution. Manufacturers were slowly gaining prominence in the British society, economy and polity. The emerging importance of manufacturers as creators of wealth and employment to masses resulted in the birth of manufacturers as a new power elite in Britain. The ideas propagated by this new power elite laid the foundation of a new economic philosophy. This philosophy was expressed in the writings of classical political economists who advocated for laissez faire, private property rights and free trade. These new ideas replaced the worldview created by the mercantilist school. The new ideology changed the attitude of the state and society against companies like the East India Company, which was forced to cease its commercial functions in c.1833. However, it remained the agency of the British Empire until c.1857.

At the same time, several socio-economic and political forces were aligning themselves against the rule of the East India Company in the

Indian sub-continent. The rule of the company had affected the interest of different groups and strata of society in the Indian sub-continent. The collective discontent of these diverse sections of the society were expressed in the form of the Sepoy Mutiny (as described by British historians) or first war of independence (as projected by the nationalist historians), which began on May 10 c.1857. The important consequence of the first war of independence was that the rule of the East India Company came to an end and the rule of the British Crown, which is being described as the British Raj, began. Direct rule of the British Crown meant that the interests of the British manufacturers came in direct conflict with the interest of Indian traditional flexible manufacturing. This is because a buffer that existed between the conflicting interests of British and Indian manufacturers disappeared. This important transformation of circumstances necessitated the need to study the impact of the transfer of power from the East India Company to the British Raj on Indian manufacturing.

The advent of the British Raj created a new form of governance of the Indian economy. The transfer of power from the company to the Raj created a political entity called India. The entire governance structure created by the East India Company was reflected by the office of the secretary of state and office of the viceroy. In the new governance structure, the involvement of British Parliament became most important. This facilitated the role of several pressure groups that existed in the British polity to influence the process of policymaking.

At this point in time, the dominant influence on policymaking in Britain was the ideas evolved by classical economists. It was generally accepted that the understanding of principles established by classical political economy will provide appropriate answers to policymakers. As a result of this acceptance amongst policymakers, there had emerged a consensus on wider areas of governance. The objective of policymakers was to facilitate the efficient working of the market economy. To facilitate this process, the British Raj believed in (i) free trade, (ii) laissez faire, and (iii) establishment of the rule of law to protect property rights and enforce contracts.

The policy of free trade was compatible with the interest of British manufacturers because it reduced the costs of production. By reducing the prices of imported raw materials, it reduced the costs of production, and by reducing the prices of imported food grains, it helped in maintaining low wages. Influenced by the interest of British manufacturers, the British Raj followed the policy of free trade, despite the fact that at that point in time the Indian manufacturing sector needed protection from British machine-made manufactured goods.

The intentions of the British Raj became quite apparent when it initiated the process of development of railways in India after c.1845. The British government purchased all the equipment and materials to develop railways from British manufacturers. It raised capital from borrowings from the British public by giving them a guarantee of a 5 per cent return on their investments. This policy mitigated the advantages of forward and backward linkages that might have resulted from the development of railways.

However, policies based on the principle of laissez faire and free trade could not be sustained for a longer period because historical circumstances made these policies incompatible with the interests of the British Raj. The circumstances created by World War I (c.1914–c.1918) forced the British government to deviate from these policies. Subsequently, the Great Depression of the 1930s also compelled the British administration to deviate from the policies of laissez faire and free trade, when it became necessary to impose tariffs and regulations.

The end of the Great Depression coincided with the beginning of World War II in c.1939. After the entry of Japan to the war, the threat of war came much closer to Indian territories. This forced the British government to initiate the process of industrialization in India to obtain necessary supply of materials and equipment, which were necessary to fight the war. Moreover, wartime compulsions compelled the British administration to impose regulations on the Indian manufacturing sector to make it serve the wartime needs of the administration. The regulatory practices began with the Defence of India Rules, which empowered the British administration to prevent entry of Indian manufacturers in certain industries. It also gave authority to the British administration to take over existing enterprises to produce wartime supplies. The British administration took control of enterprises like Texmaco and Jay Engineering Works. Tata's steel plant began to produce shells for the bombs and Tatanagar an armoured car, which was used by the British army. The Hindustan Aircraft Manufacturing Company, which was established by the Walchand-Hirachand Group, was taken over by the British government.

We came to learn from the story that during the British Raj, the Indian manufacturing sector evolved a new form of organization and management of businesses, which was known as the *managing agency system*. This innovation in the organization of businesses occurred when agency houses, which were formed by free merchants, ex-civil servants and military officials of the East India Company to sell the products of British manufacturers in India, floated enterprises in India as joint-stock companies and acquired the right to manage them. This innovation helped in financing of manufacturing enterprises on the one hand, and helped in overcoming the shortages in managerial skill on the other.

Despite the existence of several types of implicit as well as explicit discriminations in the policies followed by the British Raj, modern manufacturing enterprises did emerge in India. The existing discriminatory policy could not prevent the birth of the modern manufacturing sector in India, because historical circumstances provided the availability of all the important preconditions that are necessary for the successful emergence of modern manufacturing. There always existed a substantial supply of entrepreneurial talent in India, in the form of several communities of merchants. These merchants had earned experience in organizing networks of craftsmen to manage traditional flexible manufacturing. Some of them had also accumulated substantial financial resources from long-distance trade to invest in capital needed to establish modern manufacturing enterprises.

Communities of merchants like Parsis, Gujaratis, Bhattias, Bohras, Khojas, Memons and Marwaris had migrated from Surat and other parts of Gujarat to Bombay Presidency and had accumulated wealth by trading in cotton and opium. There also existed a large concentration of entrepreneurial talent in the form of agency houses in Calcutta, which were established by expatriate British merchants. These managing houses had also accumulated vast financial resources by exporting jute to Dundee mills.

These two groups of merchants operating at two different locations initiated two different modern manufacturing industries in India. The Bombay-based communities of merchants floated several enterprises to manufacture cotton textiles, and agency houses of Calcutta established enterprises to manufacture jute textiles. Both sets of merchants used the managing agency system to organize production. The concurrent growth of two textile-manufacturing companies in two different places has been described by scholars as '*dual dominance*'. We learnt from the story that these locations represented two different modes of growth of manufacturing. The cotton textile mills located at Bombay were promoted by native merchants, while jute textile mills of Calcutta were opened by British expatriate merchants. Due to this difference, the attitude of British policymakers was also different towards each of them. Moreover, while native merchants were manufacturing cotton textiles, which were substitutes for Manchester- and Lancashire-made cotton textiles, the jute textiles produced by the British expatriate merchants were complementary to the manufacturing interests. Both of these manufacturing industries began in the year c.1854, when Cowasjee Davar established Bombay Weaving and Spinning Mill at Bombay, and George Aukland started Aukland & Company to manufacture jute textiles at Serampore near Calcutta.

It is a historical fact that most of the cotton textiles mills located at Bombay were initially promoted by Parsi merchants like Maneckji Petit, Dinshaw Petit, Bomanji Wadia, Merwanji Bhavnagar, Tata and Kapadia. The growth of cotton textile mills was arrested by the American Civil War (c.1861–c.1865), which had suspended the availability of cotton to British cotton textile manufacturing mills. This led to massive export of Indian cotton to Britain. Despite the fact that the Indian cotton traders made substantial amount of profits, most of these profits were lost on speculation, which created a financial crisis.

The new investments in cotton textiles were made only after c. 1873. In fact, the new investment opportunities led to the entry of new entrepreneurs belonging to different communities of merchants in the manufacturing of cotton textiles. Kutchi Bhatia merchants like Morarji Gokuldas, Maneckji Khatau, Kowasji Naik, British entrepreneurs like Jams Greaves, George Cotton, Charles Killick and Presto Nixon and Baghdadi Jew, Albert Sasoon, promoted several cotton textile manufacturing units at Bombay. Ranchhodlal Chhotalal opened the first cotton textile mill at Ahmedabad.

The cotton textile mills that were located at Bombay were initially spinning mills, which were exporting cotton yarn to the markets of Southeast Asian countries, especially to cater to the markets of Japan and China. Therefore, at that point in time, cotton textile manufacturing was export oriented. However, after the adoption of modern equipment by Japanese cotton textile manufacturers after c.1860s, and Japanese occupation of China after war c.1894–1895, Indian cotton textile manufactures lost these markets for exporting yarn.

The new circumstances forced Indian cotton textile mills to restructure themselves. The restructuring reoriented the strategy of Indian cotton textile manufacturers towards a domestic market. Thus, cotton textile manufacturers became inward looking. As a consequence of the reorientation of cotton textile manufacturing, Indian mills adopted alternative strategies like (i) supplying cotton yarn to Indian handloom weavers, (ii) beginning the process of vertical integration by establishing composite textile mills, and (iii) spreading cotton textile manufacturing to new locations wherever cotton was cultivated, in the Indian sub-continent. The cumulative effect of these changes was that cotton textile manufacturing became an import-substituting industry from an export-orienting one. The rapid expansion of cotton textile manufacturing in the Indian sub-continent, and its emergence as an important import substitute for cotton textiles produced in Britain, invited hostility of British cotton textile manufacturers. This led to the imposition of several kinds of discriminatory regulations on the manufacturing of cotton textiles in India. The British administration introduced discriminating tariffs and excise duties on the manufacturers of Indian cotton textiles. Perhaps due to the experience of discriminating tariffs by cotton textile manufacturers in India, these manufacturers actively participated in the struggle for India's independence.

There existed several preconditions that are necessary for the successful promotion of manufacturing of jute textile in Calcutta. Despite the existence of these preconditions, native entrepreneurs ignored this industry. Therefore, manufacturing of jute textile was initiated by Scot entrepreneurs. The Scot merchants were involved in exporting jute from India, and had made the necessary capital required to enter into manufacturing of jute textiles by acting as agency houses. Scot entrepreneurs like George Aukland, who started Aukland & Company, and George Henderson, promoted Borneo & Company to manufacture jute textiles in Calcutta Presidency. These entrepreneurs demonstrated to other Scottish merchants the advantages of setting up jute textile manufacturing firms. Following the footsteps of these promoters of jute textile manufacturing units, several agency houses established mills to manufacture jute textiles in Calcutta Presidency. Some of these agency houses were Kettle Buller & Company, Jardine Skinner & Company, Andrew Yule & Company, and James Finlay & Company. Jute textiles manufactured in India remained largely for exports.

This book told us that during this period, Marwari merchants of Bengal were dominant players in procurement and supplying of raw jute to jute textile mills. Being an important part of the jute textile manufacturing economy of Bengal, they had established a strong association with the managing agency houses that were managing jute textile mills. These Marwari traders had accumulated substantial financial resources through trade in jute. Slowly these merchants took advantage of their knowledge about jute textile manufacturing and began to enter into the manufacturing of jute textiles. Marwari merchants followed two different strategies to enter into the business of manufacturing of jute textiles, where expatriate managing agency houses dominated. The first strategy was by lending to British managing agency houses to meet their working capital needs against collateral of blocks of shares of their jute textile manufacturing companies. This was known as the *fatka* system. Over time, these Marwari merchants became owners of a substantial percentage of stocks of the companies and became either members of the board of directors of the companies or owners of the company. Some other Marwari traders promoted their own jute textile mills. As a consequence of these alternative strategies, Marwari merchants owned twenty-two jute textile manufacturing companies, five jute textile mills owned by British managing agency houses were taken over by Marwari merchants, and in sixty-two mills owned by British agency houses, there were Marwari directors by the end of World War II.

The method of smelting iron was known to Indian craftsmen for ages. However, the manufacturing of iron and steel using the European system of manufacturing began during the rule of the East India Company (c. 1757–c. 1857), when an iron foundry called Begpore Iron Foundry was established at Beypore Inc. 1762. After this attempt, several units were started at different locations to manufacture iron and steel in India during the rule of the East India Company. These units were located at Proto-Novo (in Madras Presidency), Calcutta, Birbhum, Kumaon, Indore and Chanda. However, these units did not survive for a long period. The inventions of Bessemer converter in c.1856 and the open hearth furnace in c. 1869 changed the character of the British manufacturing of iron and steel in terms of materials required for the manufacturing of steel. These technological changes changed the location pattern of iron and steel manufacturing.

The story narrated the struggle made by Jamsetji Nusserwanji Tata to identify an appropriate location for the steel manufacturing unit that he established. He enlisted the help of Kennedy Sahlin & Company of Pittsburgh, PA to find an appropriate location for his proposed steel manufacturing firm as well as to construct his mill. This consulting firm identified the location of the proposed mill at Sakchi, which is now known as Jamshedpur. Finally, on August 26, c.1907, Tata Iron and Steel Company (TISCO) was registered at Bombay. The success of this steel mill can be attributed to the choice of suitable location, and government support in terms of purchase of substantial part of the output of the plant. Subsequently, circumstances that emerged as a consequence of the World Wars compelled the British administration to buy the entire output of steel produced by the plant. The steel manufactured by TISCO was used to construct bridges, rail roads, railway wagons and engines, bomb shells, and armoured cars known as Tatanagars, which were deployed in World War II. It is perhaps due to its appropriate location that this steel mill is still going strong. Steel mills were also started at Burnpur in Bengal (by Martin Burn & Company) and at Mysore (by Maharaja of Mysore).

It has been described in the story that the traditional flexible manufacturing produced large volume of *gur* and *khandsari*, which were consumed as sugar in the Indian sub-continent as well as in some of the neighbouring countries. British entrepreneurs introduced the cultivation of sugarcane in different British colonies with the help of indentured workers from India. British expatriate entrepreneurs began to manufacture white crystal sugar in the West Indies; the sugar that monopolized the British market. To meet the requirements of British expatriates living in the Indian sub-continent, white sugar was imported from Java, which at that time was a Dutch colony. Manufacturing of white crystal sugar began in India as a consequence of a rebellion that occurred at Hayati and San Domingo in c.1795. During this rebellion, a large proportion of the white expatriate population was killed. This rebellion paved the way for the removal of slavery in the West Indies in c.1834. Due to these reasons, supply of sugar from the West Indies to Britain was suspended. The new circumstances encouraged some of the British managing agency houses operating in India to promote sugar mills in India. Managing agency houses like Binny & Company, Parry & Company and Begg Sutherland & Company opened sugar mills in Madras Presidency and United Provinces (now known as Uttar Pradesh). However, as soon as situations in the West Indies normalized, the British government imposed discriminating tariffs in favour of sugar mills there.

Partly to prevent the import of sugar from Java, and partly in response to the demand of managing agency houses, which were engaged in the manufacturing of sugar, the British government imposed tariffs on imported sugar in India in March 1931 for the first time. The level of protection further increased following the Great Depression of the 1930s. In the new protected environment, several native entrepreneurs like Jamnalal Bajaj, Karamchand, Thapar, Gokul Chand Navrang, Ramkrishna Dalmia, Gujarmal Modi and Walchand Hirachand opened several sugar mills in different sugarcane growing areas spread across the Indian sub-continent. The process of import substitution in the manufacturing of sugar was so rapid that by the end of World War II, India was able to not only meet the demands of domestic markets, but also developed capacity to meet export demands. Apart from modernization of manufacturing of products, which were manufactured by traditional flexible manufacturing, certain entirely new manufacturing industries were also established by Indian manufacturers. These industries include paper, cement, chemicals, electric bulbs, automobiles, aircrafts, capital equipment required by cotton textile and sugar manufacturers, and chemicals that were needed by cotton textile and paper manufacturers.

We also learned that the Indian economy also attracted foreign investors. However, these investors entered the Indian markets due to different reasons. Some of the foreign investors established manufacturing units in export-oriented manufacturing industries like jute textiles, tea and sugar. Some other foreign investors entered the market to take advantage of changes in the stores policies of the British administration in India. The changes in the stores policies occurred when British government decided to procure all the requirements of their administration in India from manufacturers who had manufacturing units located in India. Companies like Alfred Herbert and Guest Kean Williams entered the Indian market especially due to this reason.

Another group of foreign manufacturing companies that established their manufacturing facilities in India were those that anticipated the requirements of plant and equipment and industrial materials by the local manufacturers. Companies like Siemens of Germany, Dunlop Rubber Company, Jenson and Nicholson, Goodlass Walls Ltd., Crompton and Parkinson, and Union Carbide entered the Indian market for this purpose. Similarly, some of the foreign investors also established their manufacturing firms to manufacture consumer goods on which the British administration had imposed tariffs. These manufacturing companies were Western India Match Company (WIMCO), Britannia Biscuit Company, Colgate–Palmolive, Lever Brothers, Bata Shoe Company, Imperial Tobacco Company, Philips, and Godfrey Phillips.

The long story of Indian manufacturing industries told us that the process of transformation of traditional flexible manufacturing into modern manufacturing occurred initially in those manufacturing industries in which India had competitive advantage. These industries were cotton textile and jute textile. Modern manufacturing also emerged in the production of sugar, cement, tea, paper, steel, safety matches, chemicals, automobiles and some of the general engineering products. We also learnt that modern manufacturing also spread to several locations in the Indian sub-continent, but it remained highly localized in Bombay and Calcutta presidencies. Madras Presidency remained the location of modern small and medium enterprises. Despite all the achievements, the size of the manufacturing sector was not sufficient to transform a continental-sized, predominantly agrarian economy into a manufacturing nation of the world. On the day of independence, only 9.6 per cent of India's working population was engaged in various forms of manufacturing activities along with modern, large, medium and small enterprises. This happened because in certain segments, traditional flexible manufacturing continued to exist, and in some others, traditional manufacturing assumed a different form of organization of production. Amongst all these different formats of manufacturing enterprises, the share of workforce engaged in large modern manufacturing firms was just 1.8 per cent, and share of all other forms of manufacturing was just 7.8 per cent.

The existence of dualistic structure in the manufacturing sector was not unique to India. Even in Britain, modern manufacturing evolved a symbiotic relationship with the earliest forms of manufacturing. However, subsequently, the entire manufacturing sector was modernized. The idea of dualism rests on the premise that there exist, concurrently, two types of technology, two different levels of economies, and two different kinds of organizational structures. The dualistic structure of the manufacturing sector emerged as a consequence of the existence of segmented markets. Segmented markets were a manifestation of social and economic segmentation of population in terms of income levels, social differentiation accompanied by religious and regional variations in consumption and tastes. The kind of consumer demographics that existed in India during this period had a very narrow base of rich and middle-income consumers, which had capability to buy standardized goods manufactured by modern manufacturing. Highly differentiated consumer demographics could not be served by products of modern manufacturing alone; they had to be complemented by other forms of manufacturing enterprises. Despite the fact there existed a narrow base of consumers who could buy the products of modern manufacturing, even this segment was made narrower by discriminatory policies of the British administration, which intended to make space for the imports of British manufactured goods. This suggests that prevalence of differentiated markets and the imposition of discriminating policies of the British Raj did not allow the modern manufacturing to be transformed into industrial revolution. These different segments of manufacturing, which co-existed concurrently, posed constraints on the process of transformation of the Indian economy into a manufacturing nation of the world, even after independence.

The analysis regarding the affairs of the Indian manufacturing sector with different political regimes that occupied political power after independence is subject matter for a separate book, which the author is writing currently.

However, one is presenting here a glimpse of the narrative as an appetizer. India achieved independence in c.1947. The independence of India was accompanied by the partitioning of the polity into India and Pakistan. Independence of India occurred in the circumstances when the global environment was highly influenced by the trauma of the Great Depression of the 1930s and the World Wars. It was the period when developed market economies were under the spell of Keynesian economics, and socialist economies of the USSR and other East European countries were following the strategy of developing their manufacturing sector within the framework of planned development.

Despite several options that were available to the political regimes that occupied political power in independent India, choices made by all of them were constrained by the prevalence of the segmented structure of the manufacturing sector. After independence, a new form of political economy emerged, where different political entities adopted different segments of the manufacturing sector for their support. Gandhians with the ruling party, which has ruled Indian polity for a substantial period of time, became supporters of traditional as well as modern small-scale enterprises. They influenced the policymakers to provide several kinds of support to these enterprises. The support was given in terms of concession in taxation and subsidies on land, fixed capital and priority sector lending. These enterprises were also facilitated by providing them well-developed infrastructure in industrial estates developed by the government. Moreover, to insulate them from the competition of large-scale enterprises, several manufacturing products were reserved especially for this sector. Most of the stringent labour laws were not imposed on this sector. The consequence of these supply-side supports was that this segment of the manufacturing sector never had incentive to transform itself into modern large-scale enterprises.

Nehruvians, within the ruling party, dedicated a very large proportion of the remaining products to be manufactured by the government enterprises under the process of planned development. Whatever residual products remained, were left for large-scale, private sector manufacturing enterprises. Instead of facilitating the growth of private sector, large-scale enterprises, government imposed several kinds of regulations that regulated all types of important managerial decisions, such as decisions regarding entry and exit, location, expansion and diversification of enterprises. These regulations not only arrested the growth of private large-scale modern manufacturing enterprises, but also strangulated the spirit of enterprise. Political regimes created hope regarding liberating the spirit of enterprise after the adoption of economic reforms. However, the reforms were too slow and did not reform labour markets, and have made acquisition of land for establishing manufacturing enterprises difficult and costly. The consequences of this strategy of development of the manufacturing sector have been that India has lost competitive advantage in the manufacturing sector.

Bibliography

Ahuja, S. P. (1992). *Paper industry in India*. New Delhi: The Institute of Economic and Market Research.

Akerlof, G. A. (1982). Labour contracts as partial gift exchange. *Quarterly Journal of Economics, 47*(4), 43–69.

Akerlof, G. A. (1983). Loyalty filters. *American Economic Renew, 73*(1), 54–63.

Akerlof, G. A. (1984). Gift exchange and efficiency wage theory four views. *American Economic Review, 74*(2), 79–83.

Alam, M. (1986). *The crisis of empire in Mughal North India-Awadh and Punjab 1707–1748*. New Delhi: Oxford University Press.

Alam, M., & Subhramanya, S. (Eds.). (1998). *The Mughal state 1526–1770*. New Delhi: Oxford University Press.

Alchian, A., & Demsetz, H. (1972). Production information costs and economics organization. *American Economic Review, 62*(5), 777–795.

Allami, A. F. (1927). *The Ain-i-Akbari* (trans: Bluchmann, H. B. & Ed. Leiut Colonel D. C. Phillot from Persian, Vols. 1 & II). New Delhi: Oriental Book Preprint Corporation.

Allami, A. F. (1956). *Akbar Nama* (trans: Bluchmann, H. B., Vol. 1). Calcutta: Bibliotheca.

Alvi, S. (Ed.). (2002). *The eighteenth century India*. New Delhi: Oxford University Press.

Ambirajan, S. (1987). *Classical political economy and British policy in India*. Cambridge: Cambridge University Press.

Amin, A. (1989). Flexible specialization and small firms in Italy—Myth and reality. *Antipode, 21*(1), 13–24.

Amin, A. (Ed.). (1995). *Post fordism—A reader*. Oxford: Blackwell.

V.K. Seth, *The Story of Indian Manufacturing*,
DOI 10.1007/978-981-10-5574-4

Amin, A., & Roberts, J. (2008). *Community economic creativity and organization.* New York: Oxford University Press.

Anderson, G. M., & Tollison, R. D. (1982). Adam Smith's analysis of joint stock companies. *Journal of Political Economy, 90*(6), 1237–1256.

Anderson, G. M., McCormick, R. E., & Tollison, R. D. (1983). The economic organization of the English East India Company. *Journal of Economic Behavior and Organization, 4*(2–3), 221–238.

Anthony, P. (2001). *People and empire—A short history of European migration, exploration and conquest: From Greece to present.* New York: Modern Library.

Aoyama, Y., & Castells, M. (2002). An empirical assessment of the information society: Employment and occupational structure of G-7 countries 1920–2000. *International Labour Review, 141*(1–2), 123–139.

Appadurai, A. (1986). *The social life of things commodities in cultural perspective.* Cambridge: Cambridge University Press.

Arasaratnam, S. (1979). Trade and political dominion in South India 1750–1790—Changing British India relationship. *Modern Asian Studies, 13,* 19–40.

Arasaratnam, S. (1980). Weavrs, merchants and company—The handloom industry in South India 1750–1790. *Indian Economic and Social History Review, 17,* 257–281.

Aristotle. (340BC/2001). *Basic works of Aristotle.* New York: Modern Library.

Arrow, K. J. (1985). The economics of agency. In J. Pratt & R. Zeckhauser (Eds.), *Principals and agents: The structure of business* (pp. 37–51). Boston: Harvard Business School Press.

Ashton, T. S. (1926). The domestic system in early Lancashire tool trade. *Economic History Review, 1*(1), 120–143.

Ashton, T. S. (1948). *Industrial revolution 1760–1800.* Oxford: Oxford University Press.

Ashton, T. S. (1960). *Industrial revolution, 1760–1800.* Oxford: Oxford University Press.

Astha, M. C. (1976). *Railway rates and fares in India 1849–1922—A historical cum analytical study.* PhD thesis, Department of Economics, Delhi School of Economics, New Delhi.

Athar Ali, M. (1975). The passing of empire—The Mughal case. *Modern Asian Studies, 9*(3), 385–396.

Athar Ali, M. (1978). Eighteenth century: An interpretation. *Indian Economic and Social History Review, 5*(1–2), 175–186.

Atthar Ali, M. (1986–1987). Recent theories of eighteenth century India. *Indian Historical Review, 13,* 102–110.

Babbage, C. (1835/1971). *Economics of machinery and manufactures.* Fairfield: A. M. Kelley.

Babur, Z. M. (1996). *Baburnama: Memoris of Muhammad Babur, Prince and Emperor* (trans: Wheeler, N. T.). New York: Modern Library.

Bagchi, A. K. (1972). *Private investment in India 1900–1939.* Cambridge: Cambridge University Press.

Bagchi, A. K. (1978). De-industrialization in India in the nineteenth century; some theoretical implications. *Journal of Development Studies, 22,* 135–164.

Bairoch, P. (1982). International industrialization levels from 1750 to 1980. *Journal of European Economic History, 11,* 10269–10333.

Baker, S. G. (1935). *Report on scientific and technical development of the jute manufacturing industry in Bengal.* Calcutta: Indian Jute Mills Association.

Baker, C. J. (1984). *An Indian rural economy 1880–1950: The Tamil Nadu country side.* New Delhi: Oxford University Press.

Bal Pours, E. (Surgeon General) (1885). *Encyclopedia of India,* Austria.

Baldwin, G. B. (1959). *Industrial growth in South India.* Glencoe: Free Press.

Ball, V. (1881). *Manual of geology of India (economic geology)* (Vol. III). Calcutta: Government of India Press.

Banarsidas. (2009). *Ardhakathanak* (trans: Chawdhury, R.). London: Penguin Books.

Banga, I. (1978). *Agrarian system of Sikhs.* New Delhi: Oxford University Press.

Basu, S. K. (1958). *The managing agency system.* Calcutta: World Press.

Bayley, C. A. (1986). The origin of swadeshi (home industry): Cloth and Indian society 1700–1930. In A. Appadurai (Ed.), *The social life of things—Commodities in cultural perspective.* Cambridge, UK: Cambridge University Press.

Bayly, C. A. (1983). *Townsman and bazars: North Indian society in the age of British expansion.* Cambridge, UK: Cambridge University Press.

Bayly, C. A. (1988a). *The British empire and the world 1780–1830.* Oxford: Oxford University Press.

Bayly, C. A. (1988b). *The new Cambridge history of India: The Indian society and the making of British empire* (Vol. II, p. 1). Cambridge, UK: Cambridge University Press.

Bayly, C. A., & Subrahmanyan, S. (1988). Portfolio capitalists and political economy of early modern India. *Indian Eco and Social History Review, 25*(4), 401–424.

Behagg, C. (1998). Mass production without factory; craft producers, guns and small firm. *Business History, 40*(3), 1–15.

Bell, D. (1973). *Coming of post-industrial society – A venture in social forecasting.* London: Arnold Heineman Publishers.

Bennett, R. B. (1980). *North India between the empires Awadh Mughals and British.* Berkley: University of California Press.

Berg, M. (1980). *Machinery question and making of political economy.* Cambridge, UK: Cambridge University Press.

Berg, M. (1993). Small producers capitalism in eighteenth century England. *Business History, 35*(l), 17–39.

Berg, M. (2005). *Luxury and pleasure in eighteenth-century Britain.* Oxford: Oxford University Press.

Berle, A. A., & Means, G. C. (1932). *Modern corporation and private property.* New York: Macmillan.

Berman, H. (1983). *Law and revolution.* Cambridge, MA: Harvard University Press.

Beveridge, H. (1886). *The trial of Maharaja Nand Kumar—A narrative of judicial murder.* Calcutta: Spock&Company/Thacker.

Beyer, H., et al. (1999). *The search for labour market flexibility—The European economies in transition.* Oxford: Clerendan Press.

Bhagwati, J. N., & Desai, P. (1970). *India: Planning for industrialization.* London: Oxford University Press. (for O.E.C.D.)

Bhardwaj, H. C. (1982). Development of iron and steel technology in India during 18th and 19th century. *Indian Journal of History of Science and Technology, 117* (2), 222–237.

Bhatacharya, S. (1982). Regional economy 1757–1857, the Eastern India. In D. Kumar & M. Desai (Eds.), *Cambridge economic history of India* (Vol. II, pp. 270–331). Hyderabad: Orient Longman.

Bhatia, B. M. (1974). *History and social development* (Vol. 1). New Delhi: Vikas.

Bhattacharya, S. (1965). Laissez Faire in India. *Indian Economic and Social History Review, 2,* 1–22.

Bhattacharya, S. (1966). Cultural and social constraint on technological innovations—Some case studies. *Indian Economic and Social History Review, 3,* 242–256.

Bired, A., & Hallow, I. (1984). *Rolls-Royce motor car.* London: B.T. Batsford.

Biswas, A. K. (2001). *Minerals and metals in pre-modern India.* New Delhi: Oxford University Press.

Blair, M. (1995). *Ownership and control—Rethinking corporate governance for twenty first century.* Washington, DC: Brooking Institute.

Blake, S. (1960). *British India centenary 1856–1956.* London: Orient Longman.

Bolt, W. (1772 [1998]). Considerations on Indian affairs. In P. Tuck (Ed.), *The East India Company 1600–1858.* London: Routledge.

Boot Arnold, W. A. (2000). Relationship banking: What do we know. *Journal of Financial Intermediation, 9,* 7–25.

Bordien, P. (2005). Principles of an economic anthropology. In N. J. Smelser & R. Swedberg (Eds.), *Handbook of economic sociology* (pp. 75–89). Princeton: Princeton University Press.

Boserup, F. (1965). *The condition of agricultural growth.* London: Allen and Union.

Boumol, W. J. (2002). Services and leaders and leaders of services. In J. Gadrey & F. Gallonz (Eds.), *Productivity, innovation and knowledge in services* (pp. 147–163). Cheltenham/Northampton: Edward Elgar.

Bowen, H. V. (1987). Lord Clive and speculation in the East India Company stocks 1766. *The Historical Journal, 30,* 890–927.

Bowen, H. V. (1989). Investment and empire in the later eighteenth century: East India stock holding 1756–1791. *Economic History Review, 42*, 186–206.

Bowen, H. V. (2006). *The business of empire—East India Company and imperial Britain 1756–1833.* Cambridge, UK: Cambridge University Press.

Bowrey, T. (1905). *A geographical account of the countries around Bay of Bengal 1669–1679.* Cambridge, UK: The Hakluyt Society.

Braverman, H. (1974). *Labor and monopoly capital – The degradation work in twentieth century.* New York: Monthly Review Press.

Braverman, H. (1989). The degradation work in twentieth century. *Monthly Review Press, 41*(5), 35–47.

Brendel, F. (1982). *Civilization and capitalism; wheels of commerce* (Vol. II). London: Collins.

Brenning, J. A. (1990). Textile producers and production in late seventeenth century Coromandel. In S. Subhramanyan (Ed.), *Merchants, markets and the state in early modern India* (pp. 66–89). New Delhi: Oxford University Press.

Brent, P. (1976). *The Mongol Empire, Genghis Khan; his triumphs and his legacy.* London: Weildenfield and Nicolson.

Brown, H. (1954). *Parry's of Madras.* Madras: Parry and Company.

Bruno, M., & Sachs, J. D. (1985). *Economics of world-wide stagflation.* Cambridge, MA: Harvard University Press.

Buchanan, D. H. (1966). *The development of capitalist enterprises in India.* London: Frank Cass.

Burke, P. (1979). *The new Cambridge modern history* (Vol. XIII). Cambridge: Cambridge University Press. (companion volume).

Burke, E. (1783 [1981], June). Ninth report of select committee. In P. J. Marshall (Ed.), *The writings and speeches of Edmund Buske* (Vol. V). Oxford: Clarendon Press.

Burt, R. (1998). Proto-industrialization and stages of growth in mining industry. *Journal of European Economic History, 27*(1), 85–104.

Burton, A. (1993). *Bukhara trade 1558–1718.* Paper on Inner Asia No. 23, Indian University Institute of Inner Asian Studies, Bloomington.

Cairncross, F. (2001). *The death of distance how the communications revolution will change our lives.* London: Texere.

Carlos, A. M. (1991). Agent opportunism and the role of Royal African companies compared. *Business and Economic History, 20*, 142–151.

Carlos, A. M. (1992a). *Bonding versus internal promotion and trust: Evidence from the Royal African Company 1672–1691.* Mimeo Department of Economics Colorado: University of Colorado.

Carlos, A. M. (1992b). Principal-agent problem in early trading companies—A tale of two firm. *American Economic Review, 82*(2), 140–145.

Carlos, A. M., & Nicholas, S. (1988). Giants of early capitalism—The chartered trading companies or modern multinational corporation. *Business History Review, 62*(3), 398–419.

Carlos, A. M., & Nicholas, S. (1990). Agency problem in early chartered companies—The case of Hudson Bay Company. *Journal of Economic History, 50*, 853–875.

Carlos, A. M., & Nicholas, S. (1993). Managing the manger: An application of principal-agent model to the Hudson Bay Company. *Oxford Economic Papers, 45* (2), 243–256.

Carlos, A. M., & Nicholas, S. (1996). Theory and history: Seventeenth century, joint-stock trading companies. *Journal of Economic History, 56*, 140–153.

Casas-Arce, P., & Santhi, H. (2004). *Job design and benifits of private trade.* Working Paper, Oxford Economics Working Paper Series Peg. 204, Oxford.

Cassels, N. G. (2010). *Social legislation of the East India Company: Public justice versus public instruction.* New Delhi: Sage.

Castells, M. (1996). *The rise of network society.* Cambridge, MA: Blackwell.

Caton, H. (1985). Pre-industrial economics of Adam Smith. *Journal of Economic History, 45*(4), 833–853.

Chadhury, S. (1972). Merchants, companies, and rulers; Bengal in eighteenth century. *Journal of Economic History of the Orient, 31*, 71–85.

Chandavarkar, R. (1994). *The origins of industrial capitalism in India. Business strategies and the working class in Bombay 1900–1940.* Cambridge, UK: Cambridge University Press.

Chandler, A. (1962). *Strategy and structure: Chapters in the history of industrial enterprise.* Cambridge, MA: MIT Press.

Chandler, A. (1977). *The visible hand; the managerial revolution in American business.* Cambridge, MA: Harvard University Press.

Chandra, S. (1998). Review of the crises of the Jagirdari system. In M. Alam & S. Subrahmanyan (Eds.), *Mughal state* (pp. 347–360). New Delhi: Oxford University Press.

Chapman, S. D. (1965). The transition to the factory system in midlands cotton spinning industry. *Economic History Review, 18*, 526–543.

Chari, S. (2004). *Fraternal capital—Peasant, workers, self-made men—Globalization in provincial India.* Ranikhet: Permanent Black.

Chatterji, M. (1983). The political economy of 'discriminating protection', the case of textiles in the 1920s. *Indian Economic and Social History Review, 20*, 239–276.

Chaudhuri, K. N. (1966). *The English East India Company—A study of an early joint stock company 1600–1640.* London: Frank Cass.

Chaudhuri, K. N. (1971). *The economic development of India under the East India Company 1814–1858.* Cambridge: Cambridge University Press.

Chaudhuri, K. N. (1974). The structure of Indian textile industry in the seventeenth and eighteen centuries. *Indian Economic and Social History Review, 11*, 23–35.

Chaudhuri, K. N. (1978). *The trading world of Asia and the East India Company 1660–1760.* Cambridge: Cambridge University Press.

Chaudhuri, K. N. (1979). Markets and traders in India during seventeenth and eighteenth centuries. In K. N. Chaudhry & C. J. Dewey (Eds.), *Economy and society-essays in Indian economic and social history.* New Delhi: Oxford University Press.

Chaudhuri, K. N. (1990). *Asia before Europe, economy and civilization of the Indian Ocean from the rise of Islam to 1750.* Cambridge: Cambridge University Press.

Chaudhury, S. (1995). *Form prosperity to decline; eighteenth century Bengal.* New Delhi: Manohar.

Chenery, H. B. (1960). Patterns of industrial growth. *American Economic Review, 53*, 220–335.

Chayanov. (1925/1968). *The theory of peasant economy* (D. Thorner, B. Kerbalay and R. E. F. Smith. Eds.). Homewood: American Economic Association.

Child, S. J. (1968). A new discourse of trade. In *Josiah child selected works 1668–1697.* Farnborough: Gregg Press/University of London.

Christian, D. (2004). *Maps of time, an introduction to big history.* London: University of California Press.

Christoplherson, S., & Storper, M. (1989). New forms of labour organizations and production policies of flexibility in specialized industry. *Industrial and Labour Relaitons Review, 42*(3), 331–347.

Cippola, C. M. (Ed.). (1974). *Fontena economic history of Europe – 16th & 17th century.* Glasgow: Collins Fontana Books.

Clapham, J. F. (1968). *Economic development of France and Germany 1815–1914.* Cambridge, MA: Cambridge University Press.

Clark, C. (1940). *The conditions of economic progress.* London: Macmillan.

Clark, C. (1950). *Conditions of economic progress.* London: Macmillan.

Clingingsmith, D., & William, J. G. (2004). *India's de-industrialization under British rule, new ideas and new evidence.* Working Paper No. 10586 National Bureau of Economic Research, Cambridge, MA.

Coleman, D. C. (1973). Textile growth. In N. B. Harte & K. G. Panting (Eds.), *Textile history and economic history.* Manchester: Manchester University Press.

Coleman, D. C. (1983). Proto-industrialization—A concept too many. *Economy History Review, 36*, 435–448.

Commander, S. (1983). Jajimani system in North India. *Modern Asian Studies, 17* (2), 283–311.

Cornaert, E. L. J. (1967). European economic institutions and the new world: The chartered companies. In E. E. Rich & C. H. Wilson (Eds.), *Cambridge economic history of Europe* (pp. 220–274). Cambridge: Cambridge University Press.

Dalrymple, W. (2006). *Last Mughal—The fall of the dynasty, Delhi 1857*. New Delhi: Penguin Viking.

Damodaran, H. (2008). *India new capitalist, caste, business and industry in modern nation*. Ranikhet: Permanent Black.

Damodaran, H. (2009). *India's new captialists—Cast, business and industry in a modern nation*. Ranikhet: Permanent Black.

Daniel, G. (1943). *Three ages*. Cambridge: Cambridge University Press.

Darling, M. (1926). *Punjab peasantry in prosperity and debt*. London: Oxford University Press.

David, S. (2002). *The Indian mutiny*. New Delhi: Penguin Books.

Davis, K. (1957). *The Royal African Company*. London: Longman Green.

Davis, G. F. (2003). Firms and environment. In N. J. Smelser & R. Swedberg (Eds.), *Handbook of economic sociology* (pp. 478–502). Princeton: Princeton University Press.

Davis, J. H., Schoorman, F. D., & Donaldson, L. (1997). Towards stewardship theory of management. *Academy of Management Review, 15*(3), 369–381.

Daw, A. (1792). *The history of Hindoostan from the death of Akbar to complete settlement of the empire under Aurangzeb*. Dublin: Luke White.

de Thevenot, J. (1687 [1949]). In S. N. Sen (Ed.), *Indian travels of Thevenot and Careri*, New Delhi: National Archives of India.

Deavy, B. (1975). *The economic development of India: A Marxist analysis*. Nottingham: Spokesman Books.

Derrett, J. D. M. (1976). Rajdharma. *Journal of Asian Studies, 35*(4), 597–609.

Desai, M. (2009). *The rediscovery of India*. New Delhi: Allen Lane and Penguin Books.

Desouza, J. D. (1961). *History of chemical industry in India*. Bombay: Technical Press.

Detmar, R. (1988). Problem of India's arrested development under British rule. In C. Dewey (Ed.), *Arrested development in India-historical dimensions* (pp. 3–11). New Delhi: Manohar.

Devra, G. S. L. (1987). The Mughal state and Indian business. In D. Tripathi (Ed.), *State and business in India—A historical perspective* (pp. 1–24). New Delhi: Manohar.

Dewey, C. (Ed.). (1988). *Arrested development in India—The historical dimension*. New Delhi: Manohar Publication.

Dey, H. L. (1933). *The Indian tariff problem in relation to industry and taxation*. London: George Allen and Unwin.

Dirk, N., & B. (2006). *The scandal of empire—India and the creation of Imperial Britain*. Ranikhet: Permanent Black.

Dittle, J. H. (1920). The house of Jagat Seth. *Bengal Past and Present, 20*, 111–120.

Dobb, M. H. (1963). *Studies in the development of capitalism*. London: Rentledge and Kegen Paul.

Dodwell, H. (1922). *Madras weavers under the company*. Proceedings of the Indian Historical Research Committee.

Donaldson, L., & Davis, J. H. (1991). Stewardship theory of agency theory-ECO, governance and shareholders returns. *Australian Journal of Management, 16*(1), 49–64.

Dore, R. (1986). *Flexible regidities*. Standford: Standford University Press.

Dosi, G. (1999). Some notes on national system of innovations and production and their implication for economic analysis. In D. Archibugi, J. Howells, & J. Michie (Eds.), *Innovation policy in global economy*. Cambridge, MA: Cambridge University Press.

Duguid, P. (2008). Community of practice then and now. In A. Amin & J. Roberts (Eds.), *Community, creativity and organization* (pp. 1–10). New York: Oxford University Press.

Dutt, R. (1906). *The economic history of India in the Victorian age*. London: Kegan Pal.

Dutt, R. C. (1956). *The economic history of India under early British rule* (p. 103). London: Rentledge and Kegan Paul.

Eaton, C., & White, W. D. (1982). Agent compensation and limits of bonding. *Economic Enquiry, 29*, 330–343.

Estall, R. G., & Buchanan, O. R. (1946). *Industrial activity and economic geography*. London: Hutchisan University Press.

Fama, E. (1980). Agency problem and the theory of firm. *Journal of Political Economy, 88*(2), 288–307.

Fama, E. F., & Jensen, M. C. (1983). Agency problems and residual claims. *Journal of Law and Economics, 26*(2, Corporations and Private Property: A Conference Sponsored by the Hoover Institution), 327–349.

Fateh, S. (2012, June 14). Tracing the origins of Pak's doom and gloom. *Times of India*.

Fei, J., & Gustav, R. (1989). Economic development in historical perspective. *American Economic Review (Papers and proceedings), 44*, 356–400.

Findley, R., & Kevin, H. O. R. (2008). *Power plenty: Trade war and the world economy in second millennium*. Princeton: Princeton University Press.

Findley, R., & Mats, L. (2006). The first globalization episode: The creation of Mongol empire or the economics of Genghis Khan. In G. Thorborn & H. H. Khonodker (Eds.), *Asia and Europe in globalization: Continents, regions and nations* (pp. 13–54). Leiden: Brill.

Fisher, M. H. (1987). *The clash of cultures: Awadh the British and Mughals*. Delhi: Oxford University Press.

Florence, P. S. (1948). *Investment location and size of plant: A realistic inquiry in to the structure of British and American industries*. Cambridge: Cambridge University Press.

Fogel, R. W. (1964). *Railroads and American growth: Essays in economic history.* Baltimonre: Johns Hopkins Press.

Foster, W. (Ed.). (1906–1972). *The English factories in India—A calendar of documents in India office* (Vol. 1–13). Oxford: British Museum and Public Records Office.

Frank, A. G. (1998). *Reorient – Global economy in the Asian age.* New Delhi: Vistar Publication.

Frank, I. M., & Brown Stone, D. M. (1986). *The silk road a history, facts.* Oxford/New York: File Publications.

Freeman, C. (1981, August). *Technological infrastructure and international competitiveness.* Draft paper submitted to OECD, Paris (Mime).

Freeman, R. E. (1984). *Strategic management: A stakeholders approach.* Boston: Pitman.

Freeman, C. (1995). The national innovation system in historical perspective. *Cambridge Journal of Economics, 19*(1), 23–45.

Freeman, C., & Louca, F. (2001). *As time goes by: From industrial revolution to the information revolution.* Oxford: Oxford University Press.

Freudenberger, H., & Redlick, F. (1964). Industrial development of Europe, reality, symbols and images. *Kyklos, 17*, 372–402.

Fruin, W. M. (1983). *Kikkoman—Company, clan and community.* Cambridge, MA: Harvard University Press.

Fruin, W. M. (2008). Business groups and inter-firm networks. In G. Jones & J. Zeitlin (Eds.), *The Oxford handbook of business history* (pp. 244–268). New York: Oxford University Press.

Fuchs, V. R. (1968). *The service economy.* New York/London: National Bureau of Economic Research.

Fukazawa, H. K. (1972). Rural servants in Maharashtrian village, demiurgic or jajmani system. *Hitotshubashi Journal of Economics, 13*, 108–124.

Fuller, C. J. (1989). Misconceiving the grain heap: A critique of the concept of jajmani system. In J. Parry & M. Bloch (Eds.), *Money and morality of exchange* (pp. 50–62). Cambridge: Cambridge University Press.

Furber, H. (1970). *John company at work: A study of European expansion in the late eighteenth century.* New York: Octagon.

Furber, H. (2004). Rival empires of trade 1600–1800. In *Maritime India.* New Delhi: Oxford University Press.

Furguson, N. (2002). *Empire: The rise and demise of the British world order and lessons for global power.* New York: Basic Books.

Furguson, N. (2004). *Empire: How Britain made the modern world.* London: Penguin.

Gadgil, D. R. (1924). *The industrial evolution of India in recent times: 1860–1939.* Oxford University Press.

Galenson, D. W. (1986). *Traders, planters and slaves: Market behavior in early English America*. Cambridge: Cambridge University Press.

Gandhi, M. P. (1930). *The Indian cotton textile industry*. Bombay.

Gandhi, M. P. (1934). *Indian sugar: Its past, present and future*. Calcutta: Government of India Press.

Gardner, B. (1971). *East India Company's history*. Report and Hart Davis, London.

Goswami, O. (1982). Collaboration and conflict—European and Indian capitalists in jute economy of Bengal 1919–1939. *Indian Economic and Social History Review, 19*(2), 141–179.

Goswami, O. (1985). Then came Marwaris: Some aspect of changes in the pattern in industrial control in Eastern India. *The Indian Economics and Social History Review, 22*(3), 225–249.

Goswami, O. (1989). Sahib, Babus and Banias; Changes in industrial control in India 1918–1950. *The Journal of Asian Studies, 48*(2), 289–309.

Goswami, O. (2011). From bazar to industry. In M. M. Kudaisya (Ed.), *Anthology of business history* (pp. 235–257). New Delhi: Oxford University Press.

Government of India. (1878). Custom notification on N043 March 18, 1978. Calcutta.

Government of India. (1939). *Defense of India act*. Shimla: Government of India Press.

Government on India. (1952). *Large industrial establishments in India in 1947*. New Delhi: Government of India Press.

Gower, L. C. B. (1969). *The principles of modern company law*. London: Stevens and Son.

Grewal, J. S. (1995). *The Sikhs of Punjab*. New Delhi: Manohar Publishers.

Grief, A. (1993). Contract enforcementability and economic institutions in early trade: The Maghribi traders coalition. *American Economic Review, 83*(3), 525–548.

Grioseclose, B. (1993). *British sculptures and Company Raj*. Newark, New Jersey: University of Delaware Press.

Groseclose, B. (1998). *British sculptures and company raj; church, monuments and public stratuary in Madras, Calcutta and Bombay to 1858*. Newark: University of Delaware Press.

Grossman, S., & Hart, O. (1983). An analysis of the principal agent problem. *Econometaica, 51*, 7–45.

Guha, S. (1989). The handloom industry of Central India 1825–1950. *Indian Economic and Social History Review, 26*(3), 355–425.

Guha, S. (2004). Civilization markets and services-villages servants from seventeenth century and twentieth century. *Indian Economic and Society History Review, 41* (1), 79–101.

Gupta, B. (1966). *Sirajuddaulah and the East India Company 1756–1757: Background to the foundation of British power in India*. London: E.J. Brill.

Gupta, A. D. (1970). The merchants of Surat c. 1700 to c. 1750. In E. Leach & S. N. Mukherjee (Eds.), *Elites in South Asia*. Cambridge: Cambridge University Press.

Gupta, R. K. (1980). Iron manufacturing industry in Birbhum. A study of growth and evolution. *Journal of Indian History, 38*, 94–106.

Gupta, A. D. (1994). *Indian merchants and the decline of Surat: 1700–1750*. New Delhi: Manohar Publishers.

Habhakuk, H. J. (1962). *American and British technology in the nineteenth century*. Cambridge, MA: Cambridge University Press.

Habib, I. (1960). Banking in the Mughal India. In T. R. Chaudri (Ed.), *Contribution to Indian economic history*. Calcutta: Firma, K.L. Mukhopadhyay.

Habib, I. (1963). *The agrarian system of Mughal India: 1506–1707*. Bombay: Asia Publishing House.

Habib, I. (1982). *An atlas of the Mughal empire*. New Delhi: Oxford University Press.

Habib, I. (1987). A system of trimetalism in the age of price revolution, effects of silver influx on the Mughal monetary system. In J. F. Richard (Ed.), *The imperial monetary system in Mughal India*. New Delhi: Oxford University Press.

Habib, I. (1990). The merchant communities in pre-colonial India. In J. D. Tracy (Ed.), *The rise of the merchant empire: Long distance trade in the early modern world 1350–1750*. Cambridge: Cambridge University Press.

Hadi, S. M., & Khan, B. (1929). *Indian sugar industry*. Bhopal: Government of Madhya Pradesh.

Hajeebu, S. (2005). Contract enforcement in English East India Company. *Journal of Economic History, 65*(2), 496–523.

Hall, H. D. (1955). *North American supply*. Oxford: Oxford University Press.

Hall, R. (1996). *Empire of monsoon*. New York: Harper Collins.

Hall, P., & Soskice, D. (2001). *Varieties of capitalism—The institutional foundation of corporate advantage*. New York: Oxford University Press.

Hambly, G. R. G. (1982). Towns and cities—Mughal India. In T. R. Chaudhari & I. Habib (Eds.), *The Cambridge economic history of India* (Vol. 1, pp. 434–457). New Delhi: Orient and Longman.

Harb, N. B. (1997). *The new draperies in low countries and England c.1300–c.1700*. Oxford: Oxford University Press.

Hardgrove, A. (2004). *Community and public culture—The Marwaris in Calcutta c.1897–c.1997*. New York: Columbia University Press.

Harnetty, P. (1991). De-industrialization revised: The handloom weavers of the central provinces of India. *Modern Asian Studies, 25*(3), 455–510.

Harris, F. R. (1954). *Jamsatji Nussarwanji Tata – A chronicle of life*. Bombay: Asia Publishing House.

Haynes, D. (1996). The logic of artisan firms in capitalist economy: Handloom weavers in Western India c.1880–1947. In B. Stein & S. Subramanyan (Eds.),

Institution and economic change in South Asia. New Delhi: Oxford University Press.

Heath, J. M. (1839). Indian iron and steel industry. *Journal of Royal Asiatic Society, 5*(3), 390–399.

Heckscher, E. R. (1935). *Mercantilism* (trans: Shapiro, M.). London: George Allen and Unwin.

Hegel, G., & Friedrich, W. (1959). *The philosophy of history* (trans: Sibree, J.). New York: Colonial Press.

Henderson, W. O. (1975). *Rise of German industrial power, 1834–1914*. Berkeley: University of Berkley.

Hibbert, C. (1978). *The great mutiny: India 1857*. London: Allen Lane.

Hicks, J. R. (1969). *A theory of economic history*. Oxford: Clarendan Press.

Hill, C. (Ed.). (1967). *Reformation to industrial revolution* (Vol. 2). Harmondsworth: Penguin Books.

Hill, C. W., & Jones, T. M. (1992). Stakeholders agency theory. *Journal of Management Studies, 29*(1), 131–154.

Hirschman, A. O. (1958). *Strategy of economic growth*. New Haven: Yale University Press.

Hobhouse, H. (1999). *Seeds of change—Six plants that transformed the mankind*. London: Papermac.

Hobsbawn, E. J. (1954). The general crises of European economy in the seventeenth century. *Past and Present, 5*, 33–53.

Hobsbawn, E. J. (1968a). *Labouring men*. London: Weidenfeld and Nicolson.

Hobsbawn, E. J. (1968b). *Industry and empire—An economic history of Britain since 1750*. London: Weiden Field and Nicolson.

Hodgson, G. H. (1938). *Thomas parry: Free merchant*. Madras: Higginbrothers.

Holmstrom, B. (1979). Moral hazard and observability. *Bell Journal of Economics, 10*, 74–91.

Holmstrom, B. (1982). Moral hazard in teams. *Bell Journal of Economics, 13*, 324–340.

Holmstrom, B., & Paul, M. (1991). Multi—Task, principal—Agent analyses: Incentives and contracts, asset ownership and job design. *Journal of Law, Economics and Organisation, 7*(1), 24–52.

Holzman, J. M. (1926). *The nabobs in England*. New York: Columbia University Press.

Hoppit, J., & Wrigley, E. A. (1994). Introduction. In J. Hoppit & E. A. Wrigley (Eds.), *The industrial revolution in Britain, the economic history society* (pp. X–XXXI). Oxford: Black Well.

Horn, J. (2007). *The industrial revolution—Milestones in business history*. London: Greenwood Press.

Housmann, H. (1996). *Ownership of enterprise*. Cambridge/New York: Harvard University Press.

Hozenetty, P. (1972). *Imperialism and free trade, Lancashire and Indian in mid-ninetieth century*. Manchester: University of Manchest Press.

Hudson, O. (1996). Proto-industrialization in England. In S. C. Ogilivie & M. Cerman (Eds.), *European proto industrialization* (pp. 49–66). Cambridge: Cambridge University Press.

Hunder, W. W. (1868). *Annal of rural Bengal*. New York: Leypoldt and Holt.

Hunter, W. W. (1874). *Famine aspects of Bengal districts*. London: Turner.

Hunter, W.W. (1888 [1974]). Statistical accounts of Bengal, Vol. IX, Concept Publishing, New Delhi.

Hurd, J. M. (1982). Railways. In D. Kumar (Ed.), *The Cambridge economic history of India* (Vol. II, pp. 737–761). Cambridge: Cambridge University Press.

Hussain, H. (1958). *The company weavers of Bengal: The East India Company and the organization of textile production in Bengal 1750–1813*. New Delhi: Oxford University Press.

Hussain, I. (Ed.). (2006). *Karl Marx on India from New York daily tribune*. New Delhi: Tulika Book.

Idrish, A.-S. (1960). *Kitab Nuzahat-ul-mushtaq* (trans: Ahmad, S. M.). Leiden: Macmillan.

Issac, T. M. T. (1984). *Class struggle and industrial sturcture—A study of coir weaving industry in Kerala 1859–1980*. PhD thesis, Centre for Development Studies, Thiruvananthapuram.

Jain, L. C. (1929). *Indigenous banking in India*. London: Macmillan.

James, L. (1997). *Raj: The making of British India*. London: Abacus.

Jenkins, D. T. (Ed.). (1994). *The textile industries*. Oxford: Blackwell Publishers.

Jensen, M. C., & William, H. M. (1976). Theory of the firm: Managerial behavior, agency costs and ownership structure. *Journal of Financial Economics, 3*(4), 305–360.

John, C. (1954). *An Indian rural economy c.1880–1955: Tamil and country*. Oxford/New Delhi: Oxford University Press.

Johnson, W. A. (1966). *The steel industry of India*. Cambridge, MA: Cambridge University Press.

Jones, S. R. H., & Ville, S. P. (1996). Efficient transactions or rent seeking monopolists? Rationale for early chartered trading companies. *Journal of Economic History, 56*(4), 898–915.

Jones, T. M., & Wicks, A. C. (1999). Convergent stakeholders theory. *Academy of Management Review, 24*(1), 200–221.

Jorgenson, D. W. (1961). The development of dual economy. *Economic Journal, 71* (3), 309–334.

Joshi, A. (1955). *Lala Shri ram—A study in entrepreneurship and industrial management*. London/Delhi: Orient Longman.

Joshi, P. C. (2007). *Rebellion 1857*. New Delhi: National Book Trust of India.

Juvani, A.-M. (1997). *Genghis Khan: The history of the world conqueror* (trans: Boyle, J. A.). Manchester: Manchester University Press.

Kahn, P. (1984). *The secret history of Mongols: The origin of Genghis Khan*. San Francisco: North Point Press.

Khan, S. A. (1923). *The East India trade in 18th century: In its political and economic aspect*. Oxford: Oxford University press.

Karen, L. (1979). The great theory of the decline of the Mughal empire. *Comparative Studies in Social History, 21*(2), 161–167.

Karen, L. (1998). The great firm theory of the decline of Mughal empire. In M. Alam & S. Subhramanyam (Eds.), *Mughal state 1526–1750* (pp. 398–148). New Delhi: Oxford University Press.

Katz, L. F. (1986). Efficiency wage theories—A partial evaluation. In S. Fisher (Ed.), *Macro-economic annual, National Bureau of Economic Research*. Cambridge, MA: MIT Press.

Keay, J. (1993). *The honorable company*. London: Harper Collins.

Kenny, M., & Florida, R. (1993). *Beyond mass production*. Oxford: Oxford University Press.

Khanolkar, G. D. (1969). *Walchand Hirachand: The man his times and achievements*. Bombay: Walchand and Company.

Kidron, M. (1965). *Foreign private investment in India*. London: Oxford University Press.

King, B. (1969). The origin of managing agency system in India. *Journal of Asian Studies, 26*, 120–130.

Kindleberger, C. P. (1976). Historical background: Adam Smith and industrial revolution. In T. Wilson & A. Skinner (Eds.), *Market and the state – Essays in honour of Adam Smith*. Oxford: Oxford University Press.

Knudsen, D. C. (1996). *The transition to flexibility*. Honwell: Kulwer Academic Press.

Koh, S. J. (1966). *Stages of industrial development in Asia – A Comparative history of cotton textiles industry in Japan, India, China and Korea*. Pennsylvania: University of Pennsylvania Press.

Kolsky, E. (2003). Confidication and rule of colonial difference: Criminal procedures in British India. *Law and History Review, 23*(3), 631–684.

Kolsky, E. (2010). *Colonial justice in British India: White violence and the rule of the law*. New Delhi: Cambridge University Press.

Kondratief, N. D. (1979). Long waves in economic life. *Review, 2*, 519–562.

Kosambi, M. (1986). *Bombay in transition – The growth and social ecology of a colonial city, 1880–1980*. Stockholm: Almqvist and Wicksell.

Kranjia, B. K. (1997). *Godrez—A hundred years* (Vol. I and II). New Delhi: Viking.

Kriedite, P., Medick, H., & Schlumbhar, J. (1981). *Industrialization before industrial revolution* (trans: Schen, B.). Cambridge: Cambridge University Press.

Krishnamurty, J. (1983). Occupational structure. In D. Kumar (Ed.), *The Cambridge history of India c. 1757–c. 1970* (Vol. II). New Delhi/Cambridge: Orient Longman/Cambridge University Press.

Kuznets, S. (1966). *Modern economic growth; rate, structure and spread.* New Haven: Yale University Press.

Lal, D. (2007). *Hindu equilibrium.* Oxford: Oxford University Press.

Lala, R. M. (1981). *The generation of wealth—The Tata story.* Bombay: IBH Publishing.

Lala, R. (2007). *Romance of Tata Steel.* New Delhi: Penguin Viking.

Lall, D. (2004). *In praise of empire—Globalization and order.* New York: Palgrave Macmillan.

Landes, D. (1994). *The wealth and poverty of nations.* New York: Abacus.

Landes, D. (1998). *The wealth and poverty of nations.* New York: Abacus.

Landes, D. S. (1969). *Unbound prometheus.* Cambridge: Cambridge University Press.

Landes, D. S. (1986). *Revolution in time: Clocks and making of modern world.* Cambridge: Cambridge University Press.

Lamb, H. (1965). Rise of Indian business class. *Pacific Affair, 23*(2), 93–126.

Latham, A. J. H. (1981). *The depression and the developing world 1914–1939.* London: Croom Helm.

Lawson, P. (1993). *The East India Company—A history.* London: Longman.

Lazomick, W. (1991). Institutional foundations of industrial dominance and decline. In W. Lazonick (Ed.), *Business organization and myth of the market economy* (pp. 23–58). Cambridge, MA: Cambridge University Press.

Lazonick, W. (1990). *Competitive advantage of the shop floor.* Cambridge, MA: Harvard University Press.

Lazonick, W. (1991). *Business organization and myth of market economy.* New York: Cambridge University Press.

La Porta, R. (1999). Corporate ownership around the world. *Journal of Finance, 54* (2), 471–517.

La Porta, R. (2003). Agency problem and dividend policies around the world. *Journal of Finance, 55*(1), 1–33.

Lebman, F. (1965). Great Britain and the supply of railway locomotives to India—A case study of economic imperialism. *Indian Economic and Social History Review, 2*(4), 297–306.

Leidholmes, C. (1972). *Indian iron and steel industry—An analysis of comparative advantage.* East Lansing: Michigan State University.

Lemire, B. (1991). *Fashion favorite: Cotton trade and consumer in Britain, 1650–1800.* Oxford: Oxford University Press.

Lemire, B. (2004). Fashion and tradition: Wearing of wool in England during consumer revolution. In G. L. Fantana & G. Gayot (Eds.), *Wool products and markets 13th to 20th centuries* (pp. 577–594). Patna: Clerandon University Press.

Levine, D. (1977). *Family formation in the age of nascent capitalism*. New York: Monthly Review Press.

Levkovsky, A. I. (1965). *Capitalism in India*. Bombay: Popular Prakashan.

Lewis, W. A. (1954). Economic development with unlimited supply of labor. *Journal of Manchester School of Economics and Social Studies, 22*, 139–192.

Lichet, W. (1995). *Industrialization of America: The nineteenth century*. Baltimore: John Hopkins Press.

List, F. (1841). *Das Nationale System der Politischen Okonomie (National systems of political economy)*. London: Longman.

Little, J. H. (1920). The house of Jagat Seth. *Bengal Past and Present, 20*, 1–119.

Liu, X. (1996). *Silk and religion*. New Delhi: Oxford University Press.

Lokanathan, P. S. (1935 [1965]). *Industrial organization of India*. London: Orient Longman.

Louca, F. (2007). Long waves, the pulsation of modern capitalism. In H. Hanusch & A. Pyka (Eds.), *Elgar companion of neo-schumipeterian economics* (pp. 766–774). Northampton: Edward Elgar.

Ludden, D. (1996). Caste society and unit of production in early modern South India. In B. Stein & S. Subrahmanyan (Eds.), *Institutions and economic change in South Asia* (pp. 105–133). New Delhi: Oxford University Press.

Lundvall, B.-A. (2007). National innovation systems; From list to freeman. In H. Hanusch & A. Pyka (Eds.), *Elgar companion to neo-Schumpeterian economics*. Cheltenham/Northampton: Edward Elgar.

Macpherson, W. J. (1955). Investment in Indian railways c. 1845–1875. *Economic History Review, VIII*, 177–187.

Maddison, A. (1971). *Class structure and economic growth: India and Pakistan*. London: Allen and Unwin.

Maddison, A. (1991). *Dynamic forces in capitalist development*. Oxford: Oxford University Press.

Maddison, A. (2003). *The world economy—A new millennial perspective*. New Delhi: Overseas Press.

Mahadevan, R. (1984). Entrepreneurship and business community in Colonial Madras c.1902–1929—Some preliminary observations. In D. Tripathi (Ed.), *Business communities of India—A historical perspective* (pp. 325–239). New Delhi: Manohar Publishing.

Mahadevan, R. (1992). Pattern of industrial control in colonial madras: Some critical observations on relative position of Indian and foreign capital 1930–1930. In A. Ghosh (Ed.), *Indian industrialization structure and policy issues* (pp. 336–364). New Delhi: Oxford University Press.

Mahadevan, R. (1999). Southern region. In FICCI (Ed.), *Footprints of Enterprise: Indian business through the ages*. New Delhi: Oxford University Press.

Maikovits, C. (2008). *Merchants, traders, entrepreneurs–Indian business in colonial era*. Renikhet: Permanent Block.

Malabari, P. D. M. (1910). *Bombay in making*. London: T. Fisher and Uniwin.
Malleson, G. B. (1901). *History of French in India c.1674 to c.1961*. Edinburg: John Grant.
Mandeville, B. (c.1714, 1989). *The fable of the bees*. Penguin Classics, Penguin Books, London.
Marco, P. (1998). *The book of Ser Marco Polo* (Ed. & trans: Yule, H.). New Delhi: Oriental Reprints (Vols. I and II).
Markovits, C. (1985). *Indian business and national politics, 1930–1939*. Cambridge: Cambridge University Press.
Markovits, C. (2000a). Colonialism and traditional crafts. *Economic and Political Weekly, 35*, 3939–3940.
Markovits, C. (2000b). *Global world of Indian merchants 1750–1947; traders of Sind from Bukhara to Panama*. Cambridge: Cambridge University Press.
Marris, R. (1964). *The economic theory of managerial capitalism*. London: Macmillan.
Marshal, P. J. (Ed.). (2003). *The eighteen century in Indian history-evolution revolution*. New Delhi: Oxford University Press.
Marshall, A. (1919). *Industry and trade*. Cambridge: Cambridge University Press.
Marshall, A. (1929). *Principles of economic analysis*. London: English Language Book Society.
Marshall, P. J. (1975a). Economic and political expansion: The case of Awadh. *Modern Asian Studies, 9*(4), 465–482.
Marshall, P. J. (1975b). Economic expansion—The case of Awadh. *Modern Asian Studies, 9*(4), 465–482.
Marshall, P. J. (1976). *East India fortunes—The British in Bengal in the eighteenth century*. Oxford: Clarendon Press.
Marshall, P. J. (1987). *Bengal: The British Brideshead, Eastern India 1740–1828*. Cambridge: Cambridge University Press.
Martin, M. (2009). Hundi/Hawala: The problem of definition. *Modern Asian Studies, 43*(4), 909–937.
Marx, K. (1853a, June 25). The Biritsh rule in India. *New York Daily Tribune*. In I. Hussain (Ed.) (2006). *Karl Marx on India* (pp. 11–17). New Delhi: Tulika Books.
Marx, K. (1853b, June 24). East India Company—Its history and results. *New York Daily Tribune*.
Marx, K. (1853c, June 24). East India Company its history and results. *New York Daily Tribune*.
Marx, K. (1853d, August 8). The future results of British rule in India, *New York Daily Tribune*. Reprinted in Husssain, I. (Ed.). (2006). *Karl Marx on India* (pp. 46–51). New Delhi: New York Daily Tribune, Aligarh History Society, Tulika Book.
Marx, K. (1867, 1978). *Capital* (Vol. I). Progress Publishers: Moscow.

Mathias, P. (1959). *Brewing industry in England, 1700–1830.* Cambridge: Cambridge University Press.
Mazumdar, S. K. (2012). *India's late, late industrial revolution—Democratization of entrepreneurship.* New Delhi: Cambridge University Press.
Mc Fetridge, D. (1995). Knowledge, market failure and multinational enterprise—A comment. *Journal of International Business Studies, 26*(2), 409–406.
Mc Graw, T. K. (1988). *Essentials of Alfred Chandler—Essays towards a historical theory of big business.* Boston: Harvard Business School Press.
McNiell, W. H. (1989). *The age of gunpowder empire: 1450–1800.* Washington, DC: American Historical Association.
Medick, H. (1994). Proto industrial family economy – The structural function of household and family during transition from rural society to industrial capitalism. In R. A. Church & E. A. Wrigley (Eds.), *Industrial revolution* (pp. 72–98). Oxford: Blackwell.
Meek, R. (1962). *Economics of physiocracy.* New York: George Allen and Unwin.
Mehta, S. D. (1954). *Cotton mills of India 1854–1954.* Bombay: Textile Association.
Mehta, M. (1991). *Indian merchants and entrepreneurship in historical perspective.* New Delhi: Academic Foundation.
Mendels, F. F. (1972). Proto-industrialization—First phase of industrialization process. *Journal of Economic History, 32*, 241–261.
Mendels, F. F. (1982). *Proto-industrialization: Theory and reality.* General Report, International Economic History Congress, Budapest.
Metcalf, B. (1993). Norms of comportment among imperial Mughal officers, in John. In F. Richards (Ed.), *Power administration and finance in Mughal India* (pp. 255–289). Eldershot: Variorum.
Meyer, P. (1993). Inventing village trading, the late 19th century origin of North Indian jajmani system. *Modern Asian Studies, 27*(2), 378–386.
Miller, J. R. (1969). A reformulation of a. V. Chayanov's theory of peasant economy. *Economic Development and Cultural Change, 28*(2), 209–220.
Mintz, W. (1986). *Sweetness and power; The place of sugar in modern history.* London: Penguin.
Mishra, B. B. (1965). *The Indian middle class.* London: Oxford University Press.
Mohan, R. (1992). Industrial policy and control. In B. Jalan (Ed.), *Indian economy, problems and prospects* (pp. 85–104). New Delhi: Vikings.
Moorland, W. H. (1923). *From Akbar to Aurangzeb.* London: Macmillan.
Moorland, W. H. (1962). *India at the death of Akbar.* New Delhi: Atma Ram and Sons.
Moorland, W. H. (1998). Ranks (Mansab) in Mughal state service. In M. Alam & S. Subramanayam (Eds.), *Mughal state 1526–1750.* New Delhi: Oxford University Press.
Moosavi, S. (1987). *The economy of the Mughal empire 1595—A statistical study.* New Delhi: Oxford University Press.

Morris, M. D. (1965). Towards to re-interpretation of nineteenth century Indian economic history. *Journal of Economic History, 23*(4), 510–540.

Morris, D. M., & Dudley, C. B. (1975). Selected railway statistics for Indian sub-continent (India, Pakistan, Bangladesh) 1853–1946. *Artha Vijnana, 17*(3), 185–298.

Morris, D. M. (1987). Indian industry and business in the age of laissez faire. In D. Tripathi (Ed.), *State and business in India* (pp. 123–156). New Delhi: Manohar Publication.

Morris, D. M., & Dudley, C. B. (1975). Selected railway statistics for Indian sub-continent (India, Pakistan, Bangladesh) c.1853–1946/7. *Artha Vijnana, 17*(3), 185–298.

Mourad, K. (2012). *In the city of gold and silver—The story of Hazarat Mahal*. New Delhi: Full Circle Publishing.

Mukherjee, R. (1974). *Rise and fall of the East India Company*. New York/London: Monthly Review Press.

Mukherjee, R. (1998 [2007]). *Spectre of violence: The 1857 Kanpur massacres*. New Delhi: Penguin Books.

Mukherjee, R. (2008). *A century of trust—A story of Tata Steel*. New Delhi: Penguin Books.

Mukherjee, M. (2010). *India in the shadow of empire: A legal and political history 1774–1950*. New Delhi: Oxford University Press.

Mukund, K. (1992). Indian textile industry in the seventeenth and eighteenth centuries: Structure organization and response. *Economic and Political Weekly, 19*, 2057–2065.

Mun, T. (1664 [1928]). *England's treasure by foreign trade*. Oxford: Basil Blackwell.

Mun, T. (1621 [1930]). *A discourse on trade from England unto East India*. New York: Facsimile Text Society.

Naoroji, D. (1901/1962). *Poverty and UnBritish rule of India*, (Reprint), New Delhi: Government of India Press.

Naqvi, H. K. (1986). *Agriculture, industrial and urban dynamics under sultans of Delhi*. New Delhi: Oxford University Press.

Naroji, D. (1901 [1962]). *Poverty and un British rule of India*. New Delhi: Reprinted by the Government of India Press.

Nayyar, D. (Ed.). (1994). *Industrial growth and stagnation, the debate in India*. New Delhi: Oxford University Press.

Neff, J. U. (1943). Industrial revolution reconsidered. *Journal of Economic History, 3*, 2–29.

Nehru, J. (1947). *The discovery of India*. London: Meridian Books.

Nigam, R. K. (1957). *Managing agency system in India: Ministry of Commerce*. New Delhi: Government of India.

Nockolds, H. (1959). *The magic of a name*. London: G.T. Foulis.

Noiel, D. (1949, 1958). *The history of sugar*. Chapman & Hall, London.
North, D. C. (1990). *Institutions, institutional change and economic performance*. Cambridge: Cambridge University Press.
North, D. C. (2006). *Understanding the process of economic change*. New Delhi: Academic Foundation.
Oglivie, S., & German, M. (Eds.). (1996). *European proto-industrialization*. Cambridge: Cambridge University Press.
Olson, M. (1965). *Logic of collective action: Public goods and theory of group*. Cambridge, MA: Harvard University Press.
Olson, M. (1982). *Rise and decline of nations; economic growth, stagflation and social regulation*. New Haven: Yale University Press.
Om Prakash (1985). *Dutch East India Company and economy of Bengal 1630–1720*. Princeton: Princeton University Press.
Om Prakash (1998). Trade and politics in eighteen country Bengal. In L. Blusse & F. Gaastra (Eds.), *On the eighteen century as a category of Asian history*. Aldershot: Ven Lein in Retrospect Ashgate.
Om Prakash (2007). From negotiation to Coersion—Textile manufacturing in the eighteen centuries. *Modern Asian Studies, 41*(6), 1331–1368.
Ovington, J. (1929). In H. G. Rawlison (Ed.), *A Voyage to Surat in the year 1689*. London: Hakluyt Society.
Parathasarthi, P. (2001). *The transition to a colonial economy-weavers, merchants and kings in South India 1720–1800*. Cambridge: Cambridge University Press.
Parker, W. N. (1982). European development in Melliennial perspective. In C. P. Kingleberger & G. di Fella (Eds.), *Economics in long view* (Vol. II). London: Macmillan.
Parthasarathi, P. (2001). *The transition to colonial economy weavers, merchants and kings in South India 1720–1800*. Cambridge: Cambridge University Press.
Parthasarathi, P. (2011). *Why Europe grew rich and Asia did not—Global divergence 1600–1850*. Cambridge: Cambridge University Press.
Pavlov, V. I. (1964). *The Indian capitalist class*. Moscow: PPH.
Pavlov, V. I. (1976). *Historical premises for India's transition to capitalism*. Moscow: Nauka Publishing House.
Pearson, M. N. (2007). Markets and merchants in the Indian Ocean. In F. Bethancourt & D. R. Curto (Eds.), *Portuguese oceanic expansion, 1400–1800* (pp. 88–108). New York: Cambridge University Press.
Perlin, F. (1983). Proto-industrialization in precolonial South Asia. *Past and Present, 98*, 30–95.
Pillai, A. S. (1930). *Monograph on Nottuketti Chettiers banking system*. Madrass Province Banking Enquiry Commission Report, Vol. III.
Piore, M. J., & Sable, C. F. (1984). *The second industrial divide*. New York: Basic Books.

Piramal, G. (1998). *Business legends: Indepth profiles of India's most daring and innovative entrepreneurs.* New Delhi: Penguin Books.

Piramal, G., & Herdeck, M. (1986). *India's industrialist.* Bombay: India Book House Distributors.

Pollard, S. (1968). *The genesis of modern management.* London: Penguin.

Porta, R. L. (2000). Agency problem and dividend policies around the world. *Journal of Finance, 55*(1), 1–33.

Powell, B. (1892). *Land-systems of British India.* Oxford: Clarendon Press.

Prescott, W. H. (1847). *History of the conquest of Mexico and Peru.* New York: Random House.

Pugh, P. (2000). *The magic of a name—The Rolls-Royce story—The first fourty years.* Cambridge/New York: Icon Books/Totem Books.

Putnam, R. (2000). *Bowling action: The collapse and revival of American community.* New York: Simon and Schuster.

Raff, D. M. G., & Summers, L. H. (1987). Did Henery Ford pay efficiency wages. *Journal of Labour Economics, 5*(4, II), 557–586.

Rajan, R., & Zingales, L. (2003). *Saving capitalism from capitalists.* Princeton: Princeton University Press.

Ramaswamy, V. (1980). Notes on textile technology in Medieval India with special reference to South. *Indian Economic and Social History Review, 17*, 227–241.

Ramaswamy, V. (2008). Traditional crafts, technology and society in pre-colonial Peninsular India. In R. Dutta (Ed.), *Rethinking a millennium, perspectives in Indian history from eight to eighteenth century* (p. 27). New Delhi: Akar.

Ramaswamy, S. (2011). *The goddess of the nation—Mapping mother India.* New Delhi: Zuban.

Raphael, S. (1994). Workshop of the world, steam power and hand technology in mid-Victorian Britain. In J. Hoppitt & E. A. Wrigley (Eds.), *Industrial revolution-economy history and society* (Vol. 3, Part-V, pp. 196–263). London: Blackwell.

Ray, R. K. (1979). *Industrialization in India: Growth and conflict in private corporate sector, 1914–1947.* New Delhi: Oxford University Press.

Ray, R. K. (1982). Pedhis and Mills—The historical integration of formal and informal sectors of the economy in Ahmedabad. *Indian Eocnomic and Social History Review, 19*(3 and 4), 239–265.

Ray, R. K. (1998). India, its society and British supremacy. In P. J. Marshall (Ed.), *Oxford history of British empire—18th century* (Vol. II, pp. 495–530). Oxford: Oxford University Press.

Ray, R. K. (2002). Indigenous banking and commission agency in India's colonial economy. In A. K. Bagchi (Ed.), *Money and credit in Indian history.* New Delhi: Tulika Books.

Raychaudhri, T. (1982). Non-agricultural production in Mughal India. In T. Raychaudhri & I. Habib (Eds.), *Cambridge economic history of India, Vol. 1 (c.1200–c.1750)* (pp. 261–307). New Delhi: Orient Longman.

Ricardo, D. (1863/1882). *Principles and political economy and taxation*. Homewood: Irwin.

Rich, E. E. (1952). *Cambridge economic history of Europe* (Vol. II). Cambridge: Cambridge University Press.

Richard, J. F. (1975). *Mughal administration in Golconda*. New Delhi: Oxford University Press.

Richard, J. F. (Ed.). (1976). *Kingship and authority in South Asia*, *Center for South Asian Studies*. Madison: University of Wisconsin.

Richard, J. F. (1983). *Precious metals in the later medieval and early modern world*. Durham: Carolina Academic Press.

Richard, R. (1996). West Africa and Pondicherry textile industry. In T. Ray (Ed.), *Cloth and commerce textiles in colonial India's*. New Delhi: Oxford University Press.

Richards, J. F., & Rao, V. N. (1998). Banditry in Mughal India: Historical and folk perception. In M. Alam & S. Subhrahmanyam (Eds.), *The Mughal state: 1526–1757* (pp. 491–519). New Delhi: Oxford University Press.

Rider, C. (Ed.). (2007). *Encylopaedia of the age of the industrial revolution, 1700–1920* (Vols. I and II). London: Greenwood Press.

Riello, G., & Ray, T. (Eds.). (2009). *How India clothed the world—The world of south Asian textiles 1500–1850*. Leiden: Brill.

Riello, G., & Parthasarathi, P. (Eds.). (2009). *The spinning world—A global history of textiles 1200–1850*. Oxford: Oxford University Press.

Rizvi, J. (2000). *Trans-Himalayan caravans: Merchant princes and peasant traders in Ladakh*. New Delhi: Oxford University Press.

Roberts, E. (1960). *Embassy to the eastern courts – Cochin, China, Saim, and Muscat in the US sloop of war during 1832–34*. Wellington: Scholarly Recourses Limited.

Robins, N. (2006). *The corporation that changed the world—How the East India Company shaped modern multinationals*. Hyderabad: Orient Longman.

Robbins, L. (1998). In S. G. Medema & W. J. Samuels (Eds.), *A history of economic thought—The LSE lectures*. Oxford: Oxford University Press.

Robinson, F. P. (1912). *The trade of East India Company form C.1709–C.1813*. Cambridge: Cambridge University Press.

Roe, M. J. (2002). The political foundations of separating ownership from control. In J. A. Mc Cathery, P. Moreland, T. R. Makero, & L. Rennebug (Eds.), *Corporate governance, regimes, convergence and diversity* (pp. 113–140). New York: Oxford University Press.

Roe, M. J. (2003). *Political determinants of corporate governance, political context, corporate impact*. New York: Oxford University Press.

Rosen, G. (1958). *Industrial change in India*. Glencoe: Jllionais.

Rosenberg, N. (1976). *Perspective on technology*. Cambridge, MA: Cambridge University Press.

Rosenberg, N. (2000). Charles Babbage in complex world. In D. Colander (Ed.), *Complexity in history of economic thought* (pp. 37–57). London/New York: Routledge.

Rostow, W. W. (1957). *Stages of economic growth—A non communist manifesto*. Cambridge, MA: Cambridge University Press.

Rostow, W. W. (Ed.). (1963). *The take-off into self sustained growth*. London: Macmillan.

Rothermund, D. (1992). *India in the great depression, 1929–1939*. New Delhi: Manohar.

Roy, T. (1993). *Artisans and industrialization—Indian weaving in twentieth century*. New Delhi: Oxford University Press.

Roy, T. (1999). *Traditional industry in the economy of colonial India*. Cambridge: Cambridge University Press.

Roy, T. (2000). *The economic history of India 1857–1947*. New Delhi: Oxford University Press.

Roy, I. (2005a). The silk industry in Bengal during colonial rule; the de-industrialization thesis revisited. *Indian Economic and Social History Review, 42*(3), 339–375.

Roy, T. (2005b). *Rethinking change in India-labour and livelihood*. London/New York: Rontledge.

Roy, T. (2006). *Economy history of India 1857–1947*. New Delhi: Oxford University Press.

Roy, T. (2010). *Company of kinsmen—Enterprise and community in South Asian history 1700–1940*. New Delhi: Oxford University Press.

Royden, H., & Zeitlin, J. (Eds.). (1985). *Division of labour, skilled workers and technological change in nineteenth century Britain*. Sussex: Harvest.

Rubin, I. I. (1979). *History of economic thought*. London: Ink-Link.

Ruderner, D. (1994). *Caste and capitalism in colonial India—The Nahokottai Chettiers*. Barkley: Barkley University Press.

Rungta, R. S. (1970). *The rise of business corporation in India, 1851–1900*. Cambridge, MA: Cambridge University Press.

Salop, S. C. (1979). Strategic entry deterrence. *American Economic Review (Paper and proceedings),69*, 305–320.

Sampson, A. (1996). *Company man: The rise and fall of corporate life*. London: Harper Collins.

Sanyal, H. R. (1968). The indigenous iron industry of Birnbaum. *Indian Economic and Social History Review, 5*(1), 102–125.

Sarkar, S. (1983). *Modern India 1885–1947*. New Delhi: Macmillan.

Sarkar, S. (2014). *Modern times: India 1880–1950, environment, economy, culture*. Ranikhet: Permanent Black.

Sarwani, A. K. (1974). *Tarikh-i-Shershahi* (trans: Ambashthya, B. P.). Patna: K. P. Jayaswal Research Institute.

Schumpeter, J. A. (1934). *The theory of economic development*. Cambridge, MA: Harvard University Press.

Schumpeter, J. A. (1939). *Business cycles a theoretical, historical and statistical analysis of capitalist process*. New York: McGraw Hill.

Schumpeter, J. A. (1947). *Capitalism socialism and democracy*. New York: Hamper.

Schumpeter, J. A. (1954). *History of economic analysis*. London: Oxford University Press.

Scitovsky, T. (1959). Growth balanced or unbalanced. In M. Abramowitz (Ed.), *Allocation of economic resources*. Standford: Stanford University Press.

Scott, W. R. (1909–1912). *The constitution and finance of the English and Irish joint-stock companies to 1720* (Vols. I–III). Cambridge: Cambridge University Press.

Sen, S. (1998). *Empire of free trade: East India Company and the making of colonial market place*. Philadelphia: University of Philadelphia Press.

Seth, V. K. (1987). *Industrialization in India spatial perspective*. New Delhi: Common Wealth Publication.

Seth, V. K. (1989). *Themes spatial perspective*. New Delhi: Common Wealth Publication.

Seth, V. K. (1994). Labour market restructing and structural adjustment programme. *Indian Journal of Labour Economics, 37*(4), 523–534.

Seth, V. K. (1998). Economic reforms and capacity utilization in Indian industry. *Productivity, 39*(3), 391–398.

Seth, V. K. (1999). *Capacity utilization in industry theory and evidence*. New Delhi: Deep and Deep Publication.

Seth, V. K. (2000). Technology and labour in retrospect and prospect. *Managment and Change, 4*(1), 61–90.

Seth, V. K. (2002). Primitive flexible to modern flexible manufacturing and the process of skilling and deskilling of labor. *Indian Journal of Labor Economics, 45* (4), 1175–1194.

Seth, V. K. (2003). The beginning and the end of primitive flexible manufacturing system. *Journal of Management Research, 3*(2), 73–86.

Seth, V. K. (2006). *Economics of services: A story of transformation of Cinderella into the queen of hearts*. New Delhi: Ane Books.

Seth, V. K. (2008). Primitive flexible manufacturing and Mughal empire. *Journal of Management Research, 8*(2), 76–95.

Seth, V. K. (2012). The East India Company—A case study in corporate governance. *Global Business Review, 13*(2), 221–238.

Seth, V. K. (2014). Debate on de-industrialization revisited: The process of decline of traditional flexible manufacturing. *Global Business Review, 15*(3), 597–610.

Seth, V. K., & Aggarwal, S. C. (2004). *The economics of labour markets policy regime changes and the process of adjustment in the organized industry in India*. New Delhi: Ane Books.

Seth, V. K., & Seth, A. (1994). *Dynamics of labour absorption in industry*. New Delhi: Deep and Deep.

Sethia, T. (1996). The rise of jute manufacturing industry in colonial India: A global perspective. *Journal of World History, 7*(1), 35–56.

Shahj, K. T. (1948). *National planning principles and administration*. Bombay: Vora.

Shavell, S. (1979). Risk sharing and incentives in principal-agent relationship. *Bell Journal of Economics, 10*, 53–73.

Shelvankar, K. S. (1940). *The problem of India*. London: Penguin.

Shrikov, G. K. (1973). *Industrialization in India*. Moscow: Progress Publishing House.

Siddiqi, A. (Ed.). (1995). *Trade and finance in colonial India* (pp. 250–264). New Delhi: Oxford University Press.

Simmons, C. (1985). "De-industrialization" industrialization and the Indian economy C.1850–1947. *Modern Asian Studies, 19*, 593–672.

Singer, C., et al. (Eds.). (1958). *History of technologies* (Vol. 2–4). Oxford: Clarendon Press.

Singh, B. (1966). *European agency houses in Bengal*. Calcutta: Firma Mukhopadhyay.

Singh, C. (1991). *Region and empire: Punjab in the seventeenth century*. Delhi: Oxford University Press.

Singh, R. (1998). *A despotism of law: Crime and justice in early colonial India*. New Delhi: Oxford University Press.

Sinha, N. K. (1968). *The economic history of Bengal*. Calcutta: Firma K. Mukhopadhyaya.

Smith, B. M. D. (1967). The Galtons of Birmingham Quacker gun merchants and bankers. *Business History, 9*(2), 135–148.

Smith, A. (1776 [2003]). Introduction. In A. B. Krueger (Ed.), *The wealth of nations* (Bantam classic). New York: Bantam Books.

Solymar, L. (1999). *Getting the message: A history of communication*. Oxford: Oxford University Press.

Stein, B. (1985). State formation and economy reconsidered. *Modern Asia Studies, 19*(3), 387–341.

Stein, B. (1998). *A history of India*. New Delhi: Oxford University Press.

Stern, P. (2004). *One body corporate and politic—The growth of East India Company state in the late seventeenth century*. Unpublished PhD thesis, Columbia University, Columbia.

Stiglitz, J. E. (1984). *Theory of wage rigidities*. Working paper no 1442, NBER. Cambridge, MA.

Stiglitz, J. E. (1988). Equilibrium wage distribution. *Economic Journal, 95*(3), 599–618.

Storper, M. (2008). Community and economics. In A. Amin & J. Roberts (Eds.), *Community, creativity and organization* (pp. 37–68). New York: Oxford University Press.

Stroper, M., & Christopherson, S. (1987). Flexible specialization and regional agglomeration. *Annal of the Association of American Geographers, 77*(1), 25–48.

Subhramanyam, S. (1995). Of 'Imarat and Tijarat': Asian merchants and state power in Western Indian Oceans—1400–1730. *Comparative Studies in Society and History, 37*(4), 225–247.

Subramaniam, L. (1987). Banias and business: The role of indigenous credit in the process of imperial expansion in Western India in the second half of eighteen century. *Modern Asian Studies, 21*(3), 455–520.

Subramaniam, L. (2012). *Three merchants of Bombay—Trawadi Arjunji, Nathji, Jamsetjee Jeejeebhoy and Premchand Roychand*. New Delhi: Penguin Books.

Subramanian, L. (1996). *Indigenous capital and imperial expansion: Bombay, Surat and the West Coast*. New Delhi: Oxford University Press.

Sung, J. K. (1966). *Stages of industrial development in Asia—A comparative history of cotton textile industry of Japan, India, China and Korea*. Philadelphia: University of Penusylvnia Press.

Sutherland, L. (1952). *The East India Company in the eighteenth century politics*. Oxford: Clarendon Press.

Tamilson, B. R. (1993). *The economy of modern India*. Cambridge: Cambridge University Press.

Taylor, F. W. (1885 [1919]). *Two papers on scientific management*. London: Routledge and Sons.

Taylor, F. W. (1947). *Scientific management*. New York: Harper and Row.

Tchitchirov, A. L. (1965). *India-changing economic structure in sixteenth to eighteenth centuries outline history of craft and trade*. New Delhi: Manohar.

Thirsk, J. (1994). Industries in countryside. In J. A. Chractre (Ed.), *Pre-industrial Britain*. Oxford: Blackwell.

Tripathi, D. (2004). *Oxford history of Indian business*. New Delhi: Oxford University Press.

Thorner, D. (1950). *Investment in empire, British railways and steam shipping enterprise 1825–1849*. Philadelphia: University of Pennsylvania Press.

Thorner, D. (1951). Great Britain and development of Indian Railway. *Journal of Economic History, XI*, 326–350.

Thorner, D. (1962). "De-industrialization" in India 1881–1931. In D. Thorner & A. Thorner (Eds.), *Land and labour in India*. New York: Asia Publishing House.

Thorner, D. (1980). *Shaping of modern India*. Bombay: Sameeksha Trust.

Tilly, C., & Tilly, R. (1971). Agenda for European economic history in 1970s. *Journal of Economic History, 31*, 176–196.

Tinberg, T. A. (1971). A study of great Marwari firm: 1860–1914. *Indian Economic and Social History Review, 8*, 267–280.

Times of India. (2010). East India company is back with Indian owner. *Times of India*, August 16.

Tod, J. (1829 [1993]). *Annels and antiquities of Rajasthan* (W. Crooke, Ed.), reprint Bibliolithic, vols 1 & 2. New Delhi.

Tomilson, B. P. (1978). Foreign private investment in India 1920–1960. *Modern Asian Studies, 12*(4), 655–677.

Tomilson, B. R. (1979). *Political economy of the raj.* London: Oxford University Press.

Tomilson, B. R. (1981). Colonial firms and decline of colonialism in eastern India 1914–1947. *Modern Asian Studies, 15*(3), 455–486.

Tomilson, B. R. (1993). *The economy of modern India 1860–1970.* Cambridge: Cambridge University Press.

Toynbee, A. (1958). *Industrial revolution.* Boston: Becon Press.

Traverse, R. (2007). *Ideology and empire in eighteen century India—The British in Bengal.* Cambridge: Cambridge University Press.

Treblicock, C. (1969). "Spin-off" in British economic history: Armaments and industry 1760–1914. *Economic History Review, 22*(3), 479–490. 2nd Session.

Tripathi, D. (1987). Business and princely states—The case of Baroda. In D. Tripathi (Ed.), *State and business in India—A historical perspective* (pp. 274–301). New Delhi: Manohar.

Tripathi, D. (Ed.). (1991). *Business and politics in India: A historical perspective.* New Delhi: Manohar Publishers.

Tripathi, D. (2004). *Oxford history of Indian business.* New Delhi: Oxford University Press.

Tripathi, D. (2011). Marwaris from community to class. In M. M. Kudaisya (Ed.), *The Oxford India anthology of business history* (pp. 133–142). New Delhi: Oxford University Press.

Tripathi, D., & Mehta, M. (1990a). The Nagarsheth of Ahmedabad: The history of urban institutions in a Gujarat city. In S. Chandra (Ed.), *Essays in medieval Indian economic history.* New Delhi: Munshi Ram Manoharlal.

Tripathi, D., & Mehta, M. (1990b). *Business houses of Western India—A study of Entreprenerurial response, 1850–1956.* New Delhi: Manohar.

Trivedi, K. K. (1998). *Agra—Economic and political profile of a Mughal Suba: 1580–1707.* Pune: Ravish Publisher.

Varshney, A. (1999). *Democracy, development and country side urban and rural struggle in India.* New York: Cambridge University Press.

Verdier, D. (2002). *Moving money, banking and finance in industrialised world.* New York: Cambridge University Press.

Verlest, H. (1776). *The view of the rise of the English government in Bengal.* London: J. Nourse.

Verma, T. (1994). *Karkhanas of Mughals.* New Delhi: Pragti Publications.

Victor, H. (1888). In J. S. Stallybrass (Ed.), *Wanderings of plants and animals from their first home.* London: Swan Sonnenshein.

Vijayabhaskar, M. (2001). *Industrial formation under conditions of flexible accumulation—The case of global knitwear node, in Southern India*. Unpublished PhD thesis, Centre of Development Studies, Thiruvananthpuram (Mimio).
von Schmoller, G. (1902). *The mercantile system and its historical significance: Illustrated chiefly from Prussian history*. New York: Macmillan.
Walker, B. (1968). *Hindu world—An encyclopedic survey Hinduism*. London: Allen and Unwin.
Wallace, D. R. (1928). *Romance of jute—A short history of the Calcutta jute mill industry 1855–1927*. London: W. Thacker.
Walton, J. K. (1989). Proto-industrialization and first industrial revolution, the case of Lancashire. In P. Hudson (Ed.), *Regions and industry—A perspective on Britain*. Cambridge: Cambridge University Press.
Washbrook, D. (1990). South Asia in the world system and world capitalism. *Journal of Asian Studies, 49*(3), 479–508.
Watson, J. F. (1879). *Report on cotton gins and cleaning quality Indian cotton*. London: Indian Office Library.
Watson, J. F. (1887). *The textile manufactures and costumes of the people of India*. London: Indian Office Library.
Weatherford, J. (2004). *Genghis khan and making of the modern world*. New York: Crown.
Weber, M. (1927). *General economic history* (trans: Knight, F.). New York: Greenberg.
Weber, A. (1929). *Theory of location of industry* (trans: Freidrich). Chicago: Chicago University Press.
Weber, M. (1958a). *Religion of India—The sociology of Hinduism and Buddhism* (Eds. & trans: Hans, H. G. & Martindale, D.). Glencoe: Free Press.
Weber, M. (1958b). *The Protestant ethic and the spirit of capitalism*. New York: Charles Scribner's Sons.
Weiss, A. (1990). *Efficiency wages—A model of unemployment lay-offs and wage dispersion*. Oxford: Clarendon Press.
Wenger, E. C. (1998). *Community of practice: Learning, meaning and identity*. Cambridge: Cambridge University Press.
West, E. J. (1976). *Adam Smith: The man and his work*. Indianapolis: Liberty Press.
Whiteway, R. S. (1893). *Rise of Portuguese power in India: 1497–1550*. Santiago De Compostela.
Wild, A. (1999). *The East India Company: Trade and conquest from 1600*. London: Harper Collins.
Williams, N. (1975). *Chronology of the modern world; 1763–1965*. Middlesex: Renguin Books.
Willam, K., et al. (1987). The end of mass production. *Economy and Society, 16*, 405–439.
Wilson, C. (1965). *England's apprenticeship 1603–1963*. London: Longman.

Wilson, B. R. (1970). *Religious sects: A sociological study*. London: Weidenfeld and Nicholson.

Wilson, C. (1994). Entrepreneurship in industrial revolution. In J. Huppit & E. A. Wriglay (Eds.), *The industrial revolution in Britain* (pp. 305–321). Oxford: Blackwell.

Winch, D. (1978). *Adam Smith's politics—An essay in historiographic revision*. Cambridge: Cambridge University Press.

Wink, A. (1986). *Land and sovereignty in India: Agrarian society and politics under the eighteen century Maratha Swarajya*. Cambridge: Cambridge University Press.

Wiser, W. H. (1969). *The Hindu jajmani system*. Lucknow: Lucknow Publishing House.

Womack, J. P., & Jones, D. J. (1996). *Lean thinking banish waste and create wealth in your corporation*. New York: Simon and Schuster.

Womack, J. P., et al. (1990). *Machines that changed the world—The story of lean production*. Temecula: Rawson Associates.

Wright, R. (2000). *Stolen continents—A conquest and resistance in the America*. London: Phoenix Press.

Wrigley, E. A. (1988). *Continuity chance and change, the character of industrial revolution in England*. Cambridge: Cambridge University Press.

Wrigley, E. A., & Roger, S. (1981). *The population history of England 1541–1871 a reconstruction*. Oxford: Oxford University Press.

Zeitlin, J. (1995). Flexibility and mass production at war-aircraft manufacturing in Britain, United States and Germany, 1939–1945. *Technology and Culture, 36*(1), 46–79.

Index

V.K. Seth, *The Story of Indian Manufacturing*,
DOI 10.1007/978-981-10-5574-4

The manufacturer's authorised representative in the EU is Springer Nature Customer Service Centre GmbH, Europaplatz 3, 69115 Heidelberg, Germany. If you have any concerns regarding our products, please contact ProductSafety@springernature.com

Printed and bound by CPI Group (UK) Ltd, Croydon, CR0 4YY
25/09/2026
02222890-0007